# Hominin Environments in the East African Pliocene:
## An Assessment of the Faunal Evidence

# Vertebrate Paleobiology and Paleoanthropology Series

Edited by

**Eric Delson**
Vertebrate Paleontology, American Museum of Natural History,
New York, NY 10024, USA
delson@amnh.org

**Ross D. E. MacPhee**
Vertebrate Zoology, American Museum of Natural History,
New York, NY 10024, USA
macphee@amnh.org

Focal topics for volumes in the series will include systematic paleontology of all vertebrates (from agnathans to humans), phylogeny reconstruction, functional morphology, paleolithic archaeology, taphonomy, geochronology, historical biogeography, and biostratigraphy. Other fields (e.g., paleoclimatology, paleoecology, ancient DNA, total organismal community structure) may be considered if the volume theme emphasizes paleobiology (or archaeology). Fields such as modeling of physical processes, genetic methodology, nonvertebrates, or neontology are out of our scope.

Volumes in the series may either be monographic treatments (including unpublished but fully revised dissertations) or edited collections, especially those focusing on problem-oriented issues, with multidisciplinary coverage where possible.

## Editorial Advisory Board

*Published and forthcoming titles in this series are listed at the end of this volume*

# Hominin Environments in the East African Pliocene:
# An Assessment of the Faunal Evidence

Edited by

**René Bobe**
*Department of Anthropology, The University of Georgia*
*Athens, USA*

**Zeresenay Alemseged**
*Department of Human Evolution*
*Max Planck Institute for Evolutionary Anthropology*
*Leipzig, Germany*

**Anna K. Behrensmeyer**
*Department of Paleobiology and*
*Evolution of Terrestrial Ecosystems Program*
*Smithsonian Institution*
*Washington, USA*

 Springer

A C.I.P. Catalogue record for this book is available from the Library of Congress.

ISBN 978-1-4020-3097-0 (HB)
ISBN 978-1-4020-3098-7 (e-book)

Published by Springer,
P.O. Box 17, 3300 AA Dordrecht, The Netherlands.

www.springer.com

Cover image:
*Exposures of the Pliocene sediments at Dikika, Ethiopia, with specimens of Homo (KNM ER 3733)
and Tragelaphus (KNM WT 18673) in the foreground.*

Cover composition and background photograph by Zeresenay Alemseged
*Tragelaphus* photograph by René Bobe
*Homo* photograph by David Brill
Fossil images: Copyrighted by the National Museums of Kenya

Printed on acid-free paper

To the people and research institutions of Africa, and to the fieldworkers and museum curators whose generosity, goodwill, time, and energy have made African fossils available to the scientific community.

# Acknowledgments

We would like to express our gratitude and appreciation to the many colleagues who have played a critical and constructive role as external reviewers of chapters in this volume: Peter Andrews, Margaret Avery, Ray Bernor, Laura Bishop, Rob Blumenschine, Christiane Denys, Craig Feibel, John Fleagle, David Fox, Henry Gilbert, Don Grayson, Yohannes Haile-Selassie, Andrew Hill, Leslea Hlusko, Clark Howell, Nina Jablonski, Martin Pickford, Kaye Reed, Martha Tappen, Blaire Van Valkenburgh, Elisabeth Vrba, Alan Walker, Alisa Winkler, and other reviewers who wish to remain anonymous. We are deeply grateful to the Smithsonian Institution and its Human Origins and Evolution of Terrestrial Ecosystems Programs, and especially the National Science Foundation and Physical Anthropology Program Director Mark Weiss, for encouragement and financial support that made the 2004 Workshop on Faunal Evidence for Hominin Paleoecology possible (NSF Grant: #0422048). Finally, we thank Ross MacPhee, Tamara Welschot, and Judith Terpos for their encouragement and support throughout the process of putting together this volume and especially Eric Delson, whose vision, patience, and persistence helped us to bring this volume to a successful completion.

# Table of Contents

# List of Contributors

**Z. Alemseged**
Department of Human Evolution
Max Planck Institute for Evolutionary Anthropology
Deutscher Platz 6
04103 Leipzig, Germany
*zeray@eva.mpg.de*

**A.K. Behrensmeyer**
National Museum of Natural History
Department of Paleobiology and Evolution
of Terrestrial Ecosystems Program
Smithsonian Institution
P.O. Box 37012, NHB MRC 121
Washington, DC 20013-7012, USA
*behrensa@si.edu*

**R. Bobe**
Department of Anthropology
The University of Georgia
Athens, GA 30602-1619, USA
*renebobe@uga.edu*

**S. Branting**
Center for Ancient Middle Eastern Landscapes
(CAMEL)
The Oriental Institute
The University of Chicago
Chicago, IL 60637, USA
*branting@uchicago.edu*

**T.G. Bromage**
Department of Biomaterials and Biomimetics
New York University College of Dentistry
345 East 24th Street
New York, NY 10010, USA
*tim.bromage@nyu.edu*

**H.B.S. Cooke**
2133 – 154th Street
White Rock, British Columbia, Canada V4A 4S5
*cookecentral@shaw.ca*

**G.G. Eck**
Department of Anthropology
University of Washington
Seattle, Washington 98195-3100, USA
*ggeck@u.washington.edu*

**S.R. Frost**
Department of Anthropology
1218 University of Oregon
Eugene, OR 97403-1218, USA
*sfrost@uoregon.edu*

**D. Geraads**
UPR 2147 CNRS-44 rue de l'Amiral Mouchez
75014 Paris, France
*dgeraads@ivry.cnrs.fr*

**B. Hallgrímsson**
Department of Cell Biology and Anatomy
University of Calgary
Calgary, Alberta, Canada T2N 1N4
*bhallgri@ucalgary.ca*

**J.M. Harris**
George C. Page Museum
5801 Wilshire Boulevard
Los Angeles, California 90036, USA
*jharris@nhm.org*

**T. Harrison**
Department of Anthropology
Center for the Study of Human Origins
New York University
25 Waverly Place
New York, NY 10003, USA
*terry.harrison@nyu.edu*

**A. Hill**
Department of Anthropology
Yale University
P.O. Box 208277
New Haven, Connecticut 06520-8277, USA
*andrew.hill@yale.edu*

**Y.M. Juwayeyi**
Department of Sociology and Anthropology
Long Island University
1 University Plaza Brooklyn
New York, NY 11201-5372, USA
*yusuf.juwayeyi@liu.edu*

**S. Killindo**
Archaeology Unit
The University of Dar es Salaam
Dar es Salaam, Tanzania
*skillindo@yahoo.uk.com*

**O. Kullmer**
Department of Paleoanthropology and Quaternary
Paleontology
Forschungsinstitut Senckenberg,
Senckenberganlage 25
60325 Frankfurt am Main, Germany
*ottmar.kullmer@senckenberg.de*

**M.E. Lewis**
Division of Natural and Mathematical Sciences
(Biology)
The Richard Stockton College of New Jersey,
P.O. Box 195
Pomona, New Jersey 08240-0195, USA
*Margaret.Lewis@stockton.edu*

**A. Mabulla**
Archaeology Unit
The University of Dar es Salaam
Dar es Salaam, Tanzania
*aumab@udsm.ac.tz*

**C. Magori**
Department of Anatomy and Histology
Bugando University College of Health Sciences
Mwanza, Tanzania
*cmagori@buchs.org*

**F. Mizambwa**
Department of Antiquities
Ministry of Tourism and Natural Resources
P.O. Box 2280, Dar es Salaam, Tanzania
*oldupai@africaonline.co.tz*

**C. Musiba**
Department of Anthropology
University of Colorado at Denver and
Health Sciences Center
Denver, Colorado 80207, USA
*charles.musiba@cudenver.edu*

**F. Ndunguru**
Department of Antiquities
Ministry of Tourism and Natural Resources
P.O. Box 2280, Dar es Salaam, Tanzania
*oldupai@africaonline.co.tz*

**R. Potts**
Human Origins Program
National Museum of Natural History
Smithsonian Institution
Washington, DC 20013-7012, USA
*pottsr@si.edu*

**D.N. Reed**
Department of Anthropology
The University of Texas at Austin
Austin, TX 78712, USA
*reedd@mail.utexas.edu*

**O. Sandrock**
Department of Geology and Paleontology
Hessisches Landesmuseum, Friedensplatz 1
64283 Darmstadt, Germany
*sandrock@hlmd.de*

**F. Schrenk**
Department of Vertebrate Paleobiology
Johann Wolfgang Goethe-University,
Siesmayerstrasse 70
60054 Frankfurt am Main, Germany
*schrenk@zoology.uni-frankfurt.de*

**T. Stein**
Department of Anatomy and Cell Biology
University of Michigan
Ann Arbor, Michigan 48109-0608, USA
*tastein@umich.edu*

**M. Stoller**
Department of Ecology and Evolution
The University of Chicago
Chicago, IL 60637, USA
*mstoller@uchicago.edu*

**D.F. Su**
Human Evolution Research Center
Department of Integrative Biology
University of California, Berkeley
Berkeley, CA 94720, USA
*denisefsu@calmail.berkeley.edu*

**R. Tuttle**
Department of Anthropology
The University of Chicago
Chicago, IL 60637, USA
*r-tuttle@uchicago.edu*

**M. Vogt**
GIS Lab, Center for Environmental
Restoration Systems
Argonne National Laboratory
Argonne, IL 60439, USA
*mvogt@anl.gov*

**L. Werdelin**
Department of Palaeozoology
Swedish Museum of Natural History
Box 50007, S-104 05 Stockholm, Sweden
*werdelin@nrm.se*

# Foreword

Hominin fossils are few and fragmentary compared with the abundant and well-preserved remains of mammals that inhabited Africa over the last seven million years. This mammalian record has been assembled from many decades of intensive field and museum work and contributes critical evidence about the evolutionary and ecological context of human evolution. With continued collecting, analysis of paleoenvironmental information, and efforts to organize the information into accessible databases, the mammalian fossil record is providing more comprehensive information on faunal change through time, regional variability, accessible levels of temporal resolution, and the impact of taphonomic and other sampling biases. This is leading to new, better supported insights and hypotheses about the interaction of environmental change and mammalian evolution, including processes that likely affected hominin evolution.

Large fossil collections and databases catalyze collaboration and, indeed, require intensive interaction among scientists, collections personnel, museum and academic administrators, and granting agencies. However, without the stimulus provided by critical questions in human evolution, it is easy to get "buried in data" and to lose sight of why we devote so much time and effort in accumulating more and more facts and fossils. In order to focus attention on some of these critical questions and to synergize the analysis of hominin paleoecology using data from the East African fossil mammal record, we organized a symposium for the April 2003 meetings of the American Association of Physical Anthropologists (in Tempe, Arizona), followed by a workshop at the Smithsonian Institution in May 2004 (in Washington, DC). The AAPA symposium, titled "Hominid Environments and Paleoecology in the East African Pliocene: an Assessment of the Faunal Evidence," provided the organizing framework for this volume. The Smithsonian "Workshop on Faunal Evidence for Hominin Paleoecology" expanded on the discussions initiated at the symposium with a broader chronological (late Cenozoic) and geographic (Africa and Eurasia) framework. The Smithsonian workshop brought together 44 scientists and students from Africa, Europe, and North America to inspire increased exchange of data and ideas, promote greater standardization and accessibility of faunal data, and lay the groundwork for future collaborations and comparative research on patterns of faunal change in the context of hominin evolution (Figure 1 (Photo)).

Workshop discussions were organized around three major issues relating to how faunal information can be used to reconstruct the changing paleoecology of the late Miocene to late Pleistocene – the time period of hominin emergence and diversification in Africa: (1) key unresolved paleoecological and paleoenvironmental issues in human evolution, (2) methodological issues in the collection and analysis of fossil data in relation to hominin paleoecology, (3) strategies for effectively storing, retrieving, and sharing the vast and rapidly expanding paleontological information now kept in many different electronic databases. The papers in this volume present new, or newly compiled, data from the mammalian fossil record that relate to all three of these topics and demonstrate the benefits as well as the challenges of handling and analyzing such data.

The book begins with a Preface by Andrew Hill and two chapters that review the major issues involved in the study of mammalian faunas in the context of human evolution in East Africa. The article by Kay Behrensmeyer, René Bobe, and Zeresenay Alemseged, "Approaches to the analysis of faunal change during the East African Pliocene," defines faunal analysis, outlines theoretical issues relating to the interpretation of faunal data, and illustrates these issues using examples drawn from published research on African mammalian faunas. The following article by Richard Potts provides a comprehensive overview of "Environmental hypotheses of Pliocene human evolution," focusing attention on the hominin record and how faunal data can be used to develop and test hypotheses about cause–effect relationships between environmental change and hominin adaptation. The following 10 chapters offer in-depth research relating to specific taxonomic groups and fossil-bearing sites. The sequence of chapters is organized generally from the northern to southern portions of the East African Rift System, where a significant proportion of the fossil record of late Cenozoic hominin evolution is preserved. Although the focus is on the Pliocene, chapter discussions and data range from the Miocene to the Pleistocene.

Steve Frost's "African Pliocene and Pleistocene cercopithecid evolution and global climatic change" provides primary data and in-depth analysis of the monkeys in light of a proposed mammalian turnover pulse between 2.8 and 2.5 Ma in Africa. Margaret Lewis and Lars Werdelin, in "Patterns of change in the Plio-Pleistocene carnivorans of eastern Africa," consider the evolution of the carnivore guild in light of the emergence of hominin hunting and scavenging. H.B.S. Cooke builds on his previous extensive work on African suids, examining taxon-specific metric patterns in "Stratigraphic variation in Suidae from the Shungura Formation and some coeval deposits," demonstrating periods of accelerated change in the different suid lineages. René Bobe and his coauthors provide a new comparison of faunal databases from different basins in "Patterns of abundance and diversity in late Cenozoic bovids from the Turkana and Hadar Basins, Kenya and Ethiopia," highlighting inter- and intrabasinal differences in contemporaneous faunas. Zeresenay Alemseged, René Bobe, and Denis Geraads continue this

theme in "Comparability of fossil data and its significance for the interpretation of hominin environments: A case study in the lower Omo Valley, Ethiopia," examining similarities and differences in the contiguous collections of the American and French expeditions. Gerald Eck focuses on the impact of different field-collecting protocols and other variables affecting the Omo Valley faunal samples in his paper on "The effects of collection strategy and effort on faunal recovery."

Studies of taphonomic processes and ecological information in modern ecosystems provide an important foundation for interpretations of East African paleoecology based on faunal data. Denné Reed's paper, "Serengeti micromammals and their implications for Olduvai paleoenvironments" provides an example of this approach and applies it to the fossil record of northern Tanzania. Working on an earlier time period in the same region, Charles Musiba and his coauthors review previous controversy and offer new interpretations in "Taphonomy and paleoecological context of the Upper Laetolil Beds (Localities 8 and 9), Laetoli in northern Tanzania." Denise Su and Terry Harrison provide an ecovariable analysis of the Laetoli mammalian faunas in "The paleoecology of the Upper Laetolil Beds at Laetoli: A reconsideration of the large mammal evidence." Insights on the mammalian record and its taphonomic biases in the southern portion of the East African rift are the focus of the paper by Oliver Sandrock and his coauthors on the "Fauna, taphonomy, and ecology of the Plio-Pleistocene Chiwondo Beds, Northern Malawi."

The epilogue by the three editors, "Finale and future: Investigating faunal evidence for hominin paleoecology in East Africa," provides a summary of the major topics and discussion points at the Workshop on Faunal Evidence for Hominin Paleoecology and indicates how the chapters in this volume begin to implement and expand upon these points. This summary includes input and recommendations from all the workshop participants on additional important issues that range from field-recording procedures and database sharing to the challenges of long-term funding for the curation of fossil collections in African museums.

The editors wish to express their thanks to the chapter authors and workshop participants – it has been a pleasure and a privilege to interact with this international community to foster new research and ideas regarding mammalian fossil records in the context of human evolution. Future progress will depend on continuing this interchange of ideas and data and increasing the number of African scholars who, in league with the international scientific community, will realize the potential of their continent's abundant fossil resources and continue to develop these resources for generations to come.

A.K. Behrensmeyer
Z. Alemseged
R. Bobe

Figure 1. Participants in the 2004 Smithsonian Faunal Workshop. Front row, left to right: Denné Reed, Zelalem Assefa, Chris Campisano, Victoria Egerton, Rick Potts, Nasser Malit, Varsha Pilbrow, Miranda Armour-Chelu. Second row, left to right: Terry Harrison, Joe Ferraro, Fred Kyalo Manthi, Katie Binetti, Kaye Reed, Samuel Ngui, Denise Su, Kay Behrensmeyer, René Bobe, Zeresenay Alemseged. Back rows, left to right: Tom Plummer, Andrew Hill, Suvi Viranta, Lars Werdelin, Margaret Lewis, Nina Jablonski, Nancy Todd, George Chaplin, Alison Brooks, Ngala Jillani, Francis Kirera, Charles Musiba, Gerry Eck, Manuel Domínguez-Rodrigo, John Yellen, Luis Alcalá, and Catherine Haradon. Steve Frost, John Harris, and Oliver Sandrock presented papers at the 2003 AAPA symposium but were not able to attend the 2004 Smithsonian workshop.

# Preface

This stimulating book has its origins in a conference session at a meeting of the American Association of Physical Anthropologists, followed by a workshop at the Smithsonian Institution. These meetings and this subsequent volume explicitly took on the problems surrounding hominin environments in eastern Africa from the time of the origin of the clade, using evidence provided by other fauna.

One major set of unresolved paleoenvironmental questions concerning hominins are those linked to hominin emergence and the early radiation of hominins through the Pliocene. Taking a historical perspective it seems that these issues are more or less the same as they were over a century ago, though we might look at them a little differently, we certainly have more data, and our provisional answers are not necessarily the same as they were in the past. The big questions are *what*? *where*? *when*?, and *why*? What particular ape evolved into a hominin? Where did it happen? When did it happen?, Why did it happen? These are all in some way paleoenvironmentally based issues.

These questions go back a long way, and Darwin is often taken as the starting point for scientific answers. In 1871 in The Descent of Man he noted that that there tended to be a relationship between extinct and living species in different regions of the world, and said: "*It is therefore probable that Africa was formerly inhabited by extinct apes closely allied to the gorilla and chimpanzee, and as these two species are now man's nearest allies, it is somewhat more probable that our early progenitors lived on the African continent than elsewhere.*" Of course, in 1871 the fossil record of our early progenitors was somewhat sparse, and so he added the following, which is less often quoted: "*But it is useless to speculate on the subject, for two or three anthropomorphous apes ... existed in Europe during the Miocene age;*

*and since so remote a period the earth has certainly undergone many great revolutions, and there has been ample time for migration on the largest scale.*"

Because of the lack of appropriate fossils other than *Dryopithecus*, the question *what* was rather unanswerable in Darwin's time – other than some anthropomorphous ape. *When* he thought was vaguely in the Miocene, or he would allow, even the Eocene. But Darwin did have something more concrete to say about *why*. Darwin saw bipedalism arising through an ape moving from the trees to the ground when, as he put it: "*some ancient member in the great series of the Primates came to be less arboreal, owing to a change in its manner of procuring subsistence, or to some change in the surrounding conditions ...*" And we have change in food, or change in immediate environment being invoked as causes.

We have a lot more fossils now than we did in Darwin's time, 135 years ago, and a lot more data of other kinds, a lot more conferences and volumes like this, but the big questions are still more or less the same, answers to them still uncertain, and these matters still debated.

In terms of *what*: we still do not have a very clear idea of the nature of the ape that led to hominins. Just like Darwin, we think it is some anthropomorphous ape. We have more than he had to choose from, but it is problematic to favor one over the other. Possibly none of them that we yet know. Partly this is due to a gap in evidence. In Africa there is a patchy but reasonable record of apes between about 23 and 14 Ma (million years ago), then hominins from 7 Ma or so onwards; but between 14 and 7 Ma, which appears to be the critical time, there are few sites or specimens. And in the context of gaps in evidence, another problem, which we all acknowledge but perhaps do not think about enough, is a geographical one, and applies more generally. Africa is approximately 4554123.6 square kilometers in area, but nearly all the relevant sites could be grouped in a box a few hundred miles on each side. So when people talk about fossil apes or hominins in "Africa" they are really talking about fossil apes or hominins from an area about 0.1% of the African continent as a whole. Clearly our current knowledge is hardly representative of what was there in the past throughout the whole of the continent; representative neither of taxa, nor of available environments. And that is something I think that should be factored more into paleoenvironmental ideas. There is too much focus on the Rift Valley; understandably, because at present there is not much information from anywhere else. But it is not necessarily the cradle of mankind. For a variety of very good geological reasons it is just where you happen to find the fossils. There are suitable environments for life, suitable environments for deposition after death in the form of lakes and their sediments, suitable volcanics that preserve the lake sediments, and a persistent tectonic regime that brings older rocks to the surface and re-exposes fossils. And *Sahelanhropus* is unlikely to be the first hominin to "have ventured out of the Rift Valley" as I read in the academic literature recently. It is just that there happened to be conditions in Chad between 6 and 7 Ma that were suitable for preserving dead hominins.

As to *where*, I think most people now consider, corresponding to Darwin's first hunch, that hominins evolved in Africa. But the deficiency of African ape fossils in the late Miocene has led some to suggest that the evolution of hominin ancestors did not take place in Africa after all. Their idea builds on Darwin's doubt, and with his observation that perhaps appropriate fossil apes exist in Europe in the Miocene, and that: "*since so remote a period the earth has certainly undergone many great revolutions, and there has been ample time for migration on the largest scale.*" They suggest that while maybe early apes evolved in Africa, at some point they all arose and moved to Asia and to Europe, where they continued evolving, before at some point getting up and moving back into Africa again. That characterization is a trifle unfair, but in opposition to this I have a conviction that fossil sites are rare accidental occurrences, and that within them, hominids are even rarer. There are very few fossiliferous exposures of the required age in Africa. Consequently there are not many large ape specimens, but apes do in fact occur in Africa during this period, in the Tugen Hills Ngorora Formation, dated at about 12 Ma, and in the Samburu Hills, also in the Rift Valley at maybe 10 Ma, and with more work I believe more will be found. However, apparently still, as in Darwin's mind 135 years ago, there remains some collective uncertainty as to where the great event happened.

*When*? Well we are much better off than formerly. *Sahelanthropus* now demonstrates that it was in fact before the Pliocene, around or before 6.5 or 7 million years ago, and that is a much better estimate than Darwin could have given. Between the time of Darwin and ourselves we have had estimates of 14 Ma based on morphology, and of 2.5 Ma from molecular work, both equally false. But now we are converging on the right number. Of course this is important paleoecologically, because we need to know with what climatic or environmental events, or with what events in other animal lineages, the speciation event might correlate. This is the reason people used to spend a lot of time trying to discover if the 14 Ma site of Fort Ternan in Kenya suggested an environment of grassland. In fact they sometimes still do, but they seem to have forgotten why.

Which brings us to the big question *why*? Why did some ape get up on its back legs and wonder what to do with its hands? Down from the trees and into the savannas of course, just like Darwin said, "*a change in its surrounding conditions.*" But the conversation is not so much in simple terms any more, and this is a good thing because the issue has been thought about too simplistically in the past. Now it is more generally suspected that hominins did not originate in the context of a savanna grassland environment. This conclusion comes partly from the fact that although we know that things have been gradually getting drier, it is hard to find evidence of good savannas at the right time. There is C3/C4 evidence from soil carbonates, some from the Baringo Project that suggests a mixture of closed and open habitats rather similar from 15.5 Ma to now. Although large forests intrude every now and again, there is no sign of a dramatic change from forest to grassland at any point, at least if I ignore my own warning and confine attention narrowly to the Rift Valley. However, evidence from a number of sources and sites show that mammals eating C4 grasses appear in increasing numbers from around 8 Ma. As additional evidence, general evaluations of the environments of the earliest hominins from a number of lines of evidence are also ambiguous with respect to a savanna environment. *Sahelanthropus* at 6–7 Ma appears to be in a very mixed situation, even near desert, based on both geological indicators and the inferred habitats of associated faunas. The site of *Ardipithecus ramidus* is reported as being wooded at 4.4 Ma – to fresh woods and postures new. *Australopithecus anamensis* occupied maybe a more open situation, but perhaps with wooded and lake margin forests at 4.2 Ma; again, most of the evidence is from fauna. So why did hominins become bipedal? Some more subtle aspect of feeding: "*a change in their manner of procuring subsistence?*" Why did other apes not become bipedal? If bipedalism was just a response to sudden aridity and the development of open conditions, why did *Sivapithecus* not become bipedal when the environment changed in Pakistan at around 7.7 Ma? Instead it just became locally extinct, or followed the retreating forests.

Other theories elaborate slightly, incorporating additional elements to the forest–savanna idea. One environmental conjecture incorporates topography and suggests that the Rift Valley proved to be a barrier for apes, vegetationally and structurally, and that this assisted in the divergence between chimpanzee and hominin lineages. Forest to the west; savanna to the east. But the recent discovery of chimpanzee fossils in the Rift puts this notion very much in question. It also demonstrates the general incompleteness of the hominid record on which we must base our theories. Despite years of work in the Rift, only one fossil chimpanzee occurrence has so far been recognized although clearly they lived there at times.

Another body of theory relates the putative forest–savanna shift to the influence of astronomical forcing on African environmental change and faunas. The possible importance of astronomical forcing has been much discussed and assumed, but only recently, from work in the Pliocene of the Tugen Hills, has the relatively great effect of such factors been actually demonstrated. It is clearly a very complicated system, but probably advances will be made through linking faunal change to predictable periods of high and low solar insolation and to the changes in rainfall and varying fragmentation of vegetation that such periods entail.

These are large fundamental paleoanthropological questions, raised here principally in the context of the origin of the hominin clade, an event that now appears to have taken place before the Pliocene. But these questions, and some more specific ones, form the overall background and context in which the following chapters can be read. Not only do some of these issues apply equally well to the Pliocene, but many extremely interesting and important events took place in that epoch. The Pliocene is a time which saw the diversification of the clade into a variety of sympatric hominin species. It saw regional differentiation into different species in different parts of Africa. There was the origin of the fascinating genus *Paranthropus*, and the one that many consider most important, the genus *Homo*. It saw the origin of systematic stone artifacts and of different modes of subsistence that their existence suggests. It is a period when global climate was changing, with inferred effects on hominins and the rest of the fauna in the interior of Africa. Obviously hominins were ecologically embedded in the faunas of which they were a part, and a close study of other mammals coexisting with them provides important evidence concerning the evolutionary changes seen in our own clade.

The following chapters amply demonstrate this. They concern aspects of faunas that range in time from the latest Miocene to the Pleistocene, and in space from the Chiwondo Beds of Malawi to the Hadar Formation in Ethiopia. Some concern whole faunas, problems of analysis, patterns in space and time. Others look at the contribution a more restricted taxon can make to our understanding of general paleoecology. Discussions range from applications on a narrow timescale, such as the horizon-specific, to the examination of broader patterns of faunal change through long periods of geological time. They provide clear evidence of the great utility of faunal approaches to questions of hominin evolution. They reinforce the need for the basic study of faunas, their taxonomy,

their adaptations, their environments, if we are fully to understand the context of our origins, and the reasons for the development of the distinctive characteristics of the human lineage during the critical phase of environmental change and faunal evolution that characterized the African Pliocene.

No longer with Darwin should we now say, *"but it is useless to speculate on the subject,"* because as this book shows, the information and evidence we now have is much more rich and reliable, and speculations and even conclusions can be justified. But such speculation should always be tempered with a keen appreciation of the limitations of that evidence; an awareness of what we do not know, as much as of what we think we do.

A. HILL
*Department of Anthropology*
*Yale University*
*P.O. Box 208277*
*New Haven, CT 06520-8277, USA*
*andrew.hill@yale.edu*

# 1. Approaches to the analysis of faunal change during the East African Pliocene

A.K. BEHRENSMEYER
*Department of Paleobiology and Evolution of Terrestrial Ecosystems Program*
*Smithsonian Institution*
*P.O. Box 37012, NHB MRC 121*
*Washington, DC 20013–7012, USA*
*behrensa@si.edu*


R. BOBE
*Department of Anthropology*
*The University of Georgia*
*Athens, GA 30602, USA*
*renebobe@uga.edu*


Z. ALEMSEGED
*Department of Human Evolution*
*Max Planck Institute for Evolutionary Anthropology*
*Deutscher Platz 6*
*04103 Leipzig, Germany*
*zeray@eva.mpg.de*

**Keywords**:   Hominin, mammals, paleoecology, paleocommunity, Africa, Turkana, Kanapoi, Omo

## Abstract

Vertebrate faunas provide important evidence for the ecological context of evolving hominins over a wide range of scales, from site-specific analysis of taxa directly associated with hominin fossils to faunal trends indicating long-term environmental change that could have affected human evolution. The foundation for all such paleoecological interpretations consists of fossil specimens in their original geological context. Study of fossils in context generates a body of "first-order" evidence consisting of taxonomic identifications of specimens and placement of these taxa in a time/space continuum. Analysis of first-order faunal data in light of additional evidence about taphonomy, sedimentology, geochemistry, and ecomorphology generates a body of "second-order" interpretations. These require additional assumptions and result in evidence for the ecological attributes of a taxon, its habitat, and its temporal and spatial relationships to other taxa. Both first- and second-order data sets can be examined for larger-scale patterns across space and through time. The validity of inferences relating faunal evidence to the ecology of a hominin species requires an additional step, i.e., careful consideration of exactly how the faunal information relates spatially and temporally to hominin remains and archeological sites. Examples of different approaches to using faunal information to infer paleoenvironmental contexts, paleoecological relationships, and long-term ecological trends highlight major issues in faunal analysis and how these relate to understanding the ecological context of human evolution.

*R. Bobe, Z. Alemseged, and A.K. Behrensmeyer (eds.) Hominin Environments in the East African*
*Pliocene: An Assessment of the Faunal Evidence, 1–24.*

## Introduction

Faunas have been an important source of information about African hominin paleoecology at least since the 1920s, when Raymond Dart began collecting fossil mammals associated with hominins during his pioneering work in the cave deposits of South Africa (Dart, 1925). Since Dart's time, and especially in the past 25 years, there has been a proliferation of goals and methods that fall under the general heading of "faunal analysis." These range from basic identification of fossils to comparisons of faunal lists in different basins and regions, to quantitative analysis of phylogenetic and ecomorphic patterns through time. Currently many of the goals of faunal analysis in the hominin fossil record are oriented toward reconstructing the paleoecology – adaptations, habitats, biogeography, and change through time – of the different hominin species as well as assembling a general picture of evolving African mammalian paleocommunities.

This paper will focus on some key issues relating to different temporal and spatial scales of faunal analysis that are used to understand hominin paleoecology and evolution. Our goal is to first review a number of conceptual and methodological issues relating to paleoecological faunal analysis and secondly to provide examples showing how different datasets and methodological approaches are used to generate paleoecological interpretations. Geological analysis also plays a critical role in paleoenvironmental reconstructions and provides evidence that can be integrated with paleoecological hypotheses based on faunas. This topic, however, is beyond the scope of our paper. The East African Pliocene is a critical time and place in the history of human origins and provides an organizing theme for the papers in this volume, but many of the problems and approaches we discuss also apply much more broadly to the vertebrate fossil record of other regions and time periods.

## Background

One of the most enticing but difficult problems in human origins research is how to relate global-scale changes in climate to shifts in species occurrence, paleoenvironments, and morphological evolution (Behrensmeyer, 2006; Potts, 2007). The question of climate and its impact on faunal change is fundamentally a paleoecological problem concerning how climate- related natural selection shaped the course of vertebrate evolution in Africa. The long-term climate records are based on marine core data from oceanic basins proximal to the African continent (e.g., deMenocal and Bloemendal, 1995; Shackleton, 1995; deMenocal, 2004; Lisiecki and Raymo, 2005), while most of the Pliocene fossil hominins and other fauna occur in the Rift Valley deposits of East Africa or in the cave deposits of South Africa. Long-term environmental records also occur in the lacustrine deposits of deep lakes in Africa (Cohen et al., 2000), but these are just beginning to provide paleoclimatic information for continental areas. The large body of site- specific faunal information from land-based fossil records is being used to address paleoecological questions at local (Cooke, 2007; Musiba et al., 2007; Su and Harrison, 2007), regional (Bobe et al., 2007; Sandrock et al., 2007), and continental scales (Frost, 2007). A major issue in faunal analysis, however, is how to take specimen-based, local-scale information from discontinuous stratigraphic records and use it to evaluate the impact of global-scale processes. Evidence from the terrestrial and marine realms has very different temporal completeness and resolution (Figure 1), and building a composite picture of faunal trends through time thus depends heavily on accurate temporal correlations between different strata and basins.

Because local tectonic and climatic processes help to control the ecology of any area, faunal patterns may reflect environments within restricted areas rather than regional or

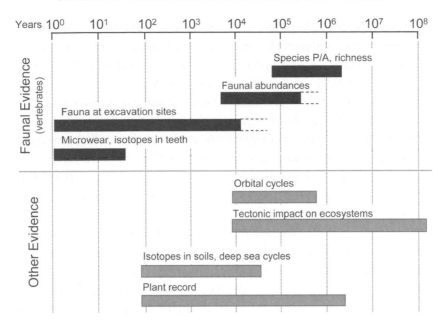

Figure 1. General temporal resolution of different types of evidence commonly used to infer hominin paleoecology, showing the approximate limits for different types of faunal and othergeological and paleontological evidence. Lower limits are set by the processes responsible for generating and preserving each type of evidence, while upper limits reflect a combination of the inherent characteristics of these processes plus decreased interest in examining their longer-term impact. Adapted from Figure 1, Kidwell and Behrensmeyer, 1993.

global-scale climate. In fact, the null hypothesis in paleoecological faunal analysis should be that smaller-scale environmental processes *do* control these patterns until proven otherwise. Therefore, before making a case for larger-scale cause and effect, researchers should ask to what extent local tectonic or climatic processes can account for the observed patterns. In the East African Rift, where there may be marked topographic differences over relatively short distances as well as frequent volcanic events, local environmental constraints may be especially important. Thus, a central problem for human origins research is how to distinguish different scales of paleoenvironmental cause and effect, from local or basin-scale controls to broader global changes that have been proposed as driving forces in human evolution (e.g., Vrba, 1988, 1995, 2000; Stanley, 1992; Behrensmeyer, 2006; Potts, 2007). Faunal evidence has been used on the one hand to support

global-scale climate forcing of East African faunal evolution (Vrba, 1995), and on the other for counterarguments and caveats based on evidence for more local controls on faunal patterns (e.g., Behrensmeyer et al., 1997; Bobe et al., 2002; deMenocal, 2004; Alemseged et al., 2007; Bobe et al., 2007).

There is no doubt that faunal evidence should play a critical role in evaluating such alternative hypotheses, and the strength of such evidence depends ultimately on the quality of the basic data and how it is analyzed. Our interest in global- scale environmental questions has emerged only recently, and much of the fossil record of eastern Africa was collected with other goals in mind, i.e., recovery and documentation of fossils in different regions and time periods (see Eck, 2007; Alemseged et al., 2007). This has resulted in a wealth of basic evidence for species presence and absence, geochronology, and paleoenvironments, much of which can be used to address larger- scale paleoecological questions. However, paleontologists probably would have structured field methods and goals somewhat differently if climate-faunal change questions had been a stronger guiding principle in their research design. Even the most sophisticated and rigorous methods of quantitative analysis (Ludwig and Reynolds, 1988; Shi, 1993) cannot overcome deficiencies in the primary data or inherent limitations of the fossil record. Researchers therefore must be aware of the weaknesses as well as the strengths of the existing body of faunal evidence and adjust their goals and methods accordingly.

## What is Faunal Change?

Faunal change, as reflected in the fossil record, consists of shifts through time in the presences or absences of taxa, relative abundances, geographic distributions, or ecological preferences, and can be measured in terms of parameters such as species richness, evenness, and turnover rates (appearances and extinctions of taxa) (e.g., Barry et al., 2002; Frost, 2007; Lewis and Werdelin, 2007). Phylogenetic relationships among taxa, including the diversification of a group or anagenetic transitions from ancestor to descendant species, also play an important role in faunal change (Cooke, 2007; Frost, 2007; Lewis and Werdelin, 2007). Faunal turnover is more specific and refers to the number of taxonomic appearances and disappearances relative to the overall taxonomic richness for sequential time intervals (e.g., Van Valkenburgh and Janis, 1993; Badgley et al., 2005). In the East African fossil record, as in any body of paleontological evidence, all of these measures of faunal change are based on interpretations of fossil specimens and their documented context in a space/time continuum, which is quite different from ecological data based on censuses of living animals. Ecologists, in fact, may use the terms "faunal change" and "faunal turnover" for differences

in faunas between contemporaneous ecosystems (Tilman and Kareiva, 1993; Russell, 1998). The many processes that intervene between a living community and the collections we use to reconstruct paleontological faunal change require a more cautious approach to evidence from the fossil record. A faunal trend is first of all a trend of fossils through time, and this may or may not be an accurate reflection of trends in the original ecosystem. Paleontologists have developed many ways to distinguish taphonomic and other biases from biological trends in the fossil record, but this remains a major focus for ongoing research (e.g., Behrensmeyer et al., 2005).

Faunal analysis may be aimed simply at documenting patterns of taxonomic biogeography, appearance, diversification, and extinction, without reference to environmental context or paleoecological implications. However, much of the faunal analysis relating to human evolution has been aimed at illuminating the ecological context of hominin taxa using the associated record of fossil mammals. Therefore, this paper will focus on selected issues relating to paleoecological faunal analysis in the African Pliocene, many of which are also represented in other contributions to this volume.

## Paleoecological Faunal Analysis

Fauna-based paleoecological information is derived either from inferences about the ecology of a single species (*autecology*) or about the ecological preferences or habitat(s) of a group of contemporaneous species (*synecology*) (Table 1). The autecology of an extinct organism depends heavily on analysis of functional morphology, dental microwear and mesowear, stable isotope analysis of tooth enamel, and how closely it is related to living species. Using these methods, a great deal of information can be gleaned about an individual, and the species to which it belonged, from a single specimen. Synecology involves ecomorphic assessment of multiple taxa that made up the original paleocommunities,

*Table 1. Types of evidence relating to terrestrial faunas, at increasing spatial/temporal scales, with examples of autecological and synecological data that can be inferred from this evidence. FAD = first appearance datum (of a taxon), LAD = last appearance datum (extinction). Entries in Column 2 constitute first-order evidence at increasing scales, while Columns 3 and 4 represent second-order evidence used for paleoecological interpretations*

| 1. Scale of evidence | 2. Type of evidence | 3. Autecology (inferences for an individual taxon) | 4. Synecology (inferences for multiple taxa) |
|---|---|---|---|
| Individual specimen | Taxonomy, skeletal and tooth morphology and wear, stable isotopes, enamel hypoplasia patterns | Taxonomic or ecomorphic relationship to extant species with known ecology, body size, diet, locomotion, etc. | N/A |
| Multiple specimens of single taxon | Variation in skeletal and tooth morphology, tooth eruption and wear, environmental context, ecomorphology | Phylogenetic relationships, demography, body size, diet, locomotion, preferred habitat, etc. | N/A |
| Multiple taxa from a single level or locality | Faunal list of species, genera, and higher taxa present in the fossil assemblage | Habitat inferred for a particular taxon based on ecomorphology and/or co-occurrence with ecological indicator taxa | Community structure and ecological preferences inferred from autecology and relative abundances of co-occurring taxa |
| Multi-locality faunal list | Analytically time-averaged compilation of taxa from a specified stratigraphic interval or area | Range of environments and co-occurring taxa indicate ecological adaptations, flexibility of taxon | Broad characterization of dominant or average community structure(s) represented by the fauna |
| Geological Formation | Sequence of faunas from members or localities in a single formation | Persistence, abundance, disappearance of individual taxa through a long stratigraphic interval | Long-term patterns of taxonomic richness, major group dominance, evenness, relationships to paleoenvironments |
| Basin (e.g., Turkana) | Biostratigraphic ranges and taxonomic dominance patterns through time | Space–time deployment, adaptations of individual taxa within a depositional basin through time | Variation in time and space of ecological parameters, correlation with different paleoenvironments |
| Region (e.g., East Africa) | Biogeography of taxa, inter- basin variation in abundance, timing of FADs and LADs for individual taxa | Biogeographic patterns, intraspecific variation in body size clines, diet, as evidence for adaptive range | Variation in community structure in different tectonic settings, latitudes, climatic zones, alpha versus beta diversity |
| Continent | Continental distribution patterns of taxa, preservational biases affecting these patterns; climate zones and biomes based on vegetation record | Evidence for ecological controls on immigration, emigration of a taxon between continents | Large-scale patterns of alpha and beta diversity, biome distributions, taxonomic dominance in relation to climate and vegetation zones |
| Global | Global FADs and LADs for taxa, marine cores as records of global-scale climate patterns | Latitudinal (temperature and moisture) controls on biogeography of individual taxa | Distribution of community types in time and across different continents |

measures of relative abundance of these taxa, sedimentological and geochemical evidence for paleoenvironments, and comparisons with ecomorphic patterns (ecological habitat spectra) in living mammal communities.

Whether one is using autecological or synecological information from fossil faunas to infer hominin paleoecology, the credibility of these inferences depends first of all on the quality of the collections- and outcrop-based information that provides the foundation for faunal analysis, and secondarily on how this information is used to infer paleoecology (i.e., methodological approaches and analysis). Following sections focus on three topics that are critical to paleoecological analysis: data quality, ecomorphological interpretation, and temporal considerations.

## DATA CATEGORIES

Data that provide the essential foundation for all faunal analyses and paleoecological interpretations based on faunal evidence can be categorized as: (1) first-order data – taxonomic identity, locality, stratigraphic position, taphonomic evidence, and geological context of each fossil occurrence in the rock record, (2) second-order data – e.g., inferred taphonomic history, original ecological habitat of a taxon and its temporal and spatial relationship to other taxa, including hominins. First-order data used in faunal analysis relies on accurate identification of specimens to genus and species and correct placement of these specimens in a time/space continuum. These types of data are now becoming increasingly accessible via computerized databases. The principles and methods of taphonomy, sedimentology, geochemistry, and ecomorphology enable paleobiologists to take first-order data and use this to build a body of second-order evidence regarding paleoecology.

First-order data encompasses much more than simple descriptions and may involve considerable research and interpretation of fragmentary fossils and complex stratigraphic relationships. However, we refer to this as "first-order" because it provides the essential foundation for all other levels of inference involved in the analysis of faunal change. Accurate taxonomic identification can be done to different levels of resolution, many of which are potentially useful in faunal analysis. The identification of bone or tooth fragments as

reptile, fish, or mammal, or at the family level as hippopotamus, elephant, or bovid (for example), usually involves minimal controversy and thus provides relatively unproblematic – and possibly very informative – taxonomic data. More intensive taxonomic analysis usually is required to accurately assign fossil specimens to a genus and species, leading to differences among workers that may affect measures of faunal diversity, provinciality, and timing of appearances and extinctions.

## ECOMORPHOLOGICAL EVIDENCE

The ecomorphology of fossil animals is characterized by their overall body shape and size, skeletal and dental anatomy, dental wear patterns, and stable isotopic signals (Andrews et al., 1979; Damuth, 1992; Plummer and Bishop, 1994; Andrews and O'Brien, 2000). Such evidence is used to infer diet, locomotion, positional and social behavior, foraging strategies, preferred habitat, and likely associations with other animals for individual taxa (autecology) and also for assemblages of taxa (synecology). Ecomorphological approaches can be applied to a single, time-specific fossil assemblage (e.g., Andrews, 1995) or to successive assemblages (e.g., Fernandez-Jalvo et al., 1998; Badgley et al., 2005), thereby providing a foundation for analyzing faunal change through long periods of time. Table 1 is arranged to show some of the different levels of resolution that are possible in paleoecological analysis incorporating both ecomorphological and contextual evidence. In general, as one scales up from specimens to locality-based faunal lists to regional and continental-scale compilations, decisions about how to use first- and second-order data become more of an issue and can lead to different interpretations based on the same evidence. Thus, it is critically important to collect, maintain, and provide access to the fossils themselves and their contextual information, the ultimate "ground truth" for resolving controversies and providing future generations of paleontologists with the opportunity to test new ideas.

## TEMPORAL FRAMEWORK

Temporal relationships among fossil mammals may be relatively straightforward when they occur in superimposed, well- dated strata, but accurate determination of absolute ages and temporal sequence of different faunal assemblages usually involves considerable interpretation, in part because datable materials often are not directly associated with fossils. It is common for paleoanthropologists and paleontologists to characterize fossils as "from the same interval or level," "associated," or "contemporaneous," without specifying exactly what this means in terms of temporal resolution. Faunal analysis could benefit greatly from more precise and accurate reporting of temporal associations among fossils and how these were determined. For some purposes,

such as general documentation of species appearances and extinctions and phylogenetic relationships, faunal lists that combine species from long stratigraphic intervals are appropriate and necessary. However, such lists may span tens to hundreds-of-thousands of years and thus provide a poorly resolved, "average" view of the ecological features of the evolving faunas. For paleoecological purposes, and also for more informative and accurate levels of resolution of biostratigraphic events, it is best to have the finest possible level of temporal and spatial resolution, i.e., site- specific contextual data for each fossil or fossil assemblage. The limits to such resolution, "taphonomic time-averaging," (Behrensmeyer and Hook, 1992) are typically set by taphonomic and depositional context (Figure 2). In practice, "analytical time- averaging" of faunas from multiple sites is often necessary to obtain an adequate sample size for paleoecological analysis. When assembled from well-documented individual sites and assemblages, however, carefully controlled analytical time-averaging can still achieve fairly high levels of temporal

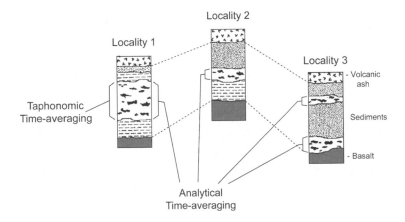

Figure 2. Different types of time-averaging of fossil remains (dark irregular shapes in white strata) that can affect paleoecological and evolutionary interpretations. Taphonomic time-averaging is the mixing of faunal remains within a single sedimentary unit; this represents the natural limit to temporal resolution of closely associated remains. Analytical time-averaging is the amount of time that results from combining faunas from different localities and strata, e.g., for a biostratigraphic faunal list. In this example, all localities occur in the interval between a volcanic ash and a lava flow (dashed lines), but their precise temporal relationships cannot be determined. Adapted from Figure 2.3 in Behrensmeyer and Hook, 1992.

resolution, and the resulting faunal lists can be used for many types of research questions.

Hominin fossils are rare components of the East African mammalian fossil assemblages and often occur as surface finds rather than in quarry situations or as *in situ* specimens. Therefore, to infer hominin paleoecology from "associated fossils," it is especially important to consider exactly how the faunal and other evidence relates spatially and temporally to the hominin remains. If hominin fossils and faunal evidence are from different stratigraphic units, or even from different localities within the same unit, there is a possibility that the two were not closely linked in the original ecosystems. Cyclical changes in climate over thousands or tens-of-thousands of years, shifting ecotonal (vegetation) boundaries (Figure 3), and biogeographical variations in vertebrate species distributions all could affect

the ecological signals in faunas from successive stratigraphic units. Taphonomic processes also can mix fossils from different times and places, even when these are buried together in the same deposit (Behrensmeyer, 1982, 1988; Reed, 2007). It is therefore critical to design analyses and to qualify inferences that take these possibilities into account.

Sampling, identification, and analytical protocols all may affect the quality and space/time resolution of paleoecological information that can be distilled from fossil assemblages, but the ecological characteristics of the original environments and faunal communities as well as taphonomic processes also impose their own limits on what we can know about hominin paleoecology. In the following section, we examine the types of ecological information that are accessible using faunal analysis.

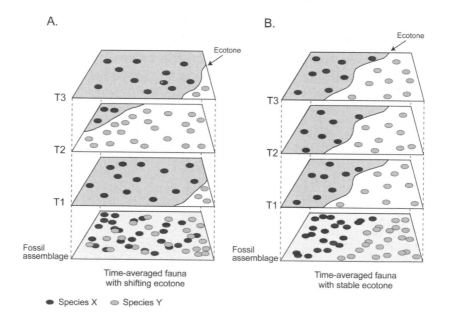

Figure 3. Hypothetical impact on autochthonous (e.g., floodplain) vertebrate fossil assemblages of shifting versus stable vegetational ecotones; gray and white backgrounds indicate different habitats, and black versus gray ovals are individuals of two different species that occur exclusively in these habitats. A. Shifts over time (T1, T2, T3) in the ecotone result in a mixed bone assemblage over much of the area represented by the fossil assemblage, and the original separation of the two species is obscured. B. Stable ecotone through T1–T3 results in two distinct faunas preserved in the same depositional unit, and information about original ecological separation of the two species can be recovered from the fossil record by samples from either side of the ecotone.

## Ecological and Taphonomic Issues in Paleoecology

Ecological variables of interest to paleoecologists include, in addition to the functional (ecomorphic) characteristics of individual taxa and their habitat preferences, the original geographic distribution and distribution of these taxa relative to different habitat types and to each other, the size of original habitat(s) and the stability of habitat boundaries over time (Reed, 2007). These variables represent ecological information that can, in theory, be preserved in rock strata and their associated fossil assemblages. However, such information is typically altered by taphonomic processes and also by analytical procedures, e.g., collection methods (Alemseged et al., 2007; Eck, 2007), the process of identifying and characterizing the taxa in a fossil assemblage, and manipulation of faunal data (e.g., time-averaging small samples to make a larger one). Taphonomic variables that must be considered when reconstructing ecological parameters include the placement of a fossil-preserving site relative to the center or boundaries of a habitat, the time over which the faunal remains accumulate (taphonomic time-averaging), effects of subsequent reworking and transport of the faunal remains to their final resting place (i.e., additional taphonomic temporal/spatial-averaging), and postburial diagenesis. Taphonomic processes and analytical procedures may, or may not, bias paleoecological information – bias must always be considered with respect to the question(s) under consideration (Behrensmeyer, 1991).

Autecological data provides ecomorphic information on the adaptations and likely habitat preferences of individual taxa, while synecological data provides information based on the patterns of occurrence and ecomorphology of multiple taxa. The following sections discuss spatial and temporal issues that affect multi-taxa occurrence patterns and thus may have important consequences for paleosynecological reconstructions based on the mammalian fossil record. Similar issues affect the use of autecological information from specific taxa to infer habitat preferences of associated hominins.

## HABITATS IN SPACE

Ecologists analyze species diversity, population abundances, etc., within the context of biomes, ecosystems, and habitats. Many studies have shown that species richness and dominance patterns are correlated with primary productivity and environmental stability, and secondarily with vegetation structure and habitat size (Pianka, 1978; Simberloff and Dayan, 1991; Tokeshi, 1993; Tilman, 1999; Waide et al., 1999; Colinvaux and De Oliveira, 2001; Ricklefs, 2004). Since paleoecology attempts to reconstruct these same parameters for ancient faunas, it is important to consider sources of variation in modern analogues and how these may affect ecological information in fossil assemblages (Behrensmeyer et al., 1979; Behrensmeyer, 1993; Reed, 2007). For example, the position of a fossil-preserving environment relative to the boundaries of a particular habitat may determine the strength of the faunal signal for that habitat in the resulting fossil assemblage. Obviously, the buried osteological record of forest fauna will be "purer" if this record accumulated in the center of a large area of forest habitat rather than near an ecotone with a different habitat. Alternatively, a fossil accumulation site in a highly fragmented mosaic of woodland, forest, and savanna can be expected to record a mixed-habitat fauna, even if animals are closely associated with one or another of these habitats. The original configuration of habitats and their animal communities thus play a critical role in shaping the fauna from a particular locality or stratum and must be kept in mind when we work back from the fossil assemblage to formulate and evaluate paleoecological hypotheses.

## HABITATS THROUGH TIME

Spatial patterning of habitats at any given time is only one variable affecting the ecological signals preserved in fossil assemblages. The nature of the stratigraphic and fossil record requires a different mind-set with regard to ecology because very long periods of ecological time usually are encompassed in the faunal samples that we use for paleoecological inference. There is a tendency to think of paleoecological reconstructions as equivalent to ecological snapshots of modern landscapes, e.g., "a mosaic open woodland with patches of grassland," etc., but in fact they may represent periods of hundreds to tens-of-thousands of years or more when grasslands and woodlands alternately dominated the landscape and the resulting faunal remains from each were mixed in a time-averaged fossil assemblage (Figure 3). If the site preserving a fossil record is near an ecotone, then even small or short-term variations in this ecotone caused by climate shifts or other variables such as river position in a fluvial system could change the types of animals being buried at a particular site (Behrensmeyer, 1982, 1993; Cutler et al., 1999). If the fossil-preserving site is near the center of a habitat, then even if boundaries shift through time, there is a higher probability of preserving an ecological signal specific to that habitat. In ecosystems where conditions are stable for long periods of time and there are few ecotones, as in continuously forested areas or open savannas, the resulting fossil record should be consistent from locality to locality and representative of these habitats. When ecosystems are characterized by a complex mosaic of grass, bush, woodland, and forest, however, then the impact of climatically induced fluctuations in these ecotones magnifies the impact of time-averaging and mixed ecological signals in the preserved faunal remains.

The above discussion assumes that faunal remains are being buried and preserved more-or-less in place, as in a floodplain soil or lake margin (Andrews, 1992). If they are reworked

and transported by fluvial or other processes, then original ecological signals are further overprinted, resulting in a larger spatial and longer temporal scale of mixing of fauna-based ecological information (Behrensmeyer, 1988, 1991; Aslan and Behrensmeyer, 1996). As previously discussed, there is the added problem of analytical time-averaging, in which fossil assemblages from different localities or levels must often be combined to provide large enough samples for analysis (Figure 2) (Behrensmeyer and Hook, 1992). Thus, it is little wonder that published interpretations of ecological signals based on faunal data often report a combination of different habitats, given the many ways that these signals may be blurred and mixed in the fossil record.

Most paleontologists understand that a fossil sample is not a one-time snapshot of the ancient community, but it is also incorrect to assume that the time-averaged faunal information represents a true "average" of the ecology of the stratigraphic interval in question. Fossils rarely occur evenly or randomly through a stratigraphic interval but are concentrated in specific zones or beds. It is theoretically possible that these fossil-bearing strata represent very specific climatic conditions – such as periods of transition from an extended dry phase to a wet one – thus biasing the ecological signal toward the drier end of the actual "average" spectrum of species in the vertebrate community. Much of the thickness in a lacustrine sequence could be taken up by deep water deposits representing wet climatic conditions, while the bulk of the land vertebrate fossil record is concentrated in shallow lacustrine to sub-aerial beds representing periods of regression of the lake caused by short periods of drier climate. It is obviously incorrect in this case to infer that the fauna represents the "average" ecological conditions for the interval. Likewise, fauna preserved in pedogenically modified fluvial silts may have been (1) buried when the silts were deposited in a floodplain pond, (2) worked in later by bioturbation associated with soil formation in a drier habitat, or (3) both. Thus, aquatic and terrestrial fauna may be mixed in the same bed. For most paleontological records in East Africa, how the preserved fauna actually relates to the different paleoenvironments within a given stratigraphic interval is currently poorly known and in need of careful examination. As a general rule, the temporal and spatial relationships between paleoecological interpretations based on sediments (including their stable isotope record) and faunas should always be demonstrated, not assumed.

## HABITATS AND LONG-TERM CLIMATE CYCLES

We know that in the historical and late Pleistocene record, climatic conditions, vegetation boundaries, and faunas have fluctuated radically over decades, centuries, and millennia (Davis, 1986; Webb, 1987; Overpeck et al., 1992; Graham, 1997; Jackson and Overpeck, 2000; deMenocal, 2004; Webb, 2004; Williams et al., 2004). If similar variability characterized the more distant past, then mixed ecological signals in both taphonomically and analytically time-averaged fossil assemblages should be the norm rather than the exception. The assumption that the past was as climatically variable as the present is likely to be incorrect, however. Increasing amplitude and frequency of climatic variability through the Plio-Pleistocene has been proposed as an important factor in human evolution (Potts, 1996). As we go back in time, the degree of habitat stability may be higher, and ecological mixing in fossil assemblages due to time-averaging consequently lower. However, testing this hypothesis is problematic because the ecological signal from any given fossil locality also depends on the proximity of this locality to an ecotone between habitats or climatic zones, as discussed above (Figure 3). Another critical issue is what portion of the climate-change curve is represented by the

interval with the fossils. Obviously, random samples of short intervals from the zigzag curve shown in Figure 4 could yield widely differing temperature conditions, even if only $10^3$–$10^4$ years apart in age, and this could translate into marked differences in vegetation and vertebrate biogeography among such samples. Since different types of depositional situations

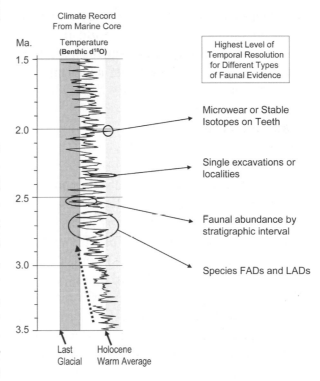

Figure 4. Examples showing how different types of faunal evidence, given their finest expected level of temporal resolution, would correspond to a Plio-Pleistocene climate record based on marine stable isotope data (Tiedemann et al., 1994). Short-term samples such as tooth microwear could represent particular points on the warm–cool climate shifts, while more poorly resolved events such as biostratigraphic appearances or extinctions (FADs or LADs) can only be linked to longer-term "packages" of warm–cool cycles. The dotted arrow indicates the well-documented late Pliocene shift from warmer, less variable climate conditions to cooler conditions with higher amplitude temperature fluctuations that occurred between ~3.5 and 2.8 Ma. Temperature curve adapted from Tiedemann et al., 1994.

as well analytical time-averaging result in fossil samples with different temporal resolution, few or many of the short-term climate cycles may be combined in the resulting faunal information (Figure 4). Furthermore, if fossils are more likely to be preserved during one phase of these cycles, e.g., during drier or more seasonal times when rates of surface decomposition are slower and/or pedogenic processes (e.g., carbonate precipitation) enhance bone permineralization, then it is possible that the resulting faunal information would over-represent ecological conditions during particular (e.g., drier) parts of the climate spectrum.

## Methodological Issues

The primary data for paleoecology are the same as for paleontological research in general – specimens and contextual information about when and where the specimens occur in the geological record. The quality of paleoecological interpretations rests on this foundation, but many fossils were not collected or documented with paleoecological questions in mind. The match between the available first-order data and the questions we seek to answer with these data is one of the most important data-related problems facing faunal analysts. For example, with some notable exceptions (Bobe and Eck, 2001), faunal evidence from published species lists or catalogued museum collections may be appropriate for determining species presence or absence but not for questions about species abundance because collecting and taxonomic procedures were inconsistent from year to year or resulted only in a record of species presence in local faunas. For these reasons, it usually is not possible to use catalogued specimens to assess the relative abundance of environmentally interesting major vertebrate groups, such as fish or reptiles versus mammals, or even the frequency of most mammal orders or families (Behrensmeyer and Barry, 2005). Only for taxa such as primates and carnivores,

for which every fragment usually is collected, can relative abundance be assessed based on catalogued specimens (e.g., Isaac and Behrensmeyer, 1997; Alemseged et al., 2007; Eck, 2007). Even in this case, however, the ability of field-workers to recognize postcranial remains as primate or carnivore may limit the accuracy of such data. If information on taxonomic abundances is necessary to address a particular paleoecological question, and data from catalogued and published specimens are not reliable, then new rounds of carefully controlled field collecting and documentation are necessary. This approach can also provide a way of measuring biases in existing collections (Behrensmeyer et al., 2004; Campisano et al., 2004; Behrensmeyer and Barry, 2005).

Examples of typical data and analysis issues relating to different types of paleoecological faunal evidence are given in Table 2. For some purposes, the simple presence of a particular taxon or ecomorph in a faunal list is sufficient to answer questions regarding habitat structure or faunal evolution. The absence of a taxon is another matter altogether, however, because treating this as an ecological or evolutionary fact (e.g., an extinction event or example of ecological exclusion) requires careful analysis to show that other possible causes of absence, such as collecting and taphonomic biases or taxonomic errors, can be eliminated or given low statistical probability.

Paleobiologists have developed a number of approaches for assessing the reality of taxonomic absence in the fossil record (Koch, 1987; Marshall, 1997; Foote, 2000; Barry et al., 2002), but these methods have yet to be widely applied to the East African fossil record. In particular, error-bars on taxonomic ranges (timing of origination and extinction) based on sample size considerations help to provide more realistic calibration of immigration events, faunal turnover patterns, shifts in paleocommunity structure, and diversification of well- represented groups such as the Bovidae (Bobe and Behrensmeyer, 2004). This in turn provides a

*Table 2. Examples of practical and theoretical issues that confront paleontologists as they work to build a solid body of evidence for faunal evolution and paleoecology*

| Type of evidence | Data issues | Analysis issues |
|---|---|---|
| Taxonomic presence and absence (major groups, tribes, genera, species) | Correct identification of fossils (up-to-date taxonomy), collecting biases, sample size, gaps in the record, taphonomic biases affecting preservation (e.g., body size, paleoenvironment), identification, chronologic and stratigraphic resolution | Characterizing the quality of record (e.g., specimen identifiability, geochronology), reconstructing taxonomic ranges (e.g., error-bars for FADs and LADs, how to treat gaps in the record) |
| Taxonomic abundance measures | Taxonomic consistency and collecting biases among sites, differential identifiability of fragmentary remains (e.g., bovid teeth versus bovid horn cores), taphonomic biases (e.g., relating to body size, depositional environment), publishing biases for or against particular taxa (e.g., primates versus elephants and hippos) | Inferences based on limited segment of the original community (e.g., Bovidae only), decisions about which abundance measures to use (e.g., locality frequency, minimum numbers of individuals, specimen numbers), biases relating to skeletal parts used as abundance proxies (e.g., teeth versus horn cores versus astragali) |
| Ecological indicators | Individual variation within species, accuracy of inferred autecology based on functional morphology, tooth wear, stable isotopes, etc., choice of modern analogues, validity of extrapolation based on phylogenetic relationships | Assumptions about relevance of modern analogues in ecological spectra analysis, impact of taphonomic biases on relative abundance of different ecomorphs, ecological flexibility of modern and fossil taxa apparently adapted to particular habitats |
| Taphonomic information from skeletal part preservation, bone modification features | Accurate representation of the proportions of different skeletal parts in the original fossil assemblages, preservational biases affecting bone modification features (e.g., tooth marks, cut marks, weathering stages) | Methods of counting skeletal part and bone modification features, minimum numbers of individuals, validity of interpretations of taphonomic processes based on limited sample of modern analogues |

more credible foundation for examining the timing of faunal change relative to regional and global-scale environmental shifts.

Beyond establishing simple presence–absence of taxa, many goals of paleoecological faunal analysis require information on relative abundance of taxa, from simple rank ordering of species or genera to percentages based on tallies of specimens (Badgley, 1986a; Badgley et al., 1995), individuals, or occurrences in multiple localities (e.g., Jernvall and Fortelius, 2002). Abundance is more likely to be biased by taphonomic processes (Badgley, 1986a, b) as well as collecting and identification practices; nevertheless, the goal of reconstructing ecological parameters for ancient faunas has generated several decades of concerted efforts to understand these biases and uncover ecological signals in abundance data. A summary of issues involved in such research is provided in Table 2. Even after taxonomic or ecomorphic abundances are established using appropriate quantitative methods (Andrews et al., 1979; Damuth, 1992; Andrews and O'Brien, 2000), the problem of how to infer ecology from these data remains. Comparisons with modern mammalian communities, and extrapolations based on taxonomic relationships (e.g., Alcelaphini = grazers) have provided the touchstones for such inferences (Andrews et al., 1979; Vrba, 1980; Bobe and Behrensmeyer, 2004). Hovering in the background, however, are assumptions that should always be kept in mind, for instance: (1) that the range of modern examples covers the range of ancient adaptations and community structure, (2) that it is valid to apply ecomorphic characteristics of modern taxa to their extinct relatives. These assumptions may be valid, but they also can limit our ability to see how the ecology of the past was different from that of today.

## Examples of Faunal Analysis and Inference

Temporal and spatial resolution affects all other types of paleoecological evidence because it determines how we order this evidence in time and relate it to environmental processes. Although temporal ordering from older to younger is usually straightforward for faunas from continuous strata, as soon as different outcrops, localities, basins, or regions are involved, temporal relationships become increasingly interpretive. The amount of time represented by a "contemporaneous fauna" is rarely discussed but also is interpretive, based on geochronology, estimates of sedimentation rates, and the depositional and taphonomic history of the fossiliferous deposit or interval. As shown in Table 1, different levels of resolution of faunal evidence relate to different scales of questions, and it is important to use evidence that is appropriate for the scale of the question at hand or to recognize the assumptions and limitations of "scale-jumping" between different levels of resolution. Following are examples of paleoecological inference based on fossil faunas that illustrate problems relating to temporal and spatial scale. Note that the intent of the discussion is not to critique these studies but to use them to illustrate general issues regarding paleoecological faunal analysis.

## EXAMPLE 1: LIMITS OF TEMPORAL RESOLUTION

In a recent study of the Kanapoi fauna of the southwestern Turkana Basin, Kenya, Harris et al. (2003) present an impressive body of faunal and ecomorphic evidence, primarily from fossils preserved in fluvial sands and paleosols deposited within a time interval between 4.17 and 4.07 Ma, i.e., approximately 100 Kyr. These deposits lie above and below a lacustrine deposit, thus the fauna as a whole represents an analytically time-averaged sample from two similar lithofacies that could be tens-of-thousands of years apart. Ecomorphic analysis of the faunas indicates that the two levels are only slightly different in terms of % terrestrial (ground-dwelling) mammals and % fresh-grass grazers (Harris et al., 2003: Figures 32 and 33), and the combined fauna is used to characterize the paleoecology of Kanapoi at the time of *Australopithecus anamensis*. This is interpreted as closed woodland based on comparisons with analogue environments and ecological structure analysis (Reed, 1997), but other lines of evidence suggest more open habitats as well, based on stable isotopic signals in tooth enamel, possible non-arboreal monkeys, and micromammals (Wynn, 2000). There is no direct evidence for the amount of time actually represented by the Kanapoi faunal samples, but it clearly was long enough by modern ecological standards to include habitat shifts across the areas of fossil accumulation (Figure 3). More open and more closed habitats could have alternated many times during hundreds to tens-of-thousands of years. Even within the time it took to form the paleosol that is the source stratum for the *A. anamensis* fossils at Kanapoi, habitats could have fluctuated. Also, the characteristics of the soil are superimposed on sediment that may represent a different ecology than occurred during the period of pedogenesis. Was *A. anamensis* buried during the sedimentary event, the early stages of soil formation, or later on in the hundreds to thousands of years represented in this unit?

Differences between the two alternative reconstructions of Kanapoi paleoecology suggested above are potentially important because the image we have of paleohabitats associated with hominins affects how we think about their adaptations. In the case of shifting habitats through time, hominin and other species could be closely tied to one habitat versus another, but still occur in a mixed-habitat fossil assemblage. In the case of a mosaic of both closed and open habitats, species would have more opportunities, and perhaps also more selective pressure, to adapt to a variety of resources and substrates. Given the limitations of the record illustrated by the Kanapoi fauna, which is one of the most age-constrained and carefully documented examples available at present, we cannot discriminate between these two alternatives. The temporal scale of our samples is too long. Mixed-habitat faunas do not necessarily mean mixed-habitat adaptations for the species on the faunal list.

The above discussion is offered as an alternative to the concept of a single complex suite of habitats that persisted throughout the time of fossil preservation. In reality, of course, we usually are dealing with varying proportions of closed versus open, or wetter versus drier habitats rather than the extremes of one or the other, and how this is recorded in the fossil record depends on the spatial scale of the sample as well as the amount of time represented. There is no simple solution to the problem of time-averaged ecological signals, but in some fossil-bearing sequences there are ways to calibrate the scale of habitat patches and evaluate the adaptations of individual species. These include higher resolution stratigraphic sampling and analysis (e.g., Fernandez-Jalvo et al., 1998), coordinated lateral sampling of faunas and paleoenvironmental variables from the same strata (e.g., stable isotopes on soil carbonate nodules, pedogenic features; Sikes et al., 1999; Blumenschine et al., 2003; Campisano et al., 2004), analysis of associations of species with known and unknown ecological preferences (Bobe and Behrensmeyer, 2004), and comparisons of tooth wear patterns and stable isotopes in the same species across space or through time (Nelson, 2003). Another strategy is to adjust the spatial and/or temporal scale of paleoecological interpretations to take account of the limitations of the record and resist the temptation to equate the ecological features of

a faunal list with snapshot (single time-plane) ecological analogues.

## EXAMPLE 2: USING DIFFERENT LEVELS OF TEMPORAL RESOLUTION

The Omo sequence of southern Ethiopia provides an example of the different levels of temporal and faunal resolution that are preserved in a faunal record spanning several million years (Coppens and Howell, 1976; de Heinzelin, 1983). The 800 m sequence of fluvial and fluvio-lacustrine deposits of the Shungura Formation is highly fossiliferous and was collected with great care and consistency by G. Eck and his team in the 1970s (Bobe and Eck, 2001; Eck, 2007). The formation is divided into members by prominent, radiometrically dated volcanic tuffs and sub-members by the fluvial cycles described by de Heinzelin (1983). Change through time in the mammalian fauna can be examined at the member or sub-member level of resolution. Data on taxonomic turnover (species appearance or disappearance), major mammalian groups, overall sample size, and proportions of different skeletal parts, resolved to the member level, provide one interpretation of faunal trends and stability through the 3 to 2 Ma time interval (Figure 5). In particular, there is little evidence for a species turnover "pulse" around 2.5 Ma in the Omo (Behrensmeyer et al., 1997; Bobe et al., 2002), as proposed by Vrba (1995, 2000, 2005) based on analysis of the Bovidae for Africa as a whole. However, at a sub-member level of resolution, using abundances of three mammal groups (bovid tribes, suid, and primate genera), another picture emerges (Figure 5). The proportions of these groups, based on chord distance comparisons of successive sub-member faunas shift from a pattern of relative stability to one of instability at about 2.5 Ma (chord distance = a measure of similarity between faunas based on numbers of specimens of each taxon; Bobe et al., 2002).

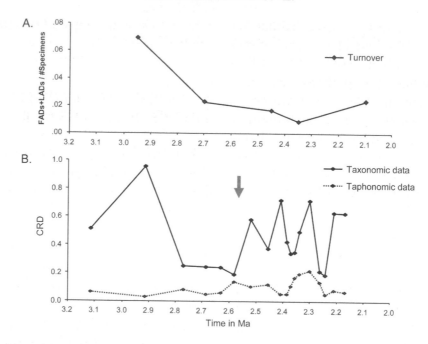

Figure 5. Comparison of patterns of faunal change analyzed at different levels of resolution for the Shungura Fm., southern Ethiopia (adapted from Figure 6 in Bobe et al., 2002). A. Plot of mammalian faunal turnover by member, showing a relatively low level of turnover between 2.8 and 2.1 Ma (original data in Behrensmeyer et al., 1997). B. Plot of abundance changes measured by chord-distance (CRD), from sub-member to sub-member, for common taxa of Bovidae, Suidae, and Primates, showing a stable interval between 2.8 and 2.55 Ma followed by 100 Kyr cycles in sub-member to sub-member faunal similarity. Gray arrow indicates approximate threshold between earlier interval of stability and subsequent cyclicity. Only sub- member intervals with 90 or more specimens were used for this plot.

A cyclic pattern of faunal similarity and dis-similarity continues to the end of the analyzed fossil record at about 2.1 Ma. There is no cor-related change in the taphonomic data (skel-etal part proportions), supporting an ecological interpretation for the shift in faunal stability.

This example demonstrates how different levels of resolution can provide very different pictures of faunal change, with added value in the comparison between them. On a continen-tal scale, based on both common and rare bovid taxa, there appears to be increased turnover between 2.8 and 2.5 Ma (Vrba, 1995, 2000, 2005), and this coincides with a global-scale environmental trend toward cooler, drier, and more variable climate. In the Omo Basin, how-ever, based on the evidence in Figure 5, the larger- scale environmental changes had minimal effect on species composition of the commu-

nity around 2.5 Ma, but these changes may be reflected in the abundances of common taxa, which began to fluctuate more frequently, sug-gesting increased ecological instability.

Elsewhere in Africa, global-scale changes may have had more or less impact on species turnover versus abundance shifts, depending on the influence of local circumstances. In the Omo example, Bobe et al. (2002, 2007) have proposed that regional and local buff-ering effects helped to reduce the impact of the larger-scale climate changes, at least for the interval between 2.8 and 2.0 Ma. This supports the interpretation that the rate and structure of faunal change varied in differ-ent regions of the African continent and indicates regional differences of hundreds-of-thousands of years in ecological responses to broad- scale climate change. In turn, this suggests that many different environmental opportunities and habitat types were available for species (including hominins) that could move from one region to another during the African Pliocene.

## EXAMPLE 3: INTERPRETING DATA REPRESENTING DIFFERENT TEMPORAL AND SPATIAL SCALES

Stable isotopes of carbon and oxygen are a major source of paleoecological information concerning diets and habitats of extinct species. In a study by Harris and Cerling (2002), isotopic analysis of the enamel of extant and Neogene suids shows that major components of $C_4$ vegetation (i.e., warm-growing season grasses) were present in extinct suid species from the Turkana Basin deposits of northern Kenya as early as about 7 Ma. The brachyodont *Kolpochoerus*, which has long been regarded as more of a browser based on dental morphology (Harris and Cerling, 2002) and a closed-habitat species based on post-cranial morphology (Bishop, 1999), plots as even more of a hyper grazer than the higher-crowned *Metridiochoerus*, based on samples of 8 *Kolpochoerus* and 7 *Metridiochoerus* specimens (Harris and Cerling, 2002: Table 6). However, *Kolpochoerus* has a lighter $\delta^{18}O$ signal, indicating it was more water-dependent, i.e., perhaps a wet-grass grazer in more mesic (moist) environments. Older suids of the *Nyanzachoerus* lineage also have carbon isotope signals indicating a $C_4$ diet.

Isotopic analysis of the enamel of suids and other species thus provides new evidence of great importance for faunal evolution in East Africa because it indicates that $C_4$ grasses formed significant biomass as early as 7 Ma (Morgan et al., 1994; Harris and Cerling, 2002), before many of the changes in dentitions linked to increased grazing and post-cranial adaptations for open habitats. Other research using pedogenic stable isotopes indicates that widespread grassland habitats were not common until the early to middle Pleistocene (Cerling, 1992; Kingston et al., 1994; Jacobs et al., 1999; Levin et al., 2005). There are obvious contradictions in these different lines of evidence that bear upon habitat reconstructions for the mammalian fauna (and associated hominins). Were these habitats very patchy, e.g., with small open areas of $C_4$ grass surrounded by dense thickets and woodland vegetation? Or is there something we are missing about the primary data that results in a biased view of the prevalence of $C_4$ diets and the importance of grasslands prior to the Pleistocene?

When contradictions emerge in paleoecology (or any other scientific field), it is always worth examining the first- and second-order data behind the competing interpretations. If nothing else, this can lead to better understanding of the complexity of the signals we are attempting to tease out of the fossil record, but at best it also can result in new interpretations and hypotheses that take us closer to the ecological reality of the past. Following are examples of questions that can be asked about the basic data used for the paleoecological inferences discussed in Example 3.

1. How well do the samples of fossil enamel represent the adaptations of the species? Each of the microsamples of enamel used for analysis represents a few weeks or months during the growth of the molar of an individual suid, and all the suid specimens are from a single basin (analytically time-averaged sample over long time-intervals for each species). Not surprisingly, the range of variation in carbon and oxygen isotopic signals for modern suid species from a large region of tropical Africa is greater than for the much smaller sample of fossil suids (Harris and Cerling, 2002: Figures 1 and 4). We do not know where the samples in question fall with respect

to regional variability in the Pliocene, and it is possible that the Turkana Basin fossils over-represent the $C_4$ end of the dietary spectrum of suids and other taxa. Proportions of grassland bovids provide supporting evidence that parts of the Turkana Basin were drier and more open than other basins during the Plio-Pleistocene (Bobe et al., 2007). Data from other basins and depositional settings could provide a more comprehensive basis for evaluating regional versus habitat- related variation in Plio-Pleistocene suid diets.

2. How do the samples of enamel correspond to the postcranial elements used to infer suid locomotion and habitat? Ideally, we would do both types of analysis on the same individuals, or samples from the same localities, but such coordination across different research approaches is rare. Evidence for locomotor adaptations is based on a limited sample of postcranial remains; nevertheless, the functional correlates of limb proportions and joint surfaces are well supported by research on modern analogues (Plummer and Bishop, 1994; Bishop, 1999). It is possible that selection pressures on suid locomotion were only weakly related to diet, i.e., an individual could depend on grazing but still retain a closed-habitat style of movement.

3. Does a $C_4$ signal in dental enamel always imply grazing? Stable isotope research is built on a solid foundation of theory and experimentation on modern analogues (e.g., Cerling and Harris, 1999), and the possibility that diagenetic overprinting could significantly affect original dietary signals is unlikely based on control samples, such as known browsing species (e.g., giraffe, deinothere) that retain $C_3$ signals. It is possible, however, that suids were eating significant amounts of non-grass $C_4$ plants (e.g., underground plant storage organs); thus the assumption that a $C_4$ signal = obligate grazing should be carefully considered. Microwear and mesowear analysis of teeth (e.g., Fortelius and Solounias, 2000) could provide tests of this hypothesis.

4. How does the diet of Plio-Pleistocene suids relate to human paleoecology and evolution? We assume for the moment that a clear isotopic signal indicating $C_4$ diets has a high probability of accurately reflecting the presence of $C_4$ habitats throughout the interval in question. Suids are abundant in the Plio-Pleistocene fossil record, and if most of them were grazers, then there must have been ample grass for them to eat. However, though this indicates one type of habitat available to Plio-Pleistocene hominins, it does not necessarily imply that the hominins frequented that habitat. Statistical analysis of associations of ecological indicator species (e.g., $C_3$ versus $C_4$) with hominins is one way to approach this problem (e.g., Bobe and Behrensmeyer, 2004). However, this is based on the assumption that co-occurrences in fossil localities reflect co-occurrences in the original communities – which may or may not be valid depending on the impact of taphonomic processes such as time-averaging and fluvial reworking.

Although in this example the clusters of points on the carbon–oxygen plots for the different fossil suid taxa provide compelling evidence for $C_4$ adaptations at the basin scale, generalizing to the level of the species, the Plio- Pleistocene, or the African continent must be done with caution, realizing that data may not be adequate to support extrapolations to much larger temporal and spatial scales. Given the proposed roles of open habitats and habitat change/increased variability in human origins (Potts, 1996), it is especially important be wary of "scale-jumping" from limited proxy samples indicating the presence of open habitats to visions of vast, Serengeti-like savannas over much of Africa. Hominin species were relatively uncommon components of the evolving faunal communities, and though we can accurately characterize general features of the ecosystems where they lived, it becomes increasingly challenging to pinpoint finer levels of their habitat preferences and adaptation.

## Conclusion

Faunal analysis has made many important contributions to understanding the history of East African mammal ecosystems and human evolution, and it will continue to expand this role in the future. Paleoecologists and paleoanthropologists are being challenged to provide high-quality data on terrestrial faunal change that can be integrated with global-scale paleoclimate research (deMenocal, 2004; Behrensmeyer, 2006). Compilation of previously published and catalogued faunal records in databases will be an increasingly important part of this effort, but ultimately the accuracy and usefulness of paleoecological interpretations depends on the quality of the fossil record and attendant documentation on which these interpretations are based. New rounds of fieldwork targeting paleoecological objectives in addition to taxonomic or biostratigraphic goals, along with increased integration of different types of paleoecological evidence, should play an important role in faunal analysis over the years to come. Meanwhile, the papers to follow in this volume showcase some of the insights and discoveries that are possible because of the creativity and long-term cooperation of researchers from around the world who have built and curated the collections and assembled an enormous body of contextual data that is now available from the East African fossil record.

## Acknowledgments

The senior author thanks her co-authors for their patience and encouragement throughout the writing of this manuscript. Many of the ideas in this paper have been simmering for years but were brought off the back burner during the Faunal Workshop in East African Faunal Analysis held in May, 2004, at the Smithsonian Institution. We thank the participants in that workshop for discussions and reactions to important issues facing us in faunal analysis and three very thoughtful reviewers who provided helpful suggestions on the initial version of this paper. AKB is grateful to the Smithsonian's Scholarly Studies Program, the National Science Foundation, and the National Museum of Natural History's Evolution of Terrestrial Ecosystems Program (ETE) and Human Origins Program (HOP) for their support of research on the paleoecological context of human evolution. The curators and staff at the National Museums of Kenya, Ethiopia, and Tanzania deserve deep appreciation for their continuing commitment to preserving the fossil heritage of their countris as a global resource. This is ETE contribution #128.

## References

Alemseged, Z., Bobe, R., Geraads, D., 2007. Comparability of fossil data and its significance for the interpretation of hominin environments: a case study in the lower Omo valley, Ethiopia. In: Bobe, R., Alemseged, Z., Behrensmeyer, A.K. (Eds.), Hominin Environments in the East African Pliocene: An Assessment of the Faunal Evidence. Springer, Dordrecht.

Andrews, P., 1992. Reconstructing past environments. Pp. 191–195. In: Jones, S., Martin, R., Pilbeam, D. (Eds.), The Cambridge Encyclopedia of Human Evolution. Cambridge University Press, Cambridge.

Andrews, P., 1995. Time resolution of the Miocene fauna from Pasalar. Journal of Human Evolution 28, 343–358.

Andrews, P., O'Brien, E.M., 2000. Climate, vegetation, and predictable gradients in mammal species richness in southern Africa. Journal of Zoology 251, 205–231.

Andrews, P., Lord, J.M., Nesbit Evans, E., 1979. Patterns of ecological diversity in fossil and modern mammalian faunas. Biological Journal of the Linnean Society 11, 177–205.

Aslan, A., Behrensmeyer, A.K., 1996. Taphonomy and time resolution of bone assemblages in a contemporary fluvial system: the East Fork River, Wyoming. Palaios 11(5), 411–421.

Badgley, C., 1986a. Counting individuals in mammalian fossil assemblages from fluvial environments. Palaios 1, 328–338.

Badgley, C., 1986b. Taphonomy of mammalian fossil remains from Siwalik rocks of Pakistan. Paleobiology 12, 119–142.

Badgley, C., Bartels, W.S., Morgan, M.E., Behrensmeyer, A.K., Raza, S.M., 1995. Taphonomy of vertebrate assemblages from the Paleogene of northwestern Wyoming and the Neogene of northern Pakistan. Palaeogeography, Palaeoclimatology, Palaeoecology 115, 157–180.

Badgley, C., Nelson, S., Barry, J.C., Behrensmeyer, A.K., Cerling, T., 2005. Testing models of faunal turnover with Neogene mammals from Pakistan. In: Lieberman, D.E., Smith, R.J., Kelley, J. (Eds.), Interpreting the Past: Essays on Human, Primate, and Mammal Evolution, in Honor of David Pilbeam. American School of Prehistoric Research Monograph Series. Brill Academic Publishers, Boston, pp. 29–44.

Barry, J.C., Morgan, M.E., Flynn, L.J., Pilbeam, D., Behrensmeyer, A.K., Raza, S.M., Khan, I.A., Badgley, C.E., Hicks, J., Kelley, J., 2002. Faunal and environmental change in the late Miocene Siwaliks of northern Pakistan. Paleobiology 28, Memoir 3 (Supplement to Number 2).

Behrensmeyer, A.K., 1982. Time resolution in fluvial vertebrate assemblages. Paleobiology 8, 211–227.

Behrensmeyer, A.K., 1988. Vertebrate Preservation in Fluvial Channels. Palaeogeography, Palaeoclimatology, Palaeoecology 63(1–3), 183–199.

Behrensmeyer, A.K., 1991. Terrestrial vertebrate accumulations. In: Allison, P.A., Briggs, D.E. (Eds.), Taphonomy: Releasing the Data Locked in the Fossil Record. Plenum Press, New York, pp. 291–335.

Behrensmeyer, A.K., 1993. The bones of Amboseli: bone assemblages and ecological change in a modern African ecosystem. National Geographic Research 9(4), 402–421.

Behrensmeyer, A.K., 2006. Climate change and human evolution. Science 311, 476–478.

Behrensmeyer, A.K., Barry, J.C., 2005. Biostratigraphic surveys in the Siwaliks of Pakistan: a method for standardized surface sampling of the vertebrate fossil record. Palaeontologica Electronica (Special issue in honor of W.R. Downs) 8(1), 15A:24p, http://palaeo-electronica.org.

Behrensmeyer, A.K., Hook, R.W., 1992. Paleoenvironmental contexts and taphonomic modes. In: Behrensmeyer, A.K., Damuth, J.D., DiMichele, W.A., Potts, R., Sues, H.D., Wing, S.L. (Eds.), Terrestrial Ecosystems through Time. University of Chicago Press, Chicago, pp. 15–136.

Behrensmeyer, A.K., Western, D., Dechant Boaz, D.E., 1979. New perspectives in paleoecology from a recent bone assemblage, Amboseli Park, Kenya. Paleobiology 5(l), 12–21.

Behrensmeyer, A.K., Todd, N.E., Potts, R., McBrinn, G.E., 1997. Late Pliocene faunal turnover in the Turkana Basin, Kenya. Science 278, 1589–1594.

Behrensmeyer, A.K., Bobe, R., Campisano, C.J., Levin, N., 2004. High resolution taphonomy and paleoecology of the Plio- Pleistocene Koobi Fora Formation, northern Kenya, with comparisons to the Hadar Formation, Ethiopia. Journal of Vertebrate Paleontology 24 (Supplement to Number 3), 38A.

Behrensmeyer, A.K., Fursich, F.T., Gastaldo, R.A., Kidwell, S.M., Kosnik, M.A., Kowaleski, M., Plotnick, R.E., Rogers, R.R., Alroy, J., 2005. Are the most durable shelly taxa also the most common in the marine fossil record? Paleobiology 31(4), 607–623.

Bishop, L., 1999. Suid paleoecology and habitat preferences at African Pliocene and Pleistocene hominid localities. In: Bromage, T.G., Schrenk, F. (Eds.), African Biogeography, Climate Change, and Human Evolution. Oxford University Press, Oxford, pp. 216–225.

Blumenschine, R.J., Peters, C.R., Masao, F.T., Clarke, R.J., Deino, A.L., Hay, R.L., Swisher, C.C., Stanistreet, I.G., Ashley, G.M., McHenry, L.J., Sikes, N.E., van der Merwe, N.J., Tactikos, J.C., Cushing, A.E., Deocampo, D.M., Njau, J.K., Ebert, J.I., 2003. Late Pliocene *Homo* and hominid land use from western Olduvai Gorge, Tanzania. Science 199, 1217–1221.

Bobe, R., Behrensmeyer, A.K., 2004. The expansion of grassland ecosystems in Africa in relation to mammalian evolution and the origin of the genus Homo. Palaeogeography, Palaeoclimatology, Palaeoecology 207, 399–420.

Bobe, R., Eck, G.G., 2001. Responses of African bovids to Pliocene climatic change. Paleobiology Memoirs 2. Paleobiology 27 (Supplement to Number 2), 1–47.

Bobe, R., Behrensmeyer, A.K., Chapman, R.E., 2002. Faunal change, environmental variability and late Pliocene hominin evolution. Journal of Human Evolution 42, 475–497.

Bobe, R., Behrensmeyer, A.K., Eck, G.G., Harris, J.M., 2007. Patterns of abundance and diversity in late Cenozoic bovids from the Turkana and Hadar Basins, Kenya and Ethiopia. In: Bobe, R., Alemseged, Z., Behrensmeyer, A.K. (Eds.), Hominin Environments in the East African Pliocene: An Assessment of the Faunal Evidence. Springer, Dordrecht.

Campisano, C.J., Behrensmeyer, A.K., Bobe, R., Levin, N., 2004. High resolution paleoenvironmental comparisons between Hadar and Koobi Fora: preliminary results of a combined geological and paleontological approach. Paleoanthropology Society 2004 Abstracts, A34.

Cerling, T.E., 1992. Development of grasslands and savannas in East Africa during the Neogene. Palaeogeography, Palaeoclimatology, Palaeoecology 97, 241–247.

Cerling, T.E., Harris, J.M., 1999. Carbon isotope fractionation between diet and bioapatite in ungulate mammals and implications for ecological and paleoecological studies. Oecologia 120, 347–363.

Cohen, A.S., Scholz, C.A., Johnson, T.C., 2000. The International Decade of East African Lakes (IDEAL) drilling initiative for the African great lakes. Journal of Paleolimnology 24, 231–235.

Colinvaux, P.A., De Oliveira, P.E., 2001. Amazon plant diversity and climate through the Cenozoic. Palaeogeography, Palaeoclimatology, Palaeoecology 166, 51–63.

Cooke, H.B.S., 2007. Stratigraphic variation in Suidae from the Shungura Formation and some coeval deposits. In: Bobe, R., Alemseged, Z., Behrensmeyer, A.K. (Eds.), Hominin Environments in the East African Pliocene: An Assessment of the Faunal Evidence. Springer, Dordrecht.

Coppens, Y., Howell, F.C., 1976. Mammalian faunas of the Omo Group: distributional and biostratigraphic aspects. In: Coppens, Y., Howell, F.C., Isaac, G.L., Leakey, R.E.F. (Eds.), Earliest Man and Environments in the Lake Rudolf Basin. University of Chicago Press, Chicago, pp. 177–192.

Cutler, A.H., Behrensmeyer, A.K., Chapman, R.E., 1999. Environmental information in a recent bone assemblage: roles of taphonomic processes and ecological change. In: Martin, R., Goldstein, S., Patterson, R.T. (Eds.), Fossil Taphonomy: Paleoenvironmental Reconstruction and Environmental Assessment. Palaeogeography, Palaeoclimatology, Palaeoecology 149, 359–372.

Damuth, J.D., 1992. Taxon-free characterization of animal communities. In: Behrensmeyer, A.K.,

Damuth, J.D., DiMichele, W.A., Potts, R., Sues, H.D., Wing, S.L. (Eds.), Terrestrial Ecosystems Through Time. University of Chicago Press, Chicago, pp. 183–203.

Dart, R.A., 1925. Australopithecus africanus: the man-ape of South Africa. Nature 155, 195–199.

Davis, M.B., 1986. Climatic instability, time lags, and community disequilibrium. In: Diamond, J., Case, T.J. (Eds.), Community Ecology. Harper and Row, New York, pp. 269–284.

de Heinzelin, J. (Ed.), 1983. The Omo Group. Musée Royale de l'Afrique Central, Tervuren, Belgique. Annales-Série in 8. Sciences Géologiques 85, Tervuren.

deMenocal, P.B., 2004. African climate change and faunal evolution during the Pliocene–Pleistocene. Earth and Planetary Science Letters 220, 3–24.

deMenocal, P.B., Bloemendal, J., 1995. Plio-Pleistocene climatic variability in subtropical Africa and the paleoenvironment of hominid evolution: a combined data-model approach. In: Vrba, E.S., Denton, G.H., Partridge, T.C., Burckle, L.H. (Eds.), Paleoclimate and Evolution, with Emphasis on Human Origins. Yale University Press, New Haven, pp. 262–288.

Eck, G.G., 2007. The effects of collection strategy and effort on faunal recovery: a case study of the American and French collections from the Shungura Formation, Ethiopia. In: Bobe, R., Alemseged, Z., Behrensmeyer, A.K. (Eds.), Hominin Environments in the East African Pliocene: An Assessment of the Faunal Evidence. Springer, Dordrecht.

Fernandez-Jalvo, Y., Denys, C., Andrews, P., Williams, T., Dauphin, Y., Humphrey, L., 1998. Taphonomy and palaeoecology of Olduvai Bed-I (Pleistocene, Tanzania). Journal of Human Evolution 34, 137–172.

Foote, M., 2000. Origination and extinction components of taxonomic diversity: general patterns. Paleobiology 26 (Supplement), 74–102.

Fortelius, M., Solounias, N., 2000. Functional characterization of ungulate molars using the abrasion-attrition wear gradient: a new method for reconstructing paleodiet. American Museum Novitates 3301, 1–36.

Frost, S.R., 2007. African Pliocene and Pleistocene cercopithecid evolution and global climatic change. In: Bobe, R., Alemseged, Z., Behrensmeyer, A.K. (Eds.), Hominin Environments in the East African Pliocene: An Assessment of the Faunal Evidence. Springer, Dordrecht.

Graham, R.W., 1997. The spatial response of mammals to Quaternary climate changes. In: Huntley,

B., Cramer, W., Morgan, A.V., Prentice, H.C., Solomon, A.M. (Eds.), Past and Future Rapid Environmental Changes: The Spatial and Evolutionary Responses of Terrestrial Biota, NATO ASI Series 1: Global Environmental Change 47:153–162.

Harris, J.M., Cerling, T.E., 2002. Dietary adaptations of extant and Neogene African suids. Journal of the Zoological Society of London 256, 45–54.

Harris, J.M., Leakey, M.G., Cerling, T.E., Winkler, A.J., 2003. Early Pliocene tetrapod remains from Kanapoi, Lake Turkana Basin, Kenya. In: Harris, J.M., Leakey, M.G. (Eds.), Contributions in Science Number 498, Geology and Vertebrate Paleontology of the Early Pliocene Site of Kanapoi, Northern Kenya. Natural History Museum, Los Angeles, pp. 39–113.

Isaac, G.Ll., Behrensmeyer, A.K., 1997. Geological context and palaeoenvironments: geology and faunas of the Koobi Fora Formation. In: Isaac, G.Ll., Isaac, B. (Eds.), Koobi Fora Research Project Volume 5: Plio-Pleistocene Archaeology. Clarendon Press, Oxford, pp. 12–32.

Jackson, S.T., Overpeck, J.T., 2000. Responses of plant populations and communities to environmental changes of the late Quaternary. Paleobiology Supplement to 26(4), 194–220.

Jacobs, B.F., Kingston, J.D., Jacobs, L.L., 1999. The origin of grass-dominated ecosystems. Annals of the Missouri Botanical Garden 86, 590–643.

Jernvall, J., Fortelius, M., 2002. Common mammals drive the evolutionary increase of hypsodonty in the Neogene. Nature 417, 538–540.

Kidwell, S.M., Behrensmeyer, A.K., 1993. Summary: estimates of time-averaging. In: Kidwell, S., Behrensmeyer, A.K. (Eds.),Taphonomic Approaches to Time Resolution in Fossil Assemblages: Short Courses in Paleontology No. 6. Paleontological Society, Knoxville, Tennessee, pp. 301–302.

Kingston, J.D., Marino, B.D., Hill, A., 1994. Isotopic evidence for Neogene hominid paleoenvironments in the Kenya Rift Valley. Science 264, 955–959.

Koch, C.F., 1987. Prediction of sample size effects on the measured temporal and geographic distribution patterns of species. Paleobiology 13(1), 100–107.

Levin, N., Cerling, T., Brown, F.H., Quade, J., 2005. The Shungura Formation Soil Carbonate Record: Evidence for an Ecosystem Buffered from Regional Environmental Change. Paleoclimate and Human Evolution Workshop, Front Royal VA (November 17–20, 2005), abstr. w/prog.

Lewis, M.E., Werdelin, L., 2007. Patterns of change in the Plio-Pleistocene carnivorans of eastern Africa: implications for hominin evolution. In: Bobe, R., Alemseged, Z., Behrensmeyer, A.K. (Eds.), Hominin Environments in the East African Pliocene: An Assessment of the Faunal Evidence. Springer, Dordrecht.

Lisiecki, L.E., Raymo, M.E., 2005. A Pliocene–Pleistocene stack of 57 globally distributed benthic $\delta^{18}O$ records. Paleoceanography 20, PA1003, doi:10.1029/2004PA001071.

Ludwig, A.J., Reynolds J.F., 1988. Statistical Ecology: A Primer on Methods and Computing. Wiley, New York.

Marshall, C.D., 1997. Confidence intervals on stratigraphic ranges with nonrandom distributions of fossil horizons. Paleobiology 23, 165–173.

Morgan, M.E., Kingston, J.D., Marino, B.D., 1994. Carbon isotopic evidence for the emergence of C4 plants in the Neogene from Pakistan and Kenya. Nature 367, 162–165.

Musiba, C., Magori, C., Stoller, M., Stein, T., Branting, S., Vogt, M., Tuttle, R., Hallgrímsson, B., Killindo, S., Mizambwa, F., Ndunguru, F., Mabulla, A., 2007. Taphonomy and paleoecological context of the Upper Laetolil Beds (Localities 8 and 9), Laetoli in northern Tanzania. In: Bobe, R., Alemseged, Z., Behrensmeyer, A.K. (Eds.), Hominin Environments in the East African Pliocene: An Assessment of the Faunal Evidence. Springer, Dordrecht.

Nelson, S.V., 2003. The Extinction of *Sivapithecus*: Faunal and Environmental Changes Surrounding the Disappearance of a Miocene Hominoid in the Siwaliks of Pakistan. Brill Academic Publishers, Boston.

Overpeck, J.T., Webb, R.S., Webb, T. III., 1992. Mapping eastern North American vegetation change of the last 18 Ka: no- analogs and the future. Geology 20, 1071–1074.

Pianka, E.R., 1978. Evolutionary Ecology. Harper and Row, New York.

Plummer, T.W., Bishop L.C., 1994. Hominid paleoecology at Olduvai Gorge, Tanzania as indicated by antelope remains. Journal of Human Evolution 27, 47–75.

Potts, R., 1996. Evolution and climate variability. Science 273, 922–923.

Potts, R., 2007. Environmental hypotheses of Pliocene human evolution. In: Bobe, R., Alemseged, Z., Behrensmeyer, A.K. (Eds.), Hominin Environments in the East African Pliocene: An Assessment of the Faunal Evidence. Springer, Dordrecht.

Reed, D., 2007. Serengeti micromammals and their implications for Olduvai paleoenvironments. In: Bobe, R., Alemseged, Z., Behrensmeyer, A.K. (Eds.), Hominin Environments in the East African Pliocene: An Assessment of the Faunal Evidence. Springer, Dordrecht.

Reed, K.E., 1997. Early hominid evolution and ecological change through the African Plio-Pleistocene. Journal of Human Evolution 32, 289–322.

Ricklefs, R.E., 2004. A comprehensive framework for global patterns in biodiversity. Ecology Letters 7(1), 1–15.

Russell, G.J., 1998. Turnover dynamics across ecological and geological scales. In: McKinney, M.L., Drake, J.A. (Eds.), Biodiversity Dynamics: Turnover of Populations, Taxa, and Communities, Columbia University Press, New York, pp. 377–404.

Sandrock, O., Kullmer, O., Schrenk, F., Juwayeyi, Y.M., Bromage, T.G., 2007. Fauna, taphonomy and ecology of the Plio- Pleistocene Chiwondo Beds, Northern Malawi. In: Bobe, R., Alemseged, Z., Behrensmeyer, A.K. (Eds.), Hominin Environments in the East African Pliocene: An Assessment of the Faunal Evidence. Springer, Dordrecht.

Shackleton, N.J., 1995. New data on the evolution of Pliocene climatic variability. In: Vrba, E.S., Denton, G.H., Partridge, T.C., Burckle, L.H. (Eds.), Paleoclimate and Evolution, with Emphasis on Human Origins. Yale University Press, New Haven, pp. 243–248.

Shi, G.R., 1993. Multivariate data analysis in palaeoecology and palaeobiology – a review. Palaeogeography, Palaeoclimatology, Palaeoecology 105, 199–234.

Sikes, N., Potts, R., Behrensmeyer, A.K., 1999. Early Pleistocene habitat in Member 1 Olorgesailie based on paleosol stable isotopes. Journal of Human Evolution 37, 721–746.

Simberloff, D., Dayan, T., 1991. The guild concept and the structure of ecological communities. Annual Review of Ecology and Systematics 22, 115–143.

Stanley, S., 1992. An ecological theory for the origin of *Homo*. Paleobiology 18, 237–257.

Su, D., Harrison, T., 2007. The paleoecology of the Upper Laetolil Beds at Laetoli: a reconsideration of the large mammal evidence. In: Bobe, R., Alemseged, Z., Behrensmeyer, A.K. (Eds.), Hominin Environments in the East African Pliocene: An Assessment of the Faunal Evidence. Springer, Dordrecht.

Tiedemann, R., Sarnthein, M., Shackleton, N.J., 1994. Astronomical timescale for the Pliocene Atlantic $\delta^{18}O$ and dust flux records of ODP Site 659. Paleoceanography 9, 619–638.

Tilman, D., 1999. The ecological consequences of biodiversity: a search for general principles. Ecology 80, 1455–1474.

Tilman, D., Kareiva, P. (Eds.), 1993. Spatial Ecology: The Role of Space in Population Dynamics and Interspecific Interactions (MPB-30). Princeton University Press, Princeton.

Tokeshi, M., 1993. Species abundance patterns and community structure. Advances in Ecological Research 24, 111–186.

Van Valkenburgh, B., Janis, C.M., 1993. Historical diversity patterns in North American large herbivores and carnivores. In: Ricklefs, R.E., Schluter, D. (Eds), Species Diversity in Ecological Communities. University of Chicago Press, Chicago, pp. 330–340.

Vrba, E.S., 1980. The significance of bovid remains as indicators of environment and predation patterns. In: Behrensmeyer, A.K., Hill, A. (Eds.), Fossils in the Making. University of Chicago Press, Chicago, pp. 247–271.

Vrba, E.S., 1988. Late Pliocene climatic events and hominid evolution. In: Grine, F.E. (Ed.), Evolutionary History of the "Robust" Australopithecines. Aldine, New York, pp. 405–426.

Vrba, E.S., 1995. The fossil record of African antelopes (Mammalia, Bovidae) in relation to human evolution and paleoclimate. In: Vrba, E.S., Denton, G.H., Partridge, T.C., Burckle, L.H. (Eds.), Paleoclimate and Evolution, with Emphasis on Human Origins. Yale University Press, New Haven, pp. 385–424.

Vrba, E.S., 2000. Major features of Neogene mammalian evolution in Africa. In: Partridge, T.C., Maud, R.R. (Eds.), The Cenozoic of Southern Africa. Oxford University Press, Oxford, pp. 277–304.

Vrba, E.S., 2005. Mass turnover and heterochrony events in response to physical change. Supplement to Paleobiology 31(2), 157–174.

Waide, R.B., Willig, M.R., Steiner, C.F., Mittelbach, G., Gough, L., Dodson, S.I., Juday, G.P., Parmenter, R., 1999. The relationship between productivity and species richness. Annual Review of Ecology and Systematics 30, 257–300.

Webb, T. III., 1987. The appearance and disappearance of major vegetational assemblages: long-term vegetational dynamics in eastern North America. Vegetation 69, 177–187.

Webb, T. III., 2004. Climatically forced vegetation dynamics in eastern North America during the late Quaternary period. In: Gillespie, A.R., Porter,

S.C., Atwater, B.F. (Eds.), The Quaternary Period in the United States. Elsevier, Amsterdam, pp. 459–478.

Williams, J.W., Shuman, B.N., Webb, T. III, Bartlein, P.J., Leduc, P.L., 2004. Late Quaternary vegetation dynamics in North America; scaling from taxa to biomes. Ecological Monographs 74(2), 309–334.

Wynn, J., 2000. Paleosols, stable carbon isotopes, and paleoenvironmental interpretation of Kanapoi, northern Kenya. Journal of Human Evolution 39, 411–432.

# 2. Environmental hypotheses of Pliocene human evolution

R. POTTS
*Human Origins Program, National Museum of Natural History*
*Smithsonian Institution*
*Washington, DC 20013-7012, USA*
*pottsr@si.edu*

**Keywords**: Human evolution, Pliocene climate, environment, fauna, climate dynamics, turnover pulse, variability selection, adaptability

## Abstract

Substantial evolutionary change in Pliocene hominins affected a suite of behaviors and anatomical features related to mobility, foraging, and diet – all related to the ways in which hominins interacted with their biotic and physical surroundings. The influence of environment on evolutionary change can be stated as a series of hypotheses. Adaptation hypotheses include the following: novel adaptations emerged in hominins and contemporaneous mammals (1) within relatively stable habitats; (2) during progressive shifts from one habitat type to another; and (3) due to significant rises in environmental variability. These ideas further suggest the "adaptability hypothesis": (4) since adaptations potentially evolved in environmentally stable, progressively changing, or highly variable periods, lineages have differed in their ability to endure environmental fluctuation. Thus, extinction of certain adaptations (and lineages) should have corresponded with heightened environmental variability, while new adaptations evolved during those periods should have enabled a lineage to persist (and spread) through a novel range of habitats. Turnover hypotheses, on the other hand, concern the timing and processes of species origination and extinction in multiple clades. These hypotheses state that species turnover; (5) was concentrated in a narrow interval of time related to a major climate shift; (6) spanned several hundred thousand years of climate change, and occurred in a predictable manner dependent on the nature of species adaptations; and (7) took place gradually over a long period as lineages originated, persisted, or went extinct within a changing mosaic of habitats. A separate, biogeographic hypothesis regarding faunal change posits that; (8) substantial climatic and tectonic disruptions resulted in multiple episodes of faunal community formation (assembly) and breakup (disassembly). This assembly–disassembly process may have had profound effects on Pliocene and Pleistocene faunas of Africa and on researchers' ability to infer significant events of faunal evolution from fossil sequences at the basin or sub-basin scale. Since all ideas about environmental effects on evolution depend on temporal correlation, an important challenge is to match faunal sampling to the precessional (~20 Kyr) and obliquity (41 Kyr) scale of Pliocene climate dynamics.

## Key Events in Pliocene Human Evolution

This chapter describes the range of hypotheses that relate early human evolution and environmental change in the Pliocene. Age-calibrated data sets now tell us a lot about African Pliocene environmental dynamics. Yet this information becomes relevant to questions of human origins only if we can identify what the pivotal events of hominin evolution were and when precisely they occurred. Likewise, evaluating how environmental change shaped

*R. Bobe, Z. Alemseged, and A.K. Behrensmeyer (eds.) Hominin Environments in the East African*
*Pliocene: An Assessment of the Faunal Evidence, 25–49.*

mammalian faunas relies on defining the precise timing and nature of species turnover and adaptive evolutionary shifts. One reason mammalian faunas are such a compelling area of inquiry is that fossil mammals comprised part of the early hominin environment, and they involved taxa whose phylogenetic, adaptive, and biogeographic trajectories were shaped as much, or as little, by environmental factors as early hominins'. The aim here, then, is to dissect the ways in which environmental change and human evolution may have coincided, with an eye on the broader question of how environmental conditions affected East African mammalian fauna. We begin here with a short list – a proposal of the main events in Pliocene human evolution.

## SPECIES ORIGINATION AND EXTINCTIONS

The oldest hominins may be late Miocene in age, with the relevant fossils from Chad, Kenya, and Ethiopia assigned to three genera – *Sahelanthropus*, *Orrorin*, and *Ardipithecus* (Haile-Selassie, 2001; Senut et al., 2001; Brunet et al., 2002; Haile-Selassie et al., 2004). The record is, at present, too incomplete to know whether the temporal range of *Sahelanthropus* or *Orrorin* extended into the Pliocene, if indeed either lineage had significant longevity. Although little is known about *Ardipithecus kadabba* (roughly 5.8 to 5.2 Ma [Haile-Selassie, 2001]), the first appearance of *Ar. ramidus* at ~4.4 Ma at Aramis, Ethiopia, implies a species origination in the early Pliocene (White et al., 1994, 1995; Renne et al., 1999). Fossils from the site of As Duma at Gona, Ethiopia, have also been attributed to *Ar. ramidus* (Semaw et al., 2005); but since these fossils, dated between 4.51 and 4.32 Ma, are chronologically indistinguishable from those of Aramis, the origination and extinction times of *Ar. ramidus* are still poorly constrained. It is an important point that when the first appearance datum

(FAD) and last appearance datum (LAD) of a taxon are essentially the same, its "lineage history" and possible environmental influences on its origination and extinction are virtually impossible to evaluate.

The genus *Australopithecus* originated during the early Pliocene based on the first appearance of *Au. anamensis* in the Turkana basin at ~4.1 Ma (Leakey et al., 1995, 1998). The FAD of *Au. afarensis* at ~3.6 Ma (Laetoli, Tanzania) – or possibly as early as ~3.85–3.89 Ma (Belohdelie, Ethiopia; [Asfaw, 1987; Renne et al., 1999]) – and its LAD at ~2.95 Ma (Hadar, Ethiopia), would appear to reflect significant evolutionary events (Leakey and Harris, 1987; Lockwood et al., 2000; White et al., 2000). *Au. afarensis* is currently one of the best-defined and best-calibrated lineages of Pliocene hominins, and shows evidence of evolutionary change late in its currently known time range (Lockwood et al., 2000).

*Au. africanus*'s first appearance, in South Africa, is less well calibrated, placed conservatively at ~2.8 Ma (Makapansgat), and possibly back to ~3.3 Ma (if the "little foot" skeleton, STW 573, from Sterkfontein Mbr 2 represents this species) (Vrba, 1995b; Clarke, 1999). The lineage persisted for at least 300 Kyr, possibly closer to 1 Myr, if the faunal age estimate of ~2.5 Ma typically linked to the *Au. africanus* sample from Sterkfontein Mbr 4 is correct (Vrba, 1988).

A fossil of *Australopithecus* (*Au. bahrelghazali*) from Chad, assigned to ~3.5 Ma (Brunet et al., 1995), shows similarities to *Au. afarensis* and raises the question of biogeographic exchange across distant regions of Africa during the Pliocene (Strait and Wood, 1999). The timings of such exchanges are difficult to pin down. The likelihood that *Australopithecus* is paraphyletic creates further uncertainty about the diversity of Pliocene hominins (Skelton and McHenry, 1992; Strait et al., 1997; Strait and Grine, 2004).

It is partly due to the paraphyly of *Australopithecus* that the new genus and species

*Kenyanthropus platyops* was named, based on an assemblage of fossils from Lomekwi in the Turkana basin, including a single, poorly preserved cranium (Leakey et al., 2001; see also White, 2003). Whether this fossil assemblage at ~3.5 Ma represents its own evolving lineage or a short-lived variation best joined with other fossils currently assigned to *Australopithecus* is not yet certain.

Two other significant Pliocene evolutionary events include, first, the origination of *Paranthropus*, based on a FAD at ~2.7 Ma (represented by L55s-33 mandibular fragment from Omo sub-Member C6) – i.e., a lineage sometimes called *P. aethiopicus* (best represented by WT-17000 cranium from West Turkana at ~2.5 Ma) – and, second, a likely phyletic transition from *P. aethiopicus* to *P. boisei* at ~2.3 Ma (Suwa et al., 1996). Although cladistic analysis supports "robust australopith" monophyly (e.g., Strait et al., 1997), the possibility of deriving *P. robustus* from *Au. africanus* in southern Africa (e.g., Rak, 1983) would require taxonomic revisions and an addition to our list of late Pliocene speciation events. Further evidence of megadontia comes from *Au. garhi*, based on fossils from Bouri, Ethiopia, at ~2.5 Ma (Asfaw et al., 1999), but it is uncertain whether this specific name refers to a sustained lineage or a short-lived variant of *Australopithecus*.

Timing of the origin of *Homo* depends on the criteria used to define the genus. Candidates believed to signal that epic event include: (1) the hypothetical coincidence of earliest *Homo* with stone toolmaking (e.g., Leakey et al., 1964), with the FAD for stone tools currently at ~2.6 Ma (Semaw et al., 2003); (2) the appearance of molars similar to those of *H. rudolfensis* in the Omo sequence at ~2.4 Ma (Suwa et al., 1996); (3) the Chemeron temporal bone fragment at ~2.4 Ma (Hill et al., 1992); (4) the AL-666 maxilla attributed to *Homo* at ~2.3 Ma (Kimbel et al., 1997); (5) neurocranial enlargement by about 1.9 Ma, indicated by crania KNM-ER 1470 and ER 1590 in the Turkana basin; (6) the appearance of African *H. erectus/ergaster* possibly by 1.9 Ma, indicated by occipital fragment KNM-ER 2598 (Wood, 1991); or (7) a reconfiguration of body proportions and size (Wood and Collard, 1999) by about 1.7 Ma (Ruff and Walker, 1993), though perhaps by ~1.9 Ma, as indicated by the large innominate KNM-ER 3228 in the Upper Burgi Member at Koobi Fora, and by reanalysis of femur length in *H. habilis* (Haeusler and McHenry, 2004).

In assessing faunal or other environmental events associated with the origin of *Homo*, therefore, it is essential to indicate exactly which earliest marker of the genus is adopted and why.

## EXPERIMENTATION IN DENTAL PROPORTIONS

One or more instances of divergence between megadont and smaller-toothed lineages took place during the Pliocene (Grine, 1988; Wood, 1991; Suwa et al., 1994; Teaford and Ungar, 2000). Greater postcanine tooth size is coincident with the origin of *Paranthropus*, and increased megadontia apparently occurred as a trend in this genus over time, although the rate and steadiness of molar and premolar size increase are not altogether clear. Megadontia in Pliocene hominins is generally correlated with cranial evidence of hypermastication, including substantial cresting along the insertions of chewing muscles and architectural modifications (e.g., substantial anterior pillars, facial "dishing") related to high-chewing forces (Rak, 1983; McCollum, 1999). The presence of prominent anterior pillars in *A. africanus* implies, however, that at least this architectural feature is not always associated with hypermastication and strong megadontia.

Smaller postcanine dentitions are typically assigned to early *Homo sensu lato*; yet megadont dentition may sometimes have been coupled with neurocranial enlargement (e.g., KNM-ER 1470) possibly indicative of *Homo*. While megadont dentition is often coupled with small canines and incisors, evidence of hominins having large anterior and postcanine teeth (Clarke, 1988; Asfaw et al., 1999) suggests that considerable experimentation in dental proportions characterized Pliocene hominins.

These aspects of dental evolution in *Australopithecus*, *Paranthropus*, and early *Homo* are suggestive of dietary and masticatory diversity, with greater or lesser emphasis on harder, more brittle foods, more abrasive foods, and eventually tougher foods when animal tissues began to contribute to the diet (Teaford and Ungar, 2000; Ungar, 2004).

## EXPERIMENTATION IN BODY PROPORTIONS

Essentially nothing is known about limb proportions in the earliest Pliocene hominins. By ~4.1 Ma, *Au. anamensis* showed broadening of the weight-bearing proximal tibia, probably related to habitual bipedal striding (Leakey et al., 1995). *Au. afarensis* (based largely on the AL-288 partial skeleton) exhibited short legs and long arms relative to estimated trunk length, suggesting a locomotor skeleton that combined arboreal climbing and terrestrial bipedality (Johanson et al., 1987). Similarly, the relationship of forelimb-to-hindlimb joint size in *Au. africanus* appears to have been ape-like (McHenry and Berger, 1998). Analysis of limb bones assigned to *Au. garhi* suggests that elongation of the femur preceded shortening of the forearm (Asfaw et al., 1999). If this interpretation is confirmed by further discoveries, it implies that an important shift in limb proportions was initiated in at least one lineage by ~2.5 Ma.

A human-like, elongated femur coupled with an ape-like, long forearm apparently also characterized *H. habilis*. Confusion arises over *H. habilis* in this regard due to earlier analyses of the OH 62 skeleton, which

indicated more ape-like limb proportions than in *Australopithecus* (Johanson et al., 1987; Hartwig-Scherer and Martin, 1991). This finding was a significant factor in the proposal to transfer *H. habilis* to the genus *Australopithecus* (Wood and Collard, 1999). Reanalysis of the fragmentary partial skeletons OH 62 and KNM-ER 3735 (~1.8 to 1.9 Ma) shows, however, that *H. habilis* possessed a modern human pattern of limb shaft proportions, an elongated hindlimb relative to *Au. afarensis* and *Au. africanus*, yet similar brachial proportions to these taxa (Haeusler and McHenry, 2004). Its elongated hindlimb suggests similarities to *H. erectus* and may imply an anatomical commitment to terrestrial bipedality over longer distances.

## NEUROCRANIAL EXPANSION

Although there is some evidence for neurocranial (braincase) expansion in hominins from early to late Pliocene, an increase in "relative" brain size is less certain. Data presented by Aiello and Wheeler (1995) and Wood and Collard (1999) indicate a gradual increase in relative brain size, yet one that extended the pattern defined by Old World monkeys and apes. Rapid encephalization was a characteristic of Pleistocene hominin evolution.

## GEOGRAPHIC SPREAD WITHIN AFRICA

Key data points in our currently impoverished understanding of Pliocene hominin biogeography include the presence of *Australopithecus* in southern and north-central Africa (*Au. bahrelghazali*) by ~3.5 Ma (Brunet et al., 1995); hominin fossils in the Malawi corridor at ~2.5–1.8 Ma (Bromage et al., 1995); and the appearance of stone tools in North Africa by ~1.8 Ma (Sahnouni et al., 2002). It is unclear if the data points in north-central and northern Africa derive from a prior hominin presence in these regions (i.e., *Sahelanthropus tchadensis*: Brunet et al., 2002) or reflect the spread of East African populations. As fossils from Malawi (and potentially Mozambique and south-central Africa) may provide evidence of exchanges between eastern and southern Africa, so Uganda and Sudan would appear to be underrepresented in our knowledge of hominin/faunal exchanges with northern and north-central Africa. Despite the inadequate fossil record, at least several episodes of Pliocene faunal and hominin exchanges among distant African regions seem likely (Strait and Wood, 1999), and may indicate significant adaptive responses to environmental events.

## DEFINITE STONE TOOLMAKING AND TRANSPORT

The earliest record of stone tools, at ~2.6 Ma, is from East Gona and Ounda Gona South, Ethiopia (Semaw et al., 1997, 2003). Even in these oldest known instances, artifacts are quite richly concentrated. Thus, although commonly thought to indicate the oldest stone tools, these sites actually provide evidence of the oldest stone-tool "accumulations." It is quite possible that flaked stone was first used and deposited very sparsely across ancient landscapes (without stone accumulation), possibly at lowland cobble sources where primary context is difficult to demonstrate, or at higher-elevation outcrops where burial did not take place. The point is that archeological visibility plays a large role in determining when regular chipping and use of stone tools began (Panger et al., 2002). The date of 2.6 Ma approximates the oldest clustering of stone artifacts brought from diverse lithic sources. This marks a key behavioral transition, the transport of materials over substantial distances to favored locations.

Making stone tools was itself a major adaptive breakthrough, as sharp flakes and pounding implements greatly extended the dental equipment available to hominins. It is still unclear, though, which (and how many) hominin species were responsible for the Pliocene artifacts, and whether these hominins depended on stone tools episodically, seasonally, or year-round. Late Pliocene toolmaking entailed the use of tools (e.g., hammerstones) to make tools, which may imply a cognitive difference with respect to the use of natural objects or hands-only toolmaking evident in chimpanzees. Further behavioral complexity is seen even in the oldest tool assemblages by the extensive reduction of stone cores and the preferential selection of certain rock types for flaking (Roche et al., 1999; Semaw et al., 2003; Delagnes and Roche, 2005).

Pliocene concentrations of stone tools and animal bones signal one of the oddest developments in human evolution – consumption of food away from its source with significant delay in eating food after acquiring it (Isaac, 1978). The ability to transport things opened up essentially all movable food resources to stone-tool processing. This meant an increase in the complexity of spatial and temporal mapping of resources, and it ultimately had enormous evolutionary consequences for hominins who adopted tool/food transport as a basic way of life. The movement of resources would have allowed Oldowan toolmakers to use tools on widely dispersed foods, to make use of foods that varied considerably in their seasonal availability, and to adjust to changing habitats and resource conditions (Potts, 1991, 1996).

## ACCESS TO LARGE MAMMAL CARCASSES

The oldest evidence of hominin interaction with large animal carcasses (size class 3: roughly 100–350 kg) coincides with the first known record of stone tools. Reported instances include a cut-marked equid calcaneum on the surface of site OGS-6 at Gona, Ethiopia, dated ~2.6 Ma (Semaw et al., 2003; Domínguez-Rodrigo et al., 2005); and cut- and percussion-marked surface bones at Bouri, Ethiopia, dated ~2.5 Ma (de Heinzelin et al., 1999). Although no tools were found with the modified bones at Bouri, sharp stone flakes and cores were directly associated with broken animal bones at Gona. This appears to establish the timing of an important dietary shift toward a high-quality food resource (high energy/protein, low digestive costs), which was patchily distributed and, in general, less predictable in time and space than plant foods. It is unknown, however, whether access to and consumption of large animal tissues (muscle, fat, and nutritious internal organs) during the late Pliocene occurred regularly throughout the year, seasonally, or rarely over a lifetime or many generations. An intriguing insight on this matter may come from the genomal analysis of tapeworms. Human-specific tapeworms had emerged (i.e., distinct from tapeworm lineages specific to African carnivores) between 1.7 and 0.7 Ma; this appears to imply that access to raw or poorly cooked meat by modern human ancestors became sufficiently regular (such that uniquely human tapeworm lineages could evolve) only after 1.7 Ma (Hoberg et al., 2001).

Even the occasional consumption of large herbivore tissues meant that certain Pliocene hominins actively entered the competitive realm of large carnivores, in which predation risks were heightened. Although much attention is paid to tissues from animals >100 kg, an archeological site at Kanjera South (~2.1 Ma) shows that under certain circumstances Pliocene toolmakers focused on smaller animals (Plummer et al., 2001; Plummer, 2004), perhaps indicative of diverse strategies of carcass acquisition dependent on environmental conditions. The sample of late Pliocene sites >2 Ma is not yet large enough to know whether the oldest instances of animal processing typically led to large quantities of meat and marrow or merely to scraps left on abandoned and largely defleshed carcasses. Early access to meat-rich bones and joints

likely had important consequences for hominin social aggregation at least over short spurts of time.

In brief, the acquisition of potentially large packages of meat and fat was one of the seminal events in the record of Pliocene hominin evolution. We do not know how dietarily or socially dependent the toolmakers were on obtaining animal tissues. Whatever the answer, certain Pliocene populations at least occasionally took on the temporal, energetic, and survival costs inherent in the carnivorous domain. This activity minimally involved carrying appropriate stones, maintaining the skill to flake sharp edges, and having the wherewithal to capture and defend animal carcasses for long enough to gain significant nutrition from them. All of the factors noted here – e.g., the means of accessing carcasses, the dietary and social payoffs, the competitive and predation costs – were almost certainly highly sensitive to local environmental conditions and the faunal context (Potts, 2003).

## SYNPOSIS OF ADAPTIVE CHANGE IN PLIOCENE HOMININS

As the preceding summary implies, key adaptive shifts in Pliocene hominin evolution occurred mainly in three interrelated domains: *mobility* (indicated, for example, by a change in limb proportions, biogeographic spread between regions, and object transport by toolmakers), *foraging* (indicated by the use and accumulation of flaked stone and hammers in acquiring food), and *diet* (indicated by change in dental proportions and the use of stone tools to access food from large animals). Organisms, in general, respond to environmental variation largely via locomotor, foraging, and dietary adaptations, especially by tracking favored resources or climatic conditions, by altering the amount of time and the strategy of finding food, and by switching to alternative foods when necessary. Environmental conditions almost certainly would have affected the expression of novel behaviors in these three domains. Thus, studies that aim to relate the Pliocene faunal record to hominin adaptive change may benefit by focusing also on evidence of changing patterns of mobility, foraging, and diet in other large mammals.

## Pliocene Environmental Dynamics

Emphasis here will be given to the nonfaunal evidence of Pliocene environments. The main data sources include deep-sea $\delta^{18}O$ records of benthic foraminifera, African dust records, Mediterranean sapropels, and stable isotope and pollen evidence obtained from hominin localities. The methods by which these data sources contribute to understanding environmental dynamics are summarized in Vrba et al. (1995), Potts (1998b), and deMenocal (2004).

## GLOBAL CLIMATE DYNAMICS

The two most significant global climate events of the Pliocene were the onset of significant Northern Hemisphere glaciation (NHG) and a shift in the dominant period of climate oscillation, both of which occurred ~2.80 to 2.75 Ma. Significant NHG was preceded by higher temperature worldwide, an era known as the early Pliocene warm period (EPW), roughly 5.3 to 3.3 Ma (Tiedemann et al., 1994; Shackleton, 1995; Ravelo et al., 2004). During the EPW, global surface temperature was ~3°C warmer than at present, sea level ~10–20 m higher, and atmospheric $CO_2$ concentrations ~30% greater (Ravelo et al., 2004). $^{18}O$ enrichment in marine benthic foraminifera beginning in the mid-Pliocene indicates decreased temperature, preferential evaporation of the lighter isotope $^{16}O$, and its retention in spreading continental ice sheets. The general $\delta^{18}O$ trend, therefore, suggests a colder, drier, more glaciated planet since about 3 Ma.

The overall trend, however, was disrupted by periodic reversals – i.e., warming, deglaciation, release of $^{16}O$ into the oceans, and sea-level rise. In fact, relative to the Oligocene and Miocene, the Pliocene was a time of considerably heightened $\delta^{18}O$ oscillation – and thus a novel degree of climate variability (Potts, 1998b). The onset of NHG was associated with a change in the period of climate oscillation from predominantly 19–23 Kyr to 41 Kyr, reflecting a shift from orbital precession to obliquity as the overarching determinant of variability in solar heating (insolation).

Tropical and subtropical climate is, of course, particularly relevant to Pliocene evolution in East Africa. According to an analysis of marine records by Ravelo et al. (2004), EPW tropical climate was (unlike today) characterized by weak east–west zonal (Walker) sea circulation, which meant essentially permanent El-Niño-like conditions. Development of strong Walker circulation took place in two steps, neither one temporally linked with the onset of NHG. The first tropical climate reorganization, between 4.5 and 4.0 Ma, was marked by altered surface water gradients and ocean circulation, possibly linked to restriction of the Panamanian and Indonesian seaways. The second, between 2.0 and 1.5 Ma, established strong Walker circulation, a steeper sea surface temperature gradient across the Pacific, and overall initiation of the modern tropical climate system. Thus, while the onset of significant NHG occurred as subtropical conditions began to cool, revisions in the Pliocene tropical climate system were independent to some degree (Ravelo et al., 2004).

Understanding the processes of climate change, especially ocean–atmosphere–land linkages, has become important in exploring the environmental events potentially related to human and faunal evolution in Africa. In particular, geo- and biochemical datasets and climate models have shown that African aridification, in general, is controlled by tropical sea surface temperature (SST), that East

African aridification beginning around 3 Ma was likely controlled by Indian Ocean SST, and that precessional variations in $C_3$ and $C_4$ plants have been controlled by changes in monsoonal precipitation driven by low-latitude insolation changes (Goddard and Graham, 1999; Philander and Fedorov, 2003; Schefuß et al., 2003). These findings imply that East African climate change has largely been governed by ocean–atmosphere linkages in the low latitudes, in close proximity to the continental basins where Pliocene hominins are known to have lived.

## NORTHEAST AFRICAN CLIMATE DYNAMICS

Despite the strong influence of tropical oceans on East African climate change, shifts in the tempo of climate variability appear to have been tightly linked across tropical-, mid-, and high-latitudes. In particular, East and West African dust records obtained from deep-sea cores (10–14°N latitude) show the shift from precessional (23 Kyr) to obliquity (41 Kyr) dominance at ~2.8 Ma, nearly concurrent with the shift in the $\delta^{18}O$ record (Tiedemann et al., 1994; deMenocal, 1995; deMenocal and Bloemendal, 1995).

Continental dust records in both regions illustrate three important aspects of Pliocene African climate change: (1) an overall increase in aridity, (2) a change in the periodicity of arid–moist cycles, both at ~2.8 Ma, and (3) the division of overall climate variability into alternating high- and low-variability packets, typically $10^3$ to $10^5$ years in duration (deMenocal, 2004). While $\delta^{18}O$ shows a significant and largely permanent rise in the amplitude of temperature and glacial–interglacial oscillation, low-latitude dust records do not clearly show this larger amplitude. What the dust records do show (which is not clearly evident in $\delta^{18}O$) is a time series of alternating intervals of high and low amplitudes. That is, inter-

vals of high aridity–moisture variability were punctuated by intervening periods of low variability, which deMenocal (2004) postulates resulted from the modulation of precession by orbital eccentricity.

Further evidence of African climate dynamics comes from the eastern Mediterranean record of sapropels, which has vastly improved since 1995 (Comas et al., 1996; Emeis et al., 1996). Sapropels are dark layers enriched in total organic carbon and certain elements such as Fe, S, Si, Ti, and Ba relative to Al (Wehausen and Brumsack, 1999). Sapropels are tied to the intensity of the African monsoon and to precessional periods of highest precipitation and discharge of the Nile into the eastern Mediterranean. The sapropel record indicates that there were peaks in African moisture every ~20 Kyr throughout the Pliocene (Emeis et al., 2000). Sapropels are not preserved in all expected intervals due to oxidation and postdepositional burn-down of organics. However, sapropel "ghosts" are now recognized, and very regular variations in Ba/Al and Ti/Al ratios indicate peaks in biological productivity associated with strongest Nile discharge (Wehausen and Brumsack, 1999, 2000). Geochemical analysis confirms, therefore, that sapropels are only the most visible indicators of a highly robust and rhythmic cyclicity in African climate that persisted through the entire Pliocene and Pleistocene. Sediments that intervene between sapropels (and other layers of high bio-productivity) show heightened levels of continental dust derived during arid intervals from the circum-Mediterranean area, including the Sahara.

Age estimates for each sapropel, sapropel "ghost", and "red interval" (in which no sapropels are preserved) for four Mediterranean cores, including ODP Site 969 dating between 5.33 Ma and 8 Ka, are given by Emeis et al. (2000), and are based on a standard 3000-year lag between mathematically predicted precession minima and sapropel midpoints (Lourens et al., 1996). Analysis of organic-carbon concentrations indicate that mid-Pliocene sapropels lasted for about 1,000 to 6,000 years, yet periods of enhanced bio-productivity (defined by Ba enrichment) lasted for 8,000 to 12,000 years (Wehausen and Brumsack, 1999). The latter range currently provides the best estimate for the duration of high African precipitation and Nile discharge during Pliocene precessional cycles. Precessional forcing of low-latitude monsoons was the leading mechanism of sapropel formation even after significant NHG began at ~2.8 Ma; however, the Mediterranean oxygen stable isotope record shows that this mechanism was more strongly affected by obliquity after 2.8 Ma and by 100-Kyr eccentricity cyclicity after 0.9 Ma (Emeis et al., 2000). The modulation of precession by obliquity and eccentricity yields a monsoon index, which has proven useful in predicting periods of highest African precipitation and sapropel formation (Rossignol-Strick, 1983).

For investigators of East African faunal and hominin evolution, the obvious question is whether the regular tempo of high precipitation and intervening aridity (as captured by the eastern Mediterranean sapropel record) and of northern African aridity–moisture cycles (as indicated by the deep-sea dust record) actually reflects the tempo and nature of climate change south of the Horn of Africa and the Sahara. That is, are the dust and sapropel findings strongly representative of Pliocene climate change associated with fossil sites in Kenya and Tanzania, for example?

This question is ultimately best answered by careful study of the timing and extent of high-moisture and high-aridity intervals in East Africa (e.g., Ashley and Hay, 2002; deMenocal, 2004). Tephrocorrelations between the Turkana basin and Gulf of Aden do suggest, though, a direct means of linking the dust record to broader East African climate change (deMenocal and Brown, 1999). Furthermore, Nile discharge is controlled by the African monsoon, in which moisture originating in the South Atlantic Ocean is captured by the Nile drain-

age. The intensity of the monsoon is known to be influenced by the meteorological equator, known as the intertropical convergence zone (ITCZ), and by trade-wind intensity – both of which influence a much broader portion of Africa than the Nile catchment (Emeis et al., 2000). For these reasons, both the sapropel and dust records appear to offer a good approximation of the tempo and degree of climate variability over East Africa, and offer specific predictions about the age of alternating arid and humid intervals throughout the region.

An intriguing hypothesis proposed recently is that precessional variation in moisture responsible for sapropel formation has also controlled depositional–erosional cycles in continental basins of northeastern Africa (Brown, 2004). Thus, the deposition of fossiliferous Member 1 in the Omo Kibish Formation, following a substantial erosional period, has been considered contemporaneous with the wet interval of sapropel S7; on this basis and Ar/Ar dating constraints, the date of this sapropel at ~195 Ka has been assigned to the early *Homo sapiens* fossils from Kibish (McDougall et al., 2005). If this hypothesis proves to be correct, it means that major depositional–erosional sequences in East Africa may be linked to climate variability. One unresolved issue is that major erosional periods, in which massive amounts of sediment must be moved, would also seem to require substantial water flow, which could be linked to significant monsoonal precipitation rather than to arid times. Nonetheless, since fossils are preserved during depositional intervals, the hypothesis linking depositional–erosional intervals with climate variability could have important implications for analyses of the fossil record.

## ENVIRONMENTS ASSOCIATED WITH PLIOCENE HOMININ LOCALITIES

Fieldwork at Pliocene sites has rarely focused on environmental dynamics; the standard goal, rather, has been to reconstruct the habitats reflected in particular strata in which fossils or archeological remains are found. For example, based on stable carbon isotopic records, fauna, and fossil seeds, WoldeGabriel et al. (1994, 2001) concluded that late Miocene and early Pliocene hominins of the Middle Awash, Ethiopia, lived in woodland and forest, and that early hominins inhabited more open vegetation only after 4.4 Ma. By contrast, a mosaic of open and wooded habitat is reconstructed by Leakey et al. (1996) for the upper Nawata Formation at Lothagam, Kenya, near the Miocene–Pliocene boundary, though hominin fossils are rare at Lothagam. Similarly, regarding the mid-Pliocene, Leakey et al. (2001) interpret the environments associated with *Australopithecus* and *Kenyanthropus* in Kenya, Tanzania, Ethiopia, and Chad as a patchwork of habitats that included open grassland, woodland, and gallery forest.

Cerling (1992; Cerling et al., 1988) and more recently Kingston (Kingston et al., 1994), Quade (Quade et al., 2004), and Wynn (2004) have developed a different approach, which is to establish a chronological record of carbon and oxygen stable isotope values for paleosol carbonates. This type of research has provided temporal sequences of vegetation structure related to carbonate precipitation in at least seasonally dry soils. The paleosol carbon stable isotope ($\delta^{13}C$) record, then, primarily detects habitats that are sufficiently arid for carbonate to precipitate and be preserved.

Wynn (2004) synthesizes $\delta^{13}C$ values beginning ~4.3 Ma for northern Kenya (Kanapoi, East and West Turkana). Although paleosols are abundant in certain time intervals, the combined stratigraphic record preserves evidence of lengthy unconformities and several lacustrine intervals, which may indicate relatively wet times. The Turkana $\delta^{13}C$ record shows, nonetheless, an overall increase in $C_4$ vegetation and aridity particularly at ~3.58–3.35, 2.52–2, and 1.81–1.58 Ma. These times of increased $\delta^{13}C$ values are associated with higher variance in $\delta^{13}C$, suggesting that an increase in vegetation spatial variability

occurred over time as climatic instability also increased.

Although paleosol carbonate $\delta^{13}C$ provides a superb environmental record, the intermittent formation of paleosols over time (i.e., periods of relative landscape stability) and the fact that carbonate precipitation depends on certain climate conditions make it impossible to obtain a time series of samples that is systematic (e.g., evenly spaced) with regard to time or stratigraphic thickness. As a result, it is difficult to place the long-term East African record of increasing aridity – essentially the story of paleosol $\delta^{13}C$ – in the context of climate oscillation – also a vital dimension of the Pliocene environmental picture. The obvious solution is to seek the ways in which oscillatory records (e.g., African dust) and progressive aridity records (such as paleosol $\delta^{13}C$) add both complementary and different kinds of information to our knowledge about Pliocene environmental change. While the terrestrial deposits in African basins offer less-complete and less-continuous environmental records, they also provide substantial evidence of regional and local tectonic activity, including volcanism, which added significantly to the environmental dynamics faced by East African hominin populations and fauna (Feibel, 1997).

Analysis of a high-resolution sequence of fossil pollen at Hadar by Bonnefille et al. (2004) provides one of the only studies of East African climate dynamics at a Pliocene hominin site. Based on detailed analysis of modern and fossil pollen samples, this study derived estimates of temperature, precipitation, humidity, and vegetation structure (biomes) associated with *Au. afarensis* at Hadar, ~3.4–3.0 Ma. The sequence of regional habitat through the Hadar Formation was reconstructed as follows, starting ~3.4 Ma and ending ~2.95 Ma: forest, wet/dry grassland, forest, woodland, wet grassland, dry grassland, and woodland. The deposits yielded, as expected, a discontinuous and uneven sequence of fossil pollen with the exception of a single ~20-Kyr interval. This densely sampled interval (approximately

1 pollen sample per 1 Kyr) recorded significant cooling (~5°C), a precipitation increase (200- to 300-mm per year), and much greater forest cover at ~3.3 Ma – i.e., contemporaneous with a global marine increase in $\delta^{18}O$ (indicative of cooling). This is the first case where habitat variability has been measured with sufficient resolution in Pliocene deposits of eastern Africa to demonstrate its correspondence with global climate variability. Bonnefille et al. (2004) note that *Au. afarensis* fossils occur in stratigraphic intervals that also record, according to the fossil pollen, substantial fluctuation in the relative extent of forest, tropical and temperate woodland, and grass cover.

While fossil mammals are often useful in deriving habitat reconstructions (e.g., Reed, 1997), they also provide insights about environmental dynamics. For example, Bobe et al. (2002) and Bobe and Behrensmeyer (2004) document a shift in late Pliocene faunal variability (the relative proportions of bovids, suids, and primates) in the Turkana basin. The shift is from relative stability (~2.8 to 2.6 Ma) to higher variability with a periodicity of about 100 Kyr, beginning ~2.5 Ma. It is not yet clear what the 100-Kyr periodicity reflects; at face value, it could indicate that faunal dynamics were more sensitive to the weak eccentricity modulation of precession than to the more rapid, precessional- and obliquity-scale variability that dominated tropical and subtropical Africa during the Pliocene. One of the interesting points of this study is that the transition to high faunal variability is associated with the first appearance of *Homo* (based on dental evidence from Omo) and stone tools in the Turkana basin (Bobe et al., 2002).

## Hypotheses Relating Environmental Dynamics, Hominin Evolution, and Faunal Change

All hypotheses that seek to relate environment and evolutionary change ultimately depend on an understanding of climate and tectonic dynamics. We have seen that Pliocene climate change involved a distinct trend – namely, cooler, more arid conditions, and significant NHG, beginning ~2.8–2.75 Ma. The Pliocene can also be divided into a period of long-term environmental stability – the EWP between 5.3 and 3.3 Ma – and a period of higher amplitude climate oscillation associated with the shift from precessional- to obliquity-dominated insolation variability after ~2.8 Ma. There is also initial evidence of alternating shorter-term packets of high- and low-climate variability throughout the Pliocene, at least in Africa.

In general, then, there are three ways in which environmental dynamics and evolutionary change may relate to one another. In Figure 1, an episode of evolutionary change (shaded vertical bars) refers to the first or last appearance of a particular adaptive character or complex, the first or last appearance of a lineage, or a well-defined increase in the rate of change in a morphological character or suite of traits. One possibility (Hypothesis A) is that such episodes of evolutionary change are unassociated with any environmental change. In this case, evolutionary events may result from internal population dynamics, ongoing resource competition, predation, disease, or other factors that operate in many different settings, and possibly at all times. Thus, evolution may occur even during times of relatively stable environment (low-climate variability). Another possibility (Hypothesis B) is that evolutionary change is correlated with, and causally related to, a progressive shift in environment. The primary evolutionary mechanisms at play in this hypothesis are directional selection and population vicariance. The former leads to the origin of novel adaptations and behaviors, the latter to a greater probability of speciation or extinction. A final possibility (Hypothesis C) is that evolutionary change is stimulated by rising environmental variability, or takes place over several periods of high-climate/resource variability interspersed with periods of low variability. The primary mechanism postulated

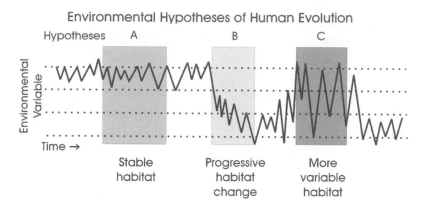

Figure 1. Three hypotheses relating evolutionary and environmental change: **Hypothesis A**: Evolutionary change (first vertical bar) is unconnected to environmental change. It takes place at any time, including intervals of environmental stability or low variability. Newly evolved traits may reflect, for instance, ongoing competition within or between species irrespective of environmental setting. **Hypothesis B**: Evolutionary change (second vertical bar) is concentrated in relatively brief periods of directional environmental change (e.g., a stepped decrease in temperature or precipitation). Newly evolved traits or taxonomic turnover directly reflects the shift from one dominant habitat type to another. **Hypothesis C**: Evolutionary change (third vertical bar) is concentrated in intervals of high environmental variability. New traits reflect greater adaptive versatility, and first/last appearances of taxa reflect increasing vicariance or resource variability during these intervals. These hypotheses are ideally testable if the record of hominin fossils and behavioral artifacts were relatively continuous and if the samples were densest around the time of a shift in the direction or variability of environmental change.

to explain this type of evolution-environment correlation is variability selection – i.e., a hypothetical process of genetic selection that results from increasing habitat/resource variability over time and results in improved environmental adaptability.

As Figure 1 implies, temporal association (coincidence in time) is the key to all tests of how instances of evolution and environmental change causally relate to one another. As much as we realize that "correlation does not equal cause," we are stuck with tests of coincidence in trying to determine the causes of events in virtually all historical sciences. Temporal association alone is insufficient, however; it is also important to specify compelling evolutionary factors and processes by which different types and tempos of environmental change may have stimulated certain types of adaptive change and speciation/extinction events. In other words, the theoretical underpinnings of an evolutionary explanation and the data pertaining to the coincidence of events (climatic and evolutionary) are both critical to consider.

For this reason, many (but not all) of the environmental hypotheses previously proposed to explain evolutionary change in Pliocene hominins and faunas have also hypothesized particular mechanisms linking the two.

Table 1 offers a synopsis of seven hypotheses (and test expectations) that have given focus to research on Pliocene evolutionary change in recent years. These ideas are divided into two types: adaptation hypotheses (which relate to the origin of adaptations) and turnover hypotheses (which relate to major turnover events – species originations and extinctions).

A variant of the turnover-pulse hypothesis, referred to as the "relay model" (Vrba, 1995a), partly derived from Vrba's (1992) "habitat theory", considers adaptation an integral part of explaining turnover. But there is an interesting difference from Vrba's original turnover-pulse hypothesis (Vrba, 1988). In the relay model, the interval of concentrated FADs and LADs may actually be quite broad, spanning several hundred thousand years, rather than essentially instantaneous. The reason is

*Table 1. Summary of hypotheses relating environmental and evolutionary change (adaptive evolution and species turnover)*

Environmental hypotheses related to adaptive evolution

1. **Habitat specific**: A specific type of habitat was necessary for a particular adaptation or suite of adaptations to emerge in one or more lineages. Examples: savanna hypothesis (e.g., Klein, 1999: 248–252), riparian woodland scavenging model (Blumenschine, 1987).

*Test expectations*: FAD for the adaptation (i.e., a functional morphological, dietary, or archeological proxy) is correlated in time and space with a specific paleohabitat. This association between the adaptation and the specific habitat is consistent in time and space and is maintained for a significant period of time. The functional or behavioral consequences of that adaptation make sense as a response to that specific habitat type.

2. **Directional change**: A particular adaptation evolved in one or more lineages as the direct result of a major progressive change in habitat – e.g., from moist forest to dry, open conditions (e.g., Vrba et al., 1989), or from warm to cool conditions (e.g., Vrba, 1994).

*Test expectations*: FAD for the adaptation is temporally constrained to a well-defined period (e.g., <100 Kyr) of significant directional environmental change, evident in one or more proxies of temperature, vegetation, etc. If the directional change reflects an ongoing environmental trend (e.g., lasting >500 Kyr), the FAD for the adaptation occurs near the initiation of that trend. The functional or behavioral shift in adaptation makes sense as a response to the newly emerging habitat.

3. **Variability selection**: A particular adaptation evolved in one or more lineages as the direct result of a significant increase in environmental variability (in time and/or space), which resulted in large variability in the adaptive conditions pertinent to those lineages (Potts, 1996, 1998a, b). Examples: earliest hominin bipedality and earliest tool/food transport emerged as adaptations to wider variability in vegetation (locomotor substrates) and in food availability (Potts, 1996); making stone tools and the adaptations of earliest *Homo* were related to heightened variability in mammalian faunas (Bobe and Behrensmeyer, 2004).

*Test expectations*: FAD for the adaptation is temporally correlated with well-defined intervals of high variability in the landscapes, food resources, or overall adaptive settings where that adaptation is first evident. The adaptive change makes sense as a response to environmental instability/uncertainty – i.e., it assists a lineage in persisting across large environmental shifts (e.g., major and repeated change in moisture and/or temperature; episodic large tephra events over broad landscapes).

3A. **Adaptability**: A particular adaptation or suite of adaptations enables a lineage to persist across larger environmental shifts than those survived by prior lineages.

*Test expectations*: A novel adaptation or suite of adaptations occurs in (or spreads to) a wider diversity of environments than prior adaptations (characteristic of earlier lineages) are known to occur. The LAD of an adaptation (or the lineage bearing it) corresponds with a significant increase in environmental variability. This represents evidence that large environmental change is a significant factor in the survival of an adaptation and the evolution of novel functions that replace or add to it.

Environmental hypotheses related to species turnover

4. **Turnover pulse**: The origination and extinction of numerous species occurs as a result of major climate change (climate forcing of evolution), which increases overall habitat and population vicariance – and thus stimulates turnover (Vrba, 1980, 1988).

*Test expectations*: Lineage FADs and LADs should be concentrated in a tightly constrained interval of climate change. The adaptations of new and terminal species during the interval of climate change should reflect the overall environmental trend (e.g., during climatic drying, there is a biased origination of arid-adapted taxa and biased extinction of moist-adapted taxa).

4A. **Relay model of turnover pulse**: Within an interval of major climate change, turnover occurs sequentially based on existing adaptations (the breadth of resource use) that prevail within clades (Vrba, 1992). That is, lineages of organisms respond to environmental change with lag effects. Thus, organisms disadvantaged by a major climate trend (e.g., cool-adapted species in a warming trend) tend to become extinct first, followed by speciation in disadvantaged lineages, followed by both extinction and speciation in organisms favored by that trend (warm-adapted species in a warming trend) (Vrba, 1995a).

*Test expectations*: As in the turnover-pulse hypothesis but with FADs and LADs distributed over a broader period, according to how the diverse adaptations of organisms relate to the overall direction of major climate change.

5. **Prolonged turnover**: Lineage turnover is not pulsed but spread out over time due to the mosaic distribution of vegetation (e.g., patches of woodland persist during the spread of arid grasslands) and the diverse adaptations of organisms. Example: even during major global climate change, the Turkana basin exhibited a prolonged turnover of species between 3 and 2 Ma, with dry- and moist-adapted mammalian species persisting throughout the interval (Behrensmeyer et al., 1997).

*Test expectations*: Even in periods of climate change, species FADs and LADs are distributed over a prolonged time, with small spikes but no single pulse. Turnover and large change in species diversity may be concentrated substantially after the onset of the climate trend if species diversity and ecological diversity are maintained within a region. Even lacking evidence of a turnover pulse, however, significant shifts in species abundance can be tightly correlated with the onset of climate change (Bobe et al., 2002).

that species that exploited different diets and resource diversity are likely to respond to environmental change at different rates, with an offset between the timing of extinction and origination events depending on those adaptations. The idea of a "pulse" that is potentially spread out over several hundred thousand years, as implied by the relay model, poses significant challenges in demonstrating a precise correlation between faunal and climate change.

The main competing explanation is what I have termed the prolonged-turnover hypothesis, the best recent example of which is given in the Turkana basin study by Behrensmeyer et al. (1997). A critical link between pulsed- and prolonged-turnover hypotheses is provided by Bobe's analysis of taxonomic abundance patterns in the Turkana basin. His study shows that the relative *abundance* of taxonomic groups may be quite sensitive to the onset of major climate change, even if species-level turnover is less so (Bobe and Eck, 2001; Bobe et al., 2002).

The variability selection (VS) hypothesis also has ramifications in the realm of faunal turnover (Bobe and Behrensmeyer, 2004), yet it was originally developed as an explanation of adaptive change (Potts, 1996, 1998a, b). The concept of VS is that environmental variability plays a substantial role in the process of evolving adaptive functions, novel behaviors, and plasticity in all biological systems (Potts, 2002). This process of originating adaptive features is a response to environmental dynamics, the degree and rate of variability in adaptive conditions as played out in a temporally continuous spectrum of environmental change – from milliseconds (in the cellular and physiological realm) to daily to seasonal to decadal, millennial, and orbital time frames.

The radical implication is that a novel trait or biological function may arise as an adaptation to resource uncertainty and unpredictability in the conditions of survival and reproduction rather than as a solution to a specific, constant environmental stimulus or adaptive challenge. The distinction, then, is

whether adaptations evolve mainly in response to specific manifestations of the environment (and overall changes in state from one manifestation to another) – or in response to environmental *dynamics* in all its messy, nested complexity. In the former case, the functioning of an organism reflects the sum or average of past habitats and resource templates. But in the latter case, the functioning of an organism reflects past environmental dynamics, which have shaped the organism's capacity to adjust to changing resource configurations, disturbances, and even novel settings – all of which define the *adaptability* of that organism.

Thus in Table 1, I define an offshoot of the VS idea, called the adaptability hypothesis. It is the flip side of the VS coin, reflecting the idea that organisms are adapted to a certain range and pace of environmental variability, expressed over the wide spectrum of time scales. The resulting dimensions of adaptability (genomal, developmental, physiological, behavioral, ecological) enable certain adaptations (and lineages) to persist across a surprising variety of environmental transitions and to disperse across a range of habitats, including settings that are unprecedented in an organism's past.

VS implies, however, that adaptive versatility can be "ratcheted up" in the face of even greater environmental variability. In this event, certain suites of adaptations can be lost, replaced by novel suites that improve the adaptable properties of later organisms. Thus, one of the test expectations of the adaptability hypothesis concerns *last* appearances. That is, the last appearance of a particular anatomical complex or behavior (e.g., combined terrestrial bipedal and arboreal activity in *Au. afarensis*) should coincide with a significant rise in environmental variability. (In this example, extinction of *Au. afarensis* may indeed take place after 2.9 Ma, associated with a rise in African climate variability [deMenocal, 2004].) This loss is then followed by the first appearance or spread of innovations that enhanced adaptive versatility. (Dietary breadth in the masticatory powerhouse *Paranthropus* and the origin of, or greater reliance on, stone tools in one or more hominin lineages may both qualify as hypothetical examples.)

The fact that certain taxa (e.g., *Au. afarensis*) and adaptive characteristics (e.g., megadontia in *Paranthropus*) endured over several hundred thousand years implies that Pliocene lineages and adaptations were typically static (or, rather, morphologically variable within certain limits) in relation to climate dynamics. The same is true for other mammalian lineages and associated sets of taxa. This observation leads to two questions that continue to challenge paleontologists: First, under what environmental conditions did *substantial reorganization* of adaptive complexes, characters, and faunal associations take place during the Pliocene? Second, why did certain adaptive traits, lineages, and faunal associations persist over periods greatly exceeding the dominant periodicities of Pliocene environmental change?

Future answers to these questions will help us better understand the processes of adaptive, phylogenetic, and ecological change – and particularly the role of environmental dynamics in deciding the balance between persistence (adaptability) and evolutionary innovation.

**Faunal Community Evolution**

A final hypothesis relevant to Pliocene human and faunal evolution comes from the study of species assembly. Species assembly concerns the co-occurrence of species populations in a given locale. The study of species assembly focuses on the processes that govern which particular taxa are found together and their relative abundances – two aspects of what is typically meant by a faunal community (Belyea and Lancaster, 1999; Hubbell, 2001). While increasingly considered in the biogeographical and ecological literature, the concept of species assembly has barely entered the realm of mammalian paleoecology.

For modern communities, the factors that govern the assembly of species in a local setting fall into three classes. The first – *environmental constraints* – concerns the fact that animals living together (and those species excluded from a given locality) are determined by climate, substrate, vegetation, and other characteristics of the local setting. The second – *internal dynamics* – highlights the role of competition, i.e., certain taxa may be recorded in a given place, and others excluded, based on the competitive ability of species populations in a given setting. The third class of factors – *dispersal constraints* – emphasizes that mobility and random geographic factors, such as habitat corridors and incumbency (those species already present in a particular place), may determine which particular species happen to become part of a local community.

To illustrate the analysis of species assembly and its potential relevance to Pliocene and Pleistocene faunal research, I will draw on an example of faunal and environmental change in the mid-Pleistocene Olorgesailie Formation, southern Kenya Rift Valley (Potts, 1996; Potts et al., 1999; Potts and Teague, 2003, in preparation). Although this faunal sequence ranges from ~1.2 to 0.5 Ma, the analysis of mammalian taxa at Olorgesailie suggests an interesting hypothesis important to examine also in Pliocene contexts.

Figure 2 shows the relative abundance of major taxonomic groups in three *in situ* mammalian fossil samples from the Olorgesailie Formation. These faunal samples were obtained by excavations and surface surveys of three widely exposed stratigraphic intervals: UM1p (upper Member 1 paleosol), LM7s (lower Member 7 sand), and M10/11 (Member 10 and lower Member 11). The stratigraphic section (Potts et al., 1999; Behrensmeyer et al., 2002) and tephra and magnetostratigraphic ages (Deino and Potts, 1990; Tauxe et al., 1992) show the temporal separation between the samples.

The Olorgesailie faunal samples are peculiar in a number of respects. First, all of them exhibit lower proportions of bovids than expected. In

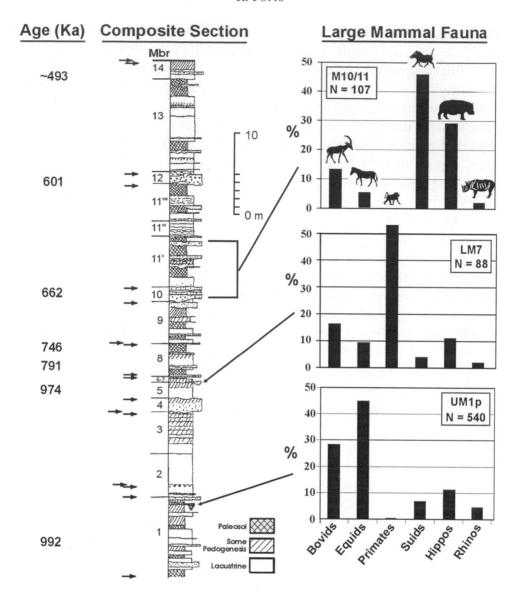

**Figure 2.** Mid-Pleistocene fauna and environmental change as recorded in the Olorgesailie basin, southern Kenya. Percentage representation (relative abundance) of major groups of large mammals is shown (right) for three fossil samples excavated from relatively narrow time intervals. The three samples are named UM1p, which is a paleosol ~990 Ka in Member 1 of the Olorgesailie Formation; LM7, which is a fluvial sand ~900 Ka in the base of Member 7; and M10/11, which is a series of sand and gravel layers ~662 to 650 Ka in Member 10 and lower Member 11. For the LM7 and M10/11 samples, *N* is based on minimum number of individuals due to the presence of partial skeletons; for UM1p, *N* is based on number of individual specimens. The horizontal arrows next to the composite section indicate the stratigraphic positions of major (basin-wide) events of landscape remodeling, which form an important context of mammal community assembly and disassembly. Composite section by A.K. Behrensmeyer and R. Potts (Potts et al., 1999; Behrensmeyer et al., 2002); age estimates based on Deino and Potts (1990) and Tauxe et al. (1992).

none of the three samples do bovids comprise the dominant group of large mammals – in contrast to most other Pleistocene and Pliocene faunal samples in East Africa. A second oddity concerns the relay of taxonomic dominance, with *Equus* spp. (especially *E. oldowayensis*) as the most abundant large mammal in UM1p giving way to *Theropithecus oswaldi* and *Equus* spp. in LM7s, followed by the dominance of *Hippopotamus* spp. (especially *H. gorgops*) and the suids *Kolpochoerus* and *Metridiochoerus* in M10/11.

The main questions are: What factors were responsible for this unusual progression? Is Olorgesailie odd relative to other mid-Pleistocene faunal samples in Africa? Was the relay of dominant taxa the result of continuous faunal change within the southern Kenya rift?

To help answer these questions, the first step was to assess how different the mammalian taxa at Olorgesailie were from other mid-Pleistocene faunal samples. Since Olorgesailie taxonomic abundances (Figure 2) are derived from detailed excavations, comparisons with other (typically surface collected) samples required working with species lists only (Potts and Teague, 2003, in preparation). Taxonomic lists from sub-Saharan fossil localities, dated ~1.4 to 0.4 Ma, were combined to establish a minimal "geographic species pool" (GSP) pertaining to the mid-Pleistocene of East Africa.

The GSP represents an estimate of the entire population (or metapopulation) from which the local combinations of mammalian species in the Olorgesailie basin were drawn.

In our initial analysis, the mid-Pleistocene GSP consists of 87 species from 17 fossil localities. Various avenues of multivariate analysis, such as detrended correspondence analysis and clustering routines, show that the combinations of taxa found in these 17 localities comprise robust geographic groupings rather than temporal groupings. Figure 3 illustrates one example from many analyses showing the four distinct geographic clusters that consistently result from this study: south-central Africa, the southern Kenya Rift Valley, the Horn of Africa, and Bed IV Olduvai.

In none of the analyses are the individual fossil samples from Olorgesailie any more unusual in terms of their member taxa than other African mid-Pleistocene samples. The

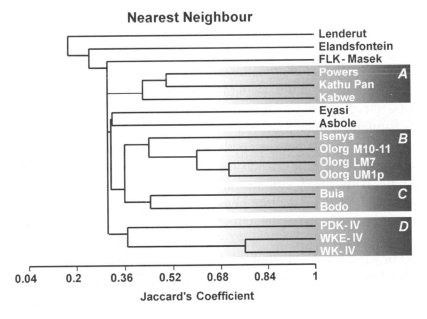

Figure 3. Cluster analysis of mid-Pleistocene mammalian fauna based on taxonomic lists from 17 localities (named on the right). The combined faunal list (*N* = 87 species) represents a minimal estimate of the geographic species pool for sub-Saharan Africa from ~1.4 to 0.4 Ma. This analysis used Jaccard's coefficient and nearest neighbor. Other distance measures (e.g., Euclidean) and clustering techniques (paired-group averages) gave similar results and consistently defined four main geographic groupings, indicated here by the shaded areas: A. South-central Africa; B. Southern Kenya Rift Valley, including Olorgesailie; C. Horn of Africa; and D. Olduvai Gorge. None of the cluster analyses produced consistent relationships between the other (nonshaded) samples, and none showed a clustering of fossil assemblages according to their temporal sequence (Potts and Teague, 2003, in preparation).

three Olorgesailie assemblages, in fact, always aggregate in the middle of each cluster analysis or correspondence analysis rather than on the periphery (Figure 3; Potts and Teague, 2003; in preparation). Olorgesailie, therefore, does not appear to be odd with regard to having unique faunal characteristics. Only when a larger number of faunal samples have been collected carefully with adequate taphonomic control, and from as wide a range of depositional contexts as each Olorgesailie sample, that this comparative analysis can be pursued further.

These initial findings suggest nonetheless that geographic factors (dispersal constraints) helped to shape each fossil assemblage in the different basins and regions. Yet this conclusion fails to explain the relay of taxonomic dominance (based on species abundance) in the Olorgesailie basin. Environmental analysis of the Olorgesailie sequence offers an illuminating clue, however: As indicated by the horizontal arrows in Figure 2, the Olorgesailie region was susceptible to repeated basin-wide revamping of the landscape and its resources. The arrows indicate the stratigraphic positions at which dramatic environmental transitions occurred, including large eruptive (tephra-depositing) events, abrupt drying of the lake, replacement of a fluvial-dominated by a lake-dominated landscape, and shifts between periods of sediment aggradation and erosion (or stability, marked by widespread soil formation). These represent the type of environmental events that would have temporarily killed off the herbaceous vegetation (in the case of the tephra events) or substantially altered local climate and the availability of water and other food resources.

I would offer, then, the following hypothesis: Basin-wide environmental events of sufficient magnitude can lead to the breakdown of mammalian communities. That is, during substantial environmental events, species populations must emigrate from the basin or die out locally, and as a result, the taxonomic community at least temporarily

dissolves – i.e., the community disassembles – followed by a subsequent period of species reassembly. Each period of reassembly is an experiment in which the factors typically thought to explain taxonomic combinations (environment, competition, and dispersal) once again intersect and play a role. Yet the instability created by climate dynamics and tectonic events is the critical factor causing recurrent episodes of community assembly and disassembly.

The focus on environmental dynamics in this community assembly–disassembly hypothesis is not meant to undermine the importance of taphonomic biases and time averaging in explaining the variations that may occur among fossil samples. Assessing such biases is a critical first step in the analysis of species co-occurrence and the differences between fossil samples. In the case of Olorgesailie, variations in species body size, skeletal part durability, and depositional environment do not play a significant role in explaining the differences between samples. Variation in time averaging also appears to play a smaller role than ecological factors in explaining the taxonomic differences (Potts and Teague, 2003, in preparation).

The point is that environmental dynamics can disrupt faunal continuity within a basin. This hypothesis challenges, therefore, the assumption that the sequence of mammalian fauna within a basin reflects *in situ* faunal turnover or a local succession of taxa due to adaptive change. Hypotheses like turnover pulse and prolonged turnover often assume the continuity of the fauna within a basin (or region) over time. The possibility of recurrent community assembly and disassembly implies, however, that faunal variation over time and space may reflect environmentally driven experimentation in how random and adaptive factors play out on a larger biogeographic scale.

The Olorgesailie example begs the question, was the Pliocene of Africa the same as the mid-Pleistocene? Given that global climate oscillation after ~900 Ka involved higher amplitude and longer periodicity, an intriguing possibility is that Pliocene faunas experienced significantly lower magnitudes of disturbance due to climate. The apparent heterogeneity of mid-Pleistocene mammalian communities in time and space may also reflect very limited connectivity among distant basins. It is possible, however, that during the Pliocene, basins in different parts of Africa would have served as viable species reservoirs for repopulating areas affected by large environmental events. This would have been possible as long as there was strong connectivity among the distant basins. An obvious test of this idea is to see whether mammalian communities were reassembled with approximately the same species at similar relative abundances before and after large events, such as the eruption of the Tulu Bor Tuff (Turkana) and Sidi Hakoma Tuff (SHT) complex (Hadar) at ~3.4 Ma.

The final point of the Olorgesailie example concerns the hominin toolmakers. The early humans responsible for making Acheulean hand axes were consistently present; they deposited their lithic debris immediately below and above almost all of the stratigraphic boundaries marking basin-wide environmental impacts. This suggests that while mammalian species, in general, had varying success in reestablishing themselves after major environmental transitions, the hominins were nearly always able to survive locally or at least to rapidly colonize the Olorgesailie basin after a disruption.

The assembly–disassembly hypothesis thus leads to interesting and relatively novel directions in mammalian paleoecology, with considerable relevance to Pliocene studies: It demands better documentation of environmental dynamics in terrestrial sequences; improved estimates of environmental impacts on local communities; new analyses of geographic species pools and fauna provinciality; and the comparison of how successful hominins and other mammals were in adapting to recurrent ecological disturbances. Eventually, this hypothesis may better our understanding of paleoecological and evolutionary processes at diverse temporal and spatial scales.

## Conclusion

The Pliocene was a 3.5-million-year period of appreciable species turnover and adaptive change apparent in hominins and other large mammals. It was also a time of marked environmental change – in particular, a substantial decrease in global temperature, the onset of NHG, heightened African aridity, a significant rise in climate oscillation, and episodic remodeling of African landscapes inhabited by hominins and other organisms. How these data sets regarding environmental and evolutionary change fit together represents a significant and challenging research agenda.

One area of this agenda needs vast improvement – namely, stratigraphic precision in correlating faunal change, evolutionary events, and environmental dynamics. Paleoenvironmental analysis of early human sites in East Africa has paid hardly any attention to environmental dynamics *per se*. Research has instead focused on the environments of particular stratigraphic intervals that also contain hominin fossils. Yet in many instances, the scientists responsible for the paleoenvironmental analysis and those responsible for the early human discoveries locate their evidence in different strata (Copes and Potts, 2005). Typically, then, there is a stratigraphic (and temporal) offset between the two lines of evidence. This offset greatly thwarts our ability to test environmental hypotheses of human evolution, which depends on examining the co-occurrence of past events in time and space.

Even where there is stratigraphic overlap, the environmental and fossil data may represent time-averaged data sets of >100 Kyr (e.g., Leakey and Harris, 1987; Bobe et al., 2002). This degree of temporal resolution inherent in most East African data sets makes it difficult, at best, to determine how precisely faunal and hominin evolutionary change matched up against specific environmental shifts and the precessional- to obliquity-scale tempo of African Pliocene climate change. A major challenge ahead is to acquire long-term evidence of faunal variation and environments from continental basins that can be compared against the ~20- and 41-Kyr periodicities of Pliocene climate variability.

A second area marked for future attention is the matter of spatial scale. Different aspects of ecological and evolutionary processes are likely to be learned by comparing diverse spatial and temporal scales (Wiens et al., 1986). Intrabasin comparison can document not only faunal heterogeneity through time but also habitat patchiness within a single narrow time interval (e.g., Potts et al., 1999). Analyses confined to a single basin can, however, lead to incorrect generalizations about preferred hominin habitat, overall taxonomic diversity and abundance, and faunal turnover during the Pliocene. Consideration of several basins within a region offers an opportunity to compare environmental, including faunal, sequences, and thus provides a way of dissecting local versus regional causes of faunal change. Interregional comparison – e.g., across eastern, southern, and north-central Africa – allows researchers to assess continent-wide influences on faunal dynamics and to examine the processes of how local communities were derived from the larger geographic species pool. Finally, comparison between continental and marine climate records represents the broadest spatial scale in which to view the environmental causes of faunal change in the Pliocene.

Most previous considerations of hominin evolution have tended to treat "the environment" in a monolithic sense. The aim of including diverse spatial and temporal scales in our analyses averts this over-simplification of the spatio-temporal complexity of environments in which Pliocene mammals, including hominins, lived and evolved. In the end, all tests of hypotheses about Pliocene environment, faunal change, and hominin evolution are matters of scale.

## Acknowledgments

The research at Olorgesailie is a collaborative project with the National Museums of Kenya, and I thank the NMK Director, I.O. Farah, and the staff of the NMK Department of Paleontology for their support. This work is funded by NSF (grant BCS-0218511) and the Smithsonian Institution's Human Origins Program. I thank Jennifer Clark for preparing the figures. Special thanks to R. Bobe, Z. Alemseged, and A.K. Behrensmeyer for organizing the workshops on African faunal change and also this volume.

## References

Aiello, L.C., Wheeler, P., 1995. The expensive tissue hypothesis. Current Anthropology 36, 199–221.

Asfaw, B., 1987. The Belohdelie frontal: new evidence of early hominid cranial morphology from the Afar of Ethiopia. Journal of Human Evolution 16, 611–624.

Asfaw, B., White, T., Lovejoy, O., Latimer, B., Simpson, S., Suwa, G., 1999. *Australopithecus garhi*: a new species of early hominid from Ethiopia. Science 284, 629–635.

Ashley, G.M., Hay, R.L., 2002. Sedimentation patterns in a Plio-Pleistocene volcaniclastic rift-platform basin, Olduvai Gorge, Tanzania. In: Renaut, R.W., Ashley, G.M. (Eds.), Sedimentation in Continental Rifts. SEPM Special Publication 73, pp. 107–122.

Behrensmeyer, A.K., Todd, N.E., Potts, R., McBrinn, G.E., 1997. Late Pliocene faunal turnover in the Turkana Basin, Kenya and Ethiopia. Science 278, 1589–1594.

Behrensmeyer, A.K., Potts, R., Deino, A., Ditchfield, P., 2002. Olorgesailie, Kenya: a million years in the life of a rift basin. In: Renaut, R.W., Ashley,

G.M. (Eds.), Sedimentation in Continental Rifts. SEPM Special Publication 73, pp. 97–106.

Belyea, L.R., Lancaster, J., 1999. Assembly rules within a contingent ecology. Oikos 86, 402–416.

Blumenschine, R.J., 1987. Characteristics of an early hominid scavenging niche. Current Anthropology 28, 383–407.

Bobe, R., Behrensmeyer, A.K., 2004. The expansion of grassland ecosystems in Africa in relation to mammalian evolution and the origin of the genus *Homo*. Palaeogeography, Palaeoclimatology, Palaeoecology 207, 399–420.

Bobe, R., Eck, G.G., 2001. Responses of African bovids to Pliocene climatic change. Paleobiology 27(suppl. no. 2). Paleobiology Memoirs 2, 1–47.

Bobe, R., Behrensmeyer, A.K., Chapman, R., 2002. Faunal change, environmental variability and late Pliocene hominin evolution. Journal of Human Evolution 42, 475–497.

Bonnefille, R., Potts, R., Chalié, F., Jolly, D., Peyron, O., 2004. High-resolution vegetation and climate change associated with Pliocene *Australopithecus afarensis*. Proceedings of the National Academy of Sciences USA 101, 12125–12129.

Bromage, T.G., Schrenk, F., Zonneveld, F.W., 1995. Paleoanthropology of the Malawi Rift: an early hominid mandible from the Chiwondo Beds, northern Malawi. Journal of Human Evolution 28, 71–108.

Brown, F.H., 2004. Geological development of the Omo-Turkana Basin during the Pliocene and Pleistocene Epochs. Geological Society of America Abstracts 36(5), 485.

Brunet, M., Beauvilain, A., Coppens, Y., Heintz, E., Moutaye, A.H.E., Pilbeam, D., 1995. The first australopithecine 2,500 kilometres west of the Rift Valley (Chad). Nature 378, 273–275.

Brunet, M., Guy, F., Pilbeam, D., Mackaye, H.T., Likius, A., Ahounta, J., Beauvillain, A., Blondel, C., Bocherens, H., Boisserie, J-R., Bonis, L.de, Coppens, Y., Dejax, J., Denys, C., Duringer, P., Eisenmann, V., Fanone, G., Fronty, P., Geraads, D., Lehmann, T., Lihoreau, F., Louchart, A., Mahamat, A., Merceron, G., Mouchelin, G., Otero, O., Campomanes, P.P., Ponce de Leon, M., Rage, J-C., Sapanet, M., Schuster, M., Sudre, J., Tassy, P., Valentin, X., Vignaud, P., Viriot, L., Zazzo, A., Zollikofer, C., 2002. A new hominid from the Upper Miocene of Chad, Central Africa. Nature 418, 145–151.

Cerling, T.E., 1992. Development of grasslands and savannas in East Africa during the Neogene. Palaeogeography, Palaeoclimatology, Palaeoecology 97, 241–247.

Cerling, T.E., Bowman, J.R., O'Neil, J.R., 1988. An isotopic study of a fluvial-lacustrine sequence: the Plio-Pleistocene Koobi Fora sequence, East Africa. Palaeogeography, Palaeoclimatology, Palaeoecology 63, 335–356.

Clarke, R., 1999. Discovery of complete arm and hand of the 3.3 million-year-old *Australopithecus* skeleton from Sterkfontein. South African Journal of Science 95, 477–480.

Clarke, R.J., 1988. A new *Australopithecus* cranium from Sterkfontein and its bearing on the ancestry of *Paranthropus*. In: Grine, F.E. (Ed.), Evolutionary History of the "Robust" Australopithecines. Aldine de Gruyter, New York, pp. 285–292.

Comas, M.C., Zahn, R., Klaus, A., et al., 1996. Proc. ODP Init. Res. 161, Ocean Drilling Program, College Station, TX.

Copes, L., Potts, R., 2005. Are hominin fossils and paleoenvironmental data precisely associated in the stratigraphic records of Turkana and Olduvai? Paleoanthropology. Online journal of the Paleoanthropology Society, p. A56.

de Heinzelin, J., Clark, J.D., White, T., Hart, W., Renne, P., WoldeGabriel, G., Beyene, Y., Vrba, E., 1999. Environment and behavior of 2.5-million-year-old Bouri hominids. Science 284, 625–629.

Deino, A., Potts, R., 1990. Single crystal $^{40}Ar/^{39}Ar$ dating of the Olorgesailie Formation, southern Kenya rift. Journal of Geophysical Research 95(B6), 8453–8470.

Delagnes, A., Roche, H., 2005. Late Pliocene hominid knapping skills: The case of Lokalalei 2C, West Turkana, Kenya. Journal of Human Evolution 48, 435–472.

deMenocal, P.B., 1995. Plio-Pleistocene African climate. Science 270, 53–59.

deMenocal, P.B., 2004. African climate change and faunal evolution during the Pliocene–Pleistocene. Earth and Planetary Science Letters 220, 3–24.

deMenocal, P.B., Bloemendal, J., 1995. Plio-Pleistocene subtropical African climate variability and the paleoenvironment of hominid evolution. In: Vrba, E.S., Denton, G.H., Partridge, T.C., Burckle, L.H. (Eds.), 1995. Paleoclimate and Evolution with Emphasis on Human Origins. Yale University Press, New Haven, CT, pp. 262–288.

deMenocal, P.B., Brown, F.H., 1999. Pliocene tephra correlations between East African hominid localities, the Gulf of Aden, and the Arabian Sea. In: Agusti, J., Rook, L., Andrews, P. (Eds.), Hominid Evolution and Climate Change in Europe, Vol. 1. Cambridge University Press, Cambridge, U.K., pp. 23–54.

Domínguez-Rodrigo, M., Pickering, T.R., Semaw, S., Rogers, M.J., 2005. Cutmarked bones from Pliocene archaeological sites at Gona, Ethiopia: implications for the function of the world's oldest stone tools. Journal of Human Evolution 48, 109–121.

Emeis, K.-C., Robertson, A.E.S., Richter, C., et al., 1996. Proc. ODP Init. Rep. 160. Ocean Drilling Program, College Station, TX.

Emeis, K.-C., Sakamoto, T., Wehausen, R., Brumsack, H.-J., 2000. The sapropel record of the eastern Mediterranean Sea – results of Ocean Drilling Program Leg. 160. Palaeogeography, Palaeoclimatology, Palaeoecology 158, 371–395.

Feibel, C.S., 1997. Debating the environmental factors in hominid evolution. Geological Society of America Today 7(3), 1–7.

Grine, F.E. (Ed.), 1988. Evolutionary History of the "Robust" Australopithecines. Aldine de Gruyter, New York.

Goddard, L., Graham, N.E., 1999. Importance of the Indian Ocean for simulating rainfall anomalies over eastern and southern Africa. Journal of Geophysical Research 104, 19099–19116.

Haile-Selassie, Y., 2001. Late Miocene hominids from the Middle Awash, Ethiopia. Nature 412, 178–181.

Haile-Selassie, Y., Suwa, G., White, T.D., 2004. Late Miocene teeth from Middle Awash, Ethiopia, and early hominid dental evolution. Science 303, 1503–1505.

Haeusler, M., McHenry, H.M., 2004. Body proportions of *Homo habilis* reviewed. Journal of Human Evolution 46, 433–465.

Hartwig-Scherer, S., Martin, R.D., 1991. Was "Lucy" more human than her "child"? Observations on early hominid postcranial skeletons. Journal of Human Evolution 21, 439–449.

Hill, A., Ward, S., Deino, A., Curtis, G., Drake, R., 1992. Earliest *Homo*. Nature 355, 719–722.

Hoberg, E.P., Alkire, N.L., de Queiroz, A., Jones, A., 2001. Out of Africa: origins of the *Taenia* tapeworms in humans. Proceedings of the Royal Society of London B. Biological Sciences 268, 781–787.

Hubbell, S.P., 2001. The Unified Neutral Theory of Biodiversity and Biogeography. Monographs in Population Biology 32, Princeton University Press, Princeton, NJ.

Isaac, G.L., 1978. The food-sharing behavior of protohuman hominids. Scientific American 238, 90–108.

Johanson, D.C., Masao, F.T., Eck, G.G., White, T.D., Walter, R.C., Kimbel, W.H., Asfaw, B., Manega, P., Ndessokia, P., Suwa, G., 1987. New partial skeleton of *Homo habilis* from Olduvai Gorge, Tanzania. Nature 327, 205–209.

Kimbel, W.H., Johanson, D.C., Rak, Y., 1997. Systematic assessment of a maxilla of *Homo* from Hadar, Ethiopia. American Journal of Physical Anthropology 103, 235–262.

Kingston, J.D., Marino, B.D., Hill, A., 1994. Isotopic evidence for Neogene hominid paleoenvironments in the Kenya Rift Valley. Science 264, 955–959.

Klein, R.G., 1999. The Human Career. University of Chicago Press, Chicago.

Leakey, L.S.B., Tobias, P.V., Napier, J.R., 1964. A new species of the genus *Homo* from Olduvai Gorge. Nature 202, 7–9.

Leakey, M.D., Harris, J.M. (Eds.), 1987. Laetoli: A Pliocene Site in Northern Tanzania. Clarendon, Oxford.

Leakey, M.G., Feibel, C.S., McDougall, I., Walker, A., 1995. New four-million-year-old hominid species from Kanapoi and Allia Bay, Kenya. Nature 376, 565–571.

Leakey, M.G., Feibel, C.S., Bernor, R.L., Harris, J.M., Cerling, T.E., Stewart, K.M., Stoors, G.W., Walker, A., Werdelin, L., Winkler, A.J., 1996. Lothagam: a record of faunal change in the Late Miocene of East Africa. Journal of Vertebrate Paleontology 16, 556–570.

Leakey, M.G., Feibel, C.S., McDougall, I., Ward, C., Walker, A., 1998. New specimens and confirmation of an early age for *Australopithecus anamensis*. Nature 393, 62–66.

Leakey, M.G., Spoor, F., Brown, F.H., Gathogo, P.N., Kiarie, C., Leakey, L.N., McDougall, I., 2001. New hominin genus from eastern Africa shows diverse middle Pliocene lineages. Nature 410, 433–440.

Lockwood, C.A., Kimbel, W.H., Johanson, D.C., 2000. Temporal trends and metric variation in the mandibles and dentititon of *Australopithecus afarensis*. Journal of Human Evolution 39, 23–55.

Lourens, L.J., Antonarakou, A., Hilgen, F.J., Van Hoof, A.A.M., Vergnaud-Grazzini, C., Zachariasse, W.J., 1996. Evaluation of the Plio-Pleistocene astronomical timescale. Paleoceanography 11, 391–413.

McCollum, M.A., 1999. The robust australopithecine face: a morphogenetic perspective. Science 284, 301–305.

McDougall, I., Brown, F.H., Fleagle, J.G., 2005. Stratigraphic placement and age of modern humans from Kibish, Ethiopia. Nature 433, 733–736.

McHenry, H.M., Berger, L.R., 1998. Body proportions in *Australopithecus afarensis* and *A. africanus* and the origin of the genus *Homo*. Journal of Human Evolution 35, 1–22.

Panger, M.A., Brooks, A.S., Richmond, B.G., Wood, B.A., 2002. Older than the Oldowan: rethinking the emergence of hominin tool use. Evolutionary Anthropology 11, 234–245.

Philander, S.G., Fedorov, A.V., 2003. Role of tropics in changing the response to Milankovitch forcing some three million years ago. Paleoceanography 18, 1045–1056.

Plummer, T., 2004. Flaked stones and old bones: Biological and cultural evolution at the dawn of technology. Yearbook of Physical Anthropology 47, 118–164.

Plummer, T., Ferraro, J., Ditchfield, P., Bishop, L., Potts, R., 2001. Late Pliocene Oldowan excavations at Kanjera South, Kenya. Antiquity 75, 809–810.

Potts, R., 1991. Why the Oldowan? Plio-Pleistocene tool-making and the transport of resources. Journal of Anthropological Reseach 47, 153–176.

Potts, R., 1996. Humanity's Descent: The Consequences of Ecological Instability. W.H. Morrow, New York.

Potts, R., 1998a. Variability selection in hominid evolution. Evolutionary Anthropology 7, 81–96.

Potts, R., 1998b. Environmental hypotheses of hominin evolution. Yearbook of Physical Anthropology 41, 93–136.

Potts, R., 2002. Complexity and adaptability in human evolution. In: Goodman, M., Moffat, A. (Eds.), Probing Human Origins. American Academy of Arts and Sciences, Cambridge, pp. 33 57.

Potts, R., 2003. Early human predation. In: Kelley, P.H., Kowalewski, M., Hansen, T. (Eds.), Predator–Prey Interactions in the Fossil Record. Kluwer, New York, pp. 359–376.

Potts, R., Behrensmeyer, A.K., Ditchfield, P., 1999. Palelandscape variation and Early Pleistocene hominid activities: Members 1 and 7, Olorgesailie Formation, Kenya. Journal of Human Evolution 37, 747–788.

Potts, R., Teague, R., 2003. Heterogeneity in large-mammal paleocommunities and hominin activities in the southern Kenya Rift Valley during the mid-Pleistocene (1.2–0.4 Ma). Paleoanthropology: Online journal of the Paleoanthropology Society.

Quade, J., Levin, N., Semaw, S., Stout, D., Renne, P., Rogers, M., Simpson, S., 2004. Paleoenvironments of the earliest stone toolmakers, Gona, Ethiopia. Geological Society of America Bulletin 116, 1529–1544.

Rak, Y., 1983. The Australopithecine Face. Academic Press, New York.

Ravelo, A.C., Andreasen, D.H., Lyle, M., Lyle, A.O., Wara, M.W., 2004. Regional climate shifts caused by gradual global cooling in the Pliocene epoch. Nature 429, 263–267.

Reed, K.E., 1997. Early hominid evolution and ecological change through the African Plio-Pleistocene. Journal of Human Evolution 32, 289–322.

Renne, P.R., WoldeGabriel, G., Hart, W.K., Heiken, G., White, T.D., 1999. Chronostratigraphy of the Miocene–Pliocene Sagantole Formation, Middle Awash Valley, Afar rift, Ethiopia. Geological Society of America Bulletin 111, 869–885.

Roche, H., Delagnes, A., Brugal, J.P., Feibel, C.S., Kibunjia, M., Mourre, B., Texier, P-J., 1999. Early hominid stone tool production and technical skill 2.34 Myr ago in West Turkana, Kenya. Nature 399, 57–60.

Rossignol-Strick, M., 1983. African monsoons, an immediate climatic response to orbital insolation forcing. Nature 303, 46–49.

Ruff, C.B., Walker, A., 1993. Body size and body shape. In: Walker, A.C., Leakey, R.E. (Eds.), The Nariokotome *Homo erectus* Skeleton. Harvard University Press, Cambridge, MA, pp. 234–265.

Sahnouni, M., Hadjouis, D., van der Made, J., Derradji, A., Canals, A., Medig, M., Belahrech, H., Harichane, Z., Rabhi, M., 2002. Further research at the Oldowan site of Ain Hanech, North-eastern Algeria. Journal of Human Evolution 43, 925–937.

Schefuß, E., Schouten, S., Jansen, J.H.F., Sinninghe Damstß, J.S., 2003. African vegetation contolled by tropical sea surface temperatures in the mid-Pleistocene period. Nature 422, 418–421.

Semaw, S., Renne, P., Harris, J.W.K., Feibel, C.S., Bernor, R.L., Fesseha, N., Mowbray, K., 1997. 2.5-Million-year-old stone tools from Gona, Ethiopia. Nature 385, 333–336.

Semaw, S., Rogers, M.J., Quade, J., Renne, P.R., Butler, R.F., Dominguez-Rodrigo, M., Stout, D., Hart, W.S., Pickering, T., Simpson, S.W., 2003. 2.6-Million-year-old stone tools and associated bones from OGS-6 and OGS-7, Gona, Afar, Ethiopia. Journal of Human Evolution 45, 169–177.

Semaw, S., Simpson, S.W., Quade, J., Renne, P.R., Butler, R.F., McIntosh, W.C., Levin, N., Dominguez-Rodrigo, M., Rogers, M.J., 2005. Early Pliocene hominids from Gona, Ethiopia. Nature 433, 301–305.

Senut, B., Pickford, M., Gommery, D., Mein, P., Cheboi, K., Coppens, Y., 2001. First hominid from the Miocene (Lukeino Formation, Kenya), C.R. Acaid. Sci. Paris 332, 137–144.

Shackleton, N.J., 1995. New data on the evolution of Pliocene climatic variability. In: Vrba, E.S., Denton, G.H., Partridge, T.C., Burckle, L.H. (Eds.), 1995. Paleoclimate and Evolution with Emphasis on Human Origins. Yale University Press, New Haven, CT, pp. 242–248.

Skelton, R.R., McHenry, H.M., 1992. Evolutionary relationships among early hominids. Journal of Human Evolution 23, 309–349.

Strait, D.S., Grine, F.E., 2004. Inferring hominoid and early hominid phylogeny using craniodental characters: the role of fossil taxa. Journal of Human Evolution 47, 399–452.

Strait, D.S., Wood, B.A., 1999. Early hominid biogeography. Proceedings of the National Academy of Sciences, USA 96, 9196–9200.

Strait, D.S., Grine, F.E., Moniz, M.A., 1997. A reappraisal of early hominid phylogeny. Journal of Human Evolution 32, 17–82.

Suwa, G., Wood, B.A., White, T.D., 1994. Further analysis of mandibular molar crown and cusp areas in Pliocene and Pleistocene hominids. American Journal of Physical Anthropology 93, 407–426.

Suwa, G., White, T.D., Howell, F.C., 1996. Mandibular postcanine dentition from the Shungura formation, Ethiopia: crown morphology, taxonomic allocations, and Plio-Pleistocene hominid evolution. American Journal of Physical Anthropology 101, 247–282.

Tauxe, L., Deino, A., Behrensmeyer, A.K., Potts, R., 1992. Pinning down the Brunhes/Matuyama and upper Jaramillo boundaries: a reconciliation of orbital and isotopic time scales. Earth and Planetary Science Letters 109, 561–572.

Teaford, M.F., Ungar, P.S., 2000. Diet and the evolution of the earliest human ancestors. Proceedings of the National Academy of Sciences USA 97, 13506–13511.

Tiedemann, R., Sarnthein, M., Shackleton, N.J., 1994. Astronomic timescale for the Pliocene Atlantic $\delta^{18}O$ and dust flux records of ODP Site 659. Paleoceanography 9, 619–638.

Ungar, P.S., 2004. Dental topography and diets of *Australopithecus afarensis* and early *Homo*. Journal of Human Evolution 46, 605–622.

Vrba, E.S., 1980. Evolution, species and fossils: how did life evolve? South African Journal of Science 76, 61–84.

Vrba, E.S., 1988. Late Pliocene climatic events and hominid evolution. In: Grine, F.E. (Ed.), Evolutionary History of the "Robust" Australopithecines. Aldine de Gruyter, New York, pp. 405–426.

Vrba, E.S., 1992. Mammals as a key to evolutionary theory. Journal of Mammalogy 73, 1–28.

Vrba, E.S., 1994. An hypothesis of heterochrony in response to climatic cooling and its relevance to early hominid evolution. In: Corruccini, R., Ciochon, R.L. (Eds.), Integrative Paths to the Past. Prentice Hall, Englewood Cliffs, NJ, pp. 345–376.

Vrba, E.S., 1995a. On the connections between paleoclimate and evolution. In: Vrba, E.S., Denton, G.H., Partridge, T.C., Burckle, L.H. (Eds.), 1995. Paleoclimate and Evolution with Emphasis on Human Origins. Yale University Press, New Haven, CT, pp. 24–45.

Vrba, E.S., 1995b. The fossil record of African antelopes (Mammalia, Bovidae) in relation to human evolution and paleoclimate. In: Vrba, E.S., Denton, G.H., Partridge, T.C., Burckle, L.H. (Eds.), 1995. Paleoclimate and Evolution with Emphasis on Human Origins. Yale University Press, New Haven, CT, pp. 385–424.

Vrba, E.S., Denton, G.H., Prentice, M.L., 1989. Climatic influences on early hominid behavior. Ossa 14, 127–156.

Vrba, E.S., Denton, G.H., Partridge, T.C., Burckle, L.H. (Eds.), 1995. Paleoclimate and Evolution with Emphasis on Human Origins. Yale University Press, New Haven, CT.

Wehausen, R., Brumsack, H.-J., 1999. Cyclical variations in the chemical composition of eastern Mediterranean Pliocene sediments: a key for understanding sapropel formation. Marine Geology 153, 161–176.

Wehausen, R., Brumsack, H.-J., 2000. The sapropel record of the eastern Mediterranean Sea – results of Ocean Drilling Program Leg 160. Palaeogeography, Palaeoclimatology, Palaeoecology 158, 371–395.

White, T., 2003. Early hominids – diversity or distortion? Science 299, 1994–1997.

White, T.D., Suwa, G., Asfaw, B., 1994. *Australopithecus ramidus*, a new species of early hominid from Aramis, Ethiopia. Nature 371, 306–312.

White, T.D., Suwa, G., Asfaw, B., 1995. Corrigendum: *Australopithecus ramidus*, a new species of early hominid from Aramis, Ethiopia. Nature 375, 88.

White, T.D., Suwa, G., Simpson, S., Asfaw, B., 2000. Jaws and teeth of *Australopithecus afarensis* from Maka, Middle Awash, Ethiopia. American Journal of Physical Anthropology 111, 45–68.

Wiens, J.A., Addicott, J.F., Case, T.J, Diamond, J., 1986. Overview: the importance of spatial and temporal scale in ecological investigations. In: Diamond, J., Case, T.J., (Eds.), Community Ecology, Harper and Row, NY, pp. 145–153.

WoldeGabriel, G., White, T.D., Suwa, G., Renne, P., de Heinzelin, J., Hart, W.K., Heiken, G., 1994. Ecological and temporal placement of early

Pliocene hominids at Aramis, Ethiopia. Nature 371, 330–333.

WoldeGabriel, G., Haile-Selassie, Y., Renne, P., Hart, W.K., Ambrose, S.H., Asfaw, B., Heiken, G., White, T., 2001. Geology and paleontology of the Late Miocene Middle Awash valley, Afar rift, Ethiopia. Nature 412, 175–177.

Wood, B.A., 1991. Koobi Fora Research Project, Vol. 4. Hominid Cranial Remains. Clarendon Press, Oxford.

Wood, B., Collard, M., 1999. The human genus. Science 284, 65–71.

Wynn, J.G., 2004. Influence of Plio-Pleistocene aridification on human evolution: evidence from paleosols of the Turkana Basin, Kenya. American Journal of Physical Anthropology 123, 106–118.

# 3. African Pliocene and Pleistocene cercopithecid evolution and global climatic change

S.R. FROST

*Department of Anthropology*
*1218 University of Oregon*
*Eugene, OR 97403-1218, USA*
*sfrost@uoregon.edu*

**Keywords**:    Habitat theory, turnover pulse, species range; taxonomic abundance; Afar Depression, Turkana Basin

## Abstract

Vrba, 1992, 1995a has put forth a series of hypotheses about the evolution of African mammals, which she has termed "habitat theory." This theory posits that changes in global climate cause "turnover pulses", within relatively short periods of time during which occur large numbers of first and last appearances of species. The present study examines whether a turnover pulse, or other forms of faunal change, occurred in the Cercopithecidae, 2.8–2.5 Ma at a time of major global cooling. The East African cercopithecid fossil record is well suited to this analysis because cercopithecids occur at most East African Pliocene and Pleistocene sites and are relatively speciose (Szalay and Delson, 1979; Delson, 1984; Jablonski, 2002). Several approaches are used including examination of cercopithecid species ranges as well as the abundance of larger taxonomic units. Both appearance and abundance data are examined for the Afar Depression and Turkana Basin, whereas only the species range data are studied for all East African and complete sub-Saharan analyses. The results provide no support for a turnover pulse 2.8–2.5 Ma. In fact, the largest number of first and last appearances are clustered around 3.4 and 2.0 Ma, with a shift in abundance at 3.4 Ma in both samples examined. These results are consistent with a relatively constant rate of turnover of cercopithecids between about 4 Ma and the Holocene.

## Introduction

Climatic change has been proposed as a cause of evolutionary process and pattern, most formally in a series of premises and hypotheses that Vrba (1992, 1995a, 1999) has called habitat theory. The first premise of habitat theory is that all animals are habitat specific. That is, they have certain temperature, moisture, and trophic requirements without which they cannot survive. For some taxa these requirements may be relatively broad, for others they may be narrower. It is further argued that natural selection will generally act to maintain this relationship between organism and habitat, resulting in morphological stasis rather than anagenetic adaptations to new habitat characteristics. As a result, the common response of most taxa to climatic change is to "passively drift" with their biome as it shifts over their continent.

A second premise of habitat theory is that allopatry is necessary for speciation, and that most allopatry is due to vicariance. Vicariance is the

*R. Bobe, Z. Alemseged, and A.K. Behrensmeyer (eds.) Hominin Environments in the East African*
*Pliocene: An Assessment of the Faunal Evidence, 51–76.*

division of a once continuous species range into two or more isolated ranges by the appearance of an isolating barrier within that range. A further assertion of habitat theory is that vicariance is most commonly caused by climatic change. In other words, biotic community interactions on their own are insufficient to cause vicariance. Speciation, extinction, and stasis are all given by Vrba as possible responses to vicariance, i.e., habitat fragmentation. Importantly, anagenesis is not suggested by Vrba as a possible evolutionary response to climatic change.

The net result of the above premises is that most if not all speciation and extinction is due to climatic change. Natural selection and biotic interactions on their own, particularly competition between species, are insufficient to cause speciation or extinction, but will instead tend to maintain an organism's adaptation to its environment. Therefore, under habitat theory, the majority of evolution occurs during periods of dramatic climatic change in significant concentrations of speciation, extinction, and migration. These evolutionary bursts are called "turnover pulses."

More specifically, Vrba (1992, 1995b, 1999, 2000) has proposed, based on data from fossil Bovidae, that global cooling between 2.8 and 2.5 million years ago (Ma) (Shackleton et al., 1984; deMenocal, 1995; Denton, 1999) caused a major turnover pulse in African mammals, as well as the origin of the hominin genera *Paranthropus* and *Homo* and the extinction of *Australopithecus*. Habitat theory, therefore, predicts that there should be a relatively large number of first and last appearances of fossil species clustered around this time interval, particularly among the more habitat specific mammals.

Several researchers have studied the temporal and geographic distributions of limited taxonomic groups in the Pliocene and Pleistocene of Africa: bovids (e.g., Vrba, 1976, 1980, 1995b; Bobe and Eck, 2001), suids (Cooke, 1978; Harris and White, 1979; White, 1995; Bishop, 1999), equids (Bernor and Armour-Chelu, 1999), cercopithecids (Delson, 1984, 1988), and hominids (e.g.,

White, 1995; Kimbel, 1995), as well as taxonomically broader faunal overviews of all Africa (Turner and Wood, 1993) or of one single region (e.g., Wesselman, 1995; Behrensmeyer et al., 1997; Bobe et al., 2002; Alemseged, 2003). There is still considerable debate as to whether a turnover pulse caused by a cooling of global climate occurred between 2.8 and 2.5 Ma (McKee, 1996, 2001; Behrensmeyer et al., 1997; Bobe et al., 2002; Alemseged, 2003).

All studies aimed at testing relationships between climate and evolution face many problems, such as those outlined by White (1995). Most of these problems relate to the quality of the data involved. There are two main categories of these problems: the biases in the fossil record, and alpha taxonomy. The biases in the fossil record that can influence such analyses include large stratigraphic gaps, preservational biases that favor particular taxa or one anatomical element over another, as well as diverse collection biases (e.g., Bobe et al., 2002). Alpha taxonomic biases are due to the fact that different paleontologists will allocate the same material to different taxa. This greatly hinders the utility of literature-based data as the species used must reflect real biological entities as closely as possible.

This study examines whether there is any evidence for a turnover pulse, or other forms of faunal change, among African cercopithecids at about the time of the 2.5 Ma global cooling event using multiple approaches. Cercopithecids were chosen as an appropriate group for such an analysis because they are relatively common in the fossil record during this period (Figure 1; Delson, 1984, 1988; Jablonski, 2002), are represented by a relatively large number of species (32 included in this analysis), and are often abundant (e.g., Kalb et al., 1982a; WoldeGabriel et al., 1994; Alemseged and Geraads, 2001; Bobe et al., 2002). In order to minimize the effects of some of the biases inherent in any analysis of this type, all of the fossils included in this analysis were examined firsthand by the author, generally in the form of original

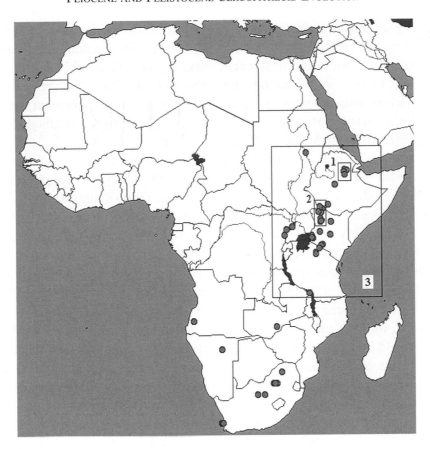

Figure 1. Map of African Pliocene and Pleistocene sites with fossil cercopithecids. Sites bounded by Boxes 1 and 2 are included in the Afar and Turkana data sets respectively. Those in Box 3 are included in the East African data set.

material, occasionally casts, or in some cases photographs. Thus, while it is impossible to be sure if the alpha taxonomy used here accurately reflects the true biological species present in the past, it is at a minimum more consistent than literature-based assignments. Some attempt has also been made to deal with the taphonomic biases in the cercopithecid record by critical analysis of the results.

## Materials and Methods

### FOSSIL SAMPLE

Included in this analysis is fossil material from 49 Pliocene and Pleistocene paleontological research areas from throughout sub-Saharan Africa. They range in extent from very large

areas consisting of many smaller collecting areas and individual localities (such as the Middle Awash, Ethiopia) to localities with only a single cercopithecid specimen (such as Kuguta, Kenya). For as many localities as possible, original material was studied directly by the author, for many of the sites where it was not possible to study the originals, casts were used. For a few others (such as Wad Medani, Sudan) only photographs were available.

For this analysis, four data sets were used. The first two represent basin-level regions of East Africa, and were studied in particular detail. These are the Afar Depression and the Turkana Basin, indicated by Boxes 1 and 2 respectively in Figure 1. These two areas were chosen because both have large and speciose collections of fossil cercopithecids and the sediments of each are tightly controlled chronologically. Furthermore, between the two of them they have yielded the majority of the East African Pliocene and Pleistocene cercopithecid fossil record. The third data set was an all East African data set, and included material from sites distributed from southern Sudan through northern Malawi. These are shown in Box 3 of Figure 1. Finally, in order to maximize the number of species that could be included in this analysis, the final data set includes material from both East and southern Africa. The sites included in this data set are shown in Figure 1. The sample was divided into different data sets for three reasons. First, because habitat theory focuses on global scale climatic change, it should affect all of Africa, and therefore the 2.5 Ma turnover pulse should be evident in each data set, although not necessarily equally given the possibility of differing environmental and other characteristics among different regions. Second, the taxonomic control differs among data sets, with the Afar data having the most and the sub-Saharan data set the least. Third, chronological control is tightest in the Afar and Turkana data sets, is also quite good in the East African data set, but is much less so for the southern African portion of the total sub-Saharan data set.

*Afar Sample*

The sample from the Afar Depression includes all of the material from the paleoanthropological collecting regions of the Middle Awash and Hadar (including nearby localities Ahmado, Leadu, and Geraru [see Kalb, 1993; Frost and Delson, 2002 for more information on these early International Afar Research Expedition sites]) that were available as of 1999, and dated to between 4.4 and approximately 0.25 Ma. The chronostratigraphy of the sediments from which this sample derives is shown in Figure 2. (Dates from: Kalb et al., 1982a, b; Walter and Aronson, 1993; White et al., 1993; Walter, 1994; Clark et al., 1994; Kimbel et al., 1996; de Heinzelin et al., 1999; Renne et al., 1999; Renne, 2000.)

The Afar sample includes over 2000 individual cercopithecid specimens representing 13 distinct species (Figure 2; Frost, 2001a, b; Frost and Delson, 2002). For this sample, it was possible to study each specimen individually, giving it the highest degree of taxonomic control.

*Turkana Sample*

The Turkana Basin sample includes material from the Omo Shungura and Usno Formations, Ethiopia, and from Koobi Fora, West Turkana, and Kanapoi, Kenya. The chronostratigraphy for the Turkana Basin sediments is shown in Figure 3. The sample of fossil cercopithecids from these sediments includes over 8000 specimens (Eck, 1977; Harris et al., 1988; 2003; personal observation) representing a total of 14 species, 3 of which are likely to be conspecific with those from the Afar Depression. These are shown in Figure 3. This sample has been published by several authors over a period of over 30 years (Patterson, 1968; Leakey and Leakey, 1973a, b, 1976; Leakey, 1976, 1982, 1987, 1993; Eck, 1976, 1977, 1987a, b; Eck and Jablonski, 1987; Harris et al., 1988, 2003; Leakey et al., 2003), and there is considerable disagreement among authorities as to the taxonomic allocation of several specimens. Similar to the Afar sample, it was possible to study almost all of the original material. However, it was not possible to study every isolated tooth from the Omo, but all of the relatively complete material (as well as many of the isolated teeth) was analyzed. The taxonomic allocation for this data set follows Frost (2001a).

*East African Sample*

The third data set is composed of all of the material that could be analyzed from East Africa. This data set includes material from 29 major collecting sites and included 23 species (shown in Figure 4 by the solid range boxes). From many sites, it was possible to study all or most of the original material. From Olduvai only that part at the British Museum was studied from the original specimens. The

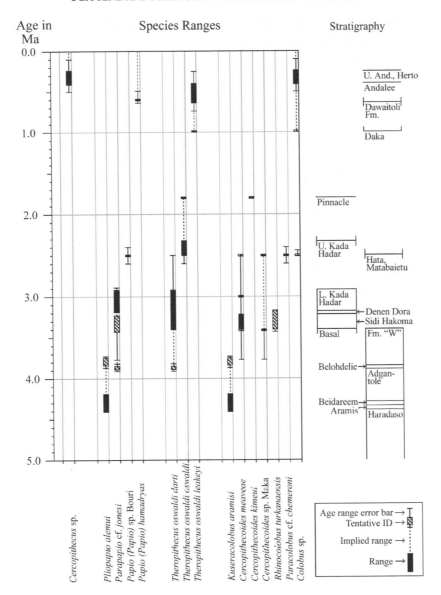

Figure 2. Temporal range of Afar Cercopithecidae. Also shown are the chronological representations of stratigraphic units. Solid boxes show the age range for species based only on confidently assigned material. Hatched boxes show ranges based on more tentatively assigned material. Solid error bars represent geochronological uncertainty. Dashed lines represent implied ranges across large gaps in the sequence.

remainder of the Olduvai sample, and the entire sample from Laetoli was analyzed from casts in the collection of Eric Delson at the AMNH. Thus, the taxonomic control for this sample is fairly tight, but not as good as that for the Afar or Turkana samples, but there are two advantages: more species that can be included in the analysis and the overall analysis should be less influenced by stratigraphic hiatuses.

*Total Sub-Saharan Sample*
The fourth and final data set was an all sub-Saharan one. In addition to the material in the East African data set, it included sites from South Africa, Zambia, Namibia, and Angola. There were two problems presented by this data set. First, none of the southern African sites are tightly controlled chronologically. Second, a smaller percentage of material

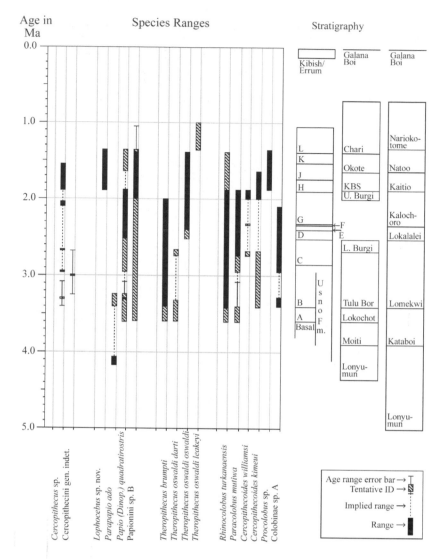

Figure 3. Temporal range of Turkana Cercopithecidae. Symbols as in Figure 2.

at these sites could be directly analyzed. In general, all of the relatively complete specimens were studied firsthand, but there was an enormous amount of more fragmentary material that could not be analyzed due to time constraints. Thus, the southern African sites are the least tightly controlled taxonomically as well. However, they add considerably to the number of species, bringing the total to 32 (Figure 4, both black and white boxes). Finally, the species ranges in this data set are the global ranges for all of the taxa involved.

## APPEARANCE DATA

For all of the material included in this analysis, each specimen was identified to anatomical element and allocated to the lowest taxonomic category possible. Often it was possible to identify specimens to species, or in the case of some *Theropithecus* specimens, to subspecies. Occasionally, this was possible entirely on the basis of the morphology preserved in a given specimen. In most cases, however, such identifications were based on the total sample from a given site. More fragmentary material was

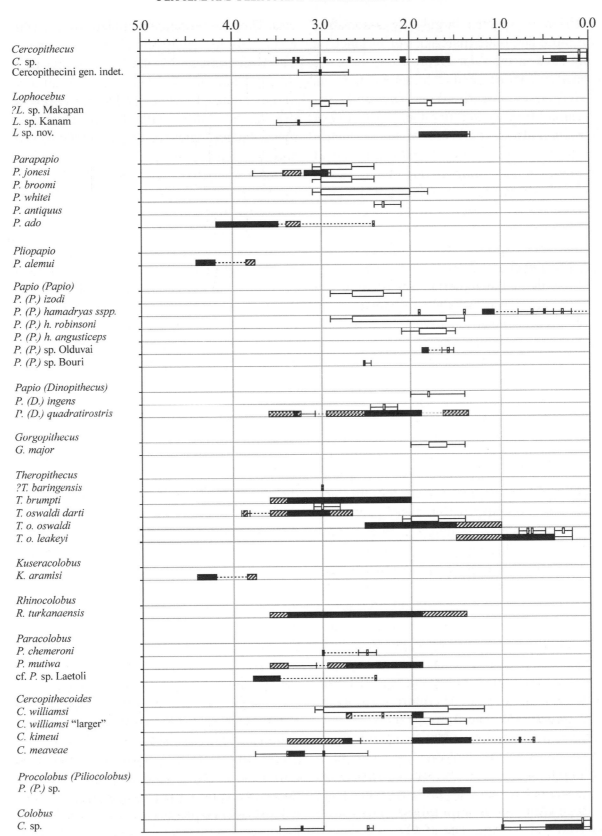

Figure 4. Temporal ranges for East and South African cercopithecid species. East African species ranges are shown by black boxes; South African ranges are shown by white boxes. Other symbols as in Figure 2.

identified to species largely by association with more complete material. For instance, at many sites in the Afar Depression, there are isolated molars or gnathic fragments that can be definitively identified as *Theropithecus*, but cannot be allocated to either *T. oswaldi* or *T. brumpti*. However, these would be identified to *T. oswaldi* in this case because all of the more complete diagnosable material from this basin represented this species, while none represented *T. brumpti*. On the other hand, similar teeth from the Omo must be left as *Theropithecus* species indeterminate because both *T. brumpti* and *T. oswaldi* are present.

Once all of the material was analyzed in this manner it was possible to construct the chronological range data for each species. These range data are affected by two types of uncertainty. The first and most obvious is uncertainty about the chronometric age of the strata from which fossils are derived. Additionally, if the provenience of a given specimen is unclear it can cause uncertainty in the age of a fossil, even in well-dated strata. In either case the effect on species ranges is the same: both reduce the precision of the age estimate for the fossil. This first type of uncertainty is indicated in the taxon range figures (Figures 2–4) by the whiskers extending from the ends of the range boxes. These are only shown where they affect the possible age range of a species. A second type of uncertainty in species ranges is due to specimens that can only be tentatively included in a given species. To ignore them may cut a range short, but to include them may stretch a range too far. Therefore, two ranges were constructed. The first one included only material that could be identified with confidence and used the geologic age estimate in the middle of the possible range. This range is shown in the range figures by either solid black or white boxes. From these ranges, first and last appearance data (FADs and LADs respectively) were determined for each spe-

cies. The second range included fossils, which could only be tentatively allocated to species. These are indicated in the range figures by the crosshatched boxes. From this material a second estimate of species FADs and LADs was determined. Additionally, for the second estimate, the geological error was added to FADs and subtracted from LADs. These second estimates are labeled FAD2 and LAD2 in all of the figures (from here on "FAD" and "LAD" will only be applied to the species ranges based on confidently assigned material, and when first and last appearances including tentative material are discussed, they will always be referred to as "FAD2" and "LAD2"). Thus, the first estimate is a conservative estimate of the taxon range, and the second produces the maximum possible species range justified by the data. Appearance data for each of the four data sets are given in Tables 1–4. These data include only the ranges of full species. In Figures 2 and 3 subspecies are sometimes shown, this is done to be more explicit about what is included as not all authors agree on the subspecific status of some of these. However, they are pooled here for all analyses.

Appearance data were grouped into 100-Kyr intervals and plotted against time (Figures 5–8) to evaluate their temporal distribution. Intervals of 100 Kyr were chosen, as this is the length suggested (Vrba, 1995a) over which the 2.8–2.5 Ma turnover pulse might occur, and shorter intervals are beyond the resolution of most of the data. Finally, longer intervals of either 400 or 500 Kyr can always be created by pooling the 100-Kyr intervals, but not the reverse.

Vrba (1995b) has provided a model for the accumulation of new distinct morphologies within a given clade. For this study, this model is applied only to the all sub-Saharan data set, as it is the only one with adequate geographic, taxonomic, and temporal coverage to provide a reasonably robust use of the model. The null hypothesis is that the exponential rate of FAD accumulation, $A$, is constant through time

*Table 1. Appearance data for the Afar data set. FADs and LADs are first and last appearance data respectively. FAD2s and LAD2s are maximum first and minimum last appearance data; see text for definitions*

| Taxon | FAD | LAD | FAD2 | LAD2 |
|---|---|---|---|---|
| *Cercopithecus* sp. | 0.40 | 0.25 | 0.40 | 0.25 |
| *Pliopapio alemui* | 4.39 | 4.19 | 4.39 | 3.75 |
| *Parapapio* cf. *jonesi* | 3.18 | 2.92 | 3.40 | 2.92 |
| *Papio* sp. Bouri | 2.50 | 2.50 | 2.60 | 2.45 |
| *Papio hamadryas* | 0.64 | 0.64 | 0.64 | 0.64 |
| *Theropithecus oswaldi* | 3.40 | 0.40 | 3.89 | 0.40 |
| *Kuseracolobus aramisi* | 4.39 | 4.19 | 4.39 | 3.75 |
| *Cercopithecoides kimeui* | 1.80 | 1.80 | 1.80 | 0.64 |
| *Cercopithecoides meaveae* | 3.40 | 3.28 | 3.40 | 2.50 |
| cf. *Cercopithecoides* sp. indet. Maka | 3.40 | 3.40 | 3.40 | 2.50 |
| cf. *Rhinocolobus turkanaensis* | 3.40 | 3.18 | 3.40 | 3.18 |
| *Paracolobus* cf. *chemeroni* | 2.50 | 2.50 | 2.60 | 2.45 |
| *Colobus* sp. | 0.40 | 0.25 | 2.50 | 0.25 |

*Table 2. Appearance data for Turkana data set. See Table 1 for description*

| Taxon | FAD | LAD | FAD2 | LAD2 |
|---|---|---|---|---|
| *Cercopithecus* sp. | 2.95 | 1.55 | 3.30 | 1.55 |
| Cercopithecini gen. et sp. indet. | 3.00 | 3.00 | 3.40 | 2.68 |
| *Lophocebus* sp. nov. | 1.88 | 1.36 | 1.88 | 1.33 |
| *Parapapio ado* | 4.17 | 4.07 | 4.17 | 3.24 |
| *Papio (Dinopithecus) quadratirostris* | 3.30 | 1.90 | 3.59 | 1.36 |
| Papionini B | 2.00 | 1.39 | 3.59 | 1.05 |
| *Theropithecus brumpti* | 3.40 | 2.00 | 3.59 | 2.00 |
| *Theropithecus oswaldi* | 3.40 | 1.00 | 3.59 | 1.00 |
| *Rhinocolobus turkanaensis* | 3.40 | 1.88 | 3.59 | 1.39 |
| *Paracolobus mutiwa* | 2.74 | 1.88 | 3.59 | 1.88 |
| *Cercopithecoides williamsi* | 2.00 | 1.88 | 2.74 | 1.88 |
| *Cercopithecoides kimeui* | 2.00 | 1.64 | 3.40 | 1.64 |
| *Procolobus* sp. | 1.88 | 1.36 | 1.88 | 1.36 |
| Colobinae sp. A | 3.40 | 2.10 | 3.40 | 2.10 |

(i.e., the dashed line in Figure 9) and modeled by the equation

$$N = N_0 e^{At} \tag{1}$$

where $N$ is the total number of accumulated FADs at time $t$, and $N_0$ is the number of taxa at the beginning of the time period. Solving Equation 1 for $A$ yields

$$A = (ln\ N_i - ln\ N_0)/(t_i - t_0) \tag{2}$$

With Equation 2, $A$ can be estimated by entering the number of accumulated FADs at time

Table 3. *Appearance data for East African data set. See Table 1 for description*

| Taxon | FAD | LAD | FAD2 | LAD2 |
|---|---|---|---|---|
| *Cercopithecus* sp. | 3.30 | 0.00 | 3.50 | 0.00 |
| Cercopithecini gen. et sp. indet. | 3.00 | 3.00 | 3.40 | 2.68 |
| *Lophocebus* sp. Kanam | 3.30 | 3.30 | 3.50 | 3.00 |
| *Lophocebus* sp. nov. | 1.88 | 1.36 | 1.88 | 1.33 |
| *Parapapio jonesi* | 3.18 | 2.92 | 3.40 | 2.92 |
| *Parapapio ado* | 4.17 | 3.49 | 4.17 | 3.24 |
| *Pliopapio alemui* | 4.40 | 4.19 | 4.40 | 3.75 |
| *Papio* sp. – Bouri | 2.50 | 2.50 | 2.60 | 2.45 |
| *Papio hamadryas ssp.* | 1.07 | 0.00 | 1.20 | 0.00 |
| *Papio (Dinopithecus) quadratirostris* | 3.30 | 1.88 | 3.59 | 1.36 |
| *?Theropithecus baringensis* | 3.00 | 3.00 | 3.00 | 3.00 |
| *Theropithecus brumpti* | 3.40 | 2.00 | 3.59 | 2.00 |
| *Theropithecus oswaldi* | 3.40 | 0.25 | 3.89 | 0.25 |
| *Kuseracolobus aramisi* | 4.40 | 4.19 | 4.40 | 3.75 |
| *Rhinocolobus turkanaensis* | 3.40 | 1.88 | 3.59 | 1.39 |
| *Paracolobus chemeroni* | 3.00 | 2.50 | 3.00 | 2.50 |
| *Paracolobus mutiwa* | 2.74 | 1.88 | 3.59 | 1.88 |
| cf. *Paracolobus* sp. | 3.79 | 3.49 | 3.79 | 2.49 |
| *Cercopithecoides williamsi* | 2.00 | 1.88 | 2.74 | 1.88 |
| *Cercopithecoides kimeui* | 2.40 | 0.80 | 3.40 | 0.64 |
| *Cercopithecoides meaveae* | 3.40 | 3.28 | 3.79 | 2.10 |
| *Colobus* sp. | 3.25 | 0.10 | 3.50 | 0.00 |
| *Procolobus* sp. | 1.88 | 1.36 | 1.88 | 1.36 |

Table 4. *Appearance data for total Sub-Saharan African data set. See Table 1 for description*

| Taxon | FAD | LAD | FAD2 | LAD2 |
|---|---|---|---|---|
| *Cercopithecus* sp. | 3.30 | 0.00 | 3.50 | 0.00 |
| Cercopithecini gen. et sp. indet. | 3.00 | 3.00 | 3.40 | 2.68 |
| *?Lophocebus* sp. Makapan | 3.00 | 3.00 | 3.20 | 1.40 |
| *Lophocebus* sp. Kanam | 3.25 | 3.25 | 3.50 | 3.00 |
| *Lophocebus* sp. nov. | 1.88 | 1.36 | 1.88 | 1.33 |
| *Parapapio jonesi* | 3.18 | 2.65 | 3.40 | 2.40 |
| *Parapapio broomi* | 3.00 | 2.65 | 3.20 | 2.40 |
| *Parapapio whitei* | 3.00 | 2.00 | 3.20 | 1.80 |
| *Parapapio antiquus* | 2.30 | 2.30 | 2.40 | 2.10 |
| *Parapapio ado* | 4.17 | 3.49 | 4.17 | 3.24 |
| *Pliopapio alemui* | 4.40 | 4.19 | 4.40 | 3.75 |
| *Papio izodi* | 2.65 | 2.30 | 2.90 | 2.10 |
| *Papio hamadryas* | 2.65 | 0.00 | 2.90 | 0.00 |
| *Papio* sp. Bouri | 2.50 | 2.50 | 2.60 | 2.45 |
| *Papio* sp. Olduvai | 1.87 | 1.70 | 1.87 | 1.52 |
| *Papio (Dinopithecus) ingens* | 1.80 | 1.80 | 2.00 | 1.40 |
| *Papio (Dinopithecus) quadratirostris* | 3.30 | 1.88 | 3.59 | 1.36 |
| *Gorgopithecus major* | 1.80 | 1.60 | 2.00 | 1.40 |
| *?Theropithecus baringensis* | 3.00 | 3.00 | 3.00 | 3.00 |
| *Theropithecus brumpti* | 3.40 | 2.00 | 3.59 | 2.00 |
| *Theropithecus oswaldi* | 3.40 | 0.25 | 3.89 | 0.25 |
| *Kuseracolobus aramisi* | 4.40 | 4.19 | 4.40 | 3.75 |

(*Continued*)

*Table 4. Appearance data for total Sub-Saharan African data set. See Table 1 for description—cont'd*

| Taxon | FAD | LAD | FAD2 | LAD2 |
|---|---|---|---|---|
| *Rhinocolobus turkanaensis* | 3.40 | 1.88 | 3.59 | 1.39 |
| *Paracolobus chemeroni* | 3.00 | 2.50 | 3.00 | 2.50 |
| *Paracolobus mutiwa* | 2.74 | 1.88 | 3.59 | 1.88 |
| cf. *Paracolobus* sp. Laetoli | 3.79 | 3.49 | 3.79 | 2.49 |
| *Cercopithecoides williamsi* | 3.00 | 1.60 | 3.20 | 1.20 |
| *Cercopithecoides williamsi* "larger" | 1.80 | 1.60 | 2.00 | 1.40 |
| *Cercopithecoides kimeui* | 2.40 | 0.80 | 3.40 | 0.64 |
| *Cercopithecoides meaveae* | 3.40 | 3.28 | 3.40 | 2.50 |
| *Procolobus* sp. | 1.88 | 1.36 | 1.88 | 1.36 |
| *Colobus* sp. | 1.00 | 0.00 | 3.50 | 0.00 |

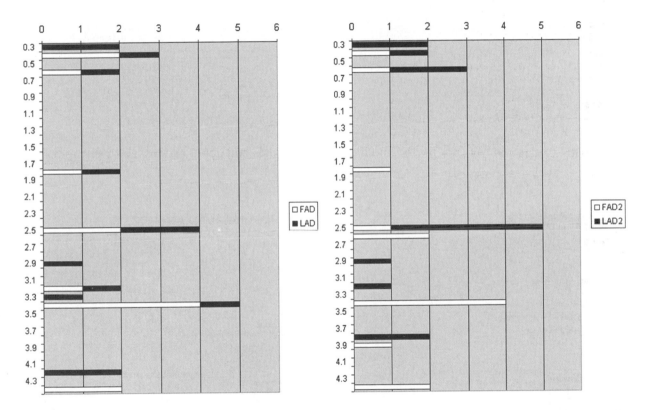

Figure 5. Frequency of appearance data for the Afar sample, with time in Ma shown on the vertical axis. First appearance data (FADs) and maximum first appearance data (FAD2s) are shown as white bars. Last appearance data (LADs) and minimum last appearance data (LAD2s) are indicated by black bars. See text for explanation of each.

$t_i$ and $t_0$. The observed FADs or FAD2s for all included species are then sorted by their rank order (Figure 9). This observed rate of accumulation can then be compared to the expected constant rate. An iterative Chi-square test is then used to test whether the observed pattern of FAD accumulation differs significantly from the null hypothesis. If there is a

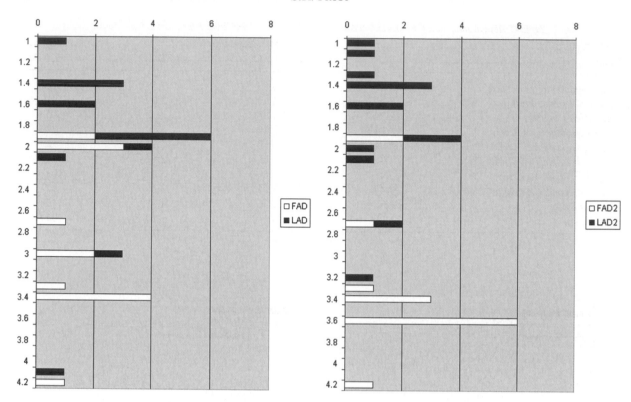

Figure 6. Appearance data for the Turkana Basin. Symbols and abbreviations as for Figure 5.

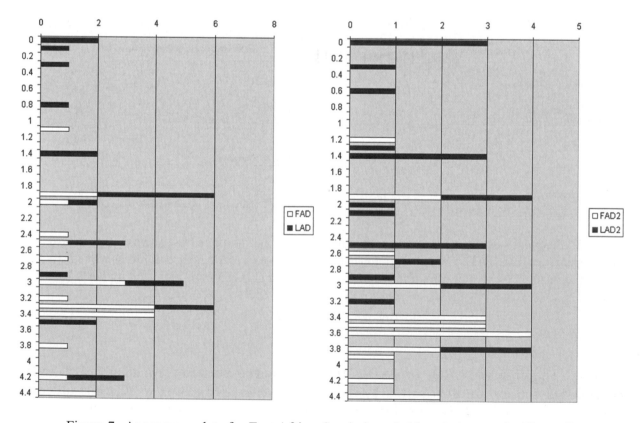

Figure 7. Appearance data for East Africa. Symbols and abbreviations as for Figure 5.

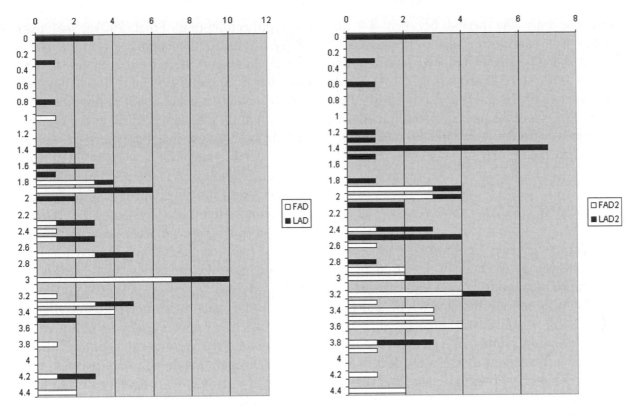

Figure 8. Appearance data for both East and South Africa. Symbols and abbreviations as for Figure 5.

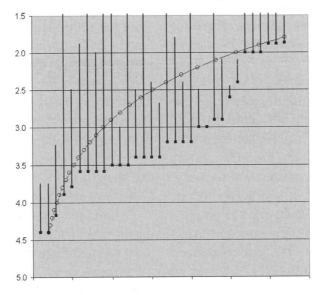

Figure 9. Durations of fossil cercopithecid species for the last 4.4 Ma plotted against expected rate of accumulation, $N = N_0 e^{1.07t}$, indicated by the dashed line. FAD2s from Table 4 are marked by the squares, and sorted by rank order. Following Vrba (1995a).

significant difference, then the period with the greatest deviation from expected is replaced by the expected value, and the test is repeated until a nonsignificant result is achieved.

Both FADs and FAD2s were examined. As the focus of this test was to examine the possibility that the number of FADs exceeded expectations at 3.5, 3.0, 2.5, and 2.0 Ma in particular only the period between 4.4 and 1.8 Ma was examined. This is because after 1.8 Ma there is greatly reduced sampling and there is only one FAD and no FAD2s. At 4.4 Ma there are two FADs and FAD2s; by 1.8 Ma there are 31 FADs and 32 FAD2s. From Equation 2, this yields expected accumulation rates of $A = 1.05$ for the FADs and 1.07 for the FAD2s.

## ABUNDANCE DATA

Abundance data are thought to be more sensitive indicators of faunal change than are appearance data, but require a higher degree of comparability among the data (Bobe and Eck, 2001; Bobe et al., 2002; Alemseged, 2003). Both taphonomic and collection biases must be taken into account to a greater degree than is the case with appearance data. Because of these considerations, only the Afar and Turkana data sets were analyzed for abundance. A large proportion of the material from both data sets is represented by isolated teeth and gnathic fragments. There is also a considerable degree of variation in the ability to definitely diagnose different species and genera based on fragmentary material. For example, *Theropithecus oswaldi leakeyi* can often be identified from isolated elements simply based upon its extreme size; however others such as *Parapapio* require relatively complete facial material to be definitely diagnosed. Therefore, in order to include the maximum amount of data, and to minimize this identification bias, the abundance data were calculated using the four groups defined by Delson (Szalay and Delson, 1979) based on molar morphology. These groupings have the advantage that fragmentary specimens can be reliably assigned. These groups are: colobines (*Colobus* and allies); cercopithecins (the guenons); the papionins other than *Theropithecus* (baboons, mangabeys, etc.); and *Theropithecus* (the gelada and extinct relatives). Except for the non-*Theropithecus* papionins, these groups represent holophyletic clades within the

Cercopithecidae (e.g., Frost, 2001a; Jablonski, 2002). It is useful to recognize *Theropithecus* as a distinct category because it is the most common fossil cercopithecid and has a uniquely derived dental morphology. It is important that while there may be overall differences between each of these groups in habitat preference, there is considerable range (and overlap) within each of them. Therefore, it is not a simple matter to translate from the abundances of these groups to paleoenvironment. However, the relative abundances of these different groups at different times and localities should indicate at least a change in overall cercopithecid fauna if not in paleoecology.

The Afar data have been organized into chronological groupings as shown in Figure 10. Data for the different formations of the Turkana Basin have not been pooled as taphonomic, sample size, and collection differences make them not entirely comparable. The Turkana Basin units have been organized into approximate chronological order in Figure 11. Proportions from the Omo Members A though H are from Bobe (1997), as well as personal observation, and only include the sample from the American contingent of the International Omo Expedition. Those from the Nachukui Formation are from Harris et al. (1988); those from Koobi Fora are from Leakey and Leakey (1973a, 1976), Leakey (1976), Delson et al. (1993), Bobe (pers. com.); and those from Kanapoi are from Harris et al. (2003); and personal observation.

Following Bobe et al. (2002), the chord distance is used to compare the abundance data from different units. Chord distance is as a measure of dissimilarity among units, based on the cosine between the vectors of abundance data for each unit. The chord distance between two samples $j$ and $k$ is computed using the formula

$$CR\,D_{jk} = \left[ 2\,(1 - ccos_{jk}) \right]^{1/2} \qquad (3)$$

When $X_{ij}$ is the abundance of the $i$th taxon in sample $j$ and $X_{ik}$ is the abundance of the $i$th

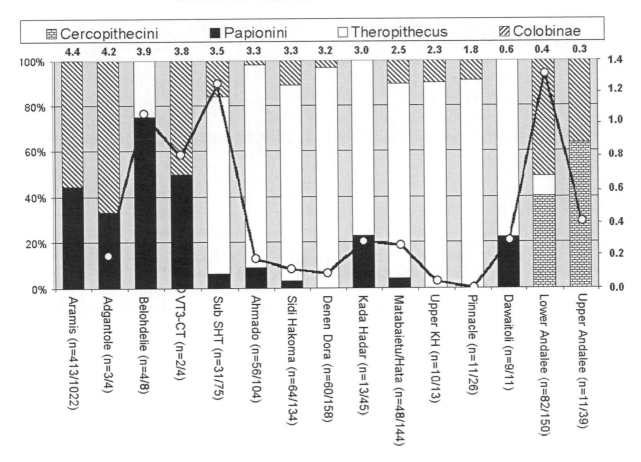

Figure 10. Relative abundance of dental groups in different stratigraphic units of the sample from the Afar Depression. Stratigraphic units arranged in approximate chronological order with ages in Ma shown across the top. White indicates *Theropithecus*, black non-*Theropithecus* papionins, bricks cercopithecins, and hatches colobines. Numbers in parentheses represent sample sizes. The numerator is the number of specimens identifiable to one of the four categories, and the denominator is the total number of specimens.

Chord distance is on the right-hand *y*-axis, line connects values for chord distance to previous unit.

taxon in sample *k*, and $X^S$ is the total number of taxa in both samples, then $ccos_{jk}$ is computed:

$$ccos_{jk} = \Sigma^S(X_{ij} \times X_{ik}) / [\Sigma^S X_{ij}^2 \Sigma^S X_{ik}^2]^{1/2} \quad (4)$$

The chord distances were then entered into a matrix with all 39 units considered in this analysis and clustered using the unweighted pair-group method, arithmetic average (UPGMA) method in NTSYSpc 2.10 (Applied Biostatistics, 2000) (Figure 12). The chord distances are also shown in Figures 10 and 11 by lines where they are used to illustrate the difference between adjacent stratigraphic levels.

**Results**

APPEARANCE DATA

When FADs and LADs for the Afar Material are plotted against geologic time (Figure 5), it appears that they are clustered around two distinct time intervals: 3.4 and 2.5 Ma, with the first being the more important. Both clusters are also present in the distribution of FAD2s and LAD2s, but the 2.5-Ma event is somewhat larger. In both cases, the 3.4-Ma event is dominated by first appearances, and more last appearances occur in the 2.5-Ma event.

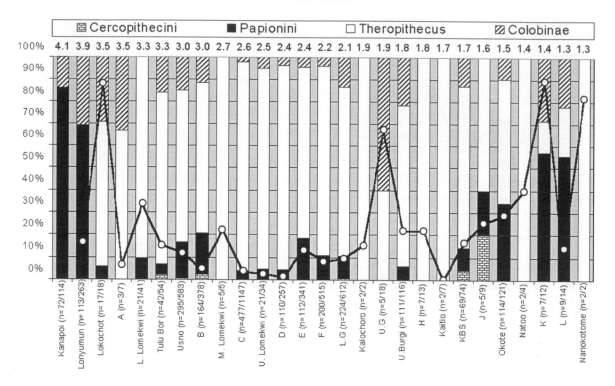

Figure 11. Relative abundance of dental groups in the different stratigraphic units of the Turkana Basin sample. Axes and symbols as for Figure 10.

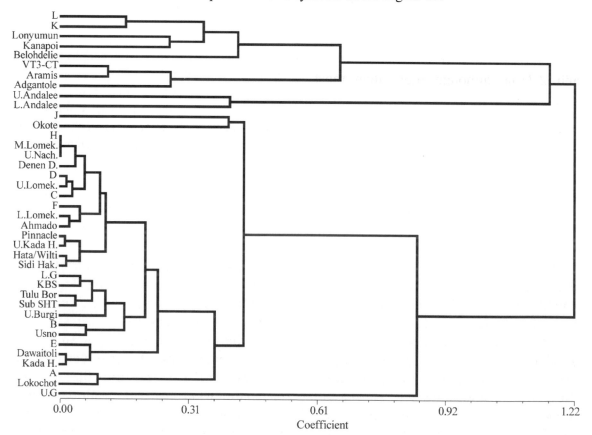

Figure 12. UPGMA cluster analysis of abundance data for stratigraphic units in Figures 10 and 11 based on chord distance.

In the Turkana Basin sample, there also appear to be two main spikes in the appearance data: one at about 3.4 to 3.6 Ma, and a second at between 1.9 and 2.0 Ma (Figure 6). In the case of the data derived only from confidently assigned material, the ca. 3.4-Ma event is entirely composed of first appearances, and the 1.9–2.0-Ma event is significantly larger. Both clusters also occur in the FAD2s and LAD2s, but the relative magnitudes are reversed. The 3.4- to 3.6-Ma event is again entirely composed of first appearances. Importantly, in the Turkana Basin data set, there are almost no FADs and LADs or FAD2s and LAD2s between 2.8 and 2.3 Ma.

For the East Africa data set, the periods of highest turnover are less pronounced than for either of the individual basin-level data sets (Figure 7). There do appear to be two main periods of relatively high turnover among the FADs and LADs, again at approximately 3.6–3.4 Ma (and possibly 3.0 Ma) and between 2.0 and 1.9 Ma. Once again there appear to be relatively fewer FADs and LADs between 2.8 and 2.3 Ma, although more than most intervals. Using the FAD2s and LAD2s, there appear to be no real peaks in the appearance data, but only a period from about 3.8 to 1.8 Ma where there is a relatively large amount of species turnover compared with earlier or later intervals.

The greatest number of FADs and LADs in the all sub-Saharan data set occurs in the 3.0-Ma time interval, with a second peak at approximately 1.9–1.8 Ma (Figure 8). The interval from 2.3 to 2.8 does not contain a relatively large cluster of first and last appearances. For the FAD2s and LAD2s, the most prominent feature is a very large number of FAD2s clustered between 1.5 and 1.2 Ma. Outside of this feature, the entire span from 3.8 to 1.9 Ma seems to have a relatively constant rate of species first and last appearances.

In both the FAD and FAD2 data sets, there are two FADs at $t_0$ and there are 31 and 32 respectively, by time $t_i$ (in this case is $t_{26}$). The

*Table 5. Significance values for the iterative Chi-square test (see Vrba, 1995b). The first column shows the time intervals that deviate most from expected. Observed – expected is number of FADs difference*

| Time interval | Observed – expected | Chi-square value |
|---|---|---|
| 2.9 | 14.09819 | 0.0000002815 |
| 3.0 | 13.09975 | 0.0000049250 |
| 2.8 | 12.98392 | 0.0000636371 |
| 3.2 | 12.80919 | 0.000554091 |
| 2.7 | 11.74427 | 0.005610215 |
| 2.6 | 11.36511 | 0.023909569 |

results of the Chi-square test for the FAD data could not reject the null hypothesis (Chi-square 0.09). For the FAD2 data set, the observed values differed significantly from expected, so the iterative procedure was conducted. This data set is summarized graphically in Figure 9. The time period with the observed value furthest from the expected was 2.9 Ma, followed by 3.0, then from there 2.8, 3.2, 2.7, and 2.6. After the value for 2.6 further intervals were not significant. These results are summarized by Table 5.

## ABUNDANCE DATA

The Afar relative abundance data show three distinctive periods (Figure 10). The earliest is a period of codominance of colobines and non-*Theropithecus* papionins. This is an unusually high abundance of colobines for a Pliocene assemblage. Importantly, *Theropithecus* is absent in this period (or at least it is very rare). This time period is essentially equivalent to that prior to the ca. 3.5-Ma cluster of FADs and LADs. The second time period is characterized by a predominance of *Theropithecus*, with its relative abundance generally approaching 80–90%. This *Theropithecus* zone continues from ca. 3.5 Ma in the sub-SHT levels through the early Middle Pleistocene. The third abundance period is only in the upper and lower

Andalee levels. This period is characterized by the very high abundance of Cercopithecini, and Colobinae.

The Turkana Basin shows a pattern of abundance that is generally similar to that of the Afar Depression (Figure 11). There is an early period prior to the ca. 3.6–3.4 Ma turnover event seen in Figure 5. It occurs at Kanapoi and Allia Bay (Harris et al., 2003; Bobe, pers. com.). In this period, non-*Theropithecus* papionins are the most common cercopithecids, colobines are rare, and *Theropithecus* and Cercopithecini are absent.

As in the Afar Depression, *Theropithecus* predominates after about 3.5 Ma, from the Lokochot Member, and especially the Tulu Bor Member through the Okote Member of the Koobi Fora Formation, the Lomekwi through Nariokotome Members of the Nachukui Formation, and Members A through upper G of the Shungura Formation (Figure 11). The relative abundances in Members J–L of the Shungura Formation may not show this predominance, but their small sample sizes argue for caution in interpreting the observed abundances. Colobines and non-*Theropithecus* papionins are generally present but of low abundance throughout the section, and guenons are very rare, and often absent.

The cluster analysis confirms that the units from both basins are similar during the so-called "*Theropithecus* zone" from approximately 3.5 through 1.5 Ma, with abundances of the genus typically comprising more than 75%. Both non-*Theropithecus* papionins and colobines are rarer, but constantly present in both basins. Cercopithecini are absent from the Afar Depression until the Middle Pleistocene (with the possible exception of a tentatively assigned humerus from Wee-ee, [Frost, 2001a]) and are extremely rare in the Turkana Basin. This dominance of *Theropithecus* for the majority of stratigraphic units in both basins is shown by the large cluster of stratigraphic groups at the bottom of Figure 12.

The distinctness of the earliest sites from the two basins is clear, given the large chord distance separating Aramis from Kanapoi and Lonyumun (Figure 12), reflecting the high proportion of colobines at Aramis and the higher frequency of papionins at Kanapoi and the Lonyumun Member.

For this analysis, samples of Middle Pleistocene age are only present in the Afar Depression. The Upper Bodo unit shows a pattern of abundance similar to other horizons dominated by *Theropithecus*, but the lower and upper Andalee deposits are clearly very different. *Theropithecus* is rare and absent (respectively) from these beds, which are dominated by colobines and cercopithecins. In fact, the Andalee deposits (and other Pleistocene sites in the Afar region, such as Asbole [Alemseged and Geraads, 2001]) may be unique in Africa for the high proportion of cercopithecins. This distinctiveness is clearly shown by the long branch separating these two from the other "low *Theropithecus*" units (Figure 12).

## Discussion

The turnover pulse hypothesis is formulated in terms of species origins, extinctions, and migrations; therefore, the appearance data are used to evaluate it. The abundance data are used as a more fine-grained lens for examining other types of faunal change to see if they are coincident with those of the appearance data or can otherwise shed light on it. Examination of the appearance data shows that while there do appear to be two distinct peaks in the data from the Afar Depression (Figure 5), there are some problems. Given that there are large stratigraphic gaps both before 2.5 Ma and after 2.3 Ma it is impossible to tell whether the peak in turnover at 2.5 Ma is really as temporally restricted as it appears, or whether the appearance data have accumulated during the stratigraphic gaps. Likewise, while there are samples from prior to 3.5 Ma from the Middle Awash, they are generally quite small between 4.4 and 3.5 Ma. Therefore, the apparent spike in first appearances at 3.5 Ma may be due to a similar stratigraphic effect. However, if either of these events reflects real increases in rates of turnover among Early to Middle Pliocene cercopithecids due to global climatic changes, then they should be reflected in the other data sets.

When the Turkana Basin data are examined, again there is ca. 3.5-Ma turnover event, apparently synchronous with that from the Afar Depression. However, prior to 3.5 Ma, the sample is very restricted, and given that this event is entirely composed of FADs or FAD2s it seems likely that it may represent stratigraphic and sampling effects. Given similar events in both the Afar and Turkana Basins, it does seem clear that between about 4.0 and 3.5 Ma there was substantial turnover of species. It is not clear, however, that this occurred in a pulse. Sampling is quite good in the Turkana Basin between 3.4 and 2.0 Ma, particularly at the Omo (Eck, 1977; Bobe et al., 2002; Alemseged, 2003) and there is no suggestion of a turnover pulse at 2.5 Ma. Vrba (e.g., 1988) has suggested that the Turkana Basin may have been a refugium during the 2.8–2.5-Ma cooling event, which could be compatible with this observation. The clustering of appearance data around 2.0 Ma may well represent a real increase in turnover rate, which would be close to an increase in FADs among bovids at 1.8 Ma (Vrba, 1995b). However, there is evidence of a major depositional change in the Omo from fluvial to lacustrine deposition at this time (de Heinzelin, 1983), which greatly decreases sampling. There is also a large sample of cercopithecids from the upper Burgi Member of the Koobi Fora Formation which is preceded by a long hiatus. Either or both of these factors could be responsible for the apparent turnover event.

The East African sample shows a pattern that more closely approximates a constant rate of turnover than do either of the individual basin-level data sets, and the FAD2s and LAD2s are more even in their distribution than those of the FADs and LADs. The interval between 3.7 and 1.5 Ma is better sampled than are those from other periods and there is in general more turnover shown during this interval, but there are also more species known, and more sites (Figure 5). Further evidence that uneven sampling may be causing this pattern is provided by the presence of the large number of LADs and LAD2s at about 1.5 Ma. In any event, there is no indication of a turnover pulse between 2.8 and 2.3 Ma, suggesting that the lack of a turnover pulse in the Turkana Basin at this time may be more indicative of East Africa overall and not necessarily because the Turkana Basin was a refuge, at least as far as the cercopithecids are concerned.

The largest cluster of FADs and LADs in the all sub-Saharan data set is at 3.0 Ma. In spite of the large number of sites included, this spike may well be the result of sampling biases. Makapansgat has a large and diverse monkey fauna, and is the oldest of the South Africa cave sites (Eisenhart, 1974; Szalay and Delson, 1979). Furthermore, the cercopithecids from South Africa are generally specifically distinct from those of the east. As a result, there are four first appearance data that occur at 3.0 Ma because of this single site as well as a last appearance, and this accounts for fully half of the cluster at this time. Outside of this time, there is also a large number of LAD2s concentrated at approximately 1.4 Ma. It is at this point that sampling from the Turkana Basin, Swartkrans, and Kromdraai greatly decreases, and, as in the case in the East African data, the increase in LAD2s at this time may simply reflect this fact. In neither the FADs and LADs nor the FAD2s and LAD2s is there any indication of a turnover pulse in cercopithecids at ca. 2.5 Ma.

While the Chi-square test of the FAD data showed no difference from the expected rate of accumulation, the FAD2 data differ most from expected at the 3-Ma interval, emphasizing the observed "Makapansgat effect" discussed above. Furthermore, many of the time periods above this are also significant. This is most likely the result of the spike at 3.2 Ma (the maximal age for Makapansgat used here) carrying over to the later intervals, as they all exceed the number of expected FAD2s. In any event, these data do not show an increased rate of turnover at ca. 2.5 Ma, but indicate that the period from 3.2 through 2.5 Ma shows an elevated rate of FAD2 accumulation. However, these data should also be regarded as less reliable than the FAD data as it includes both taxonomically less definitive diagnoses and maximal geologic ages. Regardless, this period of elevated turnover may simply reflect the fact that this is the most densely sampled and most speciose part of the cercopithecid record.

The abundance data show that prior to about 3.6 Ma, the Afar Depression and Turkana Basin are quite different. *Theropithecus* and Cercopithecini are rare to absent in both regions, but that is where the similarities end. In the Afar region, most of the 3.75 Ma and older sequences have too few specimens to yield reliable abundances, but in the Aramis Member there is a very large sample. At Aramis colobines predominate, or are at least codominant with non-*Theropithecus* papionins. The distinctiveness of the Aramis sample in its colobine proportion can be seen in Figure 12 where it is separated by a long branch from the Early Pliocene Turkana sites. Those sites that are near it have very small samples, so that their proportions are not reliable. At Kanapoi and the Lonyumun Member on the other hand, non-*Theropithecus* papionins predominate, while colobines are comparatively rare (Coffing et al., 1994; Harris et al., 2003; Bobe, pers. com.). This difference is further emphasized when the actual species are considered. The only identified colobine at Aramis is *Kuseracolobus aramisi*

(Frost, 2001b), whereas at Kanapoi colobines are divided between a medium-sized species tentatively assigned to *Cercopithecoides* and possibly two larger colobines (Harris et al., 2003). The only identified papionin at Aramis is *Pliopapio alemui* (Frost, 2001b), whereas at Kanapoi the predominant papionin is *?Parapapio* aff. *ado* (Harris et al., 2003). While the two basins appear quite different during this early time period, it should be kept in mind that this is essentially a comparison between a single site in the Afar Basin with two in the Turkana Basin. As a result, these differences may not be representative on a region-wide scale. Additionally, the Turkana sites are separated by 220 and 500 Kyr from Aramis respectively.

After approximately 3.5 Ma both basins show shifts in their relative abundances to a predominance of *Theropithecus*. In both basins *Theropithecus* is so abundant that other taxa are relegated to comparatively rare status. This predominance of *Theropithecus* seems to persist through at least the Middle Pleistocene. In fact, *Theropithecus* is not only the most common taxon in the Afar and Turkana Basins, but also at Olduvai Gorge, Kanjera, and Olorgesailie (e.g., Jolly, 1972) and may be typical of East Africa overall. This abundance in *Theropithecus* is quite different from the situation at all of the South African sites besides Hopefield (e.g., Brain, 1981; Benefit, 1999, 2000).

The abundances of upper and lower Andalee in which Colobinae and Cercopithecini predominate are clearly different from the other levels included in this analysis, most interestingly the Middle Pleistocene Dawaitoli Formation, which is similar to the other *Theropithecus*-dominated levels. This high abundance of Cercopithecini may not be typical of the Afar Depression overall during this time period. Kalb et al. (1982a) have proposed that Andalee represents a gallery forest environment adjacent to the ancient

Awash River, whereas Bodo represents a more open environment distal from the river. Presumably both habitats would have been present through the Middle Pleistocene. Thus, the difference in abundance between the Dawaitoli Formation and the Andalee Member would not be related to large-scale climatic change.

It is worth noting that while the two basins are similar in the abundance of *Theropithecus* from approximately 3.5 Ma through at least 1.5 Ma, the dominant species of *Theropithecus* in the early part of the Turkana Basin is *T. brumpti*. In the Afar Basin throughout the sequence and in the Turkana Basin after about 2.3 Ma *T. oswaldi* is the dominant species of *Theropithecus*. In fact, from Members C through lower G of the Shungura Formation there is a steady increase in the percentage of *T. oswaldi* at the expense of *T. brumpti*, with the latter species having its LAD at the top of Member G. Thus, even though both basins show predominance of *Theropithecus*, for much of the time span considered here, the predominant species differ among the basins. In fact, the taxa included in each basin are generally not shared. Of the 13 and 14 species known from the Afar and Turkana data sets respectively, only 3 are likely to be shared. Thus, even though they are similar in the patterns of overall abundance, they differ in the species included in each of the species groups used for the abundance analysis.

When the cercopithecid data are considered in light of other mammalian taxa that have been studied, some patterns emerge. Vrba (1992, 1995b, 1999) has found a turnover pulse among bovids between 2.8 and 2.5 Ma. The suids do not seem to show an increased amount of turnover at this time (White, 1995; Bishop, 1999). This discrepancy may be the result of the difference in basic biology between these groups. The Bovidae may be relatively specific in their habitat requirements dividing the ecosystem into many niches, whereas the suids may be broader in theirs, i.e., they are relatively stenobiomic and eurybiomic respectively. Many extant African cercopithecids are comparatively stenobiomic (e.g., the guenons and extant colobines), but these groups are poorly represented in the fossil record (Szalay and Delson, 1979; Jablonski, 2002). The majority of the cercopithecid fossil record consists of relatively large papionins. Additionally, many of the colobines seem to be widespread (such as *Cercopithecoides williamsi* and *C. kimeui*). Indeed, even among the fossil guenons, *C. aethiops* appears to predominate. *C. aethiops* has the broadest geographic range of any cercopithecin species, and it is one of the more terrestrial (Gebo and Sargis, 1994). Thus, like the suids the cercopithecids that predominate in the fossil record may be comparatively eurybiomic, and therefore did not participate in a possible 2.8–2.5-Ma turnover event. This distinction has direct bearing on the extinction of *Australopithecus* and the origins of *Paranthropus* and *Homo*, which are thought to occur close to this time (e.g., Kimbel, 1995; White, 1995, 2002). In order to decide if these events were ultimately the result of climatic forcing, one important question would be to decide whether these hominids were more stenobiomic or eurybiomic (e.g., Wood and Strait, 2004).

An alternative explanation as to why cercopithecids do not show a turnover pulse between 2.8 and 2.5 Ma may be that those cercopithecid taxa likely to undergo speciation and extinction as a result of global cooling had already responded to earlier cooling trends ca. 3.4 Ma so that the Cercopithecidae response to the 2.8–2.5 Ma cooling may have been dampened (Vrba, 2000).

## Conclusion

While it is not certain that species turnover among fossil cercopithecids has been relatively constant through the Pliocene and

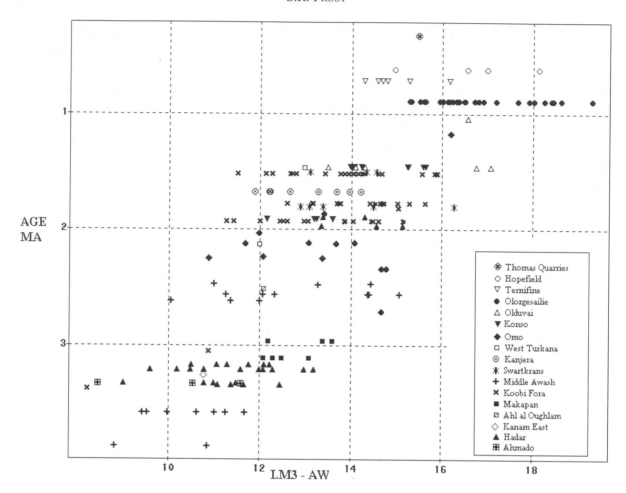

Figure 13. *x*-Axis shows M$_3$ mesial breadth of *Theropithecus oswaldi* specimens, *y*-axis is geologic time.

Pleistocene of Africa, it does seem clear that there was not a turnover pulse among cercopithecids between 2.8 and 2.3 Ma. Furthermore, all of the possible events of increased turnover, e.g., at 3.5, 3.0, 2.5, and 2.0 Ma, occur at periods where there are also possible stratigraphic causes, even in the larger data sets. Thus, the cercopithecid fossil record is consistent with a relatively constant rate of turnover. Furthermore, of these events, the only one that is also observable in the abundance data is the ca. 3.5-Ma event, which appears in both the Afar and Turkana abundance data sets.

This is not to say that climatic change has not been important in the evolution of fossil cercopithecids, as it undoubtedly has. Evidence can be seen in the anagenetic increase in size and complexity of the molar teeth of *Theropithecus oswaldi* through time

(e.g., Eck, 1987a; Leakey, 1993; Frost and Delson, 2002; Figure 13), and the diversity of relatively terrestrial large-bodied colobines during the Plio-Pleistocene. Both of these are likely adaptations to more open environments. Forest fragmentation may also have been important in speciation and diversification among the guenons (Hamilton, 1988). Thus, while climatic change may have been a significant factor in cercopithecid evolution, it does not appear that global cooling between 2.8 and 2.5 Ma caused a turnover pulse.

**Acknowledgments**

I thank René Bobe, Zeresenay Alemseged, and Kay Behrensmeyer for inviting me to participate in this symposium and write this paper. I sincerely thank the Ethiopian Center for

Research and Conservation of Cultural Heritage, National Museum of Ethiopia, the National Museums of Kenya, the Office of the President, University of the Witwatersrand Anatomy Department, the Transvaal Museum, the Natural History Museum, London, the Museum National d'Histoire Naturelle, the Hessischeslandesmuseum, Darmstadt, and the University of California, Museum of Paleontology for access to specimens in their care. I would also like to thank Eric Delson for access to his cast collection and much useful discussion. Tim White and Bill Kimbel are thanked for access to the Middle Awash and Hadar material respectively. I also want to thank Alemu Ademassu, Meave Leakey, Mary Muungu, Kevin Kuykendal, Lee Berger, Peter Andrews, Brigitte Senut, Pascal Tassy, Friedemann Schrenk, and Patricia Holroyd for much kind assistance and access to specimens housed in their respective institutions. I thank Eric Delson, Elisabeth Vrba and two anonymous reviewers for their comments. The L.S.B. Leakey Foundation, the Wenner-Gren Foundation, ETE, NYCEP, CUNY are gratefully acknowledged for financial support of this project.

# References

Alemseged, Z., 2003. An integrated approach to taphonomy and faunal change in the Shungura Formation (Ethiopia) and its implication for hominid evolution. Journal of Human Evolution 44, 451–478.

Alemseged, Z., Geraads, D., 2001. A new Middle Pleistocene fauna from the Busidima-Telalak region of the Afar, Ethiopia. Comptes rendus de l'Académie des Sciences Paris, Sciences de la Terre et des Planetes 331, 549–556.

Behrensmeyer, A.K., Todd, N.E., Potts, R., McBrinn, G.E., 1997. Late Pliocene faunal turnover in the Turkana Basin, Kenya and Ethiopia. Science 278, 1589–1594.

Benefit, B.R., 1999. Biogeography, dietary specialization, and the diversification of African Plio-Pleistocene monkeys. In: Bromage, T.G., Schrenk, F. (Eds.), African Biogeography, Climate Change, and Early Hominid Evolution. Oxford University Press, Oxford, pp. 172–188.

Benefit, B.R., 2000. Old world monkey origins and diversification: an evolutionary study of diet and dentition. In: Bromage, T.G., Schrenk, F. (Eds.), African Biogeography, Climate Change, and Early Hominid Evolution. Oxford University Press, Oxford, pp. 133–179.

Bernor, R.L., Armour-Chelu, M., 1999. Toward an revolutionary history of African hipparionine horses. In: Bromage, T.G., Schrenk, F. (Eds.), African Biogeography, Climate Change, and Hominid Evolution. Oxford University Press, Oxford, pp. 189–215.

Bishop, L.C., 1999. Reconstructing omnivore paleoecology and habitat preference in the Pliocene and Pleistocene of Africa. In: Bromage, T.G., Schrenk, F. (Eds.), African Biogeography, Climate Change, and Early Hominid Evolution. Oxford University Press, Oxford, pp. 216–225.

Bobe, R., 1997. Hominid environments in the Pliocene: an analysis of fossil mammals from the Omo Valley, Ethiopia. Ph.D. Dissertation, University of Washington.

Bobe, R., Eck, G.G., 2001. Responses of African bovids to Pliocene climatic change. Paleobiology 27 (S.2). Paleobiology Memoirs 2, 1–47.

Bobe, R., Behrensmeyer, A.K., Chapman, R.E., 2002. Faunal change, environmental variability and late Pliocene hominin evolution. Journal of Human Evolution 42, 475–497.

Brain, C.K., 1981. The Hunters or the Hunted? An Introduction to African Cave Taphonomy. University of Chicago Press, Chicago.

Clark, J.D., De Heinzelin, J., Schick, K.D., Hart, W.K., White, T.D., WoldeGabriel, G., Walter, R.S., Walter, R.C., Suwa, G., Asfaw, B., Vrba, E., Haile-Selassie, Y., 1994. African Homo erectus: old radiometric ages and young Oldowan assemblages in the Middle Awash Valley, Ethiopia. Science 264, 1907–1910.

Coffing, K., Feibel, C.S., Leakey, M.G., Walker, A., 1994. Four million year old hominids from East Lake Turkana, Kenya. American Journal of Physical Anthropology 93, 55–65.

Cooke, H.B.S., 1978. Suid evolution and correlation of African hominid localities: an alternative taxonomy. Science 201, 460–463.

Delson, E., 1984. Cercopithecid biochronology of the African Plio-Pleistocene: correlation among eastern and southern hominid-bearing localities. Courier Forschungsinstitut Senckenberg 69, 199–218.

Delson, E., 1988. Chronology of South African australopith site units. In: Grine, F.E. (Ed.), Evolutionary History of the "Robust" Australopithecines. Aldine, New York, pp. 317–324.

Delson, E., Eck, G.G., Leakey, M.G., Jablonski, N.G., 1993. A partial catalogue of fossil remains of *Theropithecus*. In: Jablonski, N.G. (Ed.), *Theropithecus*: The Rise and Fall of a Primate Genus. Cambridge University Press, Cambridge, pp. 499–525.

de Heinzelin, J., 1983. The Omo Group. Musee Royale de l'Afrique Central, Tervuren, Belgique. Annales-Serie in 8. Sciences Geologiques 85.

de Heinzelin, J., Clark, J.D., White, T.D., Hart, W., Renne, P., WoldeGabriel, G., Beyene, Y., Vrba, E., 1999. Environment and behavior of 2.5 million-year-old Bouri hominids. Science 284, 625–629.

deMenocal, P.B., 1995. Plio-Pleistocene African Climate. Science 270, 53–59.

Denton, G.H., 1999. Cenozoic climate change. In: Bromage, T.G., Schrenk, F. (Eds.), African Biogeography, Climate Change, and Early Hominid Evolution. Oxford University Press, Oxford, pp. 94–114.

Eck, G.G., 1976. Cercopithecoidea from Omo group deposits. In: Coppens, Y, Howell, F.C., Isaac, G.Ll., Leakey, R.E.F. (Eds.), Earliest Man and Environments in the Lake Rudolf Basin. University of Chicago Press, Chicago, pp. 332–344.

Eck, G.G., 1977. Diversity and frequency distribution of Omo group Cercopithecoidea. Journal of Human Evolution 6, 55–63.

Eck, G.G., 1987a. *Theropithecus oswaldi* from the Shungura Formation, Lower Omo Basin, Southwestern Ethiopia. In: Coppens, Y, Howell, F.C. (Eds.), Les faunes Plio-Pleistocenes de la Basse Vallee de l'Omo (Ethiopie). Tome 3, Cercopithecidae de la Formation de Shungura, pp. 123–139. Cahiers de Paleontologie, Travaux de Paleontologie Est-Africaine. Paris: Editions du CNRS.

Eck, G.G., 1987b. Plio-Pleistocene specimens of *Cercopithecus* from the Shungura Formation, Southwestern Ethiopia. In: Coppens, Y., Howell, F.C. (Eds.), Les faunes Plio-Pleistocenes de la Basse Vallee de l'Omo (Ethiopie). Tome 3, Cercopithecidae de la Formation de Shungura, pp. 143–146. Cahiers de Paleontologie, Travaux de Paleontologie Est-Africaine. Paris: Editions du CNRS.

Eck, G.G., Jablonski, N.G., 1987. The skull of *Theropithecus brumpti* compared with those of other species of the genus *Theropithecus*. In: Coppens, Y., Howell, F.C. (Eds.), Les faunes Plio-Pleistocenes de la Basse Vallee de l'Omo (Ethiopie). Tome 3, Cercopithecidae de la Formation de Shungura, pp. 11–122. Cahiers

de Paleontologie, Travaux de Paleontologie Est-Africaine. Paris: Editions du CNRS.

Eisenhart, W.L., 1974. The fossil cercopithecids of Makapansgat and Sterkfontein. B.A. Thesis, Harvard College.

Frost, S.R., 2001a. Fossil Cercopithecidae of the Afar Depression, Ethiopia: Species systematics and comparison with the Turkana Basin. Ph.D. Thesis, City University of New York.

Frost, S.R., 2001b. New Early Pliocene Cercopithecidae (Mammalia: Primates) from Aramis, Middle Awash Valley, Ethiopia. American Museum Novitates 3350, 1–36.

Frost, S.R., Delson, E., 2002. Fossil Cercopithecidae from the Hadar Formation and surrounding areas of the Afar Depression, Ethiopia. Journal of Human Evolution 43, 687–748.

Gebo, D.L., Sargis, E.J., 1994. Terrestrial adaptations in the postcranial skeletons of guenons. American Journal of Physical Anthropology 93, 341–371.

Hamilton, A.C., 1988. Guenon evolution and forest history. In: Gautier-Hion, A., Bourliere, F., Gautier, J-P., Kingdon, J. (Eds.), A Primate Radiation: Evolutionary Biology of the African Guenons. Cambridge University Press, Cambridge, pp. 13–34.

Harris, J.M., White, T.D., 1979. Evolution of the Plio-Pleistocene African Suidae. Transactions of the American Philosophical Society 69, 1–128.

Harris, J.M., Brown, F.H., Leakey, M.G., 1988. Stratigraphy and paleontology of Pliocene and Pleistocene localities West of Lake Turkana, Kenya. Contributions in Science 399, 1–128.

Harris, J.M., Leakey, M.G., Cerling, T.E., 2003. Early Pliocene tetrapod remains from Kanapoi, Lake Turkana, Kenya. In: Harris, J.M., Leakey, M.G. (Eds.), Geology and Vertebrate Paleontology of the Early Pliocene Site of Kanapoi, Northern Kenya, pp. 39–113. Contributions in Science 498, 1–132.

Jablonski, N.G., 2002. Fossil old world monkeys: the late Neogene. In: Hartwig, W.C. (Ed.), The Primate Fossil Record. Cambridge University Press, Cambridge, pp. 255–299.

Jolly, C.J., 1972. The classification and natural history of *Theropithecus (Simopithecus)* (Andrews, 1916), baboons of the African Plio-Pleistocene. Bulletin of the British Museum of Natural History, Geology 22, 1–122.

Kalb, J.E., 1993. Refined stratigraphy of the hominid bearing Awash Group, Middle Awash valley, Afar Depression, Ethiopia. Newsletters Stratigraphy 29, 21–62.

Kalb, J.E., Jaegar, M., Jolly, C.J., Kana, B., 1982a. Preliminary geology, paleontology and paleoecology of a Sangoan site at Andalee, Middle Awash Valley, Ethiopia. Journal of Archaeological Science 9, 349–363.

Kalb, J.E., Oswald, E.B., Mebrate, A., Tebedge, S., Jolly, C.J., 1982b. Stratigraphy of the Awash Group, Middle Awash Valley, Afar, Ethiopia. Newsletters Stratigraphy 11(3), 95–127.

Kimbel, W.H., 1995. Hominid speciation and Pliocene climatic change. In: Vrba, E.S., Denton, G.H., Partridge, T.C., Burckle, L.H. (Eds.), Paleoclimate and Evolution with an Emphasis on Human Origins. Yale University Press, New Haven, pp. 425–437.

Kimbel, W.H., Walter, R.C., Johanson, D.C., Reed, K.E., Aronson, J.L., Assefa, Z., Marean, C.W., Eck, G.G., Bobe, R., Hovers, E., Rak, Y., Vondra, C., Yemane, T., York, D., Chen, Y., Evensen, N.M., Smith, P.E., 1996. Late Pliocene *Homo* and Oldowan tools from the Hadar Formation (Kada Hadar Member), Ethiopia. Journal of Human Evolution 31, 549–561.

Leakcy, M.G., 1976. Cercopithecoidea of the East Rudolf succession. In: Coppens, Y., Howell, F.C., Isaac, G.Ll., Leakey, R.E.F. (Eds.), Earliest Man and Environments in the Lake Rudolf Basin. University of Chicago Press, Chicago, pp. 345–350.

Leakey, M.G., 1982. Extinct large colobines from the Plio-Pleistocene of Africa. American Journal of Physical Anthropology 58, 153–172.

Leakey, M.G., 1987. Colobinae (Mammalia, Primates) from the Omo Valley, Ethiopia. In: Coppens, Y., Howell, F.C. (Eds.), Les faunes Plio-Pleistocenes de la Basse Vallee de l'Omo (Ethiopie). Tome 3, Cercopithecidae de la Formation de Shungura, pp. 148–169. Cahiers de Paleontologie, Travaux de Paleontologie Est-Africaine. Paris: Editions du CNRS.

Leakey, M.G., 1993. Evolution of *Theropithecus* in the Turkana Basin. In: Jablonski, N.G. (Ed.), *Theropithecus*: The Rise and Fall of a Primate Genus. Cambridge University Press, Cambridge, pp. 15–76.

Leakey, M.G., Leakey, R.E.F., 1973a. Further evidence of *Simopithecus* (Mammalia, Primates) from Olduvai and Olorgesailie. Fossil Vertebrates of Africa 3, 101–120.

Leakey, M.G., Leakey, R.E.F., 1973b. New large Pleistocene Colobinae (Mammalia, Primates) from East Africa. Fossil Vertebrates of Africa 3, 121–138.

Leakey, M.G., Leakey, R.E.F., 1976. Further Cercopithecinae (Mammalia, Primates) from the Plio-Pleistocene of East Africa. Fossil Vertebrates from Africa 4, 121–146.

Leakey, M.G., Teaford, M.F., Ward, C.V., 2003. Cercopithecidae from Lothagam. In: Leakey, M.G., Harris, J.M. (Eds.), Lothagam: The Dawn of Humanity in East Africa. Columbia University Press, New York, pp. 201–248.

McKee, J.K., 1996. Faunal turnover patterns in the Pliocene and Pleistocene of southern Africa. South African Journal of Science 92, 111–113.

McKee, J.K., 2001. Faunal turnover rates and mammalian biodiversity of the late Pliocene and Pleistocene of eastern Africa. Paleobiology 27, 500–511.

Patterson, B., 1968. The extinct baboon, Parapapio jonesi, in the Early Pleistocene of Northwest Kenya. Breviora 282, 1–4.

Renne, P.R., WoldeGabriel, G., Hart, W.K., Heiken, G., White, T.D., 1999. Chronostratigraphy of the Mio-Pliocene Sagantole Formation, Middle Awash Valley, Afar, Ethiopia. Geological Society of America Bulletin 111(6), 869–885.

Renne P., 2000. Geochronology. In: de Heinzelin, J., Clark, J.D., Schick, K.D., Gilbert, W.H. (Eds.), The Acheulean and the Plio-Pleistocene Deposits of the Middle Awash Valley, Ethiopia, pp. 47–50. Royal Museum of Central Africa (Belgium) Annales – Sciences Geologiques 104, 1–235.

Shackleton, N.J., Backman, J., Zimmerman, H., Kent, D.V., Hall, M.A., Roberts, D.G., Schnitker, D., Baldauf, J.G., Desprairies, A., Homrighausen, R., Huddlestun, P., Keene, J.B., Kaltenback, A.J., Krumsiek, K.A.O., Morton, A.C., Murray, J.W., Westberg-Smith, J., 1984. Oxygen isotope calibration of the onset of ice-rafting and history of glaciation in the North Atlantic region. Nature 307, 620–623.

Szalay, F.S., Delson, E., 1979. Evolutionary History of the Primates. Academic Press, New York.

Turner A., Wood, B., 1993. Taxonomic and geographic diversity in robust australopithecines and other African Plio-Pleistocene larger mammals. Journal of Human Evolution 24, 147–168.

Vrba, E.S., 1976. Fossil Bovidae from Sterkfontein, Swartkrans and Kromdraai. Transvaal Museum Memoirs 21, 1–166.

Vrba, E.S., 1980. The significance of bovid remains as indicators of environment and predation patterns. In: Behrensmeyer, A.K., Hill, A.P. (Eds.), Fossils in the Making. University of Chicago Press, Chicago, pp. 247–271.

Vrba, E.S., 1988. Late Miocene climatic events and human evolution. In: Grine, F. (Ed.), Evolutionary History of "Robust" Australopithecines. Aldine de Gruyter, New York, pp. 405–426.

Vrba, E.S., 1992. Mammals as a key to evolutionary theory. Journal of Mammalogy 73(1), 1–28.

Vrba, E.S., 1995a. On the connections between paleoclimate and evolution. In: Vrba, E.S., Denton, G.H., Partridge, T.C., Burckle, L.H. (Eds.), Paleoclimate and Evolution with an Emphasis on Human Origins. Yale University Press, New Haven, pp. 24–45.

Vrba, E.S., 1995b. The fossil record of African Antelopes (Mammalia, Bovidae) in relation to human evolution and paleoclimate. In: Vrba, E.S., Denton, G.H., Partridge, T.C., Burckle, L.H. (Eds.), Paleoclimate and Evolution with an Emphasis on Human Origins. Yale University Press, New Haven, pp. 385–424.

Vrba, E.S., 1999. Habitat theory in relation to the evolution of African Neogene biota and hominids. In: Bromage, T.G., Schrenk, F. (Eds.), African Biogeography, Climate Change, and Early Hominid Evolution. Oxford University Press, Oxford, pp. 19–34.

Vrba, E.S., 2000. Major features of Neogene Mammalian Evolution in Africa. In: Partridge, T.C., Maud, R. (Eds.), Cenozoic Geology of Southern Africa. Oxford University Press, Oxford, pp. 277–304.

Walter, R.C., 1994. Age of Lucy and the First Family: single-crystal 40Ar/39Ar dating of the Denen Dora and lower Kada Hadar members of the Hadar Formation, Ethiopia. Geology 22, 6–10.

Walter, R.C., Aronson, J.L. 1993. Age and source of the Sidi Hakoma Tuff, Hadar Formation, Ethiopia. Journal of Human Evolution 25, 229–240.

Wesselman, H.B. 1995. Of mice and almost-men: regional paleoecology and human evolution in the Turkana basin. In: Vrba, E.S., Denton, G.H., Partridge, T.C., Burckle, L.H. (Eds.), Paleoclimate and Evolution with an Emphasis on Human Origins. Yale University Press, New Haven, pp. 356–368.

White, T.D., 1995. African omnivores: global climatic change and Plio-Pleistocene hominids and suids In: Vrba, E.S., Denton, G.H., Partridge, T.C., Burckle, L.H. (Eds.), Paleoclimate and Evolution with an Emphasis on Human Origins. Yale University Press, New Haven, pp. 369–384.

White, T.D., 2002. Early hominis. In: Hartwig, W.C. (Ed.), The Primate Fossil Record. Cambridge University Press, Cambridge, pp. 407–417.

White, T.D., Suwa, G., Hart, W.K., Walter, R.C., WoldeGabriel, G., de Heinzelin, J., Clark, J.D., Asfaw, B., Vrba, E., 1993. New discoveries of *Australopithecus* at Maka in Ethiopia. Nature 366, 261–265.

WoldeGabriel, G., White, T.D., Suwa, G., Renne, P., de Heinzelin, J., Hart, W.K., Heiken, G., 1994. Ecological placement of early Pliocene hominids at Aramis, Ethiopia. Nature 371, 330–333.

Wood, B., Strait, D., 2004. Patterns of resource use in early *Homo* and *Parantheopus*. Journal of Human Evolution 46, 119–162.

# 4. Patterns of change in the Plio-Pleistocene carnivorans of eastern Africa

## Implications for hominin evolution

M.E. LEWIS
*Division of Natural and Mathematical Sciences (Biology)*
*The Richard Stockton College of New Jersey*
*PO Box 195, Pomona, NJ 08240-0195, USA*
*Margaret.Lewis@stockton.edu*


L. WERDELIN
*Department of Palaeozoology*
*Swedish Museum of Natural History*
*Box 50007, S-104 05 Stockholm, Sweden*
*werdelin@nrm.se*

**Keywords**: Kleptoparasitism, extinction, origination, species richness, body mass

## Abstract

This paper uses changes in origination and extinction rates and species richness of eastern African carnivorans through time to discuss issues related to the evolution of hominin behavior. To address the question of which taxa were most likely to have had competitive interactions with hominins, modern carnivorans were sorted into size classes based on shifts in behavior, ecology, and body mass. Four size classes were created, among which the two largest (21.5–100 kg and >100 kg) include those taxa whose behavior is most relevant to the evolution of hominin dietary behavior. Fossil taxa were then assigned to these size classes. A summary of the temporal range and reconstructed behavior and ecology of fossil members of the two largest size classes is presented. We discuss the relevance of each taxon to reconstructing hominin behavior and suggest that hominins must have evolved not only successful anti-predator strategies, but also successful strategies to avoid kleptoparasitism before carcass-based resources could become an important part of the diet. Although hominins were unlikely to have been top predators upon first entrance into the carnivore guild, effective anti-predator/anti-kleptoparasitism strategies in combination with the eventual evolution of active hunting would have increased the rank of hominin species within the guild. While the appearance of stone tools at 2.6 Ma has no apparent effect upon carnivorans, the appearance of *Homo ergaster* after 1.8 Ma may have been at least partly responsible for the decrease in the carnivoran origination rate and the increase in the extinction rate at this time. The behavior of *H. ergaster*, climate change, and concomitant changes in prey species richness may have caused carnivoran species richness to drop precipitously after 1.5 Ma. In this situation, even effective kleptoparasitism by *H. ergaster* may have been enough to drive local populations of carnivorans that overlapped with hominins in dietary resources to extinction. Possibly as a result, the modern guild, which evolved within the last few hundred thousand years, is composed primarily of generalists. Although the impact of *H. sapiens* on the carnivoran guild cannot be assessed due to a lack of carnivoran fossils from this time period, one might not consider the modern carnivore guild to be complete until the appearance of our species approximately 200,000 years ago.

*R. Bobe, Z. Alemseged, and A.K. Behrensmeyer (eds.) Hominin Environments in the East African Pliocene: An Assessment of the Faunal Evidence, 77–105.*

## Introduction

Paleoanthropologists have long been interested in the transitions in diet and diet-related behavior that occurred during hominin evolution. Early work in this area envisioned hominins as top predators from whom other carnivores scavenged (e.g., Dart, 1949, 1956). Later research on the taphonomy and fauna found in South African caves suggested that early hominins were the prey and not the predators (Brain, 1969, 1981). While most researchers have accepted the hypothesis that vertebrate carcasses played an increasingly larger role in African hominin diet through time, debate over the nature of access to those carcasses and the amount of competition involved has continued for several decades (e.g., Isaac, 1971, 1978; Binford, 1981, 1986; Potts, 1984, 1988a, b; Bunn, 1986, 1991; Bunn and Kroll, 1986; Shipman, 1986a, b; Binford et al., 1988; Blumenschine, 1988, 1989, 1991; Turner, 1988; Cavallo and Blumenschine, 1989; Bunn and Ezzo, 1993; Blumenschine et al., 1994; Domínguez-Rodrigo, 1999, 2001; Treves and Naughton-Treves, 1999; Van Valkenburgh, 2001; Domínguez-Rodrigo and Pickering, 2003).

In any case, hominins, at some point, became members of the carnivore guild. As hominins became increasingly aggressive in their scavenging and/or hunting behavior, their ecological relationships with members of the order Carnivora are likely to have changed. Carnivorans, therefore, were not just potential competitors and predators, but also contributors to the structure of resources available to hominins (e.g., Schaller and Lowther, 1969; Walker, 1984; Blumenschine, 1986a,b, 1987; Turner, 1988, 1992; Cavallo and Blumenschine, 1989; Marean, 1989; Sept, 1992; Lewis, 1995b, 1997; Marean and Ehrhardt, 1995; Brantingham, 1998; Domínguez-Rodrigo, 1999; Van Valkenburgh, 2001). As such, factors affecting the partitioning of the guild

by carnivorans are relevant to the study of human evolution.

How, then, do hominins fit into the carnivore guilds of Plio-Pleistocene Africa? In other words, what niche space was available to hominin species and how did that space change through time? How might the behavior of carnivoran species and hominin species have affected one another? Before we can answer these questions, we must explore the ecological framework surrounding hominins by reconstructing the behavior and ecology of extinct carnivorans and examining patterns in the evolution of this group in Africa (Turner, 1983, 1985, 1986, 1990, 1998; Walker, 1984; Lewis, 1995b, 1997).

The present paper uses data on carnivoran originations and extinctions and species richness from a previous paper (Werdelin and Lewis, 2005) in combination with data on changes in body size through time to address theoretical issues related to the evolution of hominin dietary strategies. Specific taxa of potential relevance to hominins are discussed. Our goal is to provide a summary of up-to-date information from our research on the changing nature of the carnivoran guild through time focusing on issues of relevance to hominin behavior and evolution.

## The Importance of Carnivoran Body Size to Hominins

Many factors need to be considered to understand the complex ecological relationship between carnivorous taxa (e.g., Table 1). Part of the current study examines one of the most basic factors affecting niche partitioning of any type of guild: body size. Body size limits the choice of prey, such that larger predators capture prey with a wider range of body sizes (Gittleman, 1985). Body size is also an important determinant of interspecific rank at a carcass (Eaton, 1979; Van Valkenburgh, 2001). Sympatric carnivores that are similar in both morphology and

*Table 1. Some factors affecting carnivore guild structure*

| Factor | Definition or example |
| --- | --- |
| Habitat preference | Amount of canopy cover, underbrush, water, etc. |
| Foraging time | Diurnal, nocturnal, crepuscular, etc. |
| Prey choice | Prey size |
| Hunting method | Ambush, pack hunting, etc. |
| Scavenging | Passive, aggressive, combination, etc. |
| Food transport | Ability to remove food items from view |
| Food caching | Ability to store food safely for future use |
| Food processing | Flesh only, bone-crushing, bone-cracking, etc. |
| Level of aggression | Intra- or interspecific levels of aggression |
| Grouping behavior | Solitary, pack hunting, solitary forager in a social group, etc. |
| Individual body size | Biomass of single individual |
| Group biomass | Biomass of foraging and/or social group |

predatory behavior may reduce interspecific competition for food through differences in body size (Rosenzweig, 1966; Gittleman, 1985; Dayan and Simberloff, 1996), a pattern that has been recognized across various vertebrate taxa (e.g., Farlow and Pianka, 2002).

In carnivorous taxa, the fact that many species may hunt or at least eat in a group means that group or pack biomass must also be considered. Cooperative hunting specifically allows the capture of larger or faster prey than solitary hunting may allow (Kruuk, 1975; Packer and Ruttan, 1988), resulting in a larger prey carcass that may interest a greater range of scavengers. Grouping behavior, in general, allows larger carnivoran taxa, such as spotted hyenas (*Crocuta crocuta*) and African wild dogs (*Canis pictus*) to be more successful during interspecific competition, (e.g., at a carcass) such as that occurring over a carcass, than their body size alone would predict (Lamprecht, 1978; Eaton, 1979). Kleptoparasitism (food theft by other taxa) can still have a significant impact on groups or packs of carnivorans (e.g., Cooper, 1991; Gorman et al., 1998). For the above reasons, the potential for grouping behavior in extinct taxa must be addressed (e.g., Van Valkenburgh et al., 2003; Andersson, 2005), even though it may not be possible to positively identify specific types of grouping behavior or even grouping behavior in general in the fossil record.

## Materials and Methods

The present study is part of a larger study on the evolution of African carnivorans within a global context. The material used herein consists of many hundreds of specimens of Carnivora found at Plio-Pleistocene sites in Ethiopia, Kenya, and Tanzania (Table 2).

*Table 2. Localities included in the analyses*

| Country | Localities |
| --- | --- |
| Kenya | Allia Bay |
| | Eshoa Kakurongori |
| | Kanam East |
| | Kanapoi |
| | Koobi Fora |
| | Kosia |
| | Lainyamok |
| | Lothagam |
| | Nakoret |
| | Olorgesailie |
| | South Turkwel |
| | West Turkana (Nachukui Fm.) |
| Ethiopia | Aramis[*] |
| | Daka Mb., Bouri Fm.[*] |
| | Hadar |
| | Konso-Gardula[*] |
| | Omo |
| Tanzania | Laetoli |
| | Olduvai Gorge |

*Data from three Ethiopian localities were taken from the literature (Asfaw et al., 1992; WoldeGabriel et al., 1994; de Heinzelin et al., 1999; Gilbert, 2003). Ages of included members can be found in Table 1 of Werdelin and Lewis (2005).

Nearly all of the specimens included have been studied and identified by the authors. As a result, identifications are based on a uniform view of morphology, taxonomy, and systematics, one in which craniodental and postcranial data have been treated equally. While we do not claim to be infallible in our identifications, we believe that our analyses provide a uniform baseline for understanding patterns of change in carnivoran evolution.

All named eastern African carnivoran species known during the last 4.5 million years are included in the study. Unnamed species have been included if they have been verified by us as being different from all other known species in the material. Specimens that cannot be definitively excluded from known taxa are counted as part of known taxa. As such, we provide a minimum estimate of the number of species within the data set. Further discussion of included taxa and evidence for the suggestion that sampling overall is relatively good throughout the time period investigated can be found in our larger study of species richness and turnover in Plio-Pleistocene carnivorans of eastern Africa (Werdelin and Lewis, 2005).

Data on the behavior, ecology, and body mass of extant African carnivorans were taken from the literature (Estes, 1991; Nowak, 1991; Kingdon, 1997; Sillero-Zubiri et al., 1997). While previous studies of body mass distributions within faunas have classified carnivorans into size classes, these studies have focused on dividing carnivorans into size classes of a predetermined width (e.g., 5-kg bins in Rodriguez et al., 2004) and/or have labeled carnivorans over 20 kg (Van Valkenburgh, 2001), 40 kg (Lambert and Holling, 1998), or 45 kg (Rodriguez et al., 2004) a large carnivore or large mammal. We wanted to see if there were any natural breaks in behaviors of relevance to hominins that might not be detected by equal bin widths in body size.

Extant taxa were sorted by body mass and the behavioral and ecological data on those taxa were examined for natural breaks in the data.

In other words, we looked for changes in ecology and behavior as body size increases, with particular emphasis placed on shifts that might reflect changes in potential interactions with hominins. We identified general size classes of carnivorans based on maximum body mass such that the transition from one class to another represents a fundamental behavioral shift. While mean body mass might be a better method of assigning taxa to size classes, this information is not always available for extant taxa. Although we did examine the overall range in reported body mass for a given taxon in constructing our size classes, there were clear shifts in ecology and behavior that matched jumps in maximum body mass. While some taxa had body mass *ranges* that overlapped two categories, in all instances the *maximum* body mass fell with species most similar in behavior and ecology. Although males and females of sexually dimorphic species were considered separately during the construction of size classes, in no instance did males and females fall into separate categories. For this reason, sexes are not presented separately in this paper.

As the size classes are quite general, we did not calculate exact body masses for extinct taxa. Instead, specimens of extinct taxa were categorized based on their overall similarity in size and reconstructed behavior to extant taxa. While this study is not meant to represent the last word on the relationship between body size and behavior and ecology in carnivorans, these very general categories may prove of use in thinking about carnivoran behavior relative to hominins.

For the analyses of turnover in the carnivoran fossil record, each taxon was assigned a FAD and a LAD based on the oldest and youngest ages of the oldest and youngest sites from which it has been identified. The data were then binned into time intervals of equal length. While 500-Kyr bins are commonly used (e.g., Behrensmeyer et al., 1997), a shorter bin length (300 Kyr) provides finer grained analysis. Our previous work has demonstrated that the material was not sufficient for a shorter bin length than 300 Kyr even though we recognize that this may lump taxa together that did not actually overlap. As such, changes from bin to bin represent general changes, as bins are not equivalent to guilds. See Werdelin and Lewis (2005) for the stratigraphic ranges of all eastern African Carnivora taxa included in this study and further discussion of binning.

## Results

### BODY MASS OF EXTANT AFRICAN CARNIVORANS

Extant carnivoran taxa in Africa tend to fall into four groups based on maximum body mass, behavior, and potential relationship with hominins (Table 3). Although there is certainly overlap between groups, particularly among smaller species, the general categories hold up quite well. The characteristics of each group will be discussed in turn.

Carnivorans less than 10 kg (Size Class 1) exhibit the greatest diversity of behavior. While classification of this diverse group of taxa together might seem odd, this group is presumed to share the least potential to impact hominin behavior significantly. Members of this group today may occasionally raid human habitations, but are more easily intimidated by humans than their larger relatives. A recent study demonstrates that the carnivorans that have the highest chance of being killed by other carnivorans during interspecific encounters all belong to this group (Caro and Stoner,

Table 3. *Size classes of extant eastern African Carnivora based on maximum body mass (kg)*

| Species | Family | Common name | Min. mass | Max. mass[1] |
|---|---|---|---|---|
| **Size Class 1 (Max. mass < 10)** | | | | |
| Otocyon megalotis | Canidae | Bat-eared fox | 3.0 | 5.3 |
| Felis lybica | Felidae | Wild cat | 3.0 | 8.0 |
| Atilax paludinosus | Herpestidae | Marsh or water mongoose | 2.2 | 5.0 |
| Bdeogale crassicauda | Herpestidae | Bushy-tailed mongoose | 1.3 | 2.1 |
| Bdeogale nigripes | Herpestidae | Black-legged mongoose | 2.0 | 3.5 |
| Dologale dybowskii | Herpestidae | Savannah mongoose | 0.3 | 0.4 |
| Galerella sanguinea | Herpestidae | Slender mongoose | 0.4 | 0.8 |
| Helogale hirtula | Herpestidae | Somali dwarf mongoose | 0.2 | 0.3 |
| Helogale parvula | Herpestidae | Dwarf mongoose | 0.2 | 0.4 |
| Herpestes ichneumon | Herpestidae | Egyptian mongoose or Ichneumon | 2.2 | 4.1 |
| Ichneumia albicauda | Herpestidae | White-tailed mongoose | 1.8 | 5.2 |
| Mungos mungo | Herpestidae | Banded mongoose | 1.5 | 2.3 |
| Rhynchogale melleri | Herpestidae | Meller's mongoose | 1.7 | 3.1 |
| Ictonyx striatus | Mustelidae | Striped polecat or Zorilla | 0.4 | 1.4 |
| Lutra maculicollis | Mustelidae | African river or spot-necked otter | 4.0 | 6.5 |
| Poecilogale albinucha | Mustelidae | African striped weasel | 0.2 | 0.4 |
| Nandinia binotata | Nandiniidae | African palm civet | 1.7 | 4.7 |
| Genetta abyssinica | Viverridae | Ethiopian genet | 1.3 | 2.0 |
| Genetta genetta | Viverridae | Common genet | 1.3 | 2.3 |
| Genetta servalina | Viverridae | Servaline genet | 1.0 | 2.0 |
| Genetta tigrina | Viverridae | Blotched genet | 1.2 | 3.2 |
| **Size Class 2 (Max. mass 10–21.5)** | | | | |
| Canis adustus | Canidae | Side-striped jackal | 7.3 | 12.0 |
| Canis aureus | Canidae | Golden jackal | 6.0 | 15.0 |
| Canis mesomelas | Canidae | Black-backed jackal | 6.5 | 13.5 |
| Canis simensis | Canidae | Ethiopian wolf | 11.2 | 19.3 |
| Caracal caracal | Felidae | Caracal | 13.0 | 19.0 |

*(Continued)*

*Table 3. Size classes of extant eastern African Carnivora based on maximum body mass (kg)—cont'd*

| Species | Family | Common name | Min. mass | Max. mass[1] |
|---|---|---|---|---|
| *Leptailurus serval* | Felidae | Serval | 8.7 | 18.0 |
| *Profelis aurata* | Felidae | African golden cat | 5.3 | 16.0 |
| *Proteles cristatus* | Hyaenidae | Aardwolf | 8.0 | 14.0 |
| *Mellivora capensis* | Mustelidae | Honey badger or Ratel | 7.0 | 16.0 |
| *Civettictis civetta* | Viverridae | African civet | 7.0 | 20.0 |
| **Size Class 3 (Max. mass > 21.5–100)** | | | | |
| *Canis pictus* | Canidae | African wild dog | 18.0 | 36.0 |
| *Acinonyx jubatus* | Felidae | Cheetah | 35.0 | 72.0 |
| *Panthera pardus* | Felidae | Leopard | 28.0 | 90.0 |
| *Crocuta crocuta* | Hyaenidae | Spotted hyena | 40.0 | 90.0 |
| *Hyaena hyaena* | Hyaenidae | Striped hyena | 25.0 | 55.0 |
| *Aonyx capensis* | Mustelidae | African clawless otter | 10.0 | 28.0[2] |
| **Size Class 4 (Max mass > 100)** | | | | |
| *Panthera leo* | Felidae | Lion | 120.0 | 250.0 |

[1] Body mass and biogeographical data were taken from the literature (Estes, 1991; Nowak, 1991; Kingdon, 1997; Sillero-Zubiri et al., 1997). As data on eastern African populations were not always available, ranges represent body mass ranges reported for African populations in general. Note that eastern African populations of some species may differ in body mass range (e.g., *C. pictus*) from populations in the rest of Africa as discussed in the text.

[2] While the largest African clawless otter is reported at 34 kg, individuals of this size are not common (Estes, 1991; Kingdon, 1997).

2003). Faunivorous members of this group capture prey that is too small to be a potential source for hominin scavenging. Even the smallest hominin adult was highly unlikely to be a prey item for members of this group. Today, carnivorans of this size worldwide are more likely to be killed by humans as pests (e.g., weasels in King, 1990) or for their pelts (e.g., raccoon dogs in Novikov, 1962) than pose a threat.

Size Class 2 carnivorans (10–21.5 kg) are also quite diverse in dietary behavior. Carbone et al. (1999) have shown that of the carnivorans worldwide weighing 21.5 kg or less, only 25% feed purely on vertebrates, 45% are omnivores, 10% are purely invertebrate feeders, and 19% are mixed vertebrate/invertebrate feeders. Most importantly, among those that consume vertebrates, prey size is still too small to be a useful scavenging resource (i.e., 45% or less of their body mass; Carbone et al., 1999). Today, members of this group, such as the various jackal species, tend to be of low rank in competitive interactions at carcasses. Highly aggressive species of this size, such

as the honey badger (*Mellivora capensis*), might have been an annoying and dangerous species to encounter during foraging, particularly for any small band of hominins or solitary individuals. One would predict, however, that the potential for hominin death or serious injury at the hands of these carnivorans was less of a stressor than the potential for the same from larger carnivorans. As in the case of members of Size Class 1, most of these species are more likely to be killed by humans than actually cause the death of a human. As with the smallest size class, this group may have had less of an ecological impact on hominins than members of the two larger size classes.

Members of the third group of carnivorans (Size Class 3, >21.5–100 kg) exhibit behaviors that make them more likely than Size Class 1 or 2 carnivorans to have engaged in competitive and/or predatory encounters with hominins. Extant members are hypercarnivorous or primarily carnivorous and overlap to some degree in prey preference and other resource requirements. Terrestrial taxa in this size class prefer prey that is greater than 45% of their body mass (Carbone et al., 1999). This means that the preferred prey of Size Class 3 species would be large enough that individuals or groups might have competed with hominins for prey and/or engaged in interspecific competition at carcasses. Modern Size Class 3 species, however, do differ in behaviors associated with prey choice and procurement, carcass processing and transport, scavenging, and carcass defense (see Lewis, 1997 for a summary). Only one extant species in this class, the African clawless otter (*Aonyx capensis*), engages in behavior that is probably of little relevance to hominin evolution. We should note that all hominin species relevant to this discussion would be classified as Size Class 3.

Size Class 4 (>100 kg) today contains only the lion, *Panthera leo*. We have distinguished lions from other carnivorans due to their much larger maximum prey size (900 kg; Schaller, 1972) than other eastern African carnivorans (e.g., 300 kg for spotted hyenas; Schaller, 1972). While the most important components of the lion's diet in the Serengeti are the medium-sized (100–350 kg) wildebeest, zebra, and topi (Kruuk and Turner, 1967; Schaller, 1973), lions have also been known to prey upon elephant calves and young adult elephants (Pienaar, 1969). Lions are capable of bringing down prey in almost any habitat and are known to utilize all habitats within an area (Kruuk and Turner, 1967; Van Orsdol, 1982). Hunting success is affected by lioness group size, tactical strategy chosen by the group, prey species, and the interaction between time of the hunt and terrain (Stander and Albon, 1993). Presumably, not all extinct Size Class 4 taxa behaved like lions. However, the sheer size and robusticity of Class 4 species make them potential predators of hominins and potential competitors for prey and carcasses.

Among extant carnivorans, two Class 3 species (spotted hyenas and African wild dogs) engage in group hunting and thus are more dominant in interspecific competition

than the body mass of the average individual would predict. Interestingly, the one living Class 4 species, the lion, also engages in social grouping, thus maintaining its dominance in interspecific competitions between groups. As one can debate whether group hunting Class 3 species should be considered as Class 4 species or even whether the two size classes merit distinction, the two groups were considered both together and separately in the analyses.

## CHANGES IN DIVERSITY OF BODY MASSES THROUGH TIME

Our research indicates that both total species richness and mean standing richness across all known habitats of eastern African carnivorans reaches a peak roughly between 3.6 and 3.0 Ma (38 to 39 species) and then declines with a slight rise between 2.1 and 1.5 Ma (31 species) (see Figure 1). Unfortunately, intervals before 3.9 Ma are relatively undersampled, as are those between 0.9 Ma and the present. Nonetheless, a clear pattern can be seen during the well-sampled period between 3.9 and 0.9 Ma.

Studies of origination and extinction events in eastern Africa have indicated that there are two major peaks in the origination of new species: 3.9–3.6 Ma and 2.1–1.8 Ma (see Figure 2; also Figures 5 and 6 in Werdelin and Lewis, 2005). While these may both be sampling artifacts (the first due to poor sampling of the early Pliocene and the second due to the unique taxa found at Olduvai), it is interesting that there are absolutely no new species appearing between 3.0 and 2.4 Ma. There is a small peak in the extinction rate (3.3–3.0 Ma) that follows the first peak in origination rate. After that point, extinction rates are low and fairly constant until the extinction rate increases dramatically after 1.8 Ma. The greatest extinction rate then occurs between 1.5 and 1.2 Ma. Our data indicate that no turnover pulse occurs in carnivorans around 2.5 Ma.

One must ask whether these patterns are reflected across all carnivoran taxa. For example,

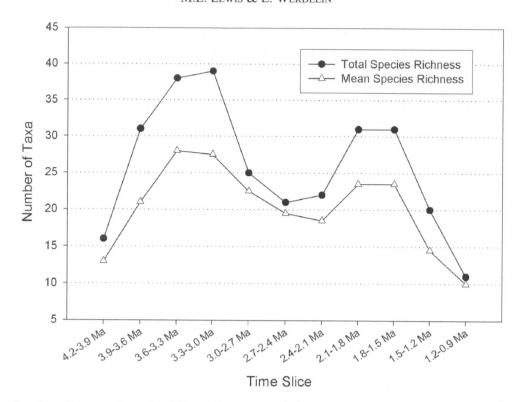

Figure 1. Species richness data for Plio-Pleistocene Carnivora of eastern Africa. Total richness and mean standing richness (MSR) are graphed in 300-Kyr bins from 4.2 to 0.9 Ma. As devised by Foote (2000), MSR = $(N_{bL} + 2N_{bt} + N_{Ft})/2$, where $N_{bL}$ = # of taxa that originate before the bin interval, but go extinct within it, $N_{bt}$ = # of taxa that originate before the interval and persist beyond it, and $N_{Ft}$ = taxa that originate in the interval and persist beyond it. Intervals 4.2–3.6 Ma and 1.5–0.9 Ma are less well sampled than the intervening intervals. Peaks before 3 Ma (higher) and after 2 Ma (lower) are evident.

how are these patterns reflected in the number of taxa large enough to have impacted hominins? When species richness of just Size Classes 3 and 4 is examined (Figure 3), the pattern seen in overall species richness is repeated: high richness between 3.6 and 3.0 Ma, then a drop, and a smaller increase around 2.1–1.5 Ma. Most importantly, there were more species in both size classes in the past until around 1.2–0.9 Ma, during which time there were still at least two Size Class 4 species (*Panthera leo* and *Dinofelis piveteaui*). While today one Size Class 4 species occurs in eastern Africa (lions), in the past, there were up to eight known species. Granted, not all of them may have been significant to hominin dietary or predator avoidance strategies as discussed below. Due to the nature of the fossil record, there is always the possibility that there were

even more species than we have detected thus far. However, such hypothetical species would most likely be close relatives of known species and thus not sympatric with them. This would mean an increase in beta- or gamma-diversity (as defined in Whittaker et al., 2001), but not the alpha-diversity that is critical for understanding the place of hominins in the carnivore guilds of the past.

## Discussion

### CARNIVORAN TAXA WITH POTENTIAL HOMININ IMPACT

To gain a full appreciation of the complexity of the changing ecological framework surrounding hominin evolution, one must take

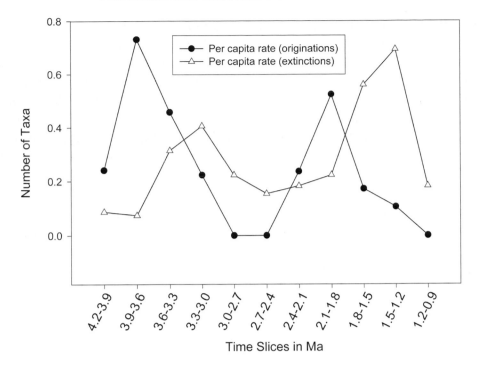

Figure 2. Per capita rates of origination and extinction for Plio-Pleistocene Carnivora of eastern Africa. Peaks in extinction follow peaks in origination, as one would expect. From 4.2 to 3.6 Ma and 1.5 to 0.9 Ma are less well sampled than the interval between them. Note the zero origination rate in the interval 3.0–2.4 Ma. No turnover pulse occurs in the 2.7–2.4 Ma interval. Per taxon rates show the same pattern as per capita rates (Werdelin and Lewis, 2005).

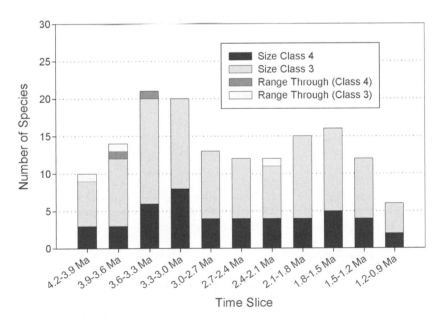

Figure 3. Changes in species richness of large-bodied carnivorans in Eastern Africa through time. These data represent minimum calculations of the number of taxa through time. Range through taxa are those that are not found during a particular time period, but are assumed to be present due to their presence in bins preceding and following a given bin. The highest overall species richness occurs between 3.6 and 3.0 Ma. A second smaller peak occurs during the 2.1–1.5 bins. The modern carnivoran fauna of eastern Africa is clearly reduced in species richness, particularly after 1.5 Ma.

an in-depth look at some of the taxa present in eastern Africa during this time period (see Table 4 for a list of Size Class 3 and 4 carnivorans present at key time periods). Unless otherwise stated, dates represent the time ranges that taxa occur in eastern Africa. In many cases, taxa may be found earlier or later in time at other localities outside of this region. Families and genera are presented in alphabetical order, for the most part, with some genera grouped by similarities in behavior and ecology or by systematic relationship. For more information on the evolution of specific African taxa or smaller members of the following families, please see Werdelin and Lewis (2005).

*Canidae*

*Canis.* Canids first appear in eastern Africa at approximately 4–4.5 Ma in the Omo (Fleagle and Brown, pers. comm.). The next oldest record occurs at Laetoli (*Nyctereutes* and cf. *Megacyon*) (Barry, 1987; Petter, 1987; Werdelin and Lewis, 2000, 2005), including the first Size Class 3 canid (cf. *Megacyon*). A third form at Laetoli has *Canis*-like features, but is smaller than the cf. *Megacyon* material (Barry, 1987; Werdelin and Lewis, 2005).

The oldest specimen of *Canis* is from South Turkwel and may have been part of a Eurasian migration of taxa into Africa at around 3.5 Ma (Werdelin and Lewis, 2000). The scant material found of this new species

*Table 4. Size Class 3 and 4 Carnivorans present at key time periods*

| Family | 3.6–3.3 Ma | 2.7–2.4 Ma | 1.8–1.5 Ma | Extant |
|---|---|---|---|---|
| Canidae | *Canis* sp. nov. A cf. *Megacyon* sp. | | *Canis lycaonoides*[1] | *Canis pictus* |
| Felidae | cf. *Acinonyx* cf. *Panthera leo*[2] cf. *Panthera pardus* | cf. *Acinonyx* cf. *Panthera pardus* | cf. *Acinonyx* *Panthera leo* *Panthera pardus* | *Acinonyx jubatus* *Panthera leo* *Panthera pardus* |
| Felidae (st) | *Dinofelis petteri* *Homotherium* sp. *Megantereon ekidoit* cf. *Megantereon whitei* | *Dinofelis aronoki* *Dinofelis petteri* *Homotherium* sp. cf. *Megantereon whitei* | *Dinofelis aronoki* *Dinofelis piveteaui* *Dinofelis* sp. B *Homotherium* sp. cf. *Megantereon whitei* | |
| Hyaenidae | *Chasmaporthetes* cf. *nitidula* *Crocuta dietrichi* cf. *Crocuta ultra* *Crocuta* sp. nov. A cf. *Hyaena makapani* *Ikelohyaena abronia* *Pachycrocuta brevirostris* *Parahyaena howelli* | *Chasmaporthetes* cf. *nitidula* *Crocuta dietrichi* cf. *Crocuta ultra* *Crocuta* sp. nov. A cf. *Hyaena makapani* | *Chasmaporthetes* cf. *nitidula* *Crocuta dietrichi* *Crocuta ultra* *Hyaena hyaena* | *Crocuta crocuta* *Hyaena hyaena* |
| Mustelidae Ursidae Viverridae | *Enhydriodon* – 4 species *Ursinae* sp. | *Enhydriodon* – 2 species | *Enhydriodon* – 1 species *Pseudocivetta ingens* *Viverridae* sp. H | *Aonyx capensis* |

st = sabertoothed felids.

[1] Formerly *Canis africanus*. See text for explanation.

[2] The attribution of the Laetoli specimens to cf. *Panthera leo* is extremely tentative. Note that no definite fossils of *Panthera leo* are known until Bed I at Olduvai Gorge. Due to the tentative attribution to *P. leo* at Laetoli, we have not scored *P. leo* as present as a range-through taxon between 3.3 and 2.1 Ma.

of *Canis* suggests that it may have been the same size as a medium-sized wolf, *C. lupus*. A fragmentary ulna from the Sidi Hakoma Member at Hadar may also belong to this species based on the size and age of the specimen, although no ulna has been found at South Turkwel to confirm this assignation. Other material at Hadar may belong to a slightly larger species of canid and two smaller species of *Canis*.

The modern species of African wild dog (subgenus *Lycaon*) is virtually unknown from eastern Africa with the earliest record occurring at Lainyamok (0.4 Ma). An earlier putative record from West Turkana (Harris et al., 1988) is an m1 trigonid that we consider indeterminate. The oldest known fossils of the extant species are from Hayonim Cave in Israel (Stiner et al., 2001). Material from South Africa suggests a relatively recent origin for African wild dogs (Lewis and Berger, 1998).

Other canid material that may be related to extant African wild dogs include *Canis africanus* from Bed II at Olduvai Gorge (Pohle, 1928; Ewer, 1965). Although the type specimen is missing, the remaining material is distinct morphologically from modern African wild dogs, but is probably part of the wild dog lineage. Martínez Navarro and Rook (2003) have synonymized *C. africanus* with *Xenocyon lycaonoides*, originally described from Gombaszög in Hungary (Kretzoi, 1938), and have suggested that this taxon is a plausible ancestor for the modern African wild dog. If this is true, leaving *Lycaon* with full generic status renders *Canis* paraphyletic. The paraphyly of *Canis* due to the nesting of *Lycaon* within *Canis* has been supported by recent systematic work on canids (Zrzavý and Řičánková, 2004). Pending revision of the genus *Canis*, we prefer to refer the wild dog lineage (e.g., *Lycaon*, *Xenocyon*) to *Canis*.

The evolution of larger canids in Africa is difficult to study due to a paucity of material. For the purposes of this paper, it is important to note that large canids were beginning around

at about 3.5 Ma, although they may have been rare. Size Class 2 canids, such as jackals, are similar in age and rarity in the fossil record of eastern Africa. Although jackal-sized *Canis* has often been referred to *C. mesomelas* or said to be like this species (Petter, 1973; Leakey, 1976; Harris et al., 1988), we have found this material to be no more similar in morphology to this species than any other extant jackal species. Dental differences among these specimens indicate that they are likely to represent more than one species. The earliest material that may belong to *C. mesomelas* is from Lainyamok (Potts et al., 1988; Potts and Deino, 1995). Other Size Class 2 canids, such as *C. simensis*, the Ethiopian wolf, have no known fossil record. In fact, the Ethiopian wolf has been suggested to be a relict population descended from a Pleistocene migration of a wolf-like canid into the formerly larger Afro-Alpine ecosystem (Wayne and Gottelli, 1997).

Extant canine canids (e.g., wolves, jackals) tend to be fairly consistent in the relationship between morphology, behavior, and body size. Although there is not enough eastern African fossil material to test hypotheses about behavior, one might suggest that the larger species of *Canis* were not too different from either modern wolves or African wild dogs. However, none of the extinct taxa show the more hypercarnivorous dental adaptations of *C. pictus*. If so, modeling these taxa as African wild dogs or even wolves may not be completely appropriate, but may give at least an approximation of the level of interaction with hominins.

Wild dogs, like cheetahs, rarely engage in scavenging (Creel and Creel, 1995). Extant African wild dogs often lose carcasses to lions and spotted hyenas and are frequently the objects of predation by these species (Fanshawe and Fitzgibbon, 1993; Mills and Gorman, 1997; Creel, 2001). However, spotted hyenas are less successful in finding wild dog kills in more wooded habitats, as in the Selous Game Reserve (Creel and Creel, 1998). In short, if extinct large-bodied species of

*Canis* behaved similarly in the past, they may have been a potential source for scavengeable carcasses. However, the fact that the dentition of the fossil forms is not as specialized as that of extant wild dogs suggests that the fossil forms are not directly equivalent ecologically to the extant forms. For example, extant African wild dogs lack an entoconid on the lower carnassial and possess an anterior cusplet on $P_4$ (Tedford et al., 1995). These features may have evolved relatively recently to facilitate even more rapid consumption of flesh before lions and spotted hyenas could steal the carcass (Lewis and Berger, 1998). The Olduvai *Canis lycaonoides*, however, retains the entoconid on $M_1$ and lacks an anterior cusplet on $P_4$.

Of course, it is likely that in open habitats hominins would have been in competition with a number of different species at any carcasses generated. If hominins developed more effective strategies at locating large canid kills rapidly in more wooded habitats than other extinct, large-bodied carnivorans and could displace the canids and defend the carcass, then these kills may have been a useful resource.

## Felidae

*Acinonyx.* The modern cheetah, *A. jubatus*, is a Size Class 3 predator adapted for short bursts of high-speed pursuit predation followed by quick consumption of the fleshy portions of carcasses. In interspecific encounters, cheetahs are of low rank and likely to be chased away from their kills, even when in coalitions (Caro, 1994). Cheetahs live at low densities and tend to be rare where the biomass of lions and spotted hyenas is high (Laurenson, 1995). Scavenging behavior in cheetahs is rare, but has been observed (e.g., Pienaar, 1969; Caro, 1994).

Unfortunately, there are few specimens of *Acinonyx* in the fossil record. A few specimens attributable to this genus first appear at Laetoli and occur infrequently, but consistently, in the Omo until about 2.5 Ma. *Acinonyx* is also known from the KBS and Okote Members at Koobi Fora and possibly earlier. These specimens clearly belong to *Acinonyx*, but are larger and morphologically distinct from the modern species. In fact, after its appearance in the Okote Member, *Acinonyx* is not found again until the present day. One can hypothesize that *Acinonyx* is present at very low levels from 3.8 Ma until the present. We consider the alternate hypothesis, that *Acinonyx* only recolonized the region fairly recently, to be less likely, but neither hypothesis can be disproved at this point. In any case, *Acinonyx* may never have been a significant competitor in the large-bodied carnivoran hierarchy of eastern Africa.

*Dinofelis.* Sabertoothed felids (subfamily Machairodontinae) in the eastern African Plio-Pleistocene include representative of three different tribes: the Metailurini (e.g., *Dinofelis*), the Homotherini (e.g., *Homotherium*), and the Smilodontini (e.g., *Megantereon*). Representatives of these tribes are quite different in morphology and presumably behavior.

The long-lived genus *Dinofelis* is found in eastern Africa from a maximum of 7.91 Ma (Lothagam) until possibly as recently as 0.9 Ma (Werdelin and Lewis, 2001b; Werdelin, 2003b). During this time interval *Dinofelis* was represented by at least five different species, many of whose ranges overlap temporally. Although an in-depth analysis and revision of this genus has been provided elsewhere (Werdelin and Lewis, 2001b), we will summarize briefly the adaptations of various species and discuss their relevance to hominin evolution.

After its occurrence at Lothagam, *Dinofelis* increases in size with the appearance of *D. petteri* (4.23–2.33 Ma). The largest eastern African form, *D. aronoki*, appears by 3.18 Ma and is found at localities in Ethiopia and Kenya until approximately 1.6 Ma (Nakoret) (Werdelin and Lewis, 2001b). *D. aronoki* may have been a little more cursorial than other members of this genus, given its reconstructed posture. Material from Bed I and II at Olduvai

Gorge indicates a unique form of *Dinofelis* (*Dinofelis* sp. B of Werdelin and Lewis, 2001b) that had extremely short, powerful forelimbs. This form has not been found elsewhere and is only known from three specimens.

A more recent species, *D. piveteaui*, was smaller than *D. aronoki* and had a more crouched posture, similar to *Megantereon*. *D. piveteaui* represents the most machairodont species of *Dinofelis* known (i.e., most flattened upper canines, reduced anterior premolars, and elongated carnassials). Although *D. piveteaui* was originally described from Southern Africa (Ewer, 1955), it is also known from Koobi Fora (Okote Mb.), Konso-Gardula, and Kanam East. If the 0.9 Ma date for Kanam East is correct (Ditchfield et al., 1999), then this would be the youngest sabertoothed felid in all of Africa.

One may note that for most of the temporal span addressed in this paper, there were two species of *Dinofelis* living at any given time in eastern Africa, although not necessarily at the same location. One species tended to be relatively larger (e.g., *D. aronoki*) and one tended to be a little smaller with a more crouched posture (e.g., *D. petteri* or *D. piveteaui*). Both types of species would have been of significance to hominins and may have preferred prey similar in size to the prey of lions.

Previous studies have demonstrated the morphological similarity between the South African species *D. barlowi* and *Panthera* (Lewis, 1995b, 1997). Despite sharing the enlarged forelimb relative to hindlimb with the genus *Megantereon* and other sabertooths, postcranial material of *Dinofelis* has been shown to be somewhat more similar to extant lions, tigers, and leopards than they are to *Megantereon* (Lewis, 1997). However, eastern African forms do not converge on modern *Panthera* in craniodental or postcranial morphology to the degree seen in *D. barlowi*.

*Dinofelis* has been suggested to have preferred dense forest (Marean, 1989) or mixed/closed habitats (Lewis, 1997). Habitat preference does not mean that a species is limited to that area, particularly as narrow categorizations of habitat preference cannot be made from carnivoran postcranial morphology (Van Valkenburgh, 1987; Taylor, 1989). Large, extant carnivorans in Africa may be found in a variety of habitats despite what their postcranial morphology might predict (e.g., lions, leopards, spotted hyenas; see review in Van Valkenburgh, 2001). Of course, it is possible that the ability of extant carnivorans to inhabit a variety of habitats successfully is a key component of the suite of adaptations that ensured their survival to the present. Yet while some species of *Dinofelis* may have preferred more open habitats (e.g., *D. aronoki*) or more closed habitats (e.g., *Dinofelis* sp. B from Olduvai Gorge), the somewhat more *Panthera*-like morphology of many species (relative to other sabertooths) suggests that species of *Dinofelis* may have ranged through a wide variety of habitats.

Suggestions that *Dinofelis* inhabited more mixed/closed habitats mean that their ability to cache carcasses in trees should be addressed. Such behavior has been rejected as being within the suite of activities open to the South African species *D. barlowi* (Lewis, 1995b, 1997). We concur with this assessment for eastern African species of *Dinofelis*, as well. Considerations of body mass aside, the smaller species certainly have the forelimb dexterity and strength to climb effectively, but possess the most machairodont canines. Tree-caching a shifting carcass would have been a risky behavior with high potential for damage to the canines. Unlike in North American taxa (Van Valkenburgh and Hertel, 1993), there is no known evidence of canine breakage in *Dinofelis*. The larger species, *D. aronoki*, had relatively less machairodont canines, but also had less rotatory ability in the forelimb. Thus, all species of *Dinofelis* appear unlikely candidates for tree-caching behavior, although they may have stolen carcasses from tree-caching leopards.

We should note that *Dinofelis* is better represented in the fossil record of eastern Africa than any of the other sabertooth taxa and yet there is no evidence of tooth breakage in *Dinofelis*. Van Valkenburgh (1988) has demonstrated that tooth breakage is associated with heavy carcass utilization and bone-eating. Based on an analogy to North American *Smilodon*, which has a large amount of tooth breakage, and the fact that modern big cats use their tongues as files to rasp flesh off bones, Van Valkenburgh (2001) has suggested that African sabertooths were probably quite capable of dismembering the carcass and engaging in bone-cracking. While this may certainly have been true, there is as yet no evidence to support the hypothesis that *Dinofelis*, or any other African sabertooth, had carcass-rendering behaviors similar to *Smilodon*. Of all of the sabertooths found in Africa during this time, some species of *Dinofelis* (e.g., *D. aronoki* or the South African *D. barlowi*) are certainly the best candidates for this behavior based on their craniodental and/or postcranial morphology. However, while some species of *Dinofelis* may converge to a small degree with *Panthera*, none of the eastern African sabertooths converges on *Smilodon* in craniodental or postcranial morphology.

*Homotherium*. *Homotherium* is present in eastern Africa from 4.35 to 1.4 Ma at a wide variety of sites. The earliest records are from Kanapoi and the Lonyumun Mb. of Koobi Fora and the most recent records are from the Okote Mb. of Koobi Fora. Previous studies have indicated that *Homotherium* was more cursorial than other sabertooths and had elongated forelimbs and relatively shortened hindlimbs (Ballesio, 1963; Beaumont, 1975; Kurtén and Anderson, 1980; Martin, 1980; Rawn-Schatzinger, 1992; Lewis, 1995b, 1996, 1997, 2001; Anyonge, 1996; Antón and Galobart, 1999). Cursorial adaptations such as the relatively (for a sabertooth) more distally elongate forelimbs and the reduction in supinatory ability of *Homotherium* are suggestive

of adaptations to more open habitats (Martin, 1980; Lewis, 1995b, 1997). However, they probably were not limited to open habitats. The slight reduction in supinatory ability suggests that members of this taxon interacted with prey in a fundamentally different manner from other sabertooths or *Panthera*. In addition, this reduction, in combination with the large body size indicates little to no scansorial ability (Lewis, 1995b, 1997, 2001). However, morphological and taphonomic studies have suggested that *Homotherium* could capture prey somewhat larger than those captured by a lion (Lewis, 1995b, 1997; Marean and Ehrhardt, 1995). If African species transported portions of carcasses, as has been suggested for North American *H. serum* (Marean and Ehrhardt, 1995), or lived in groups (Turner and Antón, 1997), then there may have been very little left for hominin scavengers. In any case, challenging this taxon was probably a very risky behavior for early (pre 2 Ma) hominins.

*Megantereon*. The smilodontin sabertooth genus *Megantereon* has been reported from Aramis (ca. 4.4 Ma; WoldeGabriel et al., 1994), although this material has not yet been described. Other than the Aramis material, the oldest material is from South Turkwel and has been referred to the new species *M. ekidoit* (Werdelin and Lewis, 2000; cf. Palmqvist, 2002; Werdelin and Lewis, 2002 for a discussion of this taxon). A second species, *M. whitei* is known from the Okote Mb. of Koobi Fora. Unfortunately, other specimens of *Megantereon* from eastern Africa (e.g., Shungura Fm. Mbs. B–G) are isolated teeth, making taxonomic identifications and behavioral reconstructions impossible. Of the three sabertooth genera in eastern Africa at this time, *Megantereon* is the most poorly represented in terms of both numbers of specimens and localities at which it is found. This genus is not found in eastern Africa after about 1.4 Ma.

Members of this genus have been shown to have extreme strength in the forelimb (Lewis, 1995a, b, 1997; Martínez-Navarro and

Palmqvist, 1996). Specimens from Kromdraai, South Africa possess a limb morphology that is more similar to that of extant jaguars than to any of the modern African taxa or other African sabertooths, although they were much more heavily muscled than jaguars (Lewis, 1995a, b, 1997). As a result, African and European *Megantereon* have been identified as potential providers of large carcasses for hominins (Lewis, 1995b, 1997; Martínez-Navarro and Palmqvist, 1996; Arribas and Palmqvist, 1999).

Like *Dinofelis*, *Megantereon* has been suggested to inhabit dense forest (Marean, 1989) or mixed/closed habitats (Lewis, 1997) based on data from European specimens. The same caveats for reconstructing habitat preferences for carnivorans that were mentioned for *Dinofelis* certainly apply to *Megantereon*. While their crouched posture is indicative of an ambush predator and their size and limb morphology suggest an ability to climb trees (Lewis, 1995a, b, 1997), this does not mean that they were tied to specific habitats. Their forelimb morphology may reflect prey grappling more than scansorial ability (Lewis, 1997). Despite being the smallest of the sabertooths at this time period, if members of this genus did climb trees, they would have been more likely to steal cached carcasses than to cache carcasses for the same reasons discussed for *Dinofelis*.

*Panthera*. The earliest specimens attributable to *Panthera* are from Laetoli and represent two species: a lion-like one and one referred to *P.* cf. *pardus* (Barry, 1987; Turner, 1990). Unfortunately, the lion-like material is not diagnostic at the species level, although we have left it as *P.* cf. *leo* (see Werdelin and Lewis, 2005 for a discussion of the attribution of these specimens). The first definite appearance of lions is at Olduvai, Bed I, leaving a large gap between Laetoli and Olduvai.

The Laetoli leopard material is better preserved and complete. Leopards and leopard-like *Panthera* are overall more common than lions or lion-like specimens. Once again, however, the first definite leopard is from Olduvai, Bed I. From Bed I to the present, the record of lions and leopards in eastern Africa is reasonably continuous.

Several lines of evidence suggest that Plio-Pleistocene leopards cached carcasses (Lewis, 1997): (1) accumulations of bones in caves that may have fallen from leopard feeding trees (Brain, 1981), (2) similarities in modern and fossil leopard postcrania, and (3) the high probability of losing a carcass on the ground due to the great number of terrestrial carnivores higher in the carnivoran hierarchy than leopards. However, if hominins began to scavenge from these cached carcasses, leopards would probably have become more diligent in guarding their carcasses (Lewis, 1997). Even though *Dinofelis* and/or *Megantereon* may not have been the best climbers, hominins may also have had to compete with them for these carcasses (Van Valkenburgh, 2001).

At present, nothing exists to suggest behaviors outside of the range of modern *Panthera* in Africa for any of these taxa although neither of the modern taxa is well represented in the eastern African fossil record.

### Hyaenidae

*Chasmaporthetes*. *Chasmaporthetes*, like *Acinonyx*, is a Size Class 3 carnivore with adaptations for hypercarnivory. The oldest specimen of *Chasmaporthetes* in eastern Africa is from Allia Bay (ca. 3.9–3.7 Ma). *C. nitidula* is known from diagnostic material from Hadar and Olduvai Gorge (Bed I). Although *C. silberbergi* has been reported from Laetoli (Turner, 1990), we are not convinced of this species attribution (see Werdelin and Lewis, 2005).

*Chasmaporthetes*, often called the "hunting hyena," has been reconstructed as being similar to cheetahs in locomotor behavior (Brain, 1981). However, our preliminary analyses of postcranial material from Hadar indicate that the eastern African species may not be quite like Eurasian or North American forms in its locomotor capabilities. Postcranial material is unknown from other eastern African localities. The dentition of eastern Africa *Chasmaporthetes*, however, is highly adapted for flesh-slicing, as has been found in material from other continents (Galiano and Frailey, 1977; Berta, 1981; Kurtén and Werdelin, 1988; Werdelin et al., 1994). While our analyses of the Hadar specimens are not complete, we can state that *Chasmaporthetes* in eastern Africa cannot be considered directly equivalent to *Acinonyx* in any behavior except for flesh-slicing ability. More likely, *Chasmaporthetes* filled a more large-bodied *Canis*-like niche, including the long-distance running seen in the African wild dog, *Canis pictus*. This hypothesis, however, is still being tested.

With respect to hominin behavior, single individuals of *Acinonyx* and *Chasmaporthetes* probably posed little threat to a group of hominins, even in the smallest, earliest species. If *Chasmaporthetes* fulfilled a more *C. pictus*-like niche and hunted in packs, the risk from predation and interspecific encounters at carcasses would have been increased. Given the lack of *Chasmaporthetes* specimens and the evolutionary distance between this taxon and modern hyenas, hypotheses of grouping behavior in *Chasmaporthetes* remain purely speculative at this point.

*Hyaena, Ikelohyaena, and Parahyaena.* Specialized, bone-cracking hyenas are absent from the Miocene and earliest Pliocene of Africa (Werdelin and Turner, 1996; Turner, 1999). The earliest species in this category is *Parahyaena howelli* from Kanapoi and Laetoli (Barry, 1987; Werdelin and Solounias, 1996; Werdelin, 2003a; Werdelin and Lewis, 2005), an early relative of the living brown hyena. Although the current distribution of the modern species, *P. brunnea*, is limited to southern and southwestern Africa, this species is known from middle Pleistocene of Kenya (Werdelin and Barthelme, 1997).

*Parahyaena howelli* differs from the modern form in both craniodental and postcranial morphology (Werdelin, 2003a). As the early form is less derived than the modern form, they cannot be considered directly equivalent in behavior or ecology, although they are certainly similar. However, on a very general level, both taxa share bone-cracking teeth in combination with gracile (for a hyaenid) postcrania. As such, the behavior of the earliest *Parahyaena* may not have been too dissimilar from the modern form, particularly with respect to behaviors that may have impacted on hominins.

The earliest close relative of the modern striped hyena found in eastern Africa is *Ikelohyaena abronia*. This species has been found at Laetoli and possibly at Lothagam and Hadar (Sidi Hakoma Mb.). The oldest specimens of *Hyaena* found in eastern Africa occur at Koobi Fora (Lokochot Mb.). The earliest species of *Hyaena*, *H. makapani*, is smaller and more gracile than the living species. This species differs from *Ikelohyaena abronia* primarily in the loss of m2 and M2. Neither *I. abronia* nor *H. makapani* have the adaptations for bone-cracking seen in the modern striped hyena, *H. hyaena*. This fact, in combination with their gracile postcrania and smaller body size, suggests that they overlapped ecologically with hominins to a much lesser degree than *H. hyaena*. The living striped hyena, *Hyaena hyaena*, appears at about 1.9 Ma and is found from that time period to the present day.

Modern striped and brown hyenas are solitary foragers with a very omnivorous diet. While fecal analyses of brown hyenas indicate a large quantity of large mammal remains (Owens and Owens, 1978), behavioral studies indicate that in both brown and striped hyenas, only a little over a third of food items eaten were mammalian (Kruuk, 1976; Mills, 1978, 1982). Wild fruits, insects, bird eggs, and small reptiles and birds comprise the rest of the diet. Mammals consumed by striped hyenas (e.g., hyraxes, rodents, ground-living birds, hares, and ibex) (MacDonald, 1978) tend to be much smaller than the prey of brown hyenas (e.g., wildebeest, gemsbok, springbok, steenbok, and small canids) (Mills, 1982). Under natural conditions, mammalian remains fed upon by striped hyenas do not provide a meal for more than one individual (Mills, 1978). On the other hand, studies indicate that the vast majority of vertebrate consumption by brown hyenas results from scavenging, not hunting (Mills, 1978, 1982; Rautenbach and Nel, 1978).

Striped and brown hyenas probably evolved bone-cracking as a means of maximizing yield from scavenged carcasses. If the extinct form of *Parahyaena* behaved much like the modern form, it is possible that competition with early hominins may have contributed to its extinction in eastern Africa. A greater reliance on small prey, fruits, and insects may have permitted the modern striped hyena to survive while the ancestors of the modern brown hyena eventually disappeared from eastern Africa. Today brown hyenas are restricted to the South West Arid Zone and drier parts of the southern savannahs in the Southern African Subregion (Hofer and Mills, 1998). The disappearance of *Hyaena* in southern Africa suggests that there is more going on than competition with early hominins.

*Pachycrocuta and Crocuta.* The evolution of *Pachycrocuta* and *Crocuta* in Africa is quite complex. One can see from Table 4 that the number of hyaenids was much greater in the past than today. A brief discussion of our current understanding of *Crocuta* and *Pachycrocuta* evolution can be found elsewhere (Lewis and Werdelin, 1997, 2000; Werdelin, 1999; Werdelin and Lewis, 2005), although analyses of the evolution of *Crocuta* are still being completed.

Although our current understanding of the evolution of these taxa is more complex than in the past, we must reiterate that one cannot simply use modern spotted hyena behavior as a model for all *Crocuta* or *Pachycrocuta*

species in the past. The bone-cracking dentition and postcranial adaptations for heavy carcass lifting evolved at different points within the evolution of *Crocuta* (Lewis and Werdelin, 1997, 2000). The early species *C. dietrichi*, for example, was small and probably more similar in behavior to modern brown hyenas. Even *C. ultra*, which is similar in craniodental morphology to *C. crocuta*, lacks the postcranial adaptations of the modern species. Unfortunately, modern spotted hyenas, *C. crocuta*, are unknown from the fossil record of eastern Africa. Although postcranial material is not known for all extinct members of the genus *Crocuta*, our current research suggests that the suite of adaptations that define spotted hyenas today developed only within the last one million years. Based on analyses of material from Eurasia and Africa, Turner and Antón (1995) have suggested that *Pachycrocuta brevirostris* was a group-living species that behaved somewhat similarly to modern spotted hyenas, but was less cursorial due to differences in body proportions. Unfortunately, there are few specimens of *Pachycrocuta* in eastern Africa.

While no single species (with the possible exception of the poorly known *Pachycrocuta*) may have encompassed the wide range of behaviors seen in extant spotted hyenas (bone-cracking, group hunting, confrontational scavenging, heavy carcass lifting, and transport behavior, etc.), eastern African hyaenids as a group probably covered all of these behaviors. Given the diversity of all hyaenids around 3.5 Ma, it is hard to imagine any salvageable carcasses present on the landscape at that time remaining available for a passive scavenger unless the hyaenid species were relatively rare and/or solitary at that time. *Crocuta*, as a genus, is a relatively much more common component of fossil assemblages in eastern Africa than *Pachycrocuta*. Although this could be due to differences in habitat usage or some other taphonomic bias, it is certainly possible that *Pachycrocuta* was not abundant in eastern

Africa despite being morphologically, and presumably ecologically, distinct from species of *Crocuta*.

## Mustelidae

The only Size Class 3 and 4 mustelids found in eastern Africa during the Pliocene and Pleistocene belong to extinct members of the Enhydrini, or sea otters. (Extinct relatives of the modern African clawless otter, *Aonyx capensis*, were smaller than the modern form.) The genus *Enhydriodon* first appears in the Upper Miocene and then radiates extensively in the Pliocene (Werdelin and Lewis, 2005). Members of this genus are known from the Omo, Hadar, Koobi Fora, and other localities. The last appearance is at Nakoret, possibly around 1.6 Ma. While the early forms are within the size range of modern otters, some of the last forms have limb bones the size of *Homotherium* or even modern ursids. Like modern sea otters, the teeth of this species are durophagous and show an increase in size matching that of the postcranial elements.

The relevance of *Enhydriodon* to hominin evolution is unknown. If this genus maintained its aquatic lifestyle as it grew larger, then it was not truly part of the terrestrial carnivoran guild. However, we are currently testing whether differences between the early and later forms are due to differences in body size and/or locomotor adaptations. The morphologies of the different taxa are certainly distinct. A brief discussion of this genus can be found elsewhere (Werdelin and Lewis, 2005).

## Ursidae

Although ursids were found in historic times in the Atlas Mountains of northern Africa (Nowak, 1991), the first report of this family in the Plio-Pleistocene of eastern Africa was from Aramis with the identification of *Agriotherium* (WoldeGabriel et al., 1994). This genus is also known from southern Africa (e.g., Hendey, 1980). Since the discovery at Aramis, material from the modern subfamily

Ursinae has also been reported from eastern Africa (Werdelin and Lewis, 2005). At least one species is known from Hadar (Denen Dora and Kada Hadar Mbs.), Koobi Fora (Tulu Bor Mb.), and possibly West Turkana (Lomekwi Mb.). As analyses of this material are still underway, it is too early to comment on possible behaviors of these taxa and their significance to hominins.

## Viverridae

Viverrids have an intriguing, albeit limited, fossil record in eastern Africa. At least two species of viverrids were large enough to warrant a Size Class 3 designation: *Pseudocivetta ingens* and a new species. *P. ingens* was first described from Olduvai, Bed I (Petter, 1967, 1973), but is now known from Koobi Fora (Upper Burgi and KBS Mbs.), the Shungura Fm. (Mbs. E + F and G), and Olduvai, Bed II. The second species is from the KBS Mb. at Koobi Fora and is somewhat similar to an enlarged *Civettictis*. The fossil record of smaller viverrids is described in more detail elsewhere (Werdelin and Lewis, 2005).

Both of the larger viverrids are within the lowest range of Size Class 3 and were not hypercarnivores. As the Koobi Fora viverrid is known from a single complete skull with dentition (KNM-ER 5339), little can currently be said about this species except that it was less hypercarnivorous than *Civettictis civetta*. As such, these taxa may have had little impact on hominins.

## INTERSPECIFIC COMPETITION AND THE EVOLUTION OF HOMININS

Studies have shown that some larger-bodied carnivoran species, such as lions and spotted hyenas, attain high densities where prey species are abundant (Van Orsdol et al., 1985; Stander, 1991). However, not all large-bodied carnivorans are affected positively by increases in diversity or abundance of potential prey species. African wild dog populations have declined, sometimes to the point of local extinction, when prey density was high, yet increased when prey density was moderate to low (Creel and Creel, 1998). Creel (2001) has hypothesized that an increase in prey abundance may reduce interspecific competition for live prey while increasing interspecific interference competition at carcasses. Creel argues that when there is asymmetry in size or fighting ability among competitors, engaging in kleptoparasitism (food theft) results in a higher net benefit for the stronger competitor than hunting live prey even when prey is abundant. Linnell and Strand (2000) have suggested that interference intraguild interactions may result in one species avoiding a specific habitat to reduce encounters with a more dominant carnivore. Predators may also avoid specific activity periods (e.g., night) when more dominant carnivores are foraging (Van Valkenburgh, 2001). Such avoidance behavior could have a severe impact on foraging efficiency. Interference competition involves more than just kleptoparasitism, as interspecific killing is common among carnivorans (Van Valkenburgh, 1985; Palomares and Caro, 1999; see Van Valkenburgh, 2001 for a detailed review of interference competition among predators in various habitats around the world). Thus, lower ranking carnivores, such as wild dogs or cheetahs, are better off when prey densities are at low or moderate levels so that densities of higher ranking carnivores, such as lions and spotted hyenas, remain low (Laurenson, 1995; Linnell and Strand, 2000; Creel, 2001).

While researchers may disagree over the amount of opportunistic or confrontational hunting and scavenging that early hominins engaged in, most would probably agree that when hominins first entered the carnivore guild, they were not top predators. Hominins lack the natural equipment (e.g., claws and sharp teeth) of other predators and would have had difficulty dominating even similarly sized carnivorans in one-on-one encounters (Shipman and Walker, 1989; Van Valkenburgh, 2001). As such, resources from carcasses were less likely to be important to hominins as they entered the carnivore guild due to potential damaging affects from kleptoparasitism by higher ranking, large-bodied predators. We hypothesize that in a predator and prey species-rich environment, individual hominin species could only survive if, in addition to predator avoidance strategies, they either had a primarily non-carnivorous diet and thus did not participate in the carnivore guild (e.g., *Paranthropus*), or if they evolved effective strategies to resist kleptoparasitism. In other words, the ancestors of *Homo sapiens*, whether they were hunting or functioning as kleptoparasites themselves, had to evolve strategies to reduce kleptoparasitism from higher ranking carnivorans before they could increase their dependence on carcasses as a resource.

Studies have suggested that animals are the preferred source of energy for modern hunter-gatherers (e.g., Cordain et al., 2002), even though it may not always be a reliable source (O'Connell et al., 2002). At some point, therefore, hominins did evolve effective strategies to resist kleptoparasitism. Predator avoidance and kleptoparasite resistance strategies, in combination with the evolution of behavior that increased the success of active hunting, would have led to an eventual reversal in the relative dominance of hominins and carnivorans. When hominins had evolved to this point, we predict that carnivorans utilizing the same food resources as hominins would be of lower rank than hominins. As seen in the impact of high densities of modern spotted hyenas on wild dog populations, any concomitant increases in the density of hominins with the previously described behaviors may have pushed local carnivoran populations to extinction.

When might these strategies have evolved? First, we must consider the patterns of turnover in eastern Africa. When mammals in general are considered, the number of species in the Turkana

Basin and the amount of turnover increases between 3.0 and 1.8 Ma (Behrensmeyer et al., 1997). Our study, however, has shown that after the high peak of 3.3 Ma, turnover is reduced in carnivorans during this period and that species richness decreases only to rise briefly to reach a second, lower peak between 2.1 and 1.8 Ma (Werdelin and Lewis, 2005). We have suggested elsewhere (Werdelin and Lewis, 2005) that the increase in originations and overall high species richness of carnivorans at 3.3 Ma is due to adaptive radiations in hyaenids, felids, and mustelids in combination with migrations of taxa (e.g., *Canis* and *Megantereon* from the north and *Hyaena* from the south) occurring after the Mio-Pliocene extinction event in Africa.

How might the pattern seen in carnivorans relate to hominins? The drop in numbers of carnivoran species precedes the appearance of stone tools at 2.6 Ma, yet presumably occurs while the species richness (and possibly abundance) of prey is increasing. The presence of stone tools at 2.6 Ma (Semaw et al., 2003) probably signals a shift in dietary behavior in hominins, yet that behavior does not appear to have impacted carnivorans directly. Whether the appearance of tool-using hominins was relevant to the decline in carnivoran species richness after 2.4 Ma is as yet impossible to test with our data set. However, the carnivoran origination rate increased briefly again after the appearance of stone tools. Could competitive interference initiated by changes in hominin behavior be part of what was driving speciation events during this time? Competition with hominins could be part of what prevented carnivorans from reattaining their previous level of species richness, even though overall species richness in mammals was increasing during this time.

One thing is clear from our data: carnivoran origination rate drops and extinction rate increases after 1.8 Ma. In addition, a precipitous drop in carnivoran species richness is apparent after 1.5 Ma. These changes coincide with the earliest appearance of *Homo ergaster*

(Feibel et al., 1989; Wood, 1991; Wood and Richmond, 2000) and a drop in overall species richness in eastern Africa (Behrensmeyer et al., 1997). We hypothesize that the changes in carnivorans are due to the appearance of new dietary strategies and behaviors in *H. ergaster* in combination with an increase in open habitats and concomitant decline in prey species richness occurring at the end of the Pliocene (Behrensmeyer et al., 1997). However, assuming that the drop in carnivoran species richness was due in part to *H. ergaster*, this decrease should not be seen as a confirmation of *H. ergaster* as active hunters. Van Valkenburgh (2001) has provided a compelling argument based on interference competition for why species of *Homo* functioning as a kleptoparasites must have engaged in confrontational scavenging. As seen in the complex web of interactions among extant African carnivorans, effective kleptoparasitism can negatively impact populations of species of lower interspecific rank. We suggest that effective confrontational scavenging by *H. ergaster* may have been enough to push some members of the carnivore guild to local extinction.

## WHEN DID THE MODERN GUILD EVOLVE?

Does the drop after the 1.8–1.5 Ma interval represent the final reduction to the low number of large-bodied carnivorans seen today? The fact that sabertoothed felids such as *Megantereon* and *Homotherium* persist until after 1.5 Ma suggests that the modern guild structure had not yet appeared. In fact, the last sabertooth found in eastern Africa, *D. piveteaui*, is found at Kanam East, approximately 0.9 Ma. Unfortunately, the absolute number of taxa present at any time after 0.9 Ma is difficult to assess due to the low number of fossil localities with carnivorans dated later than this time.

Only three Size Class 3 and 4 carnivorans in the modern guild have a definite fossil record in eastern Africa: lions, leopards, and striped hyenas. Although small carnivorans have not been the focus of this paper, the same pattern occurs in those taxa (Werdelin and Lewis, 2005). Fossil evidence from South Africa and Israel suggests a relatively recent origin for wild dogs (Lewis and Berger, 1998; Stiner et al., 2001). The fossil record of *Crocuta* also indicates a recent origin for the modern species. In short, the modern guild appeared less than one million years ago and possibly relatively recently.

We have hypothesized that specialist carnivorans (e.g., sabertooths and giant otters) were more likely to become extinct than generalists during the last three million years and that the last one million years of carnivore evolution in Africa has been the age of the generalist (Werdelin and Lewis, 2001a, 2005; Peters et al., in press). A similar pattern has been shown for carnivorans in the North American Pleistocene (Van Valkenburgh and Hertel, 1998; Wang et al., 2004). The recent origin of many of the modern species may reflect environmental pressures (including those placed on them by hominins) driving the evolution of larger niche spaces (Lewis, 1995b). In any case, the few specialists that remain (e.g., cheetahs, wild dogs) are relatively low ranking and in danger of extinction.

Similar suggestions have been made for hominins. Environmental fluctuation and a subsequent decoupling of Pleistocene hominins from adaptation to one specific habitat is an important component of Potts's variability selection hypothesis (Potts, 1998). When considering the origins of the modern carnivore guild, one might not consider the modern eastern African carnivore guild complete until the appearance of the supreme generalist species, *Homo sapiens*.

## Conclusion

Carnivorans reached their maximum species richness in eastern Africa approximately 3.3 Ma. Although we cannot accurately assess the numbers of Size Class 1 taxa or even those

of Size Class 2, the two largest size classes (Size Classes 3 and 4) included a much larger group of species than today, with a much broader range of adaptations. The greater taxonomic diversity of large-bodied carnivorans in the past suggests a more complex structuring of niche space within carnivoran guilds. As a result, interspecific competition was probably greater in the past, although the appearance of a taxon in the fossil record does not indicate the abundance of that taxon.

Turner (1988) has stated that the extinction of the large-bodied taxa provided a catalyst for hominin evolution, a view echoed elsewhere (Lewis, 1995b). As more recent finds have extended the known temporal range of some large-bodied taxa (e.g., *Dinofelis*), we modify Turner's hypothesis to state that the decline in carnivoran species richness after 3.0 Ma provided niche space for hominins to enter the carnivore guild. At that point, hominins were of relatively low rank within the guild hierarchy. Support for this low ranking comes from the lack of impact that the appearance of stone tools in the fossil record has on species richness of carnivorans, even when just the large-bodied taxa are considered. This lack of impact may be due in part to the relatively localized behavior of early toolmakers (i.e., hominin populations may have been geographically restricted and/or the use or non-use of tools may have varied among populations as described by McGrew, 2004, for modern chimpanzees). In fact, the decrease in carnivoran species richness begins after 3.0 Ma, but well before 2.6 Ma, suggesting that hominins were not responsible for this decline. In any case, early hominins, such as *Ardipithecus*, *Australopithecus*, and even *Homo habilis* were more likely to be prey than competitors with the Size Class 3 and 4 carnivorans. If carcass-based resources were to become a relied-upon food source, hominins needed to evolve effective anti-predator/anti-kleptoparasitism strategies. These strategies, in combination with the eventual evolution of active hunting, would

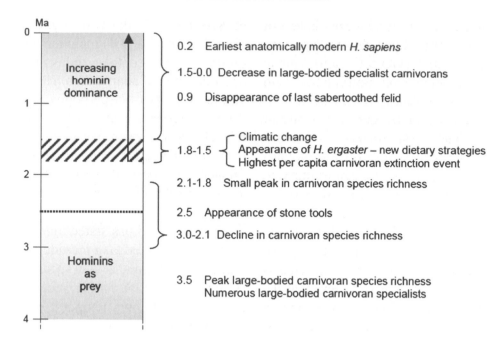

Figure 4. Key events in carnivoran and hominin evolution. Key events in the evolution of the two groups are presented side-by-side for comparison. Note that the appearance of stone tools has little effect on carnivoran species richness. After 1.5 Ma, hominins become increasingly dominant on the landscape and carnivorans decrease in species richness.

have increased the rank of hominin species within the guild (Figure 4).

The decline to the present depauperate group of carnivoran taxa began after 1.8 Ma and may be directly related to the appearance of new patterns of environmental use and dietary behavior by large-brained hominins (e.g., *Homo ergaster*) in combination with environmental change that began at the end of the Pliocene. While active hunting by hominins would have had an effect on carnivorans targeting the same prey species, active hunting is not the only means by which the carnivore guild could have been destabilized. Kleptoparasitism by high ranking carnivores has been shown to drive populations of lower ranking taxa to local extinction (Linnell and Strand, 2000; Creel, 2001). Effective kleptoparasitic strategies (i.e., confrontational scavenging) by *H. ergaster* could have had the same impact on some carnivoran taxa, although it is probably not a sufficient explanation for the extinction of all the species that become extinct during the early Pleistocene.

The question of when the modern guild first appears is difficult to answer at present due to the paucity of fossil sites known in eastern Africa from after 0.9 Ma. While a few taxa appeared quite early (e.g., lions, leopards, and striped hyenas), the rest of the large-bodied carnivorans probably appeared fairly recently. This lack of data also means that the impact of *Homo sapiens* on the carnivore guild during the last 0.2 million years cannot be assessed at present.

Finally, we must reiterate that with a few exceptions (e.g., *Panthera* and *Hyaena hyaena*), there is no morphological evidence that genera present today were morphologically (and presumably behaviorally) equivalent to their extinct congeners. While studying the present provides key insights into interpreting past behavior, any inferences drawn from studies of modern animals and environments must be treated as hypotheses and tested in the fossil record. For example, if the extant taxa were not present, were there any taxa that could have behaved in the hypothesized manner? This paper set out to address what carnivoran taxa were present in the past and provide a summary of our current understanding of behaviors relevant to the reconstructing the ever-changing ecological framework surrounding hominin evolution. While our understanding of these taxa is still incomplete, we believe that our continued analyses can only shed more light on the factors that drove a mammal from a primarily non-carnivorous order to become the dominant predator not only in eastern Africa, but across the world.

## Acknowledgments

We would like to thank the ARCCH of Ethiopia and the Governments of Ethiopia, Kenya, and Tanzania for permission to study material housed in Addis Ababa, Ethiopia and Nairobi, Kenya. For permission to study material in their care we thank Z. Alemseged, W.D. Heinrich, A. Hill, J.J. Hooker, F.C. Howell, M.G. Leakey, E. Mbua, M. Muungu, T. Plummer, R. Potts, and K.E. Reed. We are grateful for the invitation from A.K. Behrensmeyer, R. Bobe, and Z. Alemseged to participate in the "Workshop on Faunal Evidence for Hominin Paleoecology in East Africa" at the National Museum of Natural History, Smithsonian Institution in Washington, DC and for the invitation to be a part of this volume. We would also like to thank C.K. Brain whose work on bovid size classes inspired this research. The quality of this paper was significantly improved by comments from M. Lague and critical reviews from R. Blumenschine, B. Van Valkenburgh, and A. Walker. This research was supported by a grant from the L.S.B. Leakey Foundation to M.L. and K.E. Reed, Stockton Distinguished Faculty Fellowships to M.L., and Swedish Research Council Grants to L.W.

## References

Andersson, K., 2005. Were there pack-hunting canids in the Tertiary, and how can we know? Paleobiology 31, 56–72.

Antón, M., Galobart, A., 1999. Neck function and predatory behavior in the scimitar toothed cat *Homotherium latidens* (Owen). Journal of Vertebrate Paleontology 19, 771–784.

Anyonge, W., 1996. Locomotor behaviour in Plio-Pleistocene sabre-tooth cats: a biomechanical analysis. Journal of Zoology London 238, 395–413.

Arribas, A., Palmqvist, P., 1999. On the ecological connection between sabre-tooths and hominids: fauna dispersal events in the Lower Pleistocene and a review of the evidence for the first human arrival in Europe. Journal of Archaeological Science 26, 571–585.

Asfaw, B., Beyene, J., Suwa, G., Walter, R.C., White, T.D., WoldeGabriel, G., Yemane, T., 1992. The earliest Acheulian from Konso-Gardula. Nature 360, 732–735.

Ballesio, R., 1963. Monographie d'un *Machairodus* du Gisement villafranchien de Senèze: *Homotherium crenatidens* Fabrini. Travaux du Laboratoire de Géologie de la Faculté des Sciences de Lyon 9, 1–129.

Barry, J.C., 1987. Large carnivores (Canidae, Hyaenidae, Felidae) from Laetoli. In: Leakey, M., Harris, J. (Eds.), Laetoli: A Pliocene Site in Northern Tanzania. Clarendon Press, Oxford, pp. 235–258.

Beaumont, G.D., 1975. Recherches sur les félidés (Mammifères, Carnivores) du Pliocène inférieur des sables à *Dinotherium* des environs D'Eppelsheim (Rheinhessen). Archives des Sciences, Genève 28, 369–405.

Behrensmeyer, A.K., Todd, N.E., Potts, R., McBrinn, G.E., 1997. Late Pliocene faunal turnover in the Turkana Basin of Kenya and Ethiopia. Science 278, 1589–1594.

Berta, A., 1981. The Plio-Pleistocene hyaena *Chasmaporthetes ossifragus* from Florida. Journal of Vertebrate Paleontology 1, 341–356.

Binford, L.R., 1981. Bones: Ancient Men and Modern Myths. Academic Press, New York.

Binford, L.R., 1986. Reply to H.T. Bunn and E.M. Kroll's "Systematic Butchery by Plio-Pleistocene Hominids at Olduvai Gorge, Tanzania". Current Anthropology 27, 444–446.

Binford, L.R., Mills, M.G.L., Stone, N.M., 1988. Hyena scavenging behavior and its implications for the interpretation of faunal assemblages from FLK

22 (the Zinj floor) at Olduvai Gorge. Journal of Anthropology and Archeology 7, 99–135.

Blumenschine, R.J., 1986a. Carcass consumption sequences and the archaeological distinction of scavenging and hunting. Journal of Human Evolution 15, 639–660.

Blumenschine, R.J., 1986b. Early hominid scavenging opportunities; implications of carcass availability in the Serengeti and Ngororongoro ecosystems. British Archaeological Reports. International Series, Oxford.

Blumenschine, R.J., 1987. Characteristics of an early hominid scavenging niche. Current Anthropology 28, 383–407.

Blumenschine, R.J., 1988. An experimental model of the timing of hominid and carnivore influence on archaeological bone assemblages. Journal of Archeological Research 15, 483–502.

Blumenschine, R.J., 1989. A landscape taphonomic model of the scale of prehistoric scavenging opportunities. Journal of Human Evolution 18, 345–372.

Blumenschine, R.J., 1991. Hominid carnivory and foraging strategies and the socio-economic function of early archaeological sites. Philosophical Transactions of the Royal Society of London 334, 211–221.

Blumenschine, R.J., Cavallo, J.A., Capaldo, S.D., 1994. Competition for carcasses and early hominid behavioral ecology: a case study and conceptual framework. Journal of Human Evolution 27, 197–213.

Brain, C.K., 1969. The probable role of leopards as predators of the Swartkrans Australopithecines. South African Archeology Bulletin 24, 52–55.

Brain, C.K., 1981. The Hunters or the Hunted? An Introduction to African Cave Taphonomy. University of Chicago Press, Chicago.

Brantingham, P.J., 1998. Hominid-carnivore coevolution and invasion of the predatory guild. Journal of Anthropology and Archeology 17, 327–353.

Bunn, H.T., 1986. Patterns of skeletal representation and hominid subsistence activities at Olduvai Gorge, Tanzania, and Koobi Fora, Kenya. Journal of Human Evolution 15, 673–690.

Bunn, H.T., 1991. A taphonomic perspective on the archaeology of human origins. Annual Review of Anthropology 20, 433–467.

Bunn, H.T., Ezzo, J.A., 1993. Hunting and scavenging by Plio-Pleistocene hominids: nutritional constraints, archaeological patterns, and behavioral implications. Journal of Archaeological Science 20, 365–398.

Bunn, H.T., Kroll, E.M., 1986. Systematic butchery by Plio/Pleistocene hominids at Olduvai Gorge, Tanzania. Current Anthropology 27, 431–452.

Carbone, C., Mace, G.M., Roberts, S.C., Macdonald, D.W., 1999. Energetic constraints on the diet of terrestrial carnivores. Nature 402, 286–288.

Caro, T.M., 1994. Cheetahs of the Serengeti Plains: Group Living in an Asocial Species. University of Chicago Press, Chicago.

Caro, T.M., Stoner, C.J., 2003. The potential for interspecific competition among African carnivores. Biological Conservation 110, 67–75.

Cavallo, J.A., Blumenschine, R.J., 1989. Tree-stored leopard kills: expanding the hominid scavenging niche. Journal of Human Evolution 18, 393–400.

Cooper, S.M., 1991. Optimal hunting group size: the need for lions to defend their kills against loss to spotted hyaenas. African Journal of Ecology 29, 130–136.

Cordain, L., Eaton, S.B., Miller, J.B., Mann, N., Hill, K., 2002. The paradoxical nature of hunter-gatherer diets: meat-based, yet non-atherogenic. European Journal of Clinical Nutrition 56 (Supplement 1), S42–S52.

Creel, S., 2001. Four factors modifying the effect of competition on carnivore population dynamics as illustrated by African wild dogs. Conservation Biology 15, 271–274.

Creel, S., Creel, N.M., 1995. Communal hunting and pack size in African wild dogs, Lycaon pictus. Animal Behavior 50, 1325–1339.

Creel, S., Creel, N.M., 1998. Six ecological factors that may limit African wild dogs. Animal Conservation 1, 1–9.

Dart, R.A., 1949. The predatory implemental technique of Australopithecus. American Journal of Physical Anthropology 7, 1–16.

Dart, R.A., 1956. The myth of the bone-accumulating hyena. American Anthropologist 58, 40–62.

Dayan, T., Simberloff, D., 1996. Patterns of size separation in carnivore communities. In: Gittleman, J.L. (Ed.), Carnivore Behavior, Ecology, and Evolution, Volume 2. Cornell University Press, Ithaca, pp. 243–266.

de Heinzelin, J., Clark, J.D., White, T.D., Hart, W.K., Renne, P., WoldeGabriel, G., Beyene, Y., Vrba, E.S., 1999. Environment and behavior of 2.5-million-year-old Bouri hominids. Science 284, 625–629.

Ditchfield, P., Hicks, J., Plummer, T.W., Bishop, L.C., Potts, R., 1999. Current research on the Late Pliocene and Pleistocene deposits north of Homa Mountain southwestern Kenya. Journal of Human Evolution 36, 123–150.

Domínguez-Rodrigo, M., 1999. Flesh availability and bone modifications in carcasses consumed by lions: palaeoecological relevance in hominid foraging patterns. Palaeogeography, Palaeoclimatology, Palaeoecology 149, 373–388.

Domínguez-Rodrigo, M., 2001. A study of carnivore competition in riparian and open habitats of modern savannas and its implications for hominid behavioral modelling. Journal of Human Evolution 40, 77–98.

Domínguez-Rodrigo, M., Pickering, T.R., 2003. Early hominid hunting and scavenging: a zooarchaeological review. Evolutionary Anthropology 12, 275–282.

Eaton, R.L., 1979. Interference competition among carnivores: a model for the evolution of social behavior. Carnivore 2, 9–16.

Estes, R.D., 1991. The Behavior Guide to African Mammals, Including Hoofed Mammals, Carnivores, Primates. University of California Press, Berkeley.

Ewer, R.F., 1955. The fossil carnivores of the Transvaal caves: Machairodontinae. Proceedings of the Zoological Society, London 125, 587–615.

Ewer, R.F., 1965. Carnivora, preliminary notes. In: Leakey, L.S.B. (Ed.), Olduvai Gorge 1951–1961, Volume 1, A Preliminary Report on the Geology and Fauna. Cambridge University Press, Cambridge, pp. 19–23.

Fanshawe, J.H., FitzGibbon, C.D., 1993. Factors influencing the hunting success of an African wild dog pack. Animal Behavior 45, 479–490.

Farlow, J.O., Pianka, E.R., 2002. Body size overlap, habitat partitioning and living space requirements of terrestrial vertebrate predators: implications for the palocecology of large theropod dinosaurs. History of Biology 16, 21–40.

Feibel, C.S., Brown, F.H., McDougall, I., 1989. Stratigraphic context of fossil hominids from the Omo Group deposits: northern Turkana basin, Kenya and Ethiopia. American Journal of Physical Anthropology 78, 595–622.

Foote, M., 2000. Origination and extinction components of taxonomic diversity: general problems. Paleobiology 26 (Supplement), 74–102.

Galiano, H., Frailey, D., 1977. Chasmaporthetes kani, a new species from China, with remarks on phylogenetic relationships of genera within the Hyaenidae (Mammalia, Carnivora). American Museum Novitates 2632, 1–16.

Gilbert, W.H., 2003. The Daka Member of the Bouri Formation, Middle Awash Valley, Ethiopia: Fauna, Stratigraphy, and Environment. Ph.D. thesis, University of California, Berkeley, Berkeley.

Gittleman, J.L., 1985. Carnivore body size: ecological and taxonomic correlates. Oecologia 67, 540–554.

Gorman, M.L., Mills, M.G.L., Raath, J.P., Speakman, J.R., 1998. High hunting costs make African wild dogs vulnerable to kleptoparasitism by hyaenas. Nature 391, 479–481.

Harris, J., Brown, F., Leakey, M., 1988. Stratigraphy and Palaeontology of Pliocene and Pleistocene localities west of Lake Turkana, Kenya. Contributions in Science, Natural History Museum of Los Angeles 399, 1–128.

Hendey, Q., 1980. Agriotherium (Mammalia, Ursidae) from Langebaanweg, South Africa, and relationships of the genus. Annals of South African Museum 81, 1–109.

Hofer, H., Mills, M.G.L., 1998. Worldwide Distribution of Hyaenas. In: Mills, M.G.L., Hofer, H. (Eds.), Hyaenas. Status Survey and Conservation Action Plan. IUCN/SSC Hyaena Specialist Group. IUCN. Gland, Switzerland and Cambridge, UK, pp. 39–63.

Isaac, G.L., 1971. The diet of early man: aspects of archaeological evidence from Lower and Middle Pleistocene sites in East Africa. World Archaeology 2, 279–298.

Isaac, G.L., 1978. The food-sharing behavior of protohuman hominids. Scientific American 238, 90–108.

King, C.M., 1990. The Natural History of Weasels and Stoats. Cornell University Press, Ithaca.

Kingdon, J., 1997. The Kingdon Field Guide to African Mammals. Academic Press, San Diego.

Kretzoi, M., 1938. Die raubtiere von Gombaszög nebst einer Übersicht der Gesamtfauna. Annales Museum Nationale Hungaricam 31, 89–157.

Kruuk, H., 1975. Functional aspects of social hunting by carnivores. In: Baerends, G., Beer, C., Manning, A. (Eds.), Function and Evolution in Behaviour. Clarendon Press, Oxford, pp. 119–141.

Kruuk, H., 1976. Feeding and social behavior of the striped hyaena (Hyaena vulgaris Desmarest). East African Wildlife Journal 14, 91–111.

Kruuk, H., Turner, M., 1967. Comparative notes on predation by lion, leopard, cheetah and wild dog in the Serengeti area, East Africa. Mammalia 31, 1–27.

Kurtén, B., Anderson, E., 1980. Pleistocene Mammals of North America. Columbia University Press, New York.

Kurtén, B., Werdelin, L., 1988. A review of the genus Chasmaporthetes Hay, 1921 (Carnivora, Hyaenidae). Journal of Vertebrate Paleontology 8, 46–66.

Lambert, W.D., Holling, C.S., 1998. Causes of ecosystem transformation at the end of the Pleistocene: evidence from mammal body-mass distributions. Ecosystems 1, 157–175.

Lamprecht, J., 1978. The relationship between food competition and foraging group size in some larger carnivores: a hypothesis. Zeitschrift fur Tierpsychologie 46, 337–343.

Laurenson, M.K., 1995. Implications of high offspring mortality for cheetah population dynamics. In: Sinclair, A.R.E., Arcese, P. (Eds.), Serengeti II: Dynamics, Management, and Conservation of an Ecosystem. University of Chicago Press, Chicago, pp. 385–399.

Leakey, M.G., 1976. Carnivora of the East Rudolf Succession. In: Coppens, Y., Howell, F.C., Isaac, G.L., Leakey, R.E.F. (Eds.), Earliest Man and Environments in the Lake Rudolf Basin. University of Chicago Press, Chicago, pp. 302–313.

Lewis, M.E., 1995a. Functional morphology of the sabertoothed felid Megantereon from Kromdraai, South Africa. Abstract. Journal of Vertebrate Paleontology 15, 40A.

Lewis, M.E., 1995b. Plio-Pleistocene Carnivoran Guilds: Implications for Hominid Paleoecology. Ph.D. thesis, State University of New York, Stony Brook.

Lewis, M.E., 1996. Ecomorphology of the African sabertoothed felid Homotherium. Abstract. Journal of Vertebrate Paleontology 16 (Supplement to No. 3), 48A.

Lewis, M.E., 1997. Carnivoran paleoguilds of Africa: implications for hominid food procurement strategies. Journal of Human Evolution 32, 257–288.

Lewis, M.E., 2001. Implications of interspecific variation in the postcranial skeleton of Homotherium (Felidae, Machairodontinae). Journal of Vertebrate Paleontology 21, 73A.

Lewis, M.E., Berger, L.R., 1998. A new species of wolf-like canid from Gladysvale, South Africa. Abstract. Journal of Vertebrate Paleontology 18, 59A.

Lewis, M.E., Werdelin, L., 1997. Evolution and ecology of the genus Crocuta. Abstract. Journal of Vertebrate Paleontology 17.

Lewis, M.E., Werdelin, L., 2000. The evolution of spotted hyenas (Crocuta). Hyaena Specialist Group Newsletter 7, 34–36.

Linnell, J.D.C., Strand, O., 2000. Interference interactions, co-existence and conservation of mammalian carnivores. Diversity and Distributions 6, 169–176.

MacDonald, D.W., 1978. Observations on the behaviour and ecology of the striped hyaena, Hyaena hyaena, in Israel. Israel Journal of Zoology 27, 189–198.

Marean, C.W., 1989. Sabertooth cats and their relevance for early hominid diet and evolution. Journal of Human Evolution 18, 559–582.

Marean, C.W., Ehrhardt, C.L., 1995. Paleoanthropological and paleoecological implications of the taphonomy of a sabertooth's den. Journal of Human Evolution 29, 515–547.

Martin, L.D., 1980. Functional morphology and the evolution of cats. Transactions of the Nebraska Academy of Sciences 8, 141–154.

Martínez-Navarro, B., Palmqvist, P., 1996. Presence of the African saber-toothed felid Megantereon whitei (Broom, 1937) (Mammalia, Carnivora, Machairodontinae) in Apollonia-1 (Mygdonia Basin, Macedonia, Greece). Journal of Archaeological Science 23, 869–872.

Martínez Navarro, B., Rook, L., 2003. Gradual evolution of the African hunting dog lineage: systematic implications. Comptes Rendus Palevol 2, 695–702.

McGrew, W., 2004. The Culture Chimpanzee: Reflections on Cultural Primatology. Cambridge University Press, Cambridge.

Mills, M.G.L., 1978. The comparative socio-ecology of the Hyaenidae. Carnivore 1, 1–7.

Mills, M.G.L., 1982. Factors affecting group size and territory size of the brown hyaena, Hyaena brunnea in the southern Kalahari. Journal of Zoology London 198, 39–51.

Mills, M.G.L., Gorman, M.L., 1997. Factors affecting the density and distribution of wild dogs in the Kruger National Park. Conservation Biology 11, 1397–1406.

Novikov, G.A., 1962. Carnivorous Mammals of the Fauna of the USSR. Jerusalem: Israel Program for Scientific Translation.

Nowak, R.M., 1991. Walker's Mammals of the World, Fifth Edition, Volumes 1 and 2. The Johns Hopkins University Press, Baltimore.

O'Connell, J.F., Hawkes, K., Lupo, K.D., Blurton Jones, N.G., 2002. Male strategies and Plio-Pleistocene archaeology. Journal of Human Evolution 43, 831–872.

Owens, M., Owens, D., 1978. Feeding ecology and its influence on social organization in brown hyenas (Hyaena brunnea, Thunberg) of the Central Kalahari desert. East African Wildlife Journal 16, 113–135.

Packer, C., Ruttan, L., 1988. The evolution of cooperative hunting. American Naturalist 132, 159–198.

Palmqvist, P., 2002. On the presence of Megantereon whitei at the South Turkwel Hominid Site, northern Kenya. Journal of Paleontology 76, 928–930.

Palomares, F., Caro, T.M., 1999. Interspecific killing among mammalian carnivores. American Naturalist 153, 492–508.

Peters, C.R., Blumenschine, R.J., Hay, R.L., Livingstone, D.A., Marean, C.W., Harrison, T., Armour-Chelu, M., Andrews, P., Bernor, R.L., Bonnefille, R., & Werdelin, L. In press. Paleoecology of the Serengeti-Mara ecosystem. In: Sinclair, A.R.E., Packer, C., Mduma, S.A.R. & Fryxell, J.M. (Eds.)., Serengeti III: Human Impacts on Ecosystem Dynamics. University of Chicago Press, Chicago.

Petter, G., 1967. Petits carnivores villafranchiens du Bed I d'Oldoway (Tanzanie). In: Problèmes Actuels de Paléontologie (Évolution des Vertébrés). Colloques Internationaux du CNRS, Paris, pp. 529–538.

Petter, G., 1973. Carnivores Pleistocènes du Ravin d'Olduvai (Tanzanie). In: Leakey, L., Savage, R., Coryndon, S. (Eds.), Fossil Vertebrates of Africa. Academic Press, London, pp. 43–100.

Petter, G., 1987. Small carnivores (Viverridae, Mustelidae, Canidae) from Laetoli. In: Leakey, M., Harris, J. (Eds.), Laetoli: A Pliocene Site in Northern Tanzania. Clarendon Press, Oxford, pp. 194–234.

Pienaar, U.d.V., 1969. Predator-prey relationships amongst the larger mammals of the Kruger National Park. Koedoe 12, 108 176.

Pohle, H., 1928. Die Raubtiere von Oldoway. Wissenschaftliche Ergebnisse der Oldoway-Expedition 1913 (N.F.) 3, 45–54.

Potts, R., 1984. Home bases and early hominids. American Scientist 72, 338–347.

Potts, R. 1998. Environmental hypotheses of hominin evolution. Yearbook of physical Anthropology. 41, 93–136.

Potts, R., 1988a. Early Hominid Activities at Olduvai. Aldine de Gruyter Press, New York.

Potts, R., 1988b. On an early hominid scavenging niche. Current Anthropology 29, 153–155.

Potts, R., Deino, A., 1995. Mid-Pleistocene change in large mammal faunas of eastern Africa. Quaternary Research 43, 106–113.

Potts, R., Shipman, P., Ingall, E., 1988. Taphonomy, paleoecology and hominids of Lainyamok, Kenya. Journal of Human Evolution 17, 597–614.

Rautenbach, I., Nel, J., 1978. Coexistence in Transvaal Carnivora. Bulletin of Carnegie Museum of Natural History 6, 138–145.

Rawn-Schatzinger, V., 1992. The scimitar cat Homotherium serum Cope: osteology, functional morphology, and predatory behavior. Illinois State Museum Reports of Investigations 47, 1–80.

Rodriguez, J., Alberdi, M.T., Azanza, B., Prado, J.L., 2004. Body size structure in north-western Mediterranean Plio-Pleistocene mammalian faunas. Global Ecology and Biogeography 13, 163–176.

Rosenzweig, M.L., 1966. Community structure in sympatric carnivores. Journal of Mammalogy 47, 602–612.

Schaller, G.B., 1972. The Serengeti Lion. University of Chicago Press, Chicago.

Schaller, G.B., 1973. Golden Shadows, Flying Hooves. University of Chicago Press, Chicago.

Schaller, G.B., Lowther, G., 1969. The relevance of carnivore behavior to the study of early hominids. Southwest Journal of Anthropology 25, 307–341.

Semaw, S., Rogers, M.J., Quade, J., Renne, P.R., Butler, R.F., Domínguez-Rodrigo, M., Stout, D., Hart, W.S., Pickering, T.R., Simpson, S.W., 2003. 2.6-Million-year-old stone tools and associated bones from OGS-6 and OGS-7, Gona, Afar, Ethiopia. Journal of Human Evolution 45, 169–177.

Sept, J., 1992. Archaeological evidence and ecological perspectives for reconstructing early hominid subsistence behavior. In: Schiffer, M.B. (Ed.), Archaeological Theory and Method, Vol. 4 University of Arizona Press, Tucson, pp. 1–56.

Shipman, P., 1986a. Scavenging or hunting in hominids: theoretical framework and tests. American Anthropologist 88, 27–43.

Shipman, P., 1986b. Studies of hominid–faunal interactions at Olduvai Gorge. Journal of Human Evolution 15, 691–706.

Shipman, P., Walker, A., 1989. The costs of becoming a predator. Journal of Human Evolution 18, 272–392.

Sillero-Zubiri, C., Macdonald, D.W., Group, I.S.C.S., 1997. The Ethiopian Wolf – Status Survey and Conservation Action Plan. IUCN, Gland, Switzerland.

Stander, P.E., 1991. Aspects of the Ecology and Scientific Management of Large Carnivores in Sub-Saharan Africa. Ph.D. dissertation, Cambridge University, Cambridge.

Stander, P.E., Albon, S.D., 1993. Hunting success of lions in a semi-arid environment. In: Dunstone, N., Gorman, M.L. (Eds.), Mammals as Predators: The Proceedings of a Symposium Held by The Zoological Society of London and The Mammal Society, London, 22nd and 23rd November 1991. Clarendon Press, Oxford, pp. 127–144.

Stiner, M.C., Howell, F.C., Martínez Navarro, B., Tchernov, E., Bar-Yosef, O., 2001. Outside Africa: middle Pleistocene *Lycaon* from Hayonim Cave, Israel. Bolletino della Societá Paleontologica Italiana 20, 293–302.

Taylor, M., 1993. Locomotor adaptations by carnivores. In: Gittleman, J. (Ed.), Carnivore Behavior, Ecology, and Evolution. Cornell University Press, New York, pp. 382–409.

Tedford, R.H., Taylor, B.E., Wang, X., 1995. Phylogeny of the Caninae (Carnivora: Canidae): the living taxa. American Museum Novitates 3146, 1–37.

Treves, A., Naughton-Treves, L., 1999. Risk and opportunity for humans coexisting with large carnivores. Journal of Human Evolution 36, 275–282.

Turner, A., 1983. Biogeography of Miocene-Recent larger carnivores in Africa. In: Vogel, J. (Ed.), Late Cainozoic Paleoclimates of the Southern Hemisphere. Balkema, Rotterdam, pp. 499–506.

Turner, A., 1985. Extinction, speciation, and dispersal in African larger carnivores, from the late Miocene to recent. South African Journal of Science 81, 256–257.

Turner, A., 1986. Some features of African larger carnivore historical biogeography. Palaeoecology of Africa 17, 237–244.

Turner, A., 1988. Relative scavenging opportunities for east and south African Plio-Pleistocene hominids. Journal of Archaeological Science 15, 327–341.

Turner, A., 1990. The evolution of the guild of larger terrestrial carnivores during the Plio-Pleistocene in Africa. Géobios 23, 349–368.

Turner, A., 1992. Large carnivores and earliest European hominids: changing determinants of resource availability during the lower and middle Pleistocene. Journal of Human Evolution 22, 109–126.

Turner, A., 1998. Climate and evolution: implications of some extinction patterns in African and European machairodontine cats of the Plio-Pleistocene. Estudios Geológicos 54, 209–230.

Turner, A., 1999. Evolution of the African Plio-Pleistocene mammalian fauna: correlation and causation. In: Bromage, T.G., Schrenk, F. (Eds.), African Biogeography, Climate Change and Human Evolution. Oxford University Press, Oxford, pp. 76–87.

Turner, A., Antón, M., 1995. The giant hyaena, *Pachycrocuta brevirostris* (Mammalia, Carnivora, Hyaenidae). Géobios 29, 455–468.

Turner, A., Antón, M., 1997. The Big Cats and their Fossil Relatives: An Illustrated Guide to their Evolution and Natural History. Columbia University, Press New York.

Van Orsdol, K., 1982. Ranges and food habits of lions in Rwenzori National Park, Uganda. Symposium of Zoological Society of London 49, 325–340.

Van Orsdol, K., Hanby, J.P., Bygott, J.D., 1985. Ecological correlates of lion social organization (*Panthera leo*). Journal of Zoology London 206, 97–112.

Van Valkenburgh, B., 1985. Locomotor diversity within past and present guilds of large predatory mammals. Paleobiology 11, 406–428.

Van Valkenburgh, B., 1987. Skeletal indicators of locomotor behavior in living and extinct carnivores. Journal of Vertebrate Paleontology 7, 162–182.

Van Valkenburgh, B., 1988. Incidence of tooth breakage among large, predatory mammals. American Naturalist 131, 291–300.

Van Valkenburgh, B., 2001. The dog-eat-dog world of carnivores: a review of past and present carnivore community dynamics. In: Stanford, C.B., Bunn, H.T. (Eds.), Meat-Eating and Human Evolution. Oxford University Press, Oxford, pp. 101–121.

Van Valkenburgh, B., Hertel, F., 1993. Tough times at La Brea: tooth breakage in large carnivores of the late Pleistocene. Science 261, 456–459.

Van Valkenburgh, B., Hertel, F., 1998. The decline of North American predators during the Late Pleistocene. In: Saunders, J.J., Styles, B.W., Baryshnikov, G.F. (Eds.), Quaternary Paleozoology in the Northern Hemisphere. Illinois State Museum Scientific Papers, Springfield, Illinois, pp. 357–374.

Van Valkenburgh, B., Sacco, T., Wang, X., 2003. Pack hunting in the Miocene Borophagine dogs: evidence from craniodental morphology and body size. Bulletin of the American Museum of Natural History 212, 379–397.

Walker, A., 1984. Extinction in hominid evolution. In: Nitecki, M. (Ed.), Extinctions. University of Chicago Press, Chicago, pp. 119–152.

Wang, X., Tedford, R.H., Van Valkenburgh, B., Wayne, R.K., 2004. Ancestry: evolutionary history, molecular systematics, and evolutionary ecology of Canidae. In: Macdonald, D.W., Sillero-Zubiri, C. (Eds.), The Biology and Conservation of Wild Canids. Oxford University Press, Oxford, pp. 39–54.

Wayne, R.K., Gottelli, D., 1997. Systematics, population genetics and genetic management of the Ethiopian wolf. In: Sillero-Zubiri, C., MacDonald, D.W., Group, I.S.C.S. (Eds.), The Ethiopian Wolf– Status Survey and Conservation

Action Plan. IUCN, Gland, Switzerland, pp. 43–50.

Werdelin, L., 1999. *Pachycrocuta* (hyaenids) from the Pliocene of east Africa. Paläontologisches Zeitschrift 73, 157–165.

Werdelin, L., 2003a. Carnivores from the Kanapoi hominid site, Turkana Basin, northern Kenya. Contributions in Science 498, 115–132.

Werdelin, L., 2003b. Mio-Pliocene Carnivora from Lothagam, Kenya. In: Leakey, M.G., Harris, J.M. (Eds.), Lothagam: The Dawn of Humanity in Eastern Africa. Columbia University Press, New York, pp. 261–678.

Werdelin, L., Barthelme, J., 1997. Brown hyena (*Parahyaena brunnea*) from the Pleistocene of Kenya. Journal of Vertebrate Paleontology 17, 758–761.

Werdelin, L., Lewis, M.E., 2000. Carnivora from the South Turkwel hominid site, northern Kenya. Journal of Paleontology 74, 1173–1180.

Werdelin, L., Lewis, M.E., 2001a. Diversity and turnover in Eastern African Plio-Pleistocene Carnivora. Abstract. Journal of Vertebrate Paleontology 21, 112A.

Werdelin, L., Lewis, M.E., 2001b. A revision of the genus *Dinofelis* (Mammalia, Felidae). Zoological Journal of the Linnean Society 132, 147–258.

Werdelin, L., Lewis, M.E., 2002. Species identification in *Megantereon*: a reply to Palmqvist. Journal of Paleontology 76, 931–933.

Werdelin, L., Lewis, M.E., 2005. Plio-Pleistocene Carnivora of eastern Africa: species richness and turnover patterns. Zoological Journal of the Linnean Society 144, 121–144.

Werdelin, L., Solounias, N., 1996. The evolutionary history of hyenas in Europe and western Asia during the Miocene. In: Mittmann, H.-W. (Ed.), The Evolution of Western Eurasian Neogene Mammal Faunas. Columbia University Press, New York, pp. 290–306.

Werdelin, L., Turner, A., 1996. The fossil and living Hyaenidae of Africa: present status. In: Stewart, K.M., Seymour, K.L. (Eds.), Palaeoecology and Palaeoenvironments of Late Cenozoic Mammals; Tributes to the Career of C.S. Churcher. Toronto University Press, Toronto, pp. 637–659.

Werdelin, L., Turner, A., Soulounias, N., 1994. Studies of fossil hyaenids: the genera *Hyaenictis* Gaudry and *Chasmaporthetes* Hay, with a reconsideration of the Hyaenidae of Langebaanweg, South Africa. Zoological Journal of the Linnean Society 111, 197–217.

Whittaker, R.J., Willis, K.J., Field, R., 2001. Scale and species richness: toward a general, hierarchical theory of species diversity. Journal of Biogeography 28, 453–470.

WoldeGabriel, G., White, T.D., Suwa, G., Renne, P., de Heinzelin, J., Hart, W.K., Heiken, G., 1994. Ecological and temporal placement of early Pliocene hominids at Aramis, Ethiopia. Nature 371, 330–333.

Wood, B.A., 1991. Koobi Fora Research Project, Volume 4, Hominid Cranial Remains. Clarendon Press, Oxford.

Wood, B.A., Richmond, B.G., 2000. Human evolution: taxonomy and paleobiology. Journal of Anatomy 196, 19–60.

Zrzavý, J., Řičánková, V., 2004. Phylogeny of recent Canidae (Mammalia, Carnivora): relative reliability and utility of morphological and molecular data sets. Zoologica Scripta 33, 311–333.

# 5. Stratigraphic variation in Suidae from the Shungura Formation and some coeval deposits

H.B.S. COOKE

*2133-154th Street*
*White Rock, British Columbia, Canada V4A 4S5*
*cookecentral@shaw.ca*

**Keywords**:  Hypsodonty, diet, Koobi Fora, Hadar, Olduvai

## Abstract

Metrical data for suid third molars from the Shungura Formation, the Koobi Fora Formation, Hadar, and the Olduvai Beds are analyzed for each of the main genera and are set out in diagrammatic form against an appropriate time scale to facilitate visual comparison of samples from different levels. This helps to clarify the extent of variation in the same species from different sites. The analyses suggest that species transitions occurred in several lineages at about 2.7–2.8 Ma and, less clearly, around 1.6–1.8 Ma. Increasing hypsodonty is compatible with increasing amounts of harsh diet. Whether or not these changes are triggered by some external factors needs further investigation.

## Introduction

Although it is unlikely that early hominines had a special predilection for pork, it is a fact that suid remains form a common component of the fossil faunas that occur in hominid-bearing strata. The African Pliocene and Pleistocene suids belong to a limited number of lineages that were apparently diversifying fairly rapidly and are thus potentially of value in correlation and faunal dating. A major source of suid material has been the extensive sequences of strata in the basin of the Omo River in southern Ethiopia, including the western part of Turkana, and in the Afar region of central Ethiopia as well as in the classic deposits of the Olduvai Gorge in Tanzania.

Meticulous geological work has established the stratigraphy in these areas and the occurrence of numerous tuff horizons has led generally to the labeling of each stratigraphic unit using the name of the tuff occurring at its base. Many of the tuffs have yielded good radiometric dates and trace-element "fingerprinting" has permitted firm links between some of the sequences. Figure 1 outlines the correlation between the major sites from which the suids to be discussed here have been derived. The Shungura Formation provides a useful "standard" as the fossil material collected was usually tied to a particular subunit at the time of collection. The Koobi Fora Formation in Turkana has, on the whole, yielded many much better specimens but the early collections were recorded by "areas" and preceded the establishment of the

*R. Bobe, Z. Alemseged, and A.K. Behrensmeyer (eds.) Hominin Environments in the East African*
*Pliocene: An Assessment of the Faunal Evidence, 107–127.*

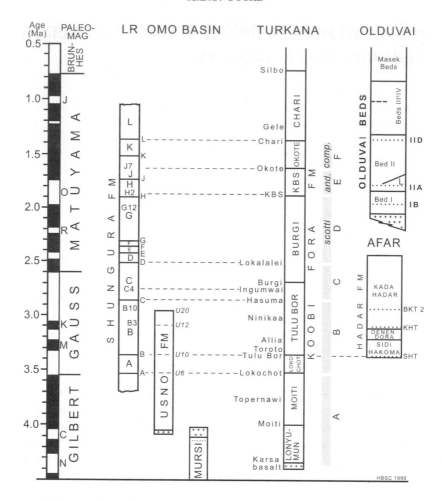

Figure 1. Summary of the stratigraphic sequences in the Lower Omo Basin, Turkana, Hadar, and Olduvai Gorge as correlated by means of radiometric dates and geochemical "fingerprinting". Dashed lines show geochemical links. For the Koobi Fora Formation, the ranges for the "collection units" are indicated: A, B, C, "*Notochoerus scotti*", "*Metridiochoerus andrewsi*", and "*Metridiochoerus compactus*".

stratigraphy, so there is sometimes uncertainty about the exact horizons for particular specimens. Faced with discrepancies between correlations based on tuffs and correlations based on faunal evidence, Harris (1978) developed the concept of informal "collection units" and made ingenious use of them as "assemblage zones", labeled "A" to "F", to suggest important amendments to the tuff correlations made before the use of trace-element "fingerprinting". Subsequently Harris (1983) proposed modifications of the zones set up by Maglio (1972) and Harris' amended zones are shown in Figure 1 – "A", "B", "C", "*Notochoerus scotti*" (= "D"), "*Metridiochoerus andrewsi*" (= "E"), and "*Metridiochoerus compactus*" (= "F"). The Hadar Formation in central

Ethiopia has furnished some radiometric dates and the SHT tuff has been correlated geochemically with the Tulu Bor tuff in Turkana and tuff B of the Shungura Formation. The Olduvai Beds are correlated by radiometric dates.

## Taxonomy

The fossil suids in Africa have been described under many different names and various attempts have been made to review and revise the nomenclature. Suids from the Shungura Formation were first described by Arambourg (1943) and almost simultaneously by Leakey (1943). East African material from many sites was described by Leakey in a monograph published in 1958a. Cooke and Maglio (1972) attempted to correlate the important stratigraphic sequences in East Africa and suggested a provisional grouping of the most important taxa of African fossil Suidae into five generic associations – *Nyanzachoerus*, *Notochoerus*, *Mesochoerus*, *Phacochoerus*, and "Aberrant Phacochoerines" (including *Metridiochoerus*). The estimated time ranges for the various species were shown graphically.

White and Harris (1977) published an important review of suid evolution and correlation of African hominid locations and presented a suggested phylogeny. This was essentially an advance summary of a monographic account that was in the press but did not appear until two years later (Harris and White, 1979). Cooke (1978a) responded with the presentation of an alternate taxonomy and phylogeny which differ in detail but are substantially similar. Basically the differences result from different taxonomic philosophies, that of Harris and White tending to regard all members of a continuous lineage as a single species while Cooke recognizes more branching and names the species primarily on morphological grounds. A comparison of the two schemes was given subsequently by Cooke (1985) when it was pointed out that the taxonomic differences have little or no impact on the correlations, which are based on the occurrence of similar morphological entities and are little affected by the labels attached to them. A critical review and revision of the African Suidae (Cooke and Wilkinson, 1978) was in press at the same time as that of Harris and White (1979) but lacks a good synonymy whereas Harris and White provide an excellent and comprehensive synonymy that is still very useful.

## Parameters

Although genera and species need to be defined on the overall morphology of the animal, especially of the skull, crania are rather rare and often incomplete so it has become necessary to rely upon the relatively abundant dental material, especially the third molars. The oft-neglected premolars also have special value in some instances. The discussions that follow are limited to the third molars. Length (L) is the basal length measured along the cingulum at the base of the crown. Breadth (B) is measured at the base of the crown, transversely to the axis of the tooth, across the cingulum at its widest point; this is usually just above the anterior roots. Crown height (H) is measured on the least worn pillar on the trigon/trigonid or on the anterior major pillars of the talon/talonid, and is measured parallel to the pillar axis from the enamel line to the apex of the pillar. If the pillar is worn, the measurement is recorded with a + sign after the figure.

The measurable crown height is affected by wear and is usually coupled with imprecise terms such as "moderately early wear" which are subjective and vary in usage by different workers. Accordingly, a recent analytical study by Kullmer (1997, 1999) is particularly welcome in establishing objective definitions for 11 "wear stages" (WS) based on the successive fusion of the cusps (enamel islands) that build the basic structure of the trigon/trigonid in all suid molars. For example, "WS3 – dentine visible in one of the second lateral pillars" or "WS7 – fusion of one of the second lateral pillars and the first central pillar". These wear stages do not imply equal increments although they could be calibrated for a species by using serial sections as proxies for normal wear surfaces.

From WS4 onwards there is progressive simplification in the degree of folding in the individual enamel islands. Kullmer (1997, 1999) has applied to suid molars a morphometric technique developed by Schmidt-Kittler (1984) for hypsodont herbivore teeth. Essentially the actual surface area of an enamel island is measured, as well as the total length of the enamel perimeter. The latter is converted into the area of a circle having the same perimeter length. The ratio of the area of this circle to the measured area provides a "density parameter", "D". The "D" values can be plotted against length/breadth ratios or against the relative crown height (H/B) to produce regression lines that show distinctive features for different evolutionary lineages. The measuring procedure is too time-consuming for this to be a practical method for assigning particular specimens to a species but Kullmer's analyses are very valuable and accord with the tooth types normally recognized as notochoerine, kolpochoerine, metridiochoerine, phacochoerine, and potamochoerine.

Length/breadth scatter diagrams for third molars have proved useful in showing the range of variation in a sample or in a species but generally the regression lines indicate that these proportions are fairly constant for a species. The relationship between crown height and basal length provides a useful visual impression of hypsodonty but a better measure is the relationship between crown height (H) and the anterior basal breadth (B). This may be defined as the "hypsodonty index" ("HI"), calculated as a percentage (100 H/B). Plots of the hypsodonty index against the basal length of the crown can be illuminating in showing a "plateau" that is close to the limiting HI value. Figure 2A shows the data for the upper third molars from Omo that have been attributed to *Metridiochoerus jacksoni* (Cooke ms) and Figure 2B is a similar plot for upper third molars from Koobi Fora, based on the tables for *M. andrewsi* published by Harris (1983), amplified by the present writer's own measurements. It is apparent that for the Omo material the hypsodonty index is close to a maximum of 150, whereas for the Koobi Fora sample it is closer to 250. As a matter of observation, the heights of unworn first and second lateral pillars are usually almost identical and even a little more wear produces only a very small change in the HI. Accordingly, the largest value for the hypsodonty index in a sample is a very useful measure for the species.

## Data Analyses

Metrical data for samples of third molars assigned to species in the three generic groups *Notochoerus* (+*Nyanzachoerus*), *Kolpochoerus*, and *Metridiochoerus* from a number of sites have been analyzed and the results are summarized in tables in Appendix A to this account. In the diagrams that follow, the data for samples from different stratigraphic horizons are plotted against a timescale to facilitate comparison between sites. A horizontal bar shows the measured range in length, the mean, and one standard deviation on either side of the mean. An abbreviation for the Formation is shown, followed by a number for the sample size. The hypsodonty index is not shown in the diagrams but is given in the tables. The appropriate species names are indicated by stipples. Comments on the diagrams for each generic group are given below.

## *NOTOCHOERUS* (AND *NYANZACHOERUS*) (TABLES 1 AND 2, FIGURES 3 AND 4)

A few specimens attributable to the genus *Nyanzachoerus* (probably *Ny. kanamensis*) are found alongside the early *Notochoerus* from Shungura Member A, the Usno Formation, and the Hadar Formation. The hypsodonty index for the *Nyanzachoerus* material is under 100 for upper M3 and a maximum of 105 for lower M3. Two specimens from Member A of the Shungura Formation have longer crowns and possibly belong to *Ny. jaegeri*.

Two species of *Notochoerus* are generally recognized, *N. euilus* and *N. scotti*, differentiated primarily by the greater hypsodonty of the latter and an increase in the number of pillars in the talon/talonid. The "typical" representatives of *N. euilus* are fairly abundant in Member B of the Shungura Formation and in the contemporary levels of the Usno Formation. *N. euilus* is also plentiful in the lower part of the Hadar Formation. A single

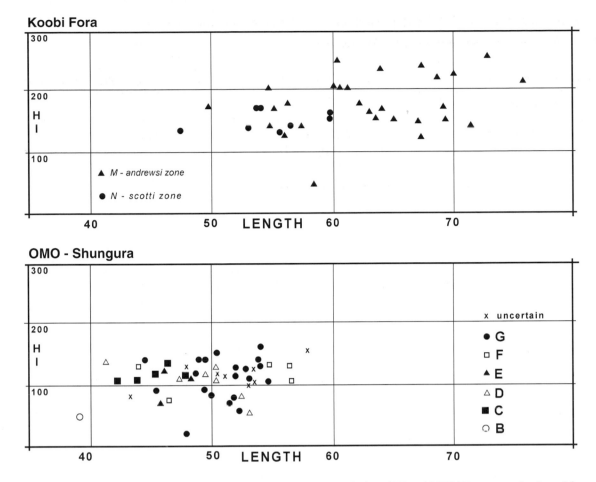

Figure 2. Scatter diagrams showing plots of the hypsodonty index (HI = 100 H/B) versus the basal length of the crowns in upper third molars assigned to *Metridiochoerus*. A is for material from various horizons in the Shungura Formation and shows a "plateau" for the HI at about 170. B is for material from the Koobi Fora Formation and shows a "plateau" at about 260 for the sample from the *Metridiochoerus andrewsi* zone but may indicate a lower value for the small sample from the *Notochoerus scotti* zone.

upper M3 of *Notochoerus* from Zone A of the Koobi Fora Formation has an HI of 98 but in Shungura B and at Hadar the index for *Notochoerus* ranges up to 144 for upper M3 and as high as 171 for lower M3. The material from Member C through lower Member G of the Shungura Formation differs slightly in that the crowns tend to be a little shorter but narrower and relatively higher crowned (HI 151 to 210 for upper and up to 184 for lower M3); these teeth are now identified as *N. clarki* (White and Suwa, 2004). An upper

M3 with the longer crown and the distinctive morphology of *N. scotti* first appears in Zone "C" of the Koobi Fora Formation, although the HI is only 139+. In the Shungura Formation Member C to lower G, molars with *N. scotti* characteristics occur alongside those regarded as *N. clarki*, but with a generally higher HI of 170–260 for upper M3's and 250–260 for lowers. At Koobi Fora *N. scotti* is fairly common in the zone that bears its name, showing an HI of 220 for upper M3's and 295 for lowers. Two teeth from the *Metridiochoerus* zone

Table 1. *Nyanzachoerus and Notochoerus upper M3*

| Formation | Unit or "zone" | N | Length | | | | Breadth | | | | Hypsodonty index |
| | | | Min | Max | Mean | SD | Min | Max | Mean | SD | |
|---|---|---|---|---|---|---|---|---|---|---|---|
| *Nyanzachoerus kanamensis* | | | | | | | | | | | |
| Hadar | Sidi Hakoma | 10 | 48.0 | 55.0 | 51.36 | 2.60 | 28.6 | 34.4 | 31.63 | 1.90 | 70 |
| | Denen Dora | 1 | | | 52.7 | | | | 34.6 | | 42++ |
| Mursi | | 4 | 46.9 | 50.0 | 48.48 | 1.30 | 28.5 | 30.7 | 29.28 | 0.84 | 80+ |
| Usno | | 3 | 43.1 | 50.9 | 46.33 | 3.32 | 27.7 | 28.5 | 28.10 | 0.37 | 85+ |
| Shungura | A | 3 | 47.6 | 51.3 | 49.30 | 1.53 | 29.0 | 31.2 | 30.17 | 0.99 | 70+ |
| *Notochoerus euilus* | | | | | | | | | | | |
| Hadar | Sidi Hakoma | 9 | 64.2 | 85.0 | 74.32 | 6.50 | 26.4 | 35.1 | 31.24 | 2.68 | 144 |
| | Denen Dora | 34 | 63.0 | 81.4 | 71.07 | 4.72 | 26.0 | 35.4 | 30.81 | 2.08 | 126 |
| | Kada Hadar | 1 | | | 76.7 | | | | 31.4 | | 108 |
| Usno | | 25 | 61.5 | 80.0 | 70.52 | 4.93 | 26.2 | 34.5 | 29.92 | 1.92 | 136 |
| Shungura | B | 14 | 72.2 | 85.9 | 78.11 | 3.50 | 25.2 | 34.8 | 30.49 | 1.94 | 142 |
| Koobi Fora | "A" | 1 | | | 68.0 | | | | 30.9 | | 98+ |
| | "B" | 5 | 69.2 | 76.5 | 72.74 | 3.29 | 30.4 | 33.2 | 31.78 | 1.20 | 105+ |
| *Notochoerus clarkii* | | | | | | | | | | | |
| Shungura | C | 3 | 56.0 | 78.8 | 67.40 | 9.31 | 24.0 | 29.2 | 26.90 | 2.16 | 151 |
| | F | 1 | | | 66.7 | | | | 24.2 | | 87+ |
| | Lower G | 10 | 53.8 | 81.5 | 65.80 | 7.35 | 22.0 | 25.9 | 24.86 | 1.99 | 210 |
| *Notochoerus scotti* | | | | | | | | | | | |
| Shungura | C | 4 | 83.6 | 95.0 | 87.98 | 4.40 | 27.4 | 31.5 | 29.15 | 1.80 | 155+ |
| | D | 3 | 95.8 | 109.6 | 102.4 | 5.65 | 29.8 | 33.0 | 31.72 | 1.39 | 195+ |
| | E | 1 | | | 90.5 | | | | 29.1 | | 201 |
| | F | 2 | 91.0 | 99.0 | 95.0 | 4.10 | 26.0 | 29.0 | 27.50 | 1.50 | 235 |
| | Lower G | 6 | 81.0 | 106.5 | 94.65 | 7.48 | 27.6 | 32.8 | 29.25 | 1.69 | 261 |
| Koobi Fora | "C" | 1 | | | 96.0 | | | | 31.0 | | 139+ |
| | "N. scotti" | 9 | 85.0 | 111.4 | 96.68 | 8.38 | 24.0 | 31.6 | 28.40 | 2.47 | 211 |
| | "M. andrewsi" | 2 | 76.0 | 96.8 | 86.40 | 14.14 | 26.6 | 27.5 | 27.05 | 0.64 | 219 |

N = sample size; SD = standard deviation; hypsodonty index = 100 × height/breadth, maximum in sample; Measurements in mm. Length and breadth are measured along and across the cingulum.

have shorter crowns and an HI of 220+ but morphologically seem to belong to *N. scotti*. The highest occurrence of *N. scotti* in the Shungura Formation is a well worn lower M3 from Member H.

The data suggest that at about 2.7 Ma the "typical" *N. euilus* is replaced by the longer and higher crowned *N. scotti* while, at the same time, the "typical" *N euilus* is succeeded by the smaller but more hypsodont variety tentatively labeled *N. clarki*. Both changes favor molars with longer lives in dealing with harsh diets. The simpler *N. clarki* persists to about 2.2 Ma while *N. scotti* ranges up to 1.7–1.8 Ma.

## KOLPOCHOERUS (TABLES 3 AND 4, FIGURES 5 AND 6)

The genus *Kolpochoerus* was formerly better known as *Mesochoerus* and is divided here into four species, *afarensis*, *heseloni*, *olduvaiensis*, and *majus*, primarily on the basis of the development of the talon/talonid. Harris and White consider *heseloni* and *olduvaiensis* to be part of a single lineage and do not recognize the separate status of *olduvaiensis*, which has additional pillars in the talon/talonid. The characteristic species *heseloni* was formerly well known

*Table 2. Nyanzachoerus and Notochoerus lower M3*

| Formation | Unit or "zone" | N | Length | | | | Breadth | | | | Hypsodonty index |
|---|---|---|---|---|---|---|---|---|---|---|---|
| | | | Min | Max | Mean | SD | Min | Max | Mean | SD | |
| Hadar | Sidi Hakoma | 26 | 53.3 | 61.0 | 57.25 | 2.15 | 23.7 | 28.1 | 25.40 | 1.20 | 103 |
| Mursi | | 3 | 55.2 | 61.6 | 58.17 | 2.63 | 20.0 | 24.8 | 23.20 | 2.26 | 80+ |
| Usno | | 3 | 51.6 | 55.6 | 53.67 | 1.64 | 24.0 | 26.1 | 24.87 | 0.90 | 84+ |
| *Notochoerus euilus* | | | | | | | | | | | |
| Hadar | Sidi Hakoma | 27 | 60.3 | 91.0 | 75.69 | 8.39 | 22.0 | 30.7 | 25.70 | 2.77 | 144 |
| | Denen Dora | 37 | 69.8 | 84.0 | 76.92 | 4.13 | 22.6 | 30.2 | 25.91 | 4.58 | 137 |
| | Kada Hadar | 3 | 66.0 | 83.0 | 71.83 | 9.67 | 22.4 | 25.4 | 23.43 | 1.70 | 81+ |
| Usno | | 25 | 68.0 | 81.0 | 73.34 | 3.47 | 22.0 | 28.7 | 24.07 | 1.49 | 146 |
| Shungura | B | 17 | 67.5 | 85.9 | 77.49 | 4.85 | 21.0 | 26.0 | 23.55 | 1.48 | 171 |
| Koobi Fora | "B" | 6 | 73.0 | 92.0 | 80.25 | 6.64 | 22.1 | 26.4 | 23.70 | 1.76 | 145 |
| | "C" | 7 | 73.0 | 90.8 | 82.11 | 6.57 | 21.4 | 27.6 | 24.09 | 2.19 | 180 |
| *Notochoerus clarkii* | | | | | | | | | | | |
| Shungura | C | 3 | 63.6 | 85.2 | 74.83 | 8.84 | 22.2 | 25.1 | 23.67 | 1.18 | 97+ |
| | F | 2 | 62.0 | 63.0 | 62.50 | 0.50 | 20.5 | 22.0 | 21.25 | 0.75 | 78++ |
| | Lower G | 8 | 60.7 | 83.0 | 70.69 | 6.78 | 19.6 | 25.0 | 21.63 | 1.78 | 184 |
| *Notochoerus scotti* | | | | | | | | | | | |
| Shungura | C | 5 | 85.0 | 101.4 | 94.34 | 5.95 | 23.5 | 25.4 | 24.38 | 0.76 | 252 |
| | D | 4 | 89.0 | 103.5 | 94.18 | 5.73 | 23.0 | 25.0 | 24.43 | 0.83 | 262 |
| | E | 6 | 85.1 | 108.0 | 95.45 | 7.95 | 20.0 | 25.5 | 23.55 | 2.16 | 252 |
| | F | 5 | 85.6 | 106.7 | 96.96 | 8.80 | 20.8 | 26.3 | 23.55 | 2.06 | 204+ |
| | Lower G | 10 | 89.2 | 112.0 | 102.00 | 6.38 | 21.0 | 26.0 | 23.93 | 1.42 | 258 |
| | H | 1 | | | 99.0 | | | | 24.7 | | 150++ |
| Koobi Fora | "N. scotti" | 15 | 91.0 | 121.0 | 102.98 | 9.95 | 19.8 | 25.8 | 22.36 | 1.82 | 211 |

under the name *Mesochoerus limnetes* but Pickford (1994) indicated that the East African material differed from the Kaiso Type of *Sus limnetes* and suggested that this specific name be confined to the Kaiso Type specimen. Cooke (1995) proposed that the common East African form be referred to *K. heseloni*, the type species of which came from the Shungura Formation (Leakey, 1943). The teeth of *K. afarensis* are best known from Hadar (Cooke, 1978b) and are more *Potamochoerus*-like than is the case with *K. heseloni*; indeed Harris and White (1979) initially regarded *afarensis* as a synonym of *Potamochoerus porcus*. *K. afarensis* is a relatively rare element in Shungura Members B and C. What is probably the ancestor of *K. afarensis* is *K. deheinzelini* from the Sagantole Formation near Aramis in the Middle Awash in Ethiopia, dated at close to 4.39 Ma (Brunet and White, 2001). These authors also

record the smallest *Kolpochoerus*, *K. cookei* as coming from Shungura Member B 10; no dimensions are given but the present writer's measurements for the type RUM3 are: length 22.9 mm, breadth 11.9 mm, and height 12.0 mm, outside the plotting limits for Figure 5.

*K. olduvaiensis* was first described by Leakey (1942) from Olduvai Gorge, where it appears in Bed I and is well represented in upper Bed II and in Beds III/IV. It is suspected that the Koobi Fora material from the "*M. andrewsi*" zone may be mixed but that from the "*M. compactus*" zone fits well with the Olduvai material. Hopwood (1934) described a new species *Koiropotamus majus* based on a partial mandibular ramus from Beds III/IV and two upper tusks from Bed I. The relatively simple cheek teeth contrast strikingly with contemporary *Kolpochoerus* material but the species is rare and not well known. Somewhat surprisingly there is not

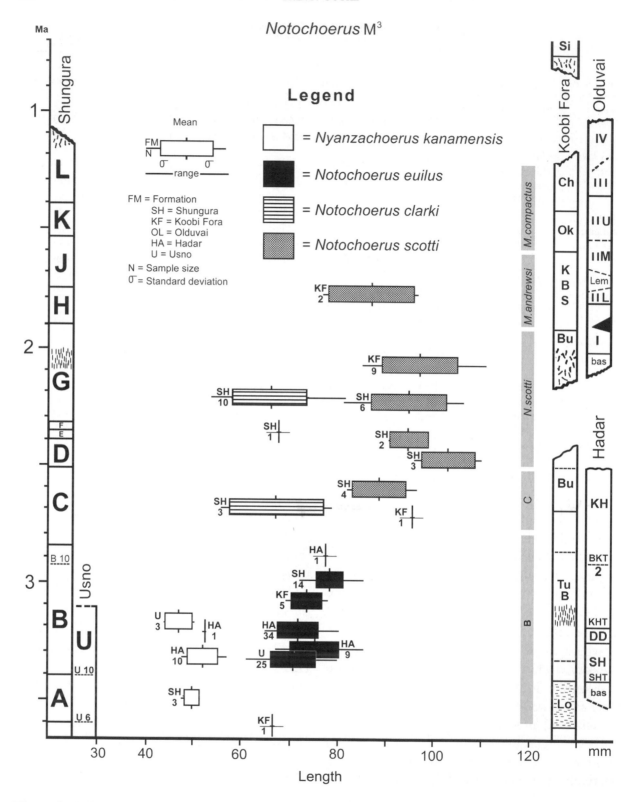

Figure 3. *NOTOCHOERUS*. Diagram showing the observed ranges in the basal lengths of upper third molars of *Nyanzachoerus*, *Notochoerus euilus*, and *No. scotti* at different horizons in the stratigraphic units shown. A key to symbols is given.

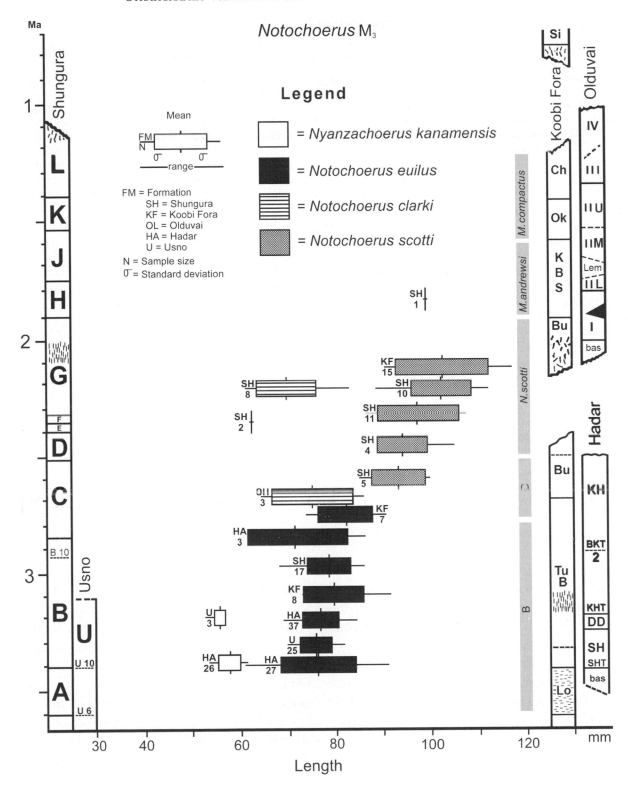

Figure 4. *NOTOCHOERUS*. As for Figure 3 but for lower third molars.

Table 3. *Kolpochoerus upper M3*

| Formation | Unit or "zone" | N | Length | | | | Breadth | | | | Hypsodonty index |
|---|---|---|---|---|---|---|---|---|---|---|---|
| | | | Min | Max | Mean | SD | Min | Max | Mean | SD | |
| *Kolpochoerus afarensis* | | | | | | | | | | | |
| Hadar | Sidi Hakoma | 4 | 32.9 | 35.6 | 34.18 | 1.18 | 19.6 | 21.0 | 20.15 | 0.62 | 91+ |
| | Denen Dora | 7 | 33.7 | 38.5 | 35.83 | 1.68 | 19.4 | 23.3 | 22.24 | 1.32 | 79+ |
| | Kada Hadar | 1 | | | 41.9 | | | | 24.2 | | 66+ |
| Usno | | 2 | 34.0 | 35.7 | 34.65 | 0.85 | 20.0 | 21.5 | 20.75 | 0.75 | 102 |
| Shungura | B | 5 | 32.6 | 41.8 | 38.56 | 3.12 | 20.0 | 24.0 | 22.72 | 1.45 | 80+ |
| | C | 2 | 37.7 | 40.2 | 38.95 | 1.25 | 20.8 | 20.9 | 20.85 | 0.05 | 85+ |
| *Kolpochoerus heseloni* | | | | | | | | | | | |
| Shungura | C | 6 | 42.1 | 50.2 | 45.62 | 2.99 | 23.5 | 27.5 | 24.92 | 1.46 | 76+ |
| | D | 6 | 42.5 | 52.5 | 47.65 | 3.00 | 22.2 | 25.4 | 23.65 | 0.96 | 120 |
| | E | 14 | 40.3 | 49.2 | 46.23 | 2.52 | 21.0 | 26.5 | 23.08 | 1.33 | 84+ |
| | F | 6 | 41.2 | 48.0 | 44.80 | 2.05 | 21.5 | 25.3 | 23.52 | 1.21 | 73+ |
| | G | 37 | 42.5 | 55.4 | 50.17 | 3.06 | 21.8 | 27.6 | 24.37 | 1.54 | 101 |
| Koobi Fora | "*N. scotti*" | 28 | 42.3 | 53.2 | 48.09 | 3.59 | 21.8 | 28.9 | 24.42 | 1.62 | 100 |
| | "*M. andrewsi*" | 28* | 45.2 | 67.0 | 56.27 | 6.02 | 19.0 | 33.0 | 25.22 | 2.48 | 110 |
| Olduvai | Bed I | 25 | 43.8 | 54.8 | 49.03 | 3.15 | 22.5 | 27.2 | 24.87 | 1.26 | 108 |
| | Bed II Lower | 6 | 44.0 | 47.6 | 47.45 | 2.08 | 20.6 | 24.5 | 22.87 | 1.39 | 111+ |
| *Kolpochoerus olduvaiensis* | | | | | | | | | | | |
| Shungura | H | 2 | 77.6 | 78.9 | 78.25 | 0.92 | 30.0 | 31.6 | 30.80 | 1.13 | 127 |
| | K | 3 | 69.7 | 73.8 | 72.00 | 1.71 | 25.2 | 29.2 | 27.67 | 1.76 | 119 |
| | L | 1 | | | 70.8 | | | | 26.0 | | 90+ |
| Koobi Fora | "*M. compactus*" | 4 | 54.4 | 67.9 | 62.08 | 5.89 | 21.0 | 27.3 | 25.07 | 3.53 | 140 |
| Olduvai | Bed II upper | 5 | 60.1 | 66.2 | 62.96 | 2.42 | 23.5 | 28.7 | 25.74 | 2.26 | 110 |
| | Bed III/IV | 2 | 53.5 | 58.5 | 56.00 | 3.54 | 24.5 | 24.6 | 24.55 | 0.07 | 118 |
| *Kolpochoerus majus* | | | | | | | | | | | |
| Olduvai | Bed II upper | 1 | | | 44.7 | | | | 23.0 | | 50+ |
| | Bed III/IV | 3 | 41.0 | 47.0 | 44.83 | 3.33 | 22.5 | 25.2 | 23.73 | 1.37 | 102 |

*The samples from this zone are suspected to be a mixture of *K. heseloni* and *K. olduvaiensis*.

much variation in the hypsodonty index in the three species suggesting that *Kolpochoerus* followed a bush pig diet of relatively soft tissues such as roots, tubers, and fruits.

*K. afarensis* ranges up to 2.7 Ma, *K. heseloni* from 2.7 to 1.7 Ma, overlapping with *K. olduvaiensis*, which ranges from 1.8 to 1.0+ Ma; *K. majus* also occurs from 1.8 to 1.0+ Ma.

## *METRIDIOCHOERUS* (TABLES 5 AND 6, FIGURES 7 AND 8)

The genus *Metridiochoerus* was established by Hopwood (1926) as *Metridiochoerus andrewsi* on the basis of a worn upper third molar from Homa Mountain, on the Winam Gulf of Lake Victoria. A maxilla fragment from Kagua, in the same general area, was described and figured by Leakey in 1958. In 1943 Leakey created a new genus and species from the Shungura Formation, *Pro-notochoerus* (sic) *jacksoni*; essentially similar material from the Shungura Formation was referred to *Metridiochoerus andrewsi* by Arambourg (1947). Harris and White (1979) recognized four species of the genus *Metridiochoerus* – *M. andrewsi*, *M. hopwoodi*, *M. modestus*, and *M. compactus*. *M. andrewsi* was divided into three stages, I, II, and III, and they regarded the differences between them as progressive changes within a single lineage. Their Stage I was

*Table 4. Kolpochoerus lower M3*

| Formation | Unit or "zone" | N | Length | | | | Breadth | | | | Hypsodonty index |
|-----------|----------------|---|-----|-----|------|-----|-----|-----|-------|------|------------------|
| | | | Min | Max | Mean | SD | Min | Max | Mean | SD | |
| *Kolpochoerus afarensis* | | | | | | | | | | | |
| Hadar | Sidi Hakoma | 12 | 33.9 | 40.0 | 37.39 | 1.87 | 16.5 | 20.8 | 18.89 | 1.03 | 104 |
| | Denen Dora | 18 | 32.6 | 41.6 | 38.51 | 2.34 | 16.5 | 21.4 | 18.89 | 1.09 | 90+ |
| | Kada Hadar | 1 | | | 37.7 | | | | 19.0 | | 79+ |
| Usno | | 1 | | | 36.5 | | | | 21.0 | | 31++ |
| Shungura | B | 6 | 33.7 | 42.9 | 39.33 | 2.84 | 15.9 | 20.2 | 19.03 | 1.46 | 124 |
| *Kolpochoerus heseloni* | | | | | | | | | | | |
| Shungura | C | 13 | 44.5 | 52.0 | 49.18 | 2.29 | 19.3 | 22.8 | 20.80 | 1.01 | 105 |
| | D | 6 | 45.5 | 58.5 | 50.88 | 5.31 | 19.3 | 25.1 | 21.43 | 2.04 | 58+ |
| | E | 30 | 44.7 | 54.5 | 49.09 | 2.40 | 18.2 | 23.2 | 20.60 | 1.25 | 100+ |
| | F | 7 | 43.0 | 59.0 | 49.79 | 5.54 | 17.7 | 23.0 | 20.31 | 1.61 | 108 |
| | Lower G | 52 | 45.3 | 61.2 | 54.75 | 3.73 | 17.5 | 24.2 | 21.60 | 1.89 | 111 |
| | Upper G | 11 | 53.0 | 63.5 | 57.05 | 4.06 | 19.4 | 24.0 | 21.28 | 1.39 | 109 |
| Koobi Fora | "*N. scotti*" | 41 | 41.0 | 65.4 | 51.94 | 5.13 | 18.0 | 24.8 | 20.99 | 1.69 | ?104 |
| | "*M. andrewsi*" | 46* | 48.2 | 72.0 | 59.15 | 4.77 | 19.0 | 29.7 | 21.87 | 1.82 | ?109 |
| Olduvai | Bed I | 19 | 46.1 | 58.9 | 52.81 | 3.29 | 18.0 | 23.5 | 21.58 | 1.50 | 117 |
| | Bed II Lower | 2 | 46.0 | 48.0 | 47.00 | 1.41 | 18.7 | 19.6 | 19.15 | 0.64 | 86+ |
| *Kolpochoerus olduvaiensis* | | | | | | | | | | | |
| Shungura | H | 1 | | | 70.0 | | | | 24.7 | | 33++ |
| | J | 2 | 62.9 | 70.3 | 66.60 | 3.70 | 21.0 | 21.0 | 21.00 | 0.00 | 114 |
| | K | 1 | | | 74.0 | | | | 25.0 | | 70+ |
| | L | 3 | 72.5 | 83.7 | 77.10 | 4.69 | 20.5 | 27.5 | 23.77 | 2.88 | 107 |
| Koobi Fora | "*M. compactus*" | 7 | 56.2 | 72.6 | 64.79 | 7.72 | 21.5 | 25.3 | 22.87 | 2.25 | 140 |
| Olduvai | Bed II Upper | 9 | 59.0 | 71.0 | 64.74 | 3.93 | 21.0 | 25.8 | 23.55 | 1.55 | 126 |
| | Bed III/IV | 3 | 66.0 | 74.0 | 68.83 | 4.48 | 21.5 | 22.0 | 21.82 | 0.29 | 109+ |
| *Kolpochoerus majus* | | | | | | | | | | | |
| Olduvai | Bed I upper | 1 | | | 42.5 | | | | 20.7 | | 34++ |
| | Bed II lower | 1 | | | 43.5 | | | | 20.3 | | 77+ |
| | Bed II upper | 3 | 45.0 | 50.0 | 48.3 | 2.86 | 21.7 | 22.0 | 21.83 | 0.15 | 118 |
| | Bed III/IV | 1 | | | 43.0 | | | | 19.7 | | 75+ |

*The samples from this zone are suspected to be a mixture of *K. heseloni* and *K. olduvaiensis*.

typified by specimens from Makapansgat in South Africa ("*Potamochoeroides shawi*") and from Members B–D of the Shungura Formation. Stage II included material from Shungura Members E–H, Olduvai Bed I and the Koobi Fora Formation at and below the KBS tuff (i.e., in the "*Notochoerus scotti*" zone). Stage III included the holotype from Homa Mountain, specimens from the KBS Member (i.e., the "*Metridiochoerus andrewsi*" zone), a skull from lower middle Bed II at Olduvai and some teeth from Swartkrans Pink Breccia in South Africa. Cooke and Wilkinson

(1978) considered the Makapansgat suid as a distinct species, *Potamochoeroides shawi* but here it is placed in *Metridiochoerus*; *M. jacksoni* is regarded as separable from *M. andrewsi*. *M. modestus* is a smaller species with columns that tend to resemble those in *Phacochoerus*. *M. hopwoodi* is a somewhat problematical entity in which the enamel pattern is more bilaterally symmetrical than in *M. andrewsi* or *M. compactus*. The latter embraces third molars that possess very high crowns with a hypsodonty index from 375 to over 500 in both the uppers and lowers and

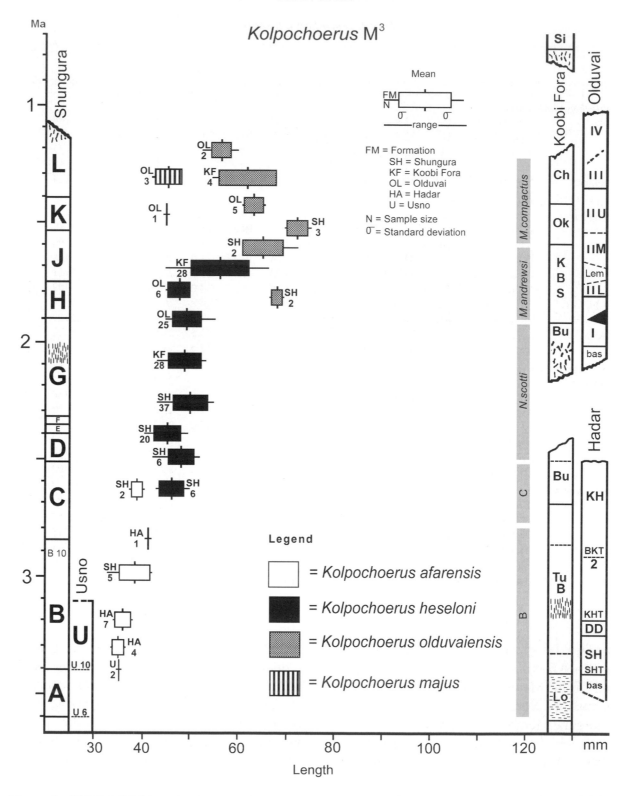

Figure 5. *KOLPOCHOERUS*. Observed ranges in the basal lengths of upper third molars of *Kolpochoerus afarensis*, *K. heseloni*, *K. olduvaiensis*, and *K. majus* at different horizons in the stratigraphic units shown. A key to symbols is given.

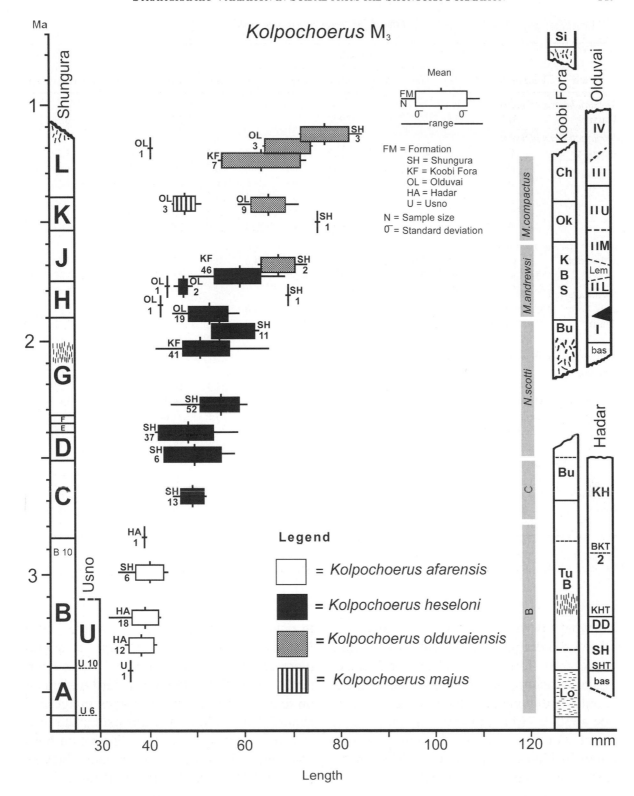

Figure 6. *KOLPOCHOERUS*. As for Figure 5 but for lower third molars.

Table 5. *Metridiochoerus upper M3*

| Formation | Unit or "zone" | N | Length | | | | Breadth | | | | Hypsodonty index |
|---|---|---|---|---|---|---|---|---|---|---|---|
| | | | Min | Max | Mean | SD | Min | Max | Mean | SD | |
| *Metridiochoerus* cf *shawi* | | | | | | | | | | | |
| Shungura | B11 | 1 | | | 39.5 | | | | 22.8 | | 68+ |
| *Metridiochoerus jacksoni* | | | | | | | | | | | |
| Shungura | C | 6 | 42.5 | 48.1 | 45.58 | 2.00 | 23.0 | 24.7 | 23.73 | 0.61 | 126 |
| | D | 5 | 47.2 | 53.0 | 50.40 | 2.06 | 24.0 | 28.7 | 26.96 | 1.64 | 119+ |
| | E | 4 | 45.8 | 48.0 | 46.67 | 0.94 | 24.5 | 26.5 | 25.55 | 0.95 | 122 |
| | F | 3 | 44.1 | 56.8 | 51.57 | 5.32 | 24.1 | 26.2 | 25.27 | 0.87 | 131 |
| | {E+F}* | 7 | 44.1 | 56.8 | 48.77 | 4.63 | 24.1 | 26.5 | 25.43 | 0.70 | 131 |
| | G | 15 | 44.8 | 54.5 | 50.80 | 2.87 | 22.3 | 29.9 | 25.34 | 2.08 | 152 |
| *Metridiochoerus andrewsi* | | | | | | | | | | | |
| Koobi Fora | "N. scotti" | 12 | 51.9 | 76.0 | 59.78 | 6.29 | 24.6 | 28.9 | 25.98 | 1.22 | 206 |
| | "M. andrewsi" | 30 | 50.2 | 82.3 | 65.38 | 6.50 | 21.1 | 30.9 | 25.75 | 2.33 | 257 |
| *Metridiochoerus "hopwoodi"* | | | | | | | | | | | |
| Koobi Fora | "M. compactus" | 2 | 51.0 | 65.0 | 58.00 | 9.90 | 24.0 | 25.4 | 24.70 | 0.99 | 217+ |
| Olduvai | Bed II Upper | 6 | 52.8 | 61.5 | 58.48 | 3.41 | 14.8 | 22.5 | 18.49 | 2.84 | 292 |
| | Bed III/IV | 3 | 61.0 | 71.5 | 65.93 | 5.33 | 18.7 | 26.1 | 23.50 | 4.18 | 282+ |
| *Metridiochoerus compactus* | | | | | | | | | | | |
| Shungura | J | 1 | | | 87.0 | | | | 24.4 | | 215+ |
| | L | 2 | 63.4 | 68.9 | 66.15 | 3.89 | 18.0 | 21.3 | 19.65 | 2.33 | 372+ |
| Koobi Fora | "M. compactus" | 5 | 81.1 | 117.7 | 94.26 | 15.00 | 25.5 | 29.6 | 26.86 | 1.68 | 332+ |
| Olduvai | Bed II upper | 14 | 60.4 | 100.0 | 77.03 | 9.29 | 18.8 | 24.5 | 21.64 | 1.88 | 445+ |
| | Bed III/IV | 12 | 60.0 | 86.9 | 71.63 | 10.50 | 17.4 | 27.3 | 19.99 | 3.23 | 474+ |
| Ternifine** | | 6 | 72.0 | 102.5 | 86.00 | 13.22 | 20.0 | 26.5 | 23.03 | 2.74 | 398 |
| *Metridiochoerus modestus* | | | | | | | | | | | |
| Shungura | Lower G | 3 | 42.5 | 46.0 | 44.83 | 2.02 | 18.1 | 20.0 | 19.03 | 0.95 | 170 |
| Koobi Fora | "M. compactus" | 1 | | | 51.2 | | | | 18.7 | | 205+ |
| Olduvai | Bed II | 2 | 38.0 | 47.6 | 42.95 | 6.58 | 18.0 | 19.9 | 18.95 | 1.34 | 101+ |
| | Bed III/IV | 2 | 45.5 | 56.5 | 51.00 | 7.75 | 16.5 | 17.0 | 16.75 | 0.35 | 279 |

*Samples from these two units are combined in order to plot them clearly within the timescale.
**Ternifine (Palikao, Tighenif) is faunally dated as about 0.7 Ma but is plotted here at 1.0 Ma for convenience.

somewhat resemble large warthog molars but with more complex enamel patterns. A good sample from Ternifine (Tighenif or Palikao) in Algeria belongs to *M. compactus*, which is the only suid in the collection. Its age is stated to be about 0.7 Ma but there is no radiometric control. For convenience of comparison it is plotted in Figures 7 and 8 as 1.0 Ma and the bar is indicated by the abbreviation TER.

A specimen from upper Member B in the Shungura Formation is inseparable from *M. shawi* from Makapansgat and its age is close to 3.0 Ma. The Makapansgat material exhibits relatively low crowns with an HI of up to 120 in upper third molars and 130 in the lowers. Material attributed to *M. jacksoni* has slightly more columnar crowns than in *M. shawi* with the HI up to 170 in the upper third molars and up to 209 in the lowers. It ranges in age from about 2.6 to 2.1 Ma. *M. andrewsi* has not been recorded from the Shungura Formation but is common in the Koobi Fora Formation between 2.1 and 1.7 Ma, exhibiting an HI up to 260 in the upper and 325 in the lower third molars. *M. compactus* appears at about 1.7 Ma and ranges through 1.1 Ma. Material attributed to *M. hopwoodi* is found at Olduvai and Koobi Fora at a level close to 1.5 Ma. *M. modestus*

*Table 6. Metridiochoerus lower M3*

| Formation | Unit or "zone" | N | Length | | | | Breadth | | | | Hypsodonty index |
|---|---|---|---|---|---|---|---|---|---|---|---|
| | | | Min | Max | Mean | SD | Min | Max | Mean | SD | |
| *Metridiochoerus jacksoni* | | | | | | | | | | | |
| Shungura | C | 5 | 44.3 | 57.9 | 49.92 | 6.36 | 18.3 | 22.5 | 19.46 | 1.53 | 166 |
| | D | 2 | 45.0 | 59.0 | 52.00 | 7.00 | 19.2 | 22.5 | 20.85 | 1.65 | 173 |
| | E | 6 | 49.0 | 58.0 | 53.73 | 2.89 | 20.5 | 23.5 | 22.43 | 1.21 | 181+ |
| | F | 4 | 50.6 | 62.2 | 56.33 | 5.24 | 17.8 | 23.2 | 20.95 | 2.61 | 172 |
| | {E+F}* | 10 | 49.0 | 62.2 | 54.77 | 4.06 | 17.8 | 23.5 | 21.84 | 1.96 | 181+ |
| | G | 21 | 46.0 | 64.4 | 56.99 | 5.00 | 18.0 | 24.9 | 20.99 | 2.17 | 209 |
| *Metridiochoerus andrewsi* | | | | | | | | | | | |
| Koobi Fora | "N. scotti" | 5 | 53.0 | 64.5 | 58.78 | 4.44 | 19.0 | 25.0 | 22.86 | 2.43 | 184 |
| | "M. andrewsi" | 32 | 56.5 | 84.3 | 66.32 | 10.92 | 18.2 | 24.8 | 20.13 | 3.94 | 328 |
| *Metridiochoerus "hopwoodi"* | | | | | | | | | | | |
| Koobi Fora | "M. andrewsi" | 4 | 53.2 | 72.0 | 59.95 | 8.28 | 14.9 | 20.4 | 17.90 | 2.38 | 249 |
| Olduvai | Bed II | 9 | 58.2 | 77.0 | 64.23 | 6.05 | 14.8 | 22.0 | 17.51 | 1.93 | 422 |
| *Metridiochoerus compactus* | | | | | | | | | | | |
| Koobi Fora | "M. compactus" | 10 | 80.5 | 101.5 | 88.79 | 6.89 | 18.0 | 23.6 | 20.81 | 1.45 | 411 |
| Olduvai | Bed II Upper | 21 | 60.0 | 110.0 | 79.38 | 13.47 | 14.0 | 22.7 | 18.65 | 2.30 | 463 |
| | Bed III/IV | 8 | 60.0 | 86.5 | 76.44 | 8.64 | 13.0 | 21.6 | 18.15 | 3.04 | 512 |
| Ternifine** | | 9 | 78.0 | 105.0 | 92.67 | 7.96 | 19.5 | 21.5 | 21.21 | 1.33 | 373+ |
| *Metridiochoerus modestus* | | | | | | | | | | | |
| Shungura | Lower G | 2 | 39.8 | 45.0 | 42.40 | 3.68 | 14.5 | 16.4 | 15.45 | 1.34 | 152+ |
| Koobi Fora | "M. andrewsi" | 3 | 45.7 | 48.6 | 47.27 | 1.46 | 13.0 | 14.8 | 13.93 | 0.90 | 267 |
| Olduvai | Bed I | 6 | 45.0 | 54.5 | 48.92 | 3.23 | 15.0 | 17.3 | 15.75 | 0.88 | 280 |
| | Bed II | 1 | | | 54.0 | | | | 16.2 | | 216+ |
| | Bed IV | 1 | | | 55.8 | | | | 14.0 | | 307+ |

*Samples from these two units are combined in order to plot them clearly within the timescale.
**Ternifine (Palikao, Tighenif) is faunally dated as about 0.7 Ma but is plotted here at 1.0 Ma for convenience.

appears at about 2.2 Ma and continues to 1.0 Ma and probably later.

## Status of *Metridiochoerus compactus*

A distinctive feature of *M. compactus* is the unusual nature of the canines and their orientation. In 1942 Leakey described as *Afrochoerus nicoli* from Bed II a suid with third molars almost twice the size of those of the warthog *Phacochoerus africanus* but with a somewhat similar structure. Associated canines were very specialized, having an oval cross section and exhibiting a core of cancellous osteo-dentine in the interior of the tusk. A substantial amount of additional material was described in 1958, mostly from Olduvai Bed II, some from Bed IV, and an almost complete mandible from Kanjera with the third molars intact and the sockets for the canines preserved. In the same monograph Leakey (1958) named a new genus and species *Orthostonyx brachyops* on the basis of four specimens from Olduvai Bed II. The teeth are somewhat like those of *Phacochoerus* but larger and all belong to immature individuals. Two maxilla fragments show that the upper canines rose diagonally outwards and upwards with a slight backward inclination, quite unlike most normal suids. A good third lower molar in very early wear resembles the teeth of *Afrochoerus* and this

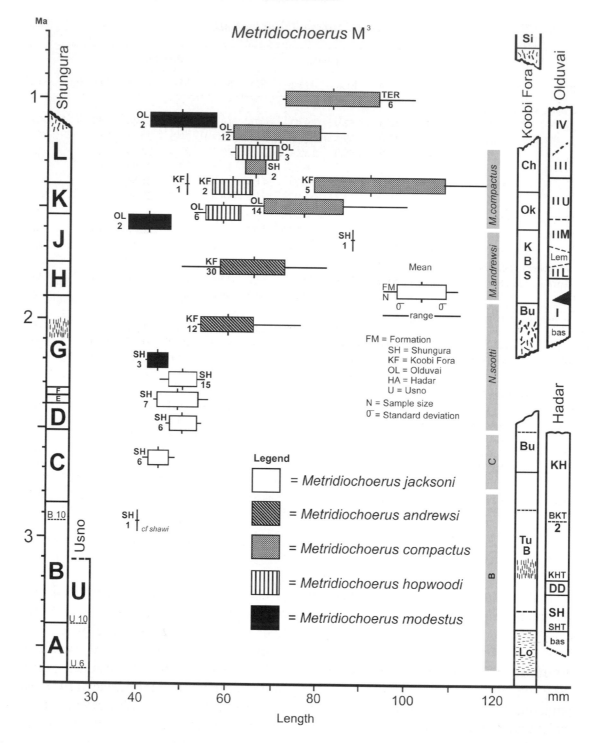

Figure 7. *METRIDIOCHOERUS*. Observed ranges in the basal lengths of upper third molars of *Metridiochoerus jacksoni*, *M. andrewsi*, *M. compactus*, *M. "hopwoodi"*, and *M. modestus* at different horizons in the stratigraphic units shown. A key to symbols is given. The single specimen from Shungura B is similar to the species *"shawi"* from Makapansgat in South Africa.

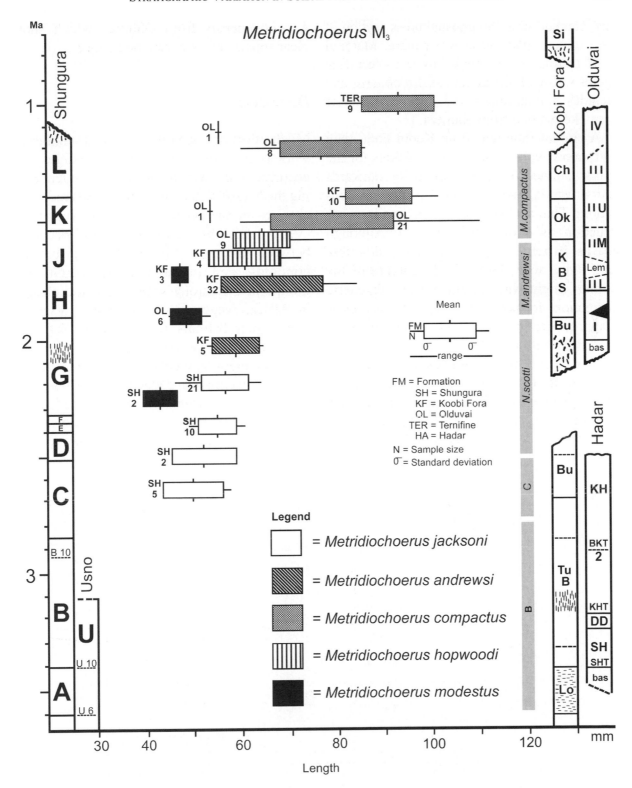

Figure 8. *METRIDIOCHOERUS*. As for Figure 7 but for lower third molars.

fact, coupled with the unusual cross section of the canines, links these two genera. Material of *Metridiochoerus compactus* from Ternifine agrees well with the molars of *Afrochoerus* and includes similar tusks and a good mandible very like the one from Kanjera.

A cranial fragment from Koobi Fora, with both upper M3's intact has the sockets for the upper canines directed diagonally outwards and upwards, with a slight posterior inclination. This makes possible a good reconstruction as shown in Figure 9. Although not as extreme as in *Babirussa*, this feature, along with the unusual structure of the canines might warrant generic or subgeneric distinction, for which the name *Stylochoerus* is appropriate. The difference in the canines is consistent with the idea that *Orthostonyx* was the female and *Afrochoerus* the male. Figure 9 shows plots of measurements on canines of *M. compactus* from Olduvai and supports this interpretation. Also shown in the figure are similar plots for *Kolpochoerus heseloni* and

*K. olduvaiensis* from Olduvai, which show clear separation between the genders.

## Discussion

Much effort has been devoted by a number of workers towards elucidating the changes that occurred in the African paleoenvironment during the Plio-Pleistocene. Vrba (1980) recorded changes in the bovid spectra and suggested (Vrba, 1993) that "turnover pulses" occurred between 2.5 and 2.0 Ma affecting bovids, hominids, and probably other mammalian groups in synchrony with a climatic change in African vegetation from more closed to more open habitats (Vrba, 1995a, b). She considered this to be linked to global climatic changes associated with the onset of Northern Hemisphere glacial conditions. The simplicity of this idea makes it attractive but it has not found universal acceptance and is debated at length in a remarkable volume

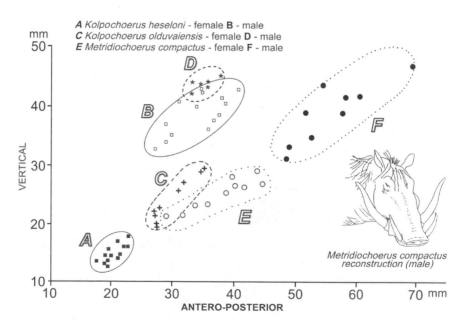

Figure 9. Scatter diagram for measurements on the cross section of upper canines from Olduvai Gorge to illustrate sexual dimorphism in the three species shown in the key on the diagram. A reconstruction of *Metridiochoerus compactus*, based primarily on a partial cranium from Koobi Fora, shows the unusual orientation of the canines.

"Paleoclimate and Evolution with emphasis on Human Origins" (Vrba et al., 1995). For the Shungura Formation the inferences based on soils, pollen, and micromammals have been summarized recently by Bobe and Eck (2001) and they present new correspondence analyses of the abundant bovid material to elucidate changes in bovid abundances through time. They observed an increase in species richness and an episode of rapid change in taxonomic abundance at $2.8 \pm 0.1$ Ma, followed by gradual and prolonged changes in abundance between 2.8 and 2.0 Ma. As environmental indicators, the bovids show a transition in the Omo area from closed and wet conditions in Member B to closed but dry conditions in Member C. The dry trend intensified between about 2.5 and 2.3 Ma, followed by an increase in bovid abundance and diversity to 2.1 Ma, possibly due to greater environmental heterogeneity. The date of 2.8 Ma corresponds closely with changes in climate recorded in terrigenous sediments off the West and North African coasts (deMenocal and Bloemendal, 1995). Bobe et al. (2002) undertook a fine scale analysis of bovids, suids, and primates in the Turkana Basin that shows a peak of faunal change at about $2.8 \pm 0.1$ Ma, followed by a stable interval between 2.7 and 2.5 Ma, not apparently responding to the global changes manifest in the marine record. Alemseged (2003) has examined the whole spectrum of mammals in the French collections from the Shungura Formation, using an integrated approach to investigate taphonomic and faunal change patterns. He found a major faunal change around the base of Member G (ca. 2.3 Ma), characterized by a change to open and edaphic grassland as a dominant type of environment. However, he did not find evidence for any rapid change from 2.9 to 2.7 Ma.

In a valuable analysis of the first appearance and last appearance data for suids, Tim White (1995) endeavored to screen out data that were judged to be "artifacts" of nomenclature and recognized two species originations at ca. 3.0–2.8 Ma and two extinctions at 2.7 and 2.5 Ma.

He notes that the most dramatic faunal turnover during the period occurred between 2.2 and 1.6 Ma. Considering the different approaches, these estimates are not widely different from those derived here. There is thus some degree of conflict between the inferences made by Vrba and by others but, although the analyses have been based on different criteria, the evidence seems to be converging on a critical span between 2.9 and 2.5 Ma.

The dietary preferences of the living suids were discussed at some length by Cooke (1985), together with some comments on the possible habits of the extinct forms. From an ecological viewpoint, the morphological changes in the cheek teeth are consistent with the concept of more closed habitats before about 2.8–2.5 Ma, changing to more open habitats thereafter. The relatively low crowned *Notochoerus euilus* probably favored the *Hylochoerus* habitat of forest and forest glade. It is replaced at about 2.7 Ma by *N. scotti* with an increase in hypsodonty that would prolong the life of the molars in the face of an increasingly abrasive diet in the grassland savanna. In *Kolpochoerus* there is little change in the hypsodonty through time, but the length of the crown increases at about 2.7–2.8 Ma and the essentially bunodont crown becomes a little more columnar, providing an effective increase in enamel area. A similar change occurs at about 1.8–1.9 Ma. It is interesting that *K. majus*, with its small and simple teeth occurs as a rare element alongside the larger and higher crowned *K. olduvaiensis*, suggesting that patches of bush with water persisted in the savanna-like environment. The same relationship is displayed by *Metridiochoerus*, with the relatively simple *M. modestus* occurring alongside the very hypsodont and warthog-like teeth of *M. compactus*.

It is suggested that the presentation of metrical data in diagrammatic form, plotted against a stratigraphic time scale, has value in the integration of data from different suites to provide at least a visual impression of variation within the species itself. This may help in allowing temporal ranges to be estimated more accurately, although it may not resolve the question of how lineages should be divided into species. Although suid teeth reflect in their morphology the diet available to them as a result of environmental shifts due to climate change, it is not yet clear that the climatic changes are the driving force of speciation. More data and analyses are needed.

## Acknowledgments

The writer is greatly indebted to several museums that house the collections discussed here and to their Directors for permission to study them. Financial support from the Wenner Gren Foundation for Anthropological Research and from the Canadian National Research Council made travel possible and is gratefully acknowledged. Comments by referees on the draft of this paper were most valuable. A particular debt is owed to Dr. Laura Bishop and to James McClosky of the John Moores University in Liverpool for volunteering to tabulate the statistical data and to redraw the rather messy hand-drafted figures using their computer skills and much valuable time. Thanks are also due to Dr. Kay Behrensmeyer, Dr. René Bobe, and Dr. Zeresenay Alemseged for having organized the conference to which this paper was submitted in the absence of the writer and for their advice and help.

## References

Alemseged, Z., 2003. An integrated approach to taphonomy and faunal change in the Shungura Formation (Ethiopia) and its implications for hominid evolution. Journal of Human Evolution 44, 451–478.

Arambourg, C., 1943. Observations sur les Suidés fossiles du Pléistocène d'Afrique. Bulletin du Muséum National d'Histoire Naturelle, Paris 15(2), 471–476.

Arambourg, C., 1947. Contribution a l'étude géologique et Paléontologique du bassin du Lac Rodolphe et de la basse valleé de l'Omo. Deuxième partie. Palèontologie Mission Scientifique de I'Omo, 1932–1933, Tome 1, Géologie-Anthropologie, Fascicule III Deuxième partie, Paléontologie. Muséum National d'Histoire Naturelle, Paris, 232–562.

Bobe, R., Eck, G.G., 2001. Responses of African bovids to Pliocene climatic changc. Paleobiology 27 (Supplement No 2), Paleobiology Memoirs 2, 1–47.

Bobe, R., Behrensmeyer, A.K., Chapman, R.E., 2002. Faunal change, environmental variability and late Pleistocene hominin evolution. Journal of Human Evolution 42, 475–497.

Brunet, M., White, T.D., 2001. Deux nouvelles espéces de Suiné (Mammalia, Suidae) du continent Africaine (Èthiopie; Tchad). Comptes Rendus de l Academie des Sciences. Paris. Earth and Planetary Sciences 332, 51–57.

Cooke, H.B.S., 1978a. Suid evolution and correlation of African hominid localities: an alternative taxonomy. Science 201, 460–463.

Cooke, H.B.S., 1978b. Pliocene–Pleistocene Suidae from Hadar, Ethiopia. Kirtlandia 29, 1–63.

Cooke, H.B.S., 1985. Plio-Pleistocene Suidae in relation to African hominid deposits. In: L'Environnement des Hominidés au Plio-Pléistocène. Fondation Singer-Polignac et Masson, Paris, pp. 101–117.

Cooke, H.B.S., 1995. The status of the African fossil suids *Kolpochoerus limnetes* (Hopwood, 1926), *K. phacochoeroides* (Thomas, 1884) and "*K*" *afarensis* (Cooke, 1978). Geobios-Lyon 30(1), 121–126.

Cooke, H.B.S., Maglio, V.J., 1972. Plio-Pleistocene stratigraphy in East Africa in relation to proboscidean and suid evolution. In: Bishop, W.W., Miller, J.A. (Eds.), Calibration of Hominid Evolution. Scottish Academic Press, pp. 303–329.

Cooke, H.B.S., Wilkinson, A.F., 1978. Suidae and Tayassuidae. In: Maglio, V.J., Cooke, H.B.S. (Eds.), Evolution of African Mammals. Harvard University Press, Cambridge, pp. 435–482.

deMenocal, P.B., Bloemendal, J., 1995. Plio-Pleistocene climatic variability in subtropical Africa and the paleoenvironment of hominid evolution: a combined data-model approach. In: Vrba, E.S., Denton, G.H., Partridge, T.C., Burckle, L.H. (Eds.), Paleoclimate and Evolution, with Emphasis on Human Origins. Yale University Press, New Haven, pp. 262–288.

Harris, J.M., 1978. Paleontology. In: Leakey, M.G., Leakey R.E.F. (Eds.), Koobi Fora Research Project, Volume 1, The Fossil Hominids and

An Introduction to Their Context, 1968–1974. Clarendon Press, Oxford, pp. 32–63.

Harris, J.M., 1983. Family Suidae. In: Harris, J.M. (Ed.), Koobi Fora Research Project, Volume 2, The Fossil Ungulates: Proboscidea, Perissodactyla, and Suidae. Clarendon Press, Oxford, pp. 215–302.

Harris, J.M., White, T.D., 1979. Evolution of the Plio-Pleistocene African Suidae. Transactions of the American Philosophical Society 69(2), 1–128.

Hopwood, A.T., 1926. Fossil Mammalia. The geology and palaeontology of the Kaiso Bone beds. Occ. Papers Geological Survey of Uganda Protectorate, Occasional Papers 2, 13–36.

Hopwood, A.T., 1934. New fossil mammals from Olduvai, Tanganyika Territory. Annals and Magazine of Natural History (10)14, 546–550.

Kullmer, O., 1997. Die Evolution der Suiden im Plio-Pleistozän Afrikas und ihre biostratigraphische, paläobiogeographische und paläoökologische Bedeutung. Doctoral Thesis, Johannes Gutenberg-Universität in Mainz.

Kullmer, O., 1999. Evolution of African Plio-Pleistocene suids (Artiodactyla: Suidae) based on tooth pattern analysis. Kaupia: Darmstädter Beitrage zur Naturwissenschaften 9, 1–34.

Leakey, L.S.B., 1942. Fossil Suidae from Oldoway. Journal of the East Africa Natural History Society, 16, 178–196.

Leakey, L.S.B., 1943. New Fossil Suidae from Shungura, Omo. Journal of the East Africa Natural History Society, 17, 45–61.

Leakey, L.S.B., 1958. Some East African Pleistocene Suidae. British Museum (Natural History), Fossil Mammals of Africa 14, 1–133.

Maglio, V.J., 1972. Vertebrate faunas and chronology of hominid-bearing sediments East of Lake Rudolf, Kenya. Nature 239, 379–385.

Pickford, M., 1994. Fossil Suidae of the Albertine Rift, Uganda-Zaire. In: Senut, B., Pickford, M. (Eds.), Geology and Paleobiology of the Albertine Rift Valley Uganda-Zaire, Volume 2, Paleobiology. Orleans, Centre International Pour la Formation et les Echange géologique. Occasional Publication 29, 339–374.

Schmidt-Kittler, N., 1984. Pattern analyses of occlusal surfaces in hypsodont herbivores and its bearing on morpho-functional studies. Koninklijke Nederlandsche Akademie Voor Wetenschappen, Proceedings B 87(4), 453–480.

Vrba, E., 1980. Evolution, species and fossils. South African Journal of Science 76, 61–84.

Vrba, E., 1993. Turnover-pulses, the Red Queen, and related topics. American Journal of Science 293A, 418–452.

Vrba, E., 1995a. On the connections between paleoclimate and evolution. In: Vrba, E.S., Denton, G.H., Partridge, T.C., Burckle, L.H. (Eds.), Paleoclimate and Evolution, with Emphasis on Human Origins. Yale University Press, New Haven, pp. 24–45.

Vrba, E., 1995b. The fossil record of African antelopes (Mammalia, Bovidae) in relation to human evolution and paleoclimate. In: Vrba, E.S., Denton, G.H., Partridge, T.C., Burckle, L.H. (Eds.), Paleoclimate and Evolution, with Emphasis on Human Origins. Yale University Press, New Haven, pp. 385–424.

Vrba, E., Denton, G.H., Partridge, T.C., Burckle, L.H. (Eds.), 1995. Paleoclimate and Evolution, with Emphasis on Human Origins. Yale University Press, New Haven.

White, T.D., 1995. African omnivores: global climatic change and Plio-Pleistocene hominids and suids. In: Vrba, E.S., Denton, G.H., Partridge, T.C., Burckle, L.H. (Eds.), Paleoclimate and Evolution, with Emphasis on Human Origins. Yale University Press, New Haven, pp. 369–384.

White, T.D., Harris, J.M., 1977. Suid evolution and correlation of African hominid localities. Science 198, 13–22.

White, T.D., Suwa, G., 2004. A new species of Notochoerus (Artiodactyla, Suidae) from the Pliocene of Ethiopia. Journal of Vertebrate Paleontology 24(2), 474–480.

# 6. Patterns of abundance and diversity in late Cenozoic bovids from the Turkana and Hadar Basins, Kenya and Ethiopia

R. BOBE
*Department of Anthropology*
*The University of Georgia*
*Athens, GA 30602, USA*
*renebobe@uga.edu*


A.K. BEHRENSMEYER
*Department of Paleobiology and Evolution of Terrestrial Ecosystems Program*
*National Museum of Natural History*
*Smithsonian Institution*
*Washington, DC, USA*
*behrensa@si.edu*


G.G. ECK
*Department of Anthropology*
*University of Washington*
*Seattle, Washington, USA*
*ggeck@u.washington.edu*


J.M. HARRIS
*George C. Page Museum*
*5801 Wilshire Boulevard*
*Los Angeles, California 90036, USA*
*jharris@nhm.org*

Keywords: East Africa, Omo, Pliocene, Pleistocene, Bovidae, paleoenvironments

## Abstract

After decades of fieldwork spurred by the search for human ancestors, paleontologists in East Africa are compiling networks of databases to address questions of long-term evolutionary, environmental, and ecological change. Paleontological databases from the Turkana Basin of Kenya and Ethiopia (East Turkana, West Turkana, Kanapoi, Lothagam, and Omo) and the Hadar Basin of Ethiopia's Afar region consist of nearly 70,000 specimens of fossil vertebrates (mostly mammals) that date from the late Miocene to the Pleistocene. Here we focus on the most abundant family of fossil mammals, the Bovidae (N = 8213 specimens), and illustrate patterns of taxonomic abundance and diversity from about 7 Ma (million years ago) to about 1 Ma. The key questions we address are the following: How much variation in patterns of faunal change is there within different areas of a large sedimentary basin? How

*R. Bobe, Z. Alemseged, and A.K. Behrensmeyer (eds.) Hominin Environments in the East African*
*Pliocene: An Assessment of the Faunal Evidence, 129–157.*

much variation is there between basins? How are these patterns related to broad signals of climatic change? What are the implications of the bovids for East African environments and for hominin evolution in the late Cenozoic? A correspondence analysis of bovid tribes indicates that important differences in taxonomic abundance existed among different areas of the Turkana Basin, and that some of these differences had environmental implications. The lower Omo Valley appears to have remained distinct from other parts of the Turkana Basin between 3 and 2 Ma, with consistently higher proportions of Tragelaphini and Aepycerotini, and at times of Reduncini and Bovini. These bovids are indicative of woodlands or forests (Aepycerotini and Tragelaphini) or of moist grasslands near wooded habitats (Reduncini and Bovini). An analysis of bovid tribes indicative of open and seasonally arid grasslands (Alcelaphini, Antilopini, and Hippotragini) shows relatively high proportions of these bovids in the West Turkana areas, but very low proportions in the Omo, especially prior to about 2 Ma. This indicates that the Omo remained wetter and more wooded than other parts of the Turkana Basin for much of the Plio-Pleistocene, while the West Turkana area appears to have been more open than other parts of the basin, and East Turkana had conditions intermediate between those at West Turkana and those in the Omo. Fossil bovids from the Hadar Basin suggest diverse environments including woodlands, wet grasslands, and drier savanna grasslands. An increase in the abundance of arid adapted bovids in the late Pliocene and early Pleistocene of the Turkana and Hadar Basins provides evidence that faunal changes in these different areas were driven by common factors consistent with the known record of climatic change. Analyses of species diversity among bovids show three peaks of richness in the Pliocene and Pleistocene. The first peak occurred at about 3.8–3.4 Ma, the second at 2.8–2.4 Ma, and the last from about 2.0 to 1.4 Ma. The last two of these peaks coincide with previously identified periods of high faunal turnover in East Africa. Although climate appears to have shaped major patterns in the evolution of bovids, the fact that different areas of a single sedimentary basin show distinct responses highlights the complexities involved in establishing causal links between paleoclimate and evolution.

## Introduction

A key problem in paleoanthropology is the extent to which climate change has influenced the course of early human evolution (Vrba et al., 1989; Feibel, 1997; Potts, 1998; Wynn, 2004; Behrensmeyer, 2006). Patterns of hominin evolution through the Pliocene and Pleistocene are sometimes used as evidence of climatic forcing, with key diversification events said to coincide with significant climatic change. However, hominin fossils are rare, and the pattern of hominin diversification is still poorly known. Discoveries in recent years have highlighted how little we still know about the early stages of human evolution (Asfaw et al., 1999; Leakey et al., 2001; Senut et al., 2001; Brunet et al., 2002; Haile-Selassie et al., 2004). Among those mammals that are most frequently associated with fossil hominins in the late Cenozoic of Africa, bovids provide the most comprehensive record of evolutionary and environmental change over the last several million years (Vrba, 1974, 1980, 1995a; Gentry, 1978, 1985; Shipman and Harris, 1988). A

detailed picture of evolutionary change in African bovids can help us understand how mammals respond to climatic changes as well as to local and regional alterations in habitat and can help us formulate hypotheses about how hominins may have responded to similar challenges. By focusing on the bovids, we aim to establish patterns that can be tested with other elements of the East African fauna.

Patterns of faunal change can be studied in a variety of ways, but here we focus on just two variables: relative abundance and taxonomic richness. We study these variables through time in scales of millions of years, and across geographical space including two distinct sedimentary basins in Kenya and Ethiopia. The key questions we seek to address are the following. What are the main patterns of faunal change in East Africa during the late Cenozoic? To what extent do different areas within the same sedimentary basin produce similar patterns of faunal change? To what extent do patterns of faunal change in one sedimentary basin resemble patterns in a different basin? Identifying significant patterns of faunal change must come before attempts to explain their causes. As suggested by Behrensmeyer and colleagues (2007), the null hypothesis is that local and regional tectonics and environmental processes control patterns of faunal change. However, if broad patterns of climate change are driving mammalian evolution in Africa (Vrba, 1995b; deMenocal, 2004), then we would expect to find similar patterns of faunal change across different regions.

To address these questions and hypotheses we provide an analysis of relative abundance and diversity in fossil bovids from the Turkana Basin of Kenya and Ethiopia, including the Shungura, Usno, Mursi, Nachukui, Koobi Fora, Kanapoi, and Nawata Formations, and from the Hadar Basin of Ethiopia's Afar region, including the Hadar and Busidima Formations. This chapter builds on the tremendous contribution of earlier research in the Turkana and Hadar areas to attempt a broad and integrative view of faunal change through time.

## THE TURKANA BASIN

The Lake Turkana Basin and its northern extension, the lower Omo Valley, hold one of the most complete and well studied archives of late Cenozoic faunal and environmental change on earth (Figure 1). The main geological formations from this area provide a relatively continuous record from the late Miocene to the early Pleistocene, i.e., from about 8 Ma (million years ago) to about 1 Ma. Extensive paleontological work in the Turkana Basin over the last few decades has produced a wealth of knowledge about faunal evolution in Africa (Coppens et al., 1976; Leakey and Leakey, 1978; Harris, 1983, 1991; Eck et al., 1987; Howell et al., 1987; Harris et al., 1988; Wood, 1991; Harris and Leakey, 2003; Leakey and Harris, 2003). The importance of the Turkana Basin for our understanding of human evolution and its environmental context has long been recognized (Howell, 1968; Behrensmeyer,

1975; Coppens, 1975; Boaz, 1977), and its fossil record continues to provide crucial contextual evidence for new analytical studies (Leakey, 2001; Alemseged, 2003). The geologic framework of the Turkana Basin has been extensively studied, and strong stratigraphic correlations as well as chronometric dates have been established (Brown, 1969, 1994; de Heinzelin, 1983; McDougall, 1985; Feibel et al., 1989; McDougall and Feibel, 1999; Brown et al., 2006). A summary of dating and correlations is presented in Table 1.

## HADAR

Another major area of research for human evolution in Africa has been the Hadar Basin in the Afar region of northern Ethiopia (Figure 1). Hadar provides a rich record of fossil vertebrates including early hominins, but it is more restricted than the Turkana Basin in both geographical extent and temporal depth. Nevertheless, Hadar has played a pivotal role in our understanding of early human evolution and its environmental context (Johanson and Taieb, 1976; Gray, 1980; Aronson and Taieb, 1981; White et al., 1981; Johanson et al., 1982; Kalb et al., 1982; Tiercelin, 1986; Bonnefille et al., 2004). The Hadar Formation deposits date from at least 3.4 Ma to about 2.9 Ma, and is divided into the Basal Member (>3.4 Ma), the Sidi Hakoma Member (3.4–3.22 Ma), the Denen Dora Member (3.22–3.18 Ma), and the Kada Hadar Member (3.18–2.9 Ma) (Aronson and Taieb, 1981; Walter and Aronson, 1993; Walter, 1994; Alemseged et al., 2005). A major unconformity separates the Hadar Formation from the overlying Busidima Formation as defined in the Gona area west of Hadar (Quade et al., 2004). A.L. 666 in the Maka'amitalu drainage, a locality that has yielded a maxilla attributed to early *Homo* and associated lithic artifacts (Kimbel et al., 1996), is considered here as part of the Busidima Formation, even though it has not been formally assigned to that

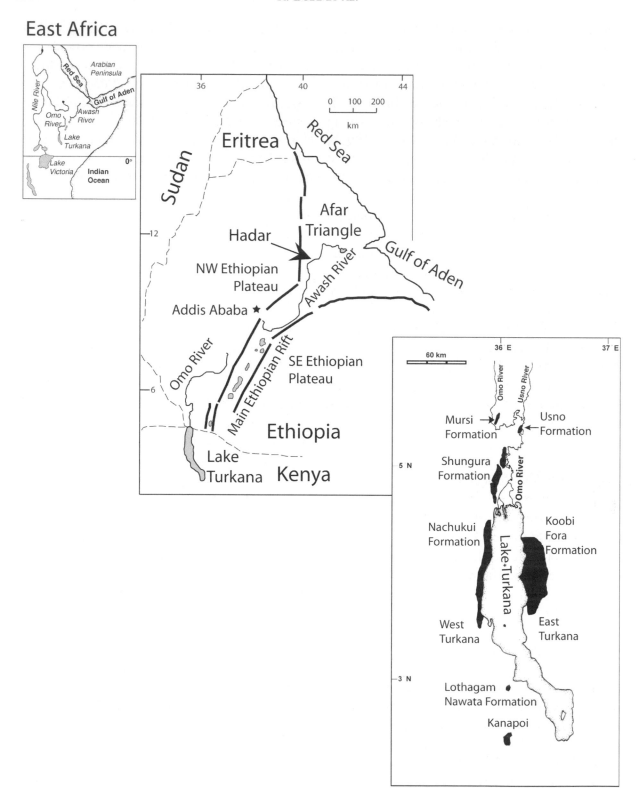

Figure 1. Map of East Africa showing the Turkana Basin and the Hadar area. Here we use the term "Turkana Basin" to refer to both the Lake Turkana area and the adjacent lower Omo Valley. The Nawata, Kanapoi, Nachukui, Koobi Fora, Shungura, Usno, and Mursi Formations from the Turkana Basin are considered in this study. The term "Hadar Basin" as used here refers to the Hadar Formation in Ethiopia's Afar region.

Table 1. Dating and correlations among formations in the Turkana Basin of Kenya and Ethiopia

| Omo | | West Turkana | | East Turkana | | GPTS | Age (Ma) | +/− | Dated unit (Fm) | Method | Refs. |
|---|---|---|---|---|---|---|---|---|---|---|---|
| Mbs | | Mbs | | Mbs | | | | | | | |
| Silbo Tuff | | Silbo Tuff | | Silbo Tuff | | | 0.74 | 0.01 | Silbo Tuff (KF) | K-Ar, Ar-Ar | 16, 24 |
| | | | | | | | 0.780 | | | Matuyama/Brunhes | 1, 10, 30 |
| | L | top Nariokotome Mb | Nariokotome | | | | 0.980 | 0.05 | | Stratigraphic scaling | 14 |
| L9 (top) | | | | | | | 0.990 | | | GPTS Jaramillo top | 1, 10 |
| | | | | | | | 1.05 | 0.12 | | Stratigraphic scaling | 14 |
| | | | | | | | 1.070 | | | GPTS Jaramillo base | 1, 10 |
| L5/L6 | | | | | | | 1.12 | 0.02 | L5/6 (Sh) | GPTS Cobb Mtn subchron (N) | 10, 14, 19 |
| | | | | Gele Tuff | | | 1.25 | 0.02 | Gele Tuff (KF) | K-Ar | 8 |
| | | L. Nariokotome Tuff | | | | | 1.33 | 0.03 | Nariokotome (Nk) | K-Ar | 8, 16 |
| Tuff L-3 | | Naito Tuff | | | | | | | | | 11, 18 |
| Tuff L | K | Chari Tuff | | Chari Tuff | | | 1.39 | 0.02 | Chari Tuff (KF) | K-Ar, Ar-Ar | 4, 24 |
| Tuff K1 | | | | Koobi Fora Tuff | | | | | | | 15 |
| | | | Natoo | MFB | | | 1.49 | 0.05 | MFB (KF) | Stratigraphic scaling | 14 |
| Tuff K (α) | | Upper Okote Tuff | | Upper Okote Tuff | Okote | | 1.53 | 0.03 | Tuff K (Sh) | Stratigraphic scaling | 7, 14 |
| Tuff J7-1 | | Black Pumice Tuff | | Black Pumice Tuff | | | 1.55 | 0.03 | BPT (KF) | Stratigraphic scaling | 5, 14, 16 |
| Tuff J6-2 | | L. Koobi Fora Tuff | | L. Koobi Fora Tuff | | | 1.62 | 0.02 | LKF Tuff (KF) | Interpolation | 14 |
| | | | | Okote Tuff | | | | | | | 6 |
| | | | | Lower Okote Tuff | | | 1.62 | 0.02 | L. Okote Tuff (KF) | Stratigraphic scaling | 14 |
| Tuff J4 | | Morutot Tuff | | Morutot Tuff | KBS | | 1.65 | 0.03 | J4 (Sh) | K-Ar | 14, 26 |
| | | | | A6 | | | 1.68 | 0.05 | A6 (KF) | Stratigraphic scaling | 6, 14 |
| | | | | White Tuff | | | 1.70 | 0.03 | WT (KF) | Stratigraphic scaling | 5, 14 |
| | | | | C6 | | | 1.73 | 0.05 | C6 (KF) | Stratigraphic scaling | 6, 14 |
| Tuff J | | | | | | | 1.74 | 0.03 | Tuff J (Sh) | Stratigraphic scaling | 14 |
| Unit H7 | | | | | | | 1.770 | | H7 (Sh) | GPTS Olduvai (top) | 1, 10, 14 |
| | | | | A2 | | | 1.78 | 0.05 | A2 (KF) | Stratigraphic scaling | 6, 14 |
| | | | | | | | 1.796 | | Pliocene/Pleistocene boundary | | 1, 10, 31 |
| Unit H5 | | | | C4 | | | 1.86 | 0.05 | C4 (KF) | Stratigraphic scaling | 3, 6, 14 |
| Tuff H4 | | Malbe Tuff | | Malbe Tuff | | | 1.86 | 0.02 | Malbe Tuff (KF) | K-Ar, Ar-Ar | 24 |
| Tuff H2 | | KBS Tuff | | KBS Tuff | | | 1.88 | 0.02 | KBS Tuff (KF) | K-Ar, Ar-Ar | 4, 24 |
| Tuff H | | | | | | | 1.90 | 0.03 | Tuff H (Sh) | Stratigraphic scaling | 14 |
| | H | | | Lorenyang Tuff | | | 1.90 | 0.05 | Lorenyang (KF) | Stratigraphic scaling | 14 |
| Unit G27 | G | | | | | | 1.950 | | Unit G27 (Sh) | GPTS Olduvai base | 1, 10, 14 |

(Continued)

*Table 1. Dating and correlations among formations in the Turkana Basin of Kenya and Ethiopia—cont'd*

| Grp | Formation | Member/Tuff (W) | Unit | Formation (E) | Tuff (E) | Age | ± | Correlation | Method | References |
|---|---|---|---|---|---|---|---|---|---|---|
| F | | | Unit G14 | Burgi | | 2.09 | 0.04 | Unit G14 (Sh) | GPTS Reunion II top | 20, 27 |
| F | | | Unit G12 | | | 2.13 | 0.04 | Unit G 12 (Sh) | GPTS Reunion II base | 20, 27 |
| F | | | Unit G9 | | | 2.17 | 0.04 | Unit G9 (Sh) | GPTS Reunion I top | 20, 27 |
| F | | | Unit G4 | | | 2.25 | 0.04 | Unit G4 (Sh) | GPTS Reunion I base | 20, 27 |
| F | | | Tuff G | | | 2.32 | 0.03 | Tuff G (Sh) | K-Ar | 8 |
| F | | | Units F3/4 | | | 2.37 | | Units F3/4 (Sh) | GPTS X(N) subchron | 14 |
| E | Kalochoro | Ekalalei Tuff | Tuff F1 | | | | | | | 14 |
| E | | Kalochoro Tuff | Tuff F | | | 2.34 | 0.05 | Tuff F (Sh) | K-Ar | 8, 14 |
| E | | Kokiselei Tuff | Tuff E | | | 2.40 | 0.05 | Tuff E (Sh) | Stratigraphic scaling | 14 |
| D | Lok | Lokalelei Tuff | Tuff D | | Lokalalei Tuff | 2.52 | 0.05 | Tuff D (Sh) | K-Ar, Ar-Ar | 8, 14, 16 |
| D | Lomekwi | Emekwi Tuff | Tuff C9 | | | 2.581 | | Emekwi (Nk) | GPTS Gauss/Matuyama | 11, 14, 14, 10 |
| C | | | Unit C9 | | Basal Burgi Mb | 2.581 | | Unit C9 (Sh) | GPTS Gauss/Matuyama | 2, 14, 10 |
| C | | | | | Burgi Tuff | 2.64 | 0.05 | Burgi Tuff (KF) | K-Ar | 14 |
| C | | | Tuff C4 | | Ingumwai Tuff | 2.74 | 0.08 | Tuff C4 (Sh) | Stratigraphic scaling | 14 |
| B | | | Tuff C (α) | Tulu Bor | Hasuma Tuff | 2.85 | 0.08 | Tuff C (Sh) | Stratigraphic scaling | 14 |
| B | | | Tuff B10 | | | 2.95 | 0.05 | Tuff B-10 (Sh) | K-Ar | 8 |
| B | | | | | | 3.04 | | | GPTS Kaena top | 1, 10, 27 |
| B | | | | | Ninikaa Tuff | 3.08 | 0.03 | Ninikaa (KF) | K-Ar, Ar-Ar | 24 |
| B | | | | | | 3.11 | | | GPTS Kaena base | 1, 10, 27 |
| B | | | U14 | | Karo | 3.2 | | U14 (Us) | Stratigraphic scaling | 9 |
| B | | | | | Allia Tuff | 3.22 | 0.05 | Allia Tuff (KF) | GPTS Mammoth top | 14 |
| B | | | Unit B2 (U12) | | | 3.22 | | Unit B2 (Sh) | GPTS Mammoth base | 2, 10, 27 |
| B | | Tulu Bor Tuff (beta) | Unit B1 (U11-12) | | | 3.33 | | Unit B1 (Sh) | | 2, 10, 27 |
| B | | | | | Toroto Tuff | 3.32 | 0.02 | Toroto (KF) | K-Ar, Ar-Ar | 24 |
| B | Kataboi | Burrowed bed | Tuff B-β (U10) | | Tulu Bor Tuff (β) | 3.40 | 0.03 | SHT (Hadar) | Ar-Ar | 2, 28, 32 |
| B | | | | | | 3.53 | | | Interpolation | 23 |
| A | Moi Lok | Lokochot Tuff | Tuff A (U6) | Moi Lok | Lokochot Tuff | 3.594 | | Tuff A (Sh) | GPTS Gilbert/Gauss | 12, 14, 29 |
| A | | | | | Wargolo Tuff | 3.77 | | VT-3 (Maka) | | 15, 33 |
| | | Moiti Tuff | Usno 1 | | Moiti Tuff | 3.94 | 0.04 | Moiti Tuff (Nk) | Ar-Ar | 23, 24, 33 |

| | | Age | ± | Unit | Method | References |
|---|---|---|---|---|---|---|
| Topernawi Tuff | Topernawi Tuff | 3.96 | 0.03 | Topernawi Tuff (Nk) | Ar-Ar | 14, 23 |
| | Kataboi Basalt | 4.05 | 0.06 | Basalt (Nk) | K-Ar | 8 |
| | Kanapoi Tuff | 4.07 | 0.02 | KT (Kp) | Ar-Ar | 21, 22 |
| Usno Basalt | | 4.10 | 0.06 | Basalt (Us) | K-Ar, Ar-Ar | 8, 11, 17 |
| Mursi Basalt | | 4.2 | 0.2 | Basalt (Mu) | K-Ar | 8, 14, 17 |
| | Kanapoi upper pumiceous tuff | 4.12 | 0.02 | tuff (Kp) | Ar-Ar | 21 |
| | Kanapoi lower pumiceous tuff | 4.17 | 0.03 | tuff (Kp) | Ar-Ar | 21 |
| | Lothagam Basalt | 4.20 | 0.03 | Lothagam Basalt (Nw) | K-Ar, Ar-Ar | 13, 17, 25 |
| | Tuff in Apak Mb | 4.22 | 0.03 | Tuff in Apak Mb (Nk) | Ar-Ar | 13, 25 |
| | Karsa Basalt | 4.35 | 0.05 | Karsa Basalt (KF) | K-Ar | 14, 17, 24 |
| | | 5.3 | | Miocene/Pliocene boundary | | 1, 10 |
| | Marker Tuff | 6.54 | 0.04 | Marker Tuff (Nw) | Ar-Ar | 13 |
| | Lower Markers | 7.44 | 0.05 | Lower Markers (Nw) | Ar-Ar | 25 |

Lonyumun

**References:**

1) Berggren et al., 1995; 2) Brown, 1983; 3) Brown, 1995; 4) Brown and Nash, 1976; 5) Brown and Feibel, 1985; 6) Brown and Feibel, 1986; 7) Brown and McDougall, 1993; 8) Brown et al., 1985; 9) Brown et al., 1992; 10) Cande and Kent, 1995; 11) de Heinzelin, 1983; 12) de Heinzelin and Haesaerts, 1983; 13) Feibel, 2003; 14) Feibel et al., 1989; 15) Haileab and Brown, 1992; 16) Haileab and Brown, 1994; 17) Haileab et al., 2004; 18) Harris et al., 1988; 19) Howell et al., 1987; 20) Lanphere et al., 2002; 21) Leakey et al., 1995; 22) Leakey et al., 1998; 23) Leakey et al., 2001; 24) McDougall, 1985; 25) McDougall and Feibel, 2003; 26) McDougall et al., 1985; 27) McDougall et al., 1992; 28) Sa:na-Wojcicki et al., 1985; 29) Shackleton, 1995; 30) Tauxe et al., 1992; 31) Van Couvering, 1996; 32) Walter and Aronson, 1993; 33) White et al., 1993.

Key to Abbreviations: Sh = Shungura Formation, Nk = Nachukui Formation, KF = Koobi Fora Formation; GPTS = Global Polarity Time Scale; black indicates intervals of normal polarity and white indicates intervals of reversed polarity.

Note: Ages given of sub-members reflect the age at the base of the unit, unless noted otherwise. For the three major regions of the Turkana Basin, alternating gray shading and white are used to separate successive members within each formation. Gray bands on right side of the table highlight the major Epoch boundaries.

sequence. Thus, the specimens analyzed here derive from the well-studied Hadar deposits north of the Awash River and the A.L. 666 locality in the Maka'amitalu drainage.

## PALEONTOLOGICAL DATABASES

One of the earliest efforts in East Africa to maintain an electronic catalogue of collected specimens was made in the 1960s and 1970s by the Omo Research Expedition (Eck, 2007). Researchers today routinely keep their paleontological records in electronic databases, but there is little standardization of fields or compatibility among the various formats that are used. Following the model of the Shungura Formation Catalogue of Fossil Specimens (compiled by G. Eck), we have developed a network of paleontological databases which, although independent of each other, maintain comparable fields of information about each fossil specimen and its geological context. The Shungura Formation American Catalogue has 22,335 records, most of which consist of fossil mammals. A separate catalogue documenting specimens from the French Shungura collection has nearly 27,000 records (Alemseged et al., 2007). There is also a Mursi Formation Catalogue, with 142 records and an Usno Formation Catalogue, with 2525 records (maintained by G. Eck and R. Bobe). For the rest of the Turkana Basin (Koobi Fora, Nachukui, Kanapoi, and Nawata Formations), we have compiled the Turkana Basin Paleontology Database, which has about 16,500 records of fossil mammals (and is maintained by R. Bobe, A.K. Behrensmeyer, E. Mbua, and M. Leakey). This database has been created by a collaborative project between the National Museums of Kenya and the Smithsonian Institution, and is scheduled to be posted online by these institutions. A separate database (created and maintained by G. Eck) archives records of fossil vertebrates from Hadar; the Catalogue of Hadar Fossil Specimens has 8131 records. This network of databases relies on FileMaker Pro software, but records can be easily exported to other formats. We use this network of databases to analyze patterns of relative abundance and diversity in bovids from the late Cenozoic of the Turkana and Hadar Basins. We exclude the French Shungura database because it is being separately analyzed by Alemseged and colleagues (2007).

## Methods

In order to assess patterns of faunal change over intervals of several million years, we use the geological member as our basic "sampling unit" (Table 2) (Ludwig and Reynolds, 1988). A few members spanning long time intervals were subdivided as long as adequate sample size could be maintained (n >50 specimens). For example, the Lomekwi Member of the Nachukui Formation is divided into lower, middle, and upper sections; we combined the lower with the middle sections, and the upper section with the overlying Lokalalei Member. Member B of the Shungura Formation has a rich fossil record in the upper sections, but a very poor record in the lower ones. The rich record of the Usno Formation (U-12) correlates with lower Member B, and therefore we separate Member B into a lower section that includes Usno, here labeled as B(L), and an upper section, B(U). Shungura Member G is divided into a fluvio-deltaic lower section, G(L), and a largely lacustrine upper section, G(U). Analyses at finer levels of temporal resolution are possible given the rich fossil record of some sections of the Turkana and Hadar deposits (Bobe et al., 2002), but the goal here is to detect the major trends over long time intervals. Geological members with small numbers of bovid specimens (<50) were combined with adjacent members (e.g., Kataboi-Kaiyumung, Upper Lomekwi-Lokalalei, and Natoo-Nariokotome) so that

*Table 2. Abundance of bovid tribes (number of specimens) across geological members from the Hadar, Busidima, Shungura, Usno, Koobi Fora, Nachukui, Kanapoi, and Nawata Formations*

| Formation | Member or unit | BOVID TRIBES | | | | | | | | |
|---|---|---|---|---|---|---|---|---|---|---|
| | | Aepycerotini | Alcelaphini | Antilopini | Boselaphini | Bovini | Hippotragini | Reduncini | Tragelaphini | Total |
| Nawata | L. Nawata | 72 | 7 | 1 | 37 | 2 | 8 | 17 | 5 | 149 |
| Nawata | U. Nawata | 58 | 25 | 2 | 11 | 3 | 8 | 31 | 3 | 141 |
| Nachukui | Apak Mb | 20 | 9 | 3 | 2 | 6 | 3 | 6 | 13 | 62 |
| Kanapoi | Kanapoi | 18 | 23 | 2 | 0 | 4 | 4 | 8 | 52 | 111 |
| Nachukui | Kataboi-Kaiyumung | 14 | 10 | 4 | 0 | 7 | 2 | 8 | 13 | 58 |
| Nachukui | Lomekwi (L-M) | 79 | 102 | 42 | 0 | 10 | 17 | 61 | 36 | 347 |
| Nachukui | Lomekwi(U)-Lokalalei | 10 | 46 | 10 | 0 | 9 | 6 | 65 | 14 | 160 |
| Nachukui | Kalochoro | 10 | 27 | 14 | 0 | 5 | 5 | 40 | 9 | 110 |
| Nachukui | Kaitio | 9 | 23 | 3 | 0 | 10 | 4 | 14 | 30 | 93 |
| Nachukui | Natoo-Nariokotome | 9 | 50 | 16 | 0 | 8 | 3 | 43 | 11 | 140 |
| Koobi Fora | Lokochot | 16 | 9 | 4 | 0 | 4 | 0 | 10 | 24 | 67 |
| Koobi Fora | Tulu Bor | 4 | 14 | 7 | 0 | 25 | 0 | 96 | 39 | 185 |
| Koobi Fora | Burgi(U) | 36 | 59 | 47 | 0 | 12 | 5 | 194 | 65 | 418 |
| Koobi Fora | KBS | 78 | 172 | 81 | 0 | 66 | 14 | 318 | 121 | 850 |
| Koobi Fora | Okote | 6 | 35 | 7 | 0 | 15 | 0 | 71 | 33 | 167 |
| Usno | B(L) | 128 | 4 | 3 | 0 | 28 | 0 | 4 | 50 | 217 |
| Shungura | B(U) | 49 | 7 | 1 | 0 | 24 | 0 | 63 | 32 | 176 |
| Shungura | C | 94 | 5 | 3 | 0 | 73 | 3 | 74 | 178 | 430 |
| Shungura | D | 56 | 6 | 1 | 0 | 15 | 0 | 30 | 66 | 174 |
| Shungura | E | 91 | 9 | 0 | 0 | 13 | 1 | 56 | 117 | 287 |
| Shungura | F | 141 | 25 | 5 | 0 | 15 | 0 | 69 | 87 | 342 |
| Shungura | G(L) | 461 | 51 | 4 | 0 | 38 | 2 | 684 | 375 | 1615 |
| Shungura | G(U) | 24 | 13 | 9 | 0 | 2 | 0 | 29 | 2 | 79 |
| Shungura | H | 7 | 12 | 3 | 0 | 2 | 0 | 125 | 4 | 153 |
| Shungura | J | 6 | 8 | 0 | 0 | 1 | 0 | 39 | 0 | 54 |
| Shungura | K | 5 | 15 | 2 | 0 | 1 | 0 | 27 | 1 | 51 |
| Shungura | L | 4 | 11 | 1 | 0 | 2 | 0 | 42 | 2 | 62 |
| Hadar | SH | 146 | 94 | 16 | 0 | 83 | 12 | 53 | 92 | 496 |
| Hadar | DD | 132 | 127 | 5 | 0 | 77 | 7 | 298 | 103 | 749 |
| Hadar | KH | 28 | 46 | 46 | 0 | 24 | 7 | 15 | 33 | 199 |
| Busidima | Maka'amitalu | 2 | 11 | 11 | 0 | 14 | 5 | 10 | 18 | 71 |
| | Total | 1813 | 1055 | 353 | 50 | 598 | 116 | 2600 | 1628 | 8213 |

the total sample size per interval exceeds 50 specimens (Table 2). Thus we have tried to maximize resolution across geological formations while maintaining samples adequate for statistical analyses.

There are 12 bovid tribes represented in the late Cenozoic fossil record of the Turkana Basin: Aepycerotini, Alcelaphini, Antilopini, Boselaphini, Bovini, Caprini, Cephalophini, Hippotragini, Neotragini, Ovibovini, Reduncini, and Tragelaphini. Boselaphini are found only in the Nawata Formation and the lowermost Nachukui Formation (Apak Member) at Lothagam (Harris, 2003). Caprini, Cephalophini, Neotragini, and Ovibovini have a sparse fossil record and therefore are not used here in the analyses of abundance, but they are used in analyses of diversity. Our analyses of taxonomic abundance are based on numbers of fossil specimens, counting multiple catalogued specimens of the same individual as one. Because of taphonomic processes, these abundances likely are a biased representation of actual live abundances in the original ecosystems. However, these biases can be controlled by restricting the analysis to the most commonly collected and identifiable elements of the bovid skeleton, primarily horn cores and teeth. We assess relative abundances with two methods: correspondence analysis and changes in tribe-specific proportions through time. Correspondence analysis provides a visual assessment of contingency tables with cells containing frequency counts (Table 2). Correspondence analysis displays graphically the relationship between two nominal variables (e.g., geological members and bovid taxa). Categories that are similar to each other appear close together in the graphic display, and those that are different occur farther apart (Benzécri,

1992; Greenacre, 1993). The graphical output (Figure 2) shows each taxon pulled toward the geological members in which the taxon has high relative abundance. Thus we may obtain associations of taxa that have high abundances in particular members. Correspondence analysis also distributes members (sampling units) in relation to the taxa they contain. Interpretations of these graphs consist of examining the spread of taxa and sampling units across each axis in search of underlying ecological or environmental features that may explain the spread of points (Greenacre and Vrba, 1984). The correspondence analysis shown in Figure 2 excludes the Nawata Formation because the high abundance of Boselaphini in the late Miocene makes this sequence distinctly different from the later deposits in the Turkana and Hadar areas. Boselaphini became rare or absent from the East African fossil record after

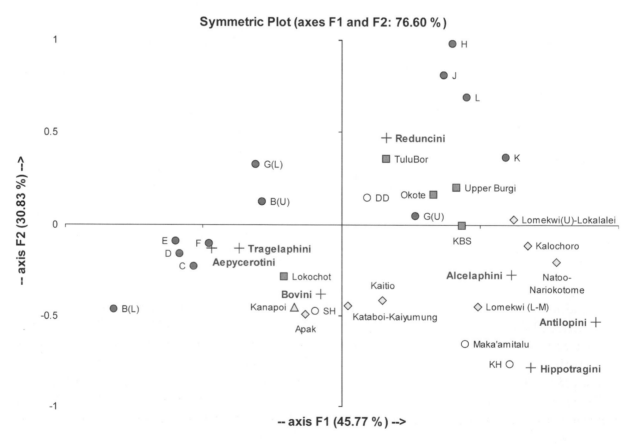

Figure 2. Correspondence analysis of bovid tribes in the Shungura, Usno, Koobi Fora, Nachukui, Kanapoi, Hadar, and Busidima Formations (Maka'amitalu drainage).

the Miocene/Pliocene boundary (Vrba, 1995a; Harris, 2003).

The correspondence analysis display (Figure 2) shows an association among Alcelaphini, Antilopini, and Hippotragini, and this association is confirmed by different methods with a cluster analysis (Figure 3). Modern Alcelaphini (wildebeests, hartebeests, and topi) are cursorial and hypsodont grazers that typically occur in open grasslands or wooded grasslands (Kingdon, 1982b; Gagnon and Chew, 2000), and their Plio-Pleistocene ancestors may have shared those habits (Kappelman et al., 1997; Spencer, 1997; Sponheimer et al., 1999). Antilopini are cursorial and hypsodont grazers or browsers in arid bushland–grasslands (Kingdon, 1982b). Although not all species of Antilopini are grazers (Sponheimer et al., 1999; Gagnon and Chew, 2000; Cerling et al., 2003), they tend to occur in arid grasslands or bushland, and even in deserts. Hippotragini are cursorial and hypsodont grazers that occur in wooded grasslands, grasslands, or semideserts (Kingdon, 1982b). The

close association among these three tribes (Alcelaphini, Antilopini, and Hippotragini) in both the correspondence and the cluster analyses suggests that they may be combined and treated as a variable. Vrba (1980) has argued that the abundance of Alcelaphini and Antilopini as a proportion of the entire bovid fauna may be used as an indication of open and arid conditions. As noted above, Hippotragini (especially the genera *Oryx* and *Addax*) are likewise well adapted to open and arid conditions (Kingdon, 1982a). Thus we use the Alcelaphini–Antilopini–Hippotragini (AAH) criterion as a variation of Vrba's (1980) Alcelaphini–Antilopini criterion of open-dry environments, usually savanna grasslands, bushland or woodland–grassland mosaics. It should be noted that Reduncini and Bovini are also grazers, but these bovids consume fresh grasses in moist habitats, woodland clearings, or near waterlogged conditions (Kingdon, 1982b), and therefore are poor indicators of seasonally arid grasslands. AAH is measured as the combined proportion of Alcelaphini, Antilopini, and Hippotragini among all bovid specimens that can be identified to tribe. Confidence limits of AAH proportions are calculated with the methods of Buzas and Hayek (Buzas, 1990; Hayek and Buzas, 1997).

There have been notable efforts to identify the feeding adaptations and habitat preferences of extinct bovids through analyses of ecomorphology (Plummer and Bishop, 1994; Kappelman et al., 1997; Spencer, 1997; Sponheimer et al., 1999; DeGusta and Vrba, 2003). These studies indicate that the habitat preferences of bovid tribes have remained relatively consistent over the last few million years, even if some taxa have changed their feeding behavior since the Pliocene. Alcelaphini and Antilopini in particular appear to have radiated into seasonally arid habitats early in their evolutionary history (Greenacre and Vrba, 1984; Vrba, 1987).

The AAH variable is not meant to provide a direct translation of habitat type at any one

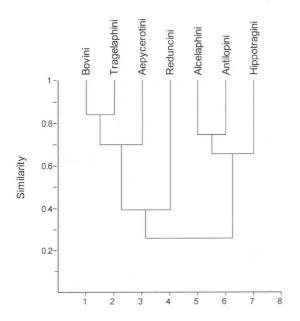

Figure 3. Cluster analysis of bovid tribes based on abundance data across geological members from the Turkana and Hadar areas. Paired groups using rho similarity measure with Past software.

time. An AAH abundance of 50% does not necessarily mean that the environment from which these bovids derive was composed of grasslands over 50% of its area. Several biases influence this proportion. For example, most collections are probably biased against small bovids such as duikers or dik-diks, which may fragment more easily or be more difficult to see in surface surveys than larger bovids. But if taphonomic and collection biases are relatively constant among sampling units, then changes in this variable over time should convey important environmental information. Thus, we assume that a significant increase in AAH means an increase in the importance of seasonally arid environments in the landscape, even though we do not know exactly the proportions of different vegetation types on that landscape.

In analyses of relative abundance in the fossil record it is important to evaluate the taphonomic context of the specimens under consideration. In previous analyses of relative abundance we have done just that (Bobe and Eck, 2001; Bobe et al., 2002): we have considered the sedimentological context and used skeletal elements as a proxy measure of taphonomic conditions. After controlling for taphonomic and sedimentologic conditions, we have found that changes in taxonomic relative abundances can provide important environmental and ecological information. In this regard it is also critical to consider collection methods in the field that may bias the proportions of the taxa being analyzed. Eck (2007) provides an important analysis of collection effort in relation to the Shungura Formation. A detailed assessment of taphonomy, sedimentology, and collection biases in the Turkana and Hadar areas is beyond the scope of this paper, although efforts in this regard are in progress (Behrensmeyer et al., 2004; Campisano et al., 2004). Thus, we assume for the time being that these processes bias relative abundances more or less equally across geological members. The results presented here would not be significantly altered as long as these biases affect the sampling units to a similar degree. However, until further research can be done to assess how taphonomic and collection factors affected the catalogued fossil collections from the Turkana and Hadar Basins, our results should be considered as hypotheses and subjected to further testing.

In the analysis of diversity we have focused on species richness, i.e., the number of bovid taxa per time interval. In the assessment of species richness over time it is desirable to consider approximately equal time intervals. We have divided the time span from 4.2 to 1 Ma into intervals of 200 Kyr (thousand years) (Table 3). Finer levels of temporal resolution are possible for the Shungura record, which often contains information at the submember level, but resolution tends to be coarser elsewhere. Many of the fossils analyzed here derive from geological units that do not fall neatly into 200-Kyr categories. In these cases we have assigned specimens to the 200-Kyr intervals encompassed by the geological unit. For example, the Kaitio Member of the Nachukui Formation spans from 1.88 Ma to about 1.6 Ma, and we have assigned taxa that occur in that member to both the 2.0–1.8 Ma interval and the 1.8–1.6 Ma interval. This procedure undoubtedly introduces an error into the richness calculations, and highlights the need to assign fossil specimens to finer levels of stratigraphic resolution. Nevertheless, by using broad time intervals we have tried to match the scale of the analysis to the quality of the record.

## Results

### CORRESPONDENCE ANALYSIS

The correspondence analysis depicted in Figure 2 shows the distribution of bovid tribes with respect to their abundance across geological members in the Shungura, Usno, Nachukui,

*Table 3. Turkana Basin Bovidae across time intervals in the late Cenozoic*

| Time in Ma | 7.44-6.54 | 6.54-5.23 | 4.5-4.22 | 4.2-4.0 | 4.0-3.8 | 3.8-3.6 | 3.6-3.4 | 3.4-3.2 | 3.2-3.0 | 3.0-2.8 | 2.8-2.6 | 2.6-2.4 | 2.4-2.2 | 2.2-2.0 | 2.0-1.8 | 1.8-1.6 | 1.6-1.4 | 1.4-1.2 | 1.2-1.0 | Total |
|---|---|---|---|---|---|---|---|---|---|---|---|---|---|---|---|---|---|---|---|---|
| Omo | | | | Mursi | | | A | B2-3-U12 | B6-B9 | B10-C2 | C3-C9 | D0-5 | E-F-G8 | G9-23 | G24-H | J | K | L0-L4 | L5-L9 | |
| West Turkana | Nawata (L) | Nawata (U) | Apak | Kanapoi | Kataboi-Kaiyumung | Kataboi-Kaiyumung | Kataboi-Kaiyumung | L. Lomekwi | L. Lomekwi | M. Lomekwi LO10 | M. Lomekwi LO9 | U. Lomekwi-Lokalalei | Kalochoro | Kalochoro | Kalochoro-Kaitio | Kaitio | Natoo | Nariokotome | Nariokotome | |
| East Turkana | | | | Lonyumun | Lonyumun-Moiti | Moiti | Lokochot | L. Tulu Bor | L. Tulu Bor | U. Tulu Bor | U. Tulu Bor | L. Burgi | Unconformity | Unconformity | U. Burgi-KBS | KBS | Okote | Chari | Chari | |
| **AEPYCEROTINI** | | | | | | | | | | | | | | | | | | | | |
| Aepyceros premelampus | 70 | 58 | 20 | 7 | 3 | 3 | 3 | 0 | 0 | 0 | 0 | 0 | 0 | 0 | 0 | 0 | 0 | 0 | 0 | 164 |
| Aepyceros shungurae-melampus | 0 | 0 | 0 | 0 | 0 | 3 | 20 | 176 | 10 | 50 | 95 | 60 | 408 | 290 | 97 | 55 | 14 | 5 | 2 | 1285 |
| Aepyceros sp. Nov. A | 0 | 0 | 0 | 0 | 0 | 0 | 0 | 0 | 0 | 0 | 0 | 0 | 6 | 0 | 0 | 0 | 0 | 0 | 0 | 6 |
| Aepyceros sp. Nov. B | 0 | 0 | 0 | 0 | 0 | 0 | 0 | 0 | 0 | 0 | 0 | 0 | 0 | 0 | 0 | 0 | 1 | 0 | 0 | 1 |
| **ALCELAPHINI** | | | | | | | | | | | | | | | | | | | | |
| Beatragus antiquus | 0 | 0 | 0 | 0 | 0 | 0 | 0 | 0 | 0 | 0 | 0 | 0 | 0 | 0 | 3 | 8 | 2 | 0 | 0 | 13 |
| Beatragus hunteri | 0 | 0 | 0 | 0 | 0 | 0 | 0 | 0 | 0 | 0 | 0 | 0 | 0 | 0 | 0 | 0 | 0 | 0 | 1 | 1 |
| Connochaetes aff. gentryi | 0 | 0 | 0 | 0 | 0 | 0 | 7 | 3 | 0 | 0 | 0 | 0 | 0 | 0 | 0 | 0 | 0 | 0 | 0 | 10 |
| Connochaetes gentryi | 0 | 0 | 0 | 0 | 0 | 0 | 0 | 0 | 0 | 0 | 0 | 2 | 0 | 0 | 13 | 16 | 9 | 0 | 0 | 40 |
| Damalacra sp. A | 4 | 23 | 4 | 0 | 0 | 1 | 0 | 0 | 0 | 0 | 0 | 0 | 0 | 0 | 0 | 0 | 0 | 0 | 0 | 32 |
| Damalacra sp. B | 1 | 0 | 2 | 0 | 0 | 0 | 0 | 0 | 0 | 0 | 0 | 0 | 0 | 0 | 0 | 0 | 0 | 0 | 0 | 3 |
| Damaliscus (Parmularius) | 0 | 0 | 0 | 0 | 0 | 0 | 0 | 0 | 0 | 0 | 5 | 0 | 0 | 0 | 33 | 33 | 7 | 0 | 0 | 78 |
| Damaliscus cf. niro | 0 | 0 | 0 | 0 | 0 | 0 | 0 | 0 | 0 | 0 | 0 | 0 | 0 | 0 | 0 | 3 | 0 | 0 | 0 | 3 |
| Damalops aff. palaeindicus | 0 | 0 | 0 | 0 | 0 | 0 | 0 | 0 | 0 | 1 | 0 | 0 | 0 | 0 | 0 | 0 | 0 | 0 | 0 | 1 |
| Megalotragus isaaci | 0 | 0 | 0 | 0 | 0 | 0 | 0 | 0 | 0 | 0 | 0 | 0 | 0 | 0 | 35 | 37 | 9 | 0 | 0 | 81 |

*(Continued)*

Table 3. Turkana Basin Bovidae across time intervals in the late Cenozoic—cont'd

| Taxon | | | | | | | | | | | | | | | |
|---|---|---|---|---|---|---|---|---|---|---|---|---|---|---|---|
| Megalotragus kattwinkeli | 7 | 1 | 1 | 0 | 0 | 1 | 1 | 0 | 0 | 0 | 0 | 0 | 0 | 0 | 0 |
| Parmularius altidens | 3 | 0 | 0 | 1 | 0 | 1 | 1 | 0 | 0 | 0 | 0 | 0 | 0 | 0 | 0 |
| cf. Parmularius angusticornis | 2 | 0 | 0 | 0 | 0 | 0 | 0 | 0 | 1 | 1 | 0 | 0 | 0 | 0 | 0 |
| **ANTILOPINI** | | | | | | | | | | | | | | | |
| Antidorcas recki | 90 | 0 | 3 | 49 | 18 | 5 | 4 | 1 | 1 | 1 | 3 | 4 | 1 | 0 | 0 |
| Antilope aff. subtorta | 2 | 0 | 0 | 0 | 0 | 0 | 0 | 0 | 2 | 0 | 0 | 0 | 0 | 0 | 0 |
| Gazella cf. granti | 4 | 0 | 1 | 1 | 0 | 0 | 0 | 1 | 0 | 1 | 0 | 0 | 1 | 0 | 0 |
| Gazella janenschi | 15 | 0 | 2 | 4 | 4 | 0 | 0 | 0 | 1 | 0 | 2 | 2 | 0 | 1 | 0 |
| Gazella praethomsoni | 52 | 0 | 5 | 11 | 19 | 2 | 4 | 1 | 1 | 1 | 2 | 2 | 4 | 0 | 0 |
| **BOSELAPHINI** | | | | | | | | | | | | | | | |
| Tragoportax cf. cyrenaicus | 10 | 0 | 0 | 0 | 0 | 0 | 0 | 0 | 0 | 0 | 0 | 0 | 0 | 2 | 1 |
| Tragoportax sp. A | 6 | 0 | 0 | 0 | 0 | 0 | 0 | 0 | 0 | 0 | 0 | 0 | 0 | 0 | 2 |
| Tragoportax sp. B | 3 | 0 | 0 | 0 | 0 | 0 | 0 | 0 | 0 | 0 | 0 | 0 | 0 | 0 | 2 |
| **BOVINI** | | | | | | | | | | | | | | | |
| Pelorovis oldowayensis | 21 | 0 | 7 | 12 | 2 | 0 | 0 | 0 | 0 | 0 | 0 | 0 | 0 | 0 | 0 |
| Pelorovis turkanensis | 68 | 1 | 8 | 53 | 6 | 0 | 0 | 0 | 0 | 0 | 0 | 0 | 0 | 0 | 0 |
| Simatherium cf. kohllarseni | 30 | 0 | 0 | 0 | 0 | 0 | 0 | 0 | 0 | 11 | 0 | 14 | 2 | 0 | 0 |
| Simatherium shungurense | 1 | 0 | 0 | 0 | 0 | 1 | 0 | 0 | 0 | 0 | 0 | 0 | 0 | 0 | 0 |
| Syncerus acoelotus | 23 | 2 | 0 | 0 | 1 | 3 | 4 | 0 | 13 | 0 | 0 | 0 | 0 | 0 | 0 |
| Syncerus cf. caffer | 2 | 0 | 0 | 0 | 0 | 0 | 0 | 1 | 0 | 0 | 0 | 0 | 0 | 0 | 0 |
| Ugandax sp. Nov. | 11 | 0 | 0 | 0 | 0 | 1 | 1 | 0 | 1 | 2 | 3 | 0 | 0 | 0 | 0 |
| **CAPRINI** | | | | | | | | | | | | | | | |
| Caprini sp. 1 | 2 | 0 | 0 | 0 | 0 | 1 | 1 | 0 | 0 | 0 | 0 | 0 | 0 | 0 | 0 |
| Caprini sp. A | 2 | 1 | 0 | 0 | 1 | 0 | 0 | 1 | 0 | 0 | 0 | 0 | 0 | 0 | 0 |
| Caprini sp. B | 1 | 0 | 1 | 0 | 0 | 0 | 0 | 0 | 0 | 0 | 0 | 0 | 0 | 0 | 0 |
| Caprini sp. C | 2 | 1 | 0 | 0 | 0 | 0 | 0 | 0 | 1 | 0 | 0 | 0 | 0 | 0 | 0 |
| **CEPHALOPHINI** | | | | | | | | | | | | | | | |
| Cephalophus | 2 | 0 | 1 | 1 | 0 | 0 | 0 | 0 | 0 | 0 | 0 | 0 | 0 | 0 | 0 |
| **HIPPOTRAGINI** | | | | | | | | | | | | | | | |
| Hippotragus sp. | 4 | 0 | 0 | 0 | 0 | 0 | 0 | 0 | 0 | 0 | 0 | 0 | 2 | 0 | 2 |
| Hippotragus gigas | 8 | 0 | 1 | 0 | 3 | 0 | 0 | 0 | 1 | 0 | 0 | 0 | 0 | 0 | 0 |
| Oryx sp. | 16 | 0 | 0 | 12 | 2 | 0 | 0 | 1 | 1 | 0 | 0 | 0 | 0 | 0 | 0 |
| Praedamalis sp. | 6 | 0 | 0 | 0 | 0 | 0 | 0 | 0 | 0 | 0 | 0 | 0 | 2 | 0 | 2 |
| **NEOTRAGINI** | | | | | | | | | | | | | | | |
| Madoqua | 9 | 0 | 0 | 0 | 2 | 0 | 0 | 0 | 0 | 0 | 0 | 0 | 1 | 1 | 1 |
| Raphicerus | 15 | 0 | 0 | 5 | 1 | 2 | 2 | 0 | 0 | 0 | 0 | 0 | 2 | 1 | 1 |
| **OVIBOVINI** | | | | | | | | | | | | | | | |
| Makapania | 3 | 0 | 0 | 0 | 1 | 0 | 1 | 0 | 0 | 0 | 0 | 0 | 0 | 0 | 0 |

| | 1 | 2 | 3 | 4 | 5 | 6 | 7 | 8 | 9 | 10 | 11 | 12 | 13 | 14 | 15 | 16 | 17 | 18 | 19 | Total |
|---|---|---|---|---|---|---|---|---|---|---|---|---|---|---|---|---|---|---|---|---|
| **REDUNCINI** | | | | | | | | | | | | | | | | | | | | |
| Kobus ancystrocera | 0 | 0 | 0 | 0 | 0 | 0 | 3 | 2 | 3 | 11 | 0 | 6 | 25 | 6 | 44 | 0 | 0 | 0 | 0 | 100 |
| Kobus ellipsiprymnus | 0 | 0 | 0 | 0 | 0 | 0 | 0 | 0 | 0 | 0 | 0 | 1 | 0 | 1 | 5 | 2 | 0 | 1 | 0 | 10 |
| Kobus cf. kob | 0 | 0 | 0 | 0 | 0 | 1 | 1 | 1 | 2 | 2 | 1 | 0 | 1 | 0 | 0 | 0 | 0 | 0 | 0 | 9 |
| Kobus kob | 0 | 0 | 0 | 0 | 0 | 0 | 0 | 0 | 0 | 0 | 0 | 0 | 0 | 0 | 29 | 65 | 6 | 0 | 0 | 100 |
| Kobus laticornis | 0 | 10 | 0 | 0 | 0 | 0 | 0 | 0 | 0 | 0 | 0 | 0 | 0 | 0 | 0 | 1 | 0 | 1 | 0 | 10 |
| Kobus aff. Leche | 0 | 0 | 0 | 0 | 0 | 0 | 0 | 0 | 0 | 0 | 0 | 0 | 0 | 3 | 15 | 0 | 0 | 0 | 0 | 20 |
| Kobus oricornus | 0 | 0 | 0 | 2 | 2 | 6 | 11 | 12 | 7 | 14 | 0 | 0 | 0 | 0 | 0 | 0 | 0 | 0 | 0 | 54 |
| Kobus presigmoidalis | 5 | 13 | 4 | 0 | 0 | 0 | 0 | 0 | 0 | 0 | 2 | 0 | 0 | 0 | 0 | 0 | 0 | 0 | 0 | 23 |
| Kobus sigmoidalis | 0 | 0 | 0 | 0 | 2 | 2 | 6 | 6 | 1 | 8 | 11 | 204 | 40 | 108 | 81 | 4 | 0 | 0 | 0 | 473 |
| Menelikia leakeyi | 4 | 3 | 0 | 0 | 13 | 0 | 4 | 4 | 5 | 11 | 1 | 0 | 0 | 0 | 0 | 0 | 0 | 0 | 0 | 45 |
| Menelikia lyrocera | 0 | 0 | 0 | 0 | 1 | 0 | 1 | 1 | 0 | 1 | 2 | 94 | 45 | 51 | 84 | 3 | 0 | 0 | 0 | 283 |
| Redunca | 0 | 0 | 0 | 0 | 0 | 0 | 0 | 0 | 1 | 2 | 0 | 2 | 0 | 0 | 0 | 1 | 1 | 0 | 0 | 7 |
| **TRAGELAPHINI** | | | | | | | | | | | | | | | | | | | | |
| Taurotragus arkelli-oryx | 0 | 0 | 0 | 0 | 0 | 0 | 0 | 0 | 0 | 0 | 0 | 0 | 0 | 0 | 0 | 1 | 1 | 0 | 0 | 2 |
| Tragelaphus gaudryi | 0 | 0 | 0 | 0 | 0 | 0 | 0 | 0 | 0 | 0 | 0 | 162 | 39 | 2 | 0 | 0 | 0 | 0 | 0 | 203 |
| Tragelaphus kyaloae | 0 | 1 | 7 | 38 | 3 | 3 | 0 | 0 | 0 | 0 | 0 | 0 | 0 | 0 | 0 | 0 | 0 | 0 | 0 | 65 |
| Tragelaphus nakuae | 0 | 0 | 1 | 0 | 2 | 19 | 60 | 18 | 31 | 176 | 62 | 301 | 93 | 21 | 5 | 1 | 0 | 0 | 0 | 794 |
| Tragelaphus pricei | 0 | 0 | 0 | 0 | 0 | 0 | 0 | 0 | 0 | 1 | 0 | 0 | 0 | 0 | 0 | 0 | 0 | 0 | 0 | 1 |
| Tragelaphus scriptus | 0 | 0 | 0 | 0 | 0 | 0 | 0 | 0 | 0 | 0 | 1 | 0 | 0 | 0 | 1 | 0 | 0 | 0 | 1 | 3 |
| Tragelaphus strepsiceros | 0 | 0 | 1 | 0 | 0 | 0 | 0 | 0 | 0 | 0 | 0 | 0 | 3 | 38 | 105 | 21 | 2 | 1 | 1 | 170 |
| Sample size | 102 | 123 | 44 | 48 | 32 | 71 | 291 | 62 | 115 | 349 | 150 | 1203 | 549 | 470 | 673 | 174 | | | | 4510 |
| Number of taxa | 12 | 13 | 9 | 5 | 11 | 14 | 15 | 12 | 12 | 21 | 17 | 18 | 14 | 24 | 26 | 23 | 20 | 11 | 9 | |

Koobi Fora, and Hadar Formations. It is noteworthy that bovid tribes in this analysis separate strongly into recognizable habitat categories. The bovid tribes Tragelaphini, Aepycerotini, and Bovini occur on the left side of the graph. These bovids are associated with woodlands or forests (Tragelaphini), woodland–grassland ecotones (Aepycerotini), or woodland–grasslands near water (Bovini). The tribes Alcelaphini, Antilopini, and Hippotragini cluster toward the right side of the graph. These bovids generally are hypsodont and cursorial grazers, although some species of recent Antilopini are browsers in semiarid bushland. Thus, the main axis of the correspondence analysis (horizontal axis in Figure 2, explaining 46% of the variation) may be interpreted as a gradient from closed and moist environments associated with the Tragelaphini–Aepycerotini–Bovini pole to dry and open environments associated with the Alcelaphini–Antilopini–Hippotragini pole. A third pole occupied by Reduncini on the upper part of the graph likely indicates wet grasslands or waterlogged conditions.

The distribution of geological units in Figure 2 shows that Omo Shungura Members B through lower G are pulled strongly toward the Tragelaphini–Aepycerotini–Bovini pole. These members are clearly separated from Shungura Members H through L, with upper Member G in an intermediate position. This separation into lower and upper members indicates a major shift in the bovid fauna through the sequence. The association of Members B through lower G with Tragelaphini–Aepycerotini–Bovini (and to some extent with Reduncini) indicates that these members were characterized by wooded and moist habitats. The later members of the Shungura Formation (H through L) have higher proportions of Alcelaphini, Antilopini, and Reduncini, and indicate a greater dominance of open grasslands during deposition of these members, even though moist grasslands and waterlogged conditions also occurred near the environments of deposition. Thus, the

distribution of Shungura members depicted in this analysis provides evidence of environmental changes in the lower Omo Valley during the Plio-Pleistocene, with increasingly open environments dominating the landscape after about 2 Ma (the age of upper Member G).

The lower members of the Shungura Formation contrast not only with the upper members, but also with most of the geological members from the Nachukui, Koobi Fora, and Hadar Formations. Only the Lokochot Member of the Koobi Fora Formation, the Apak Member of the Nachukui Formation, the Sidi Hakoma Member of the Hadar Formation, and the Kanapoi sequence approach the lower Omo members in the distribution of bovid tribes (Figure 2). These members (Lokochot, Apak, Sidi Hakoma, and Kanapoi) are older than 3.4 Ma, and they all cluster toward the closed-wooded end members of the correspondence analysis figure.

Geological units most strongly pulled toward the Alcelaphini–Antilopini–Hippotragini pole are the Kada Hadar Member of the Hadar Formation, the Maka'amitalu site (A.L. 666 locality), and the Lomekwi, Lokalalei, Kalochoro, Natoo, and Nariokotome Members of the Nachukui Formation. Thus, most of the Nachukui Formation members occur toward the open-arid pole of the correspondence analysis graph, in contrast to Shungura members which occur either near the wet-closed pole or the wet-open pole (Reduncini).

The Hadar units are spread out in the graph: the Sidi Hakoma Member is pulled in the direction of Tragelaphini–Aepycerotini–Bovini, the Denen Dora Member in the direction of Reduncini, and the Kada Hadar Member with the Maka'amitalu locality toward the Alcelaphini–Antilopini–Hippotragini pole. This would indicate a prevalence of closed environments during Sidi Hakoma times, wet grasslands during Denen Dora times, and a shift to more open environments during Kada Hadar times.

The correspondence analysis of bovid tribes across geological units from the Turkana and

Hadar Basins suggests several noteworthy points. First, the lower Shungura Formation members (B through lower G) differ from most other geological units and were characterized by a high abundance of bovids indicative of closed, wet, and wooded environments. Second, the upper members of the Shungura Formation (H through L) clearly differ from earlier members and indicate that significant environmental changes occurred in the lower Omo Valley toward more open conditions beginning with the deposition of upper Member G, i.e., around 2 Ma. Third, most of the Nachukui Formation members are pulled toward the Alcelaphini–Antilopini–Hippotragini pole and suggest that environmental conditions in the western parts of the Turkana Basin were consistently more open and arid than in the lower Omo Valley during the Plio-Pleistocene. Fourth, Koobi Fora Formation members are separated into two groups: the Lokochot Member is pulled toward the Tragelaphini–Aepycerotini–Hippotragini pole while the Tulu Bor, Upper Burgi, KBS, and Okote members are pulled toward a Reduncini–Alcelaphini pole, a pattern that suggests a greater importance of grasslands (both moist and dry) in the eastern parts of the Turkana Basin after 3.4 Ma. Finally, the Hadar sampling units are well separated from each other in the correspondence analysis figure: the Sidi Hakoma Member is dominated by bovids characteristic of wet and wooded conditions; the Denen Dora Member is dominated by bovids characteristic of wet grasslands; while the Kada Hadar Member and the Maka'amitalu locality are dominated by bovids indicative of more open and arid habitats. This distribution of Hadar units indicates a shift to drier conditions after about 3.2 Ma.

## CLUSTER ANALYSIS

The cluster analysis (rho similarity measure) presented in Figure 3 provides a grouping of bovid tribes similar to the pattern seen in the correspondence analysis figure. Correspondence analysis suggests a correlation, or covariation, of Tragelaphini and Aepycerotini which is confirmed in the cluster analysis of Figure 3. These two taxa also cluster with Bovini and Reduncini, but are distinctly separated from the cluster of Alcelaphini, Antilopini, and Hippotragini. The cluster formed by the "moist-grass grazers" (Reduncini) plus Bovini, Aepycerotini, and Tragelaphini likely represents vegetation types that depend on moister climatic or ground-water conditions with less severe dry seasons, while the cluster formed by Alcelaphini, Antilopini, and Hippotragini represents open or bushy vegetation that characterizes environments with marked dry seasons. Thus, this cluster analysis reinforces the associations of bovid tribes derived from correspondence analysis.

## HABITAT INDICATORS

The proportions of Alcelaphini–Antilopini–Hippotragini (AAH) among bovids from the Turkana and Hadar Basins during the interval from about 7 to 1 Ma are shown in Figure 4A, while Figure 4B presents the same data focused on the Turkana Basin during the interval from 4 to 1 Ma. A noteworthy aspect of this analysis is that Omo Shungura members from 3.4 to about 2 Ma show consistently low proportions of AAH. As this proportion increases after 2 Ma, Shungura members approach the levels seen in other sequences. The persistently low proportion of AAH in Members B through lower G indicates a high degree of environmental stability in the lower Omo Valley during the interval from about 3.4 to 2 Ma. This is evidence for wooded environments in the Omo during the late Pliocene, even though other parts of the Turkana Basin and other regions of Africa had more open and perhaps more unstable environments. The idea of an Omo refugium during the Pliocene has

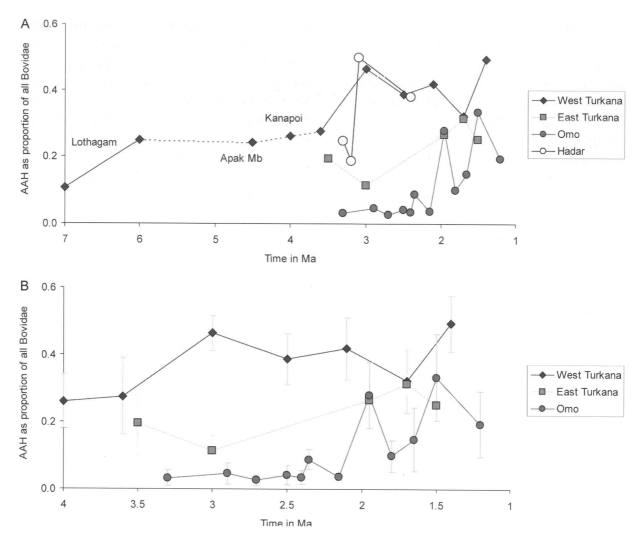

Figure 4. Abundance of Alcelaphini, Antilopini, and Hippotragini (AAH) as a proportion of all bovid specimens per geological interval.

been proposed by Vrba (1988) and is supported by this analysis.

The Nachukui Formation has significantly higher proportions of AAH than the Shungura Formation during the Plio-Pleistocene, but both sequences show a degree of convergence after 2 Ma. Although the lowest proportion of AAH in the Nachukui Formation occurs in the earliest interval (Apak Member), and the highest proportion occurs in the latest interval (Natoo-Nariokotome Members), the West Lake Turkana sequence does not show as strong a trend toward higher proportions of AAH over time as the Omo does. At East

Turkana, the earliest intervals (Lokochot and Tulu Bor) show the lowest AAH proportions and the later intervals (Upper Burgi, KBS, and Okote) the highest. Thus at East Turkana, as in the Shungura and Nachukui Formations, the trend is toward increasing proportions of open country bovids toward the latest Pliocene and earliest Pleistocene. Nevertheless, the significant differences in AAH proportions among the three main Turkana Basin areas suggest that intra-basinal faunal differences existed through the Pliocene and Pleistocene. These differences are consistent with the idea that the Shungura Formation sampled the main axis of the Plio-Pleistocene Turkana Basin, close to the course of the paleo-Omo River, whereas the Nachukui Formation sampled more marginal habitats in the basin, as proposed by Feibel et al. (1991). In this view, the Koobi Fora Formation on the eastern margin of the basin presents a more complex picture that included axial and marginal habitats at different times. This is consistent with the structural configuration of the Turkana Basin half-graben, with the major fault on the west side and a hinged platform or "ramp" on the east. This configuration allowed the paleo-Omo river to avulse occasionally into the East Turkana area, and west-flowing tributaries from areas east of the basin also could have drained into the basin across the west-sloping rift margin.

The Sidi Hakoma and Denen Dora Members of the Hadar Formation have AAH proportions similar to those of contemporaneous members in the Koobi Fora Formation. The significant increase in AAH proportions in

the Kada Hadar Member suggests environmental changes toward more open conditions in the Hadar area after 3.2 Ma, following upon a significant wet interval associated with the Denen Dora Member (Figure 2). The Maka'amitalu locality (A.L. 666) indicates that these open conditions persisted in the Hadar Basin at around 2.4–2.3 Ma, where both *Homo* and lithic artifacts occur in close association (Kimbel et al., 1996).

## SPECIES RICHNESS

Given that there are differences among Turkana Basin areas in the proportions of the different bovid tribes, we now ask whether differences exist in species richness among the different areas under consideration. Richness is defined as the total number of species (or distinct taxa) in a sample. It is well known that taxonomic richness is dependent on sample size (Figure 5).

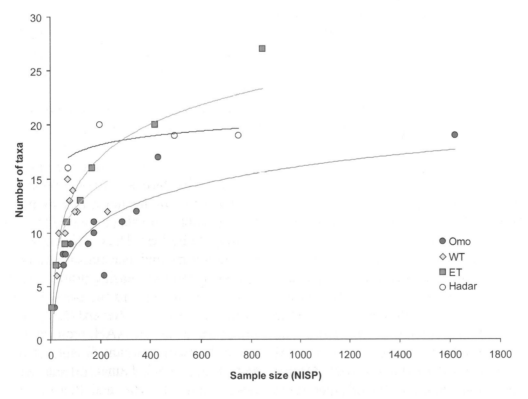

Figure 5. Bovid species richness vs. sample size, with logarithmic trend lines (rarefaction curves) highlighting the differences in this relationship for the different samples.

For a given sample size, the Shungura Formation tends to have lower richness than other parts of the Turkana Basin. The lower richness in the Shungura samples could be due to differences in taphonomic conditions, rather than to a lower number of species in the Shungura paleoecosystem. The Shungura collections include large numbers of isolated teeth, and specimens often show signs of rolling and abrasion that indicate reworking and deposition under relatively high energy fluvial conditions. The fragmentary nature of many Shungura specimens would thus tend to reduce the likelihood that these could be identified to genus or species. Better preservation in other parts of the basin would thus contribute to the greater species richness in samples from the Koobi Fora and Nachukui Formations. Two of the Hadar samples (Kada Hadar and Maka'amitalu) are richer in bovid species and a third (Sidi Hakoma) is similar in richness to samples of comparable size from the Turkana Basin. However, the largest of the Hadar samples (Denen Dora Member) appears to have low species richness, perhaps because it spans a comparatively short interval of time (40,000 years) and is strongly dominated by a single bovid tribe, the Reduncini.

As noted above, species richness is dependent on sample size, and sample size varies greatly across geological members in the Turkana Basin. Figure 6A shows the number of bovid taxa in the Turkana Basin during the 3.5 million-year interval from the Apak Member of the Nachukui Formation to the top of Member L of the Shungura Formation, and Figure 6B shows the abundance of the bovid fossil record during this time. This span of time in the Turkana Basin has a high concentration of fossils and a relatively continuous record that can be divided into roughly 200 thousand-year intervals. The dating and correlations among the different areas of the basin is based on the data provided in Table 1, and the distribution of bovid specimens identified below the tribal level is provided in Table 3.

There is a plethora of indices to measure species richness in relation to sample size (Ludwig and Reynolds, 1988; Magurran, 1988; Hayek and Buzas, 1997), but Fisher's $\alpha$ is recommended as a reliable measure when a single index is used (Hayek and Buzas, 1997). As measured by Fisher's $\alpha$ (Figure 7), bovid species richness in the Turkana Basin shows three major peaks in the time from about 4.5 to about 1 Ma. The first of these peaks occurred from 3.8 to 3.4 Ma, and includes the Moiti and Lokochot Members of the Koobi Fora Formation, the Kataboi Member of the Nachukui Formation, and Member A of the Shungura Formation. The second peak occurred from 2.8 to 2.4 Ma, including the upper Tulu Bor Member of the Koobi Fora Formation, the middle-upper Lomekwi and Lokalalei Members of the Nachukui Formation, and Members C and D of the Shungura Formation. A third and more sustained peak occurred from about 2 to 1.4 Ma, and includes the abundant record of the Upper Burgi, KBS, and Okote Members of the Koobi Fora Formation, the Kalochoro (part), Kaitio, and Natoo Members of the Nachukui Formation, and Members G (upper) to K of the Shungura Formation. These peaks of diversity in the Turkana Basin appear to come in cycles of about one million years, although the significance and causes of this pattern are not clear. The peak in diversity between 2.8 and 2.4 Ma occurred during the time of Vrba's (1995a) hypothesized turnover pulse in Africa. Elsewhere we have shown that significant faunal changes occurred at about 2.8 Ma: there is a peak of turnover at this time (although not as pronounced as a later Plio-Pleistocene peak) (Bobe and Behrensmeyer, 2004), and there are significant changes in the relative abundances of Omo bovids (Bobe and Eck, 2001). The third peak of taxonomic richness at 2.0–1.4 Ma coincides with another period of important faunal changes in the Turkana Basin. The time around the Pliocene–Pleistocene boundary (~1.8 Ma) is characterized by high rates of faunal turnover (Behrensmeyer et al., 1997)

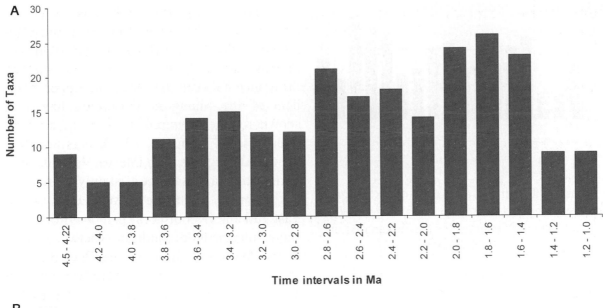

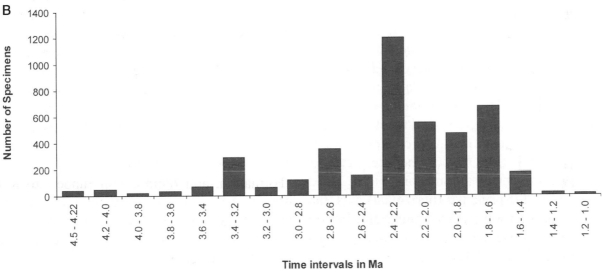

Figure 6. A) Number of bovid taxa per 200,000 year intervals in the Turkana Basin. B) Number of bovid specimens per 200,000 year intervals in the Turkana Basin.

and an increase in the importance of open grassland environments in the Turkana Basin (Bobe and Behrensmeyer, 2004).

The record of the Hadar Formation is not long enough to analyze patterns of species richness at 200-Kyr intervals. Such an analysis would include the Sidi Hakoma Member and half of the Denen Dora Member in the interval from 3.4 to 3.2 Ma, and the other half of the Denen Dora Member plus the Kada Hadar Member in the interval from 3.2 to 3.0 Ma. Such a partitioning of the Hadar record produces Fisher's $\alpha$ values of 4.8 and 5.4 for the earlier and later intervals respectively. Both values are higher than those of contemporaneous Turkana Basin intervals. Although sampling and taphonomic considerations remain to be assessed, these values confirm the results of Figure 5 that show taxonomic richness at Hadar within the upper part of the range of the Turkana Basin samples.

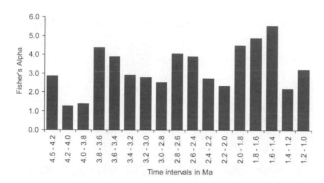

Figure 7. Fisher's α, a measure of taxonomic (species) richness relative to sample size, plotted by 200,000 year interval for the Turkana Basin bovid sample.

## Discussion and Conclusion

This analysis has sought to identify significant trends in the abundance and diversity of bovids during the late Cenozoic of East Africa (Figure 1). A correspondence analysis of bovid tribes and geological members in the Turkana and Hadar Basins produces distribution of bovid tribes that can be interpreted in terms of environmental conditions (Figure 2). At one end of the principal (horizontal) axis, the tribes Aepycerotini, Tragelaphini, and Bovini indicate mostly closed and moist environments dominated by woodlands and fresh grasses. At the other end, the tribes Alcelaphini, Antilopini, and Hippotragini indicate more open and seasonally dry environments dominated by grasslands, wooded grasslands, or bushland. A third pole toward the position of the tribe Reduncini (high on the vertical axis) suggests open but moist-soil environments dominated by fresh or edaphic grasses. This distribution of bovid tribes parallels the associations indicated by cluster analysis (Figure 3).

The distribution of geological members in the correspondence analysis diagram (Figure 2) shows that the lower Omo Valley was characterized by closed and wet environments from about 3.4 to 2.0 Ma (Usno U-12 and Members B to lower G). These environmental

characteristics contrast with those that predominated in the Omo after 2 Ma and elsewhere in the Turkana Basin for much of the late Pliocene and early Pleistocene. This marked difference between the lower and upper members of the Shungura Formation has been shown for the mammalian fauna as a whole (Geraads and Coppens, 1995). It is noteworthy that during this time interval the Nachukui Formation maintained consistently more open environments than the lower Omo Valley. East Turkana (Koobi Fora) paleoenvironments show intermediate conditions between those in the Omo and those in the West Turkana areas. The wide separation of Hadar Formation members in the correspondence analysis diagram (Figure 2) indicates that significant environmental changes occurred in the Hadar area from 3.4 to 3 Ma, as closed environments in the Sidi Hakoma Member gave way to extensive moist grasslands in the Denen Dora Member, followed by drier and more open environments in the Kada Hadar Member and the A.L. 666 locality. These results show that significant environmental changes occurred in the Turkana and Hadar Basins during the Pliocene, and that important environmental differences existed in different areas of the same sedimentary basin.

The use of Alcelaphini, Antilopini, and Hippotragini (AAH) as indicators of open and seasonally arid environments dominated by grasslands confirms that the lower Omo Valley remained significantly more wooded and moist than other parts of the Turkana Basin during the late Pliocene, from about 3.4 to 2.0 Ma (Figure 4). These results are consistent with the view that the Omo Shungura Formation sampled the axis of a sedimentary basin incised by a large meandering river, while the West Turkana deposits sampled more marginal habitats nearer the tectonic margins of the basin (Feibel et al., 1991). The Koobi Fora Formation on the east side of modern Lake Turkana sampled more complex or intermediate environments on the hinged side of a half-graben, while West Turkana sampled the fault-bounded side. The latter might be expected to subside more rapidly and be consistently wetter, but the uplifted rift shoulder to the west may have acted as a rain shadow and may have had limited runoff into the Turkana Basin (drainages on uplifted rift shoulders usually are directed away from the rift valley). In the latest Pliocene the Turkana Basin underwent significant tectonic changes; outflow of the paleo-Omo River to the Indian Ocean became blocked and this resulted in the expansion of a major lake at about 2.1 Ma. Subsequently the outflow of the river was diverted to the Nile drainage (Feibel et al., 1991). Thus, tectonism likely had a significant impact in the faunal changes that occurred in the Omo after 2 Ma.

In the Afar, the distribution of Hadar Formation samples indicates shifts from closed and wooded environments in the Sidi Hakoma Member (3.4–3.22 Ma) to moist substrate-waterlogged conditions in the Denen Dora Member (3.22–3.18 Ma), and seasonally arid grasslands in the Kada Hadar Member (3.18–3.0 Ma) and in the Maka'amitalu locality (~2.34 Ma) (Figures 2 and 4). Because of the short temporal spans of the Hadar faunal samples, the shift from woodland to moist grasslands to drier grasslands in could reflect either facies changes in the paleo-Awash drainage basin or short-term, basin-scale climate changes.

The relative abundance of Alcelaphini, Antilopini, and Hippotragini as a proportion of all bovid specimens increased in all areas toward the late Pliocene and the earliest Pleistocene, suggesting a common driving mechanism across basins (Figure 4). These bovid tribes are indicators of seasonally arid grasslands or bushland, and their increase in relative abundance during the late Cenozoic indicates an expansion of arid environments. The increase in aridity in East Africa is consistent with the known record of climatic change derived from marine sediments (deMenocal, 1995). Nevertheless, these aridity trends in the Turkana and Hadar areas show different responses at different times, evidence that local geography and tectonics played an important role in mediating environmental change.

The analysis of bovid diversity over time highlights three intervals of high species richness (Figure 7). The first one occurred from about 3.8 to 3.4 Ma, a time during which bovids such as *Kobus oricornus*, *Kobus sigmoidalis*, and *Aepyceros shungurae* first appear in the fossil record (Vrba, 1995a; Bobe and Behrensmeyer, 2004). The second interval of high species richness occurred from 2.8 to 2.4 Ma. Previous analyses have shown considerable changes in bovid relative abundances at 2.8 Ma (Bobe and Eck, 2001), a time that also corresponds to Vrba's hypothesized late Pliocene turnover event (Vrba, 1995a) and significant changes in African climate (deMenocal, 1995). The third interval of high species richness occurs across the Pliocene–Pleistocene boundary, from about 2 to 1.4 Ma. This was a time of expanding grasslands and diversification of grazing bovids, including the species *Pelorovis turkanensis*, *Pelorovis oldowayensis*, *Megalotragus isaaci*, and *Beatragus antiquus* (Vrba, 1995a; Bobe and Behrensmeyer, 2004). Previous analyses of bovid diversity using evenness and richness data have also identified the earliest Pleistocene as a time of high bovid diversity (Geraads, 1994). Bovid diversity seems to decrease significantly after 1.4 Ma, but samples are small.

Although hominins were relatively rare elements of the East African fauna during the late Pliocene, species of *Australopithecus* (*A. afarensis*), *Paranthropus*, and early *Homo* persisted through significant changes in local and regional landscapes (Bobe et al., 2002; Bonnefille et al., 2004). By the early Pleistocene, East African landscapes were becoming increasingly open and seasonal, and hominins were showing significant shifts in biological and cultural adaptations to these novel conditions (McHenry and Coffing, 2000).

Patterns of faunal change are dependent on the scale of analysis, and this analysis is meant to elucidate the main trends in bovid abundance and diversity over millions of years. Studies of the Omo fauna at finer levels of temporal resolution have shown that during the time between 3.4 and 2 Ma there were intervals of environmental stability and intervals of instability (Bobe et al., 2002). Also, different methods of assessing taxonomic abundance changes over time have shown that significant faunal shifts occurred in the Omo at about 2.8 Ma (Bobe and Eck, 2001). These patterns are obscured in the lower resolution analyses presented here, which focus on longer time intervals and broader geographic scales.

The diverse patterns of faunal abundance and richness documented above highlight the complexities involved in establishing correlations between climatic change and faunal evolution. Both within the Turkana Basin and between this basin and the Afar region there are similar but asynchronous trends that suggest increasing dry season stress on the vegetation and the faunas. This provides a measure of the capacity of local environmental "buffering" on larger-scale climatic trends. It is clear that during the critical period of human evolution between 4.0 and 2.0 Ma there were many different habitats in East Africa, but after 2.0 Ma the more vegetated and moist end of the habitat spectrum became more limited. This analysis highlights broad patterns of faunal change, but it is based on only one major group of mammals, the Bovidae. This provides a foundation for future testing with other mammals, as well as with geological, geochemical, and other types of paleontological evidence.

## Acknowledgments

For insightful comments that greatly improved this paper, we are grateful to Meave Leakey, Louise Leakey, Denis Geraads, Chris Campisano, Zeresenay Alemseged, Eric Delson, and the anonymous reviewers. The Turkana Basin Paleontology Database is a collaborative Project between the National Museums of Kenya and the Smithsonian Institution. The database was created with support from the Smithsonian Institution's Evolution of Terrestrial Ecosystems Program and Scholarly Studies Program, and from the National Science Foundation (BSC0137235).

## References

Alemseged, Z., 2003. An integrated approach to taphonomy and faunal change in Shungura Formation (Ethiopia) and its implication for hominid evolution. Journal of Human Evolution 44, 461–478.

Alemseged, Z., Wynn, J.G., Kimbel, W.H., Reed, D., Geraads, D., Bobe, R., 2005. A new hominin from the Basal Member of the Hadar Formation, Dikika, Ethiopia, and its geological context. Journal of Human Evolution 49, 499–514.

Alemseged, Z., Bobe, R., Geraads, D., 2007. Comparability of fossil data and its significance for the interpretation of hominin environments: a case study in the lower Omo valley, Ethiopia. In: Bobe, R., Alemseged, Z., Behrensmeyer, A.K. (Eds.), Hominin Environments in the East African Pliocene: An Assessment of the Faunal Evidence. Springer, Dordrecht.

Aronson, J.L., Taieb, M., 1981. Geology and paleogeography of the Hadar hominid site, Ethiopia. In: Rapp, G., Vondra, C.F. (Eds.), Hominid Sites: Their Geologic Settings. Westview Press, Boulder, pp. 165–195.

Asfaw, B., White, T.D., Lovejoy, C.O., Latimer, B.M., Simpson, S., Suwa, G., 1999. Australopithecus garhi: a new species of early hominid from Ethiopia. Science 284, 629–635.

Behrensmeyer, A.K., 1975. The taphonomy and paleoecology of Plio-Pleistocene vertebrate assemblages East of Lake Rudolf, Kenya. Bulletin of the Museum of Comparative Zoology 146(10), 473–578.

Behrensmeyer, A.K., 2006. Climate change and human evolution. Science 311, 476–478.

Behrensmeyer, A.K., Todd, N.E., Potts, R., Mcbrinn, G.E., 1997. Late Pliocene faunal turnover in the Turkana Basin, Kenya and Ethiopia. Science 278, 1589–1594.

Behrensmeyer, A.K., Bobe, R., Campisano, C.J., Levin, N., 2004. High resolution taphonomy and paleoecology of the Plio-Pleistocene Koobi Fora

Formation, northern Kenya, with comparisons to the Hadar Formation, Ethiopia. Journal of Vertebrate Paleontology 24(3 (Supplement)), 38A.

Behrensmeyer, A.K., Bobe, R., Alemseged, Z., 2007. Approaches to the analysis of faunal change during the East African Pliocene. In: Bobe, R., Alemseged, Z., Behrensmeyer, A.K. (Eds.), Hominin Environments in the East African Pliocene: An Assessment of the Faunal Evidence. Springer, Dordrecht.

Benzécri, J.-P., 1992. Correspondence Analysis Handbook. Dekker, New York.

Berggren, W.A., Kent, D.V., Aubry, M.-P., Hardenbol, J. (Eds.), 1995. Geochronology, Time Scales and Global Stratigraphic Correlation. Society for Sedimentary Geology, Tulsa, Oklahoma.

Boaz, N.T., 1977. Paleoecology of early Hominidae in Africa. The Kroeber Anthropological Society Papers 50, 37–62.

Bobe, R., Behrensmeyer, A.K., 2004. The expansion of grassland ecosystems in Africa in relation to mammalian evolution and the origin of the genus Homo. Palaeogeography, Palaeoclimatology, Palaeoecology 207, 399–420.

Bobe, R., Eck, G.G., 2001. Responses of African bovids to Pliocene climatic change. Paleobiology Memoirs, Paleobiology 27(Supplement to Number 2), 1–47.

Bobe, R., Behrensmeyer, A.K., Chapman, R.E., 2002. Faunal change, environmental variability and late Pliocene hominin evolution. Journal of Human Evolution 42(4), 475–497.

Bonnefille, R., Potts, R., Chalie, F., Jolly, D., Peyron, O., 2004. High-resolution vegetation and climate change associated with Pliocene Australopithecus afarensis. Proceedings of the National Academy of Sciences USA 101(33), 12125–12129.

Brown, F.H., 1969. Observations on the stratigraphy and radiometric age of the Omo Beds, lower Omo Basin, southern Ethiopia. Quaternaria 11, 7–14.

Brown, F.H., 1983. Chronology. In: de Heinzelin, J. (Ed.), The Omo Group. Musée Royale de l'Afrique Central, Tervuren, pp. 145–147.

Brown, F.H., 1994. Development of Pliocene and Pleistocene chronology of the Turkana Basin, East Africa, and its relation to other sites. In: Corruccini, R.S., Ciochon, R.L. (Eds.), Integrative Paths to the Past. Prentice-Hall, Englewood Cliffs, New Jersey, pp. 285–312.

Brown, F.H., 1995. The potential of the Turkana Basin for paleoclimatic reconstruction in East Africa.

In: Vrba, E.S., Denton, G.H., Partridge, T.C., Burckle, L.H. (Eds.), Paleoclimate and Evolution with Emphasis on Human Origins. Yale University Press, New Haven, pp. 319–330.

Brown, F.H., Feibel, C.S., 1985. Stratigraphical notes on the Okote Tuff Complex at Koobi Fora, Kenya. Nature 316, 794–797.

Brown, F.H., Feibel, C.S., 1986. Revision of lithostratigraphic nomenclature in the Koobi Fora region, Kenya. Journal of the Geological Society of London 143, 297–310.

Brown, F.H., McDougall, I., 1993. Geologic setting and age. In: Walker, A., Leakey, R.E. (Eds.), The Nariokotome Homo erectus Skeleton. Harvard University Press, Cambridge, pp. 9–20.

Brown, F.H., Nash, W.P., 1976. Radiometric dating and tuff mineralogy of Omo Group deposits. In: Coppens, Y., Howell, F.C., Isaac, G.L., Leakey, R.E. (Eds.), Earliest Man and Environments in the Lake Rudolf Basin. University of Chicago Press, Chicago, pp. 50–63.

Brown, F.H., McDougall, I., Davis, T., Maier, R., 1985. An integrated Plio-Pleistocene chronology for the Turkana Basin. In: Delson, E. (Ed.), Ancestors: The Hard Evidence. Alan R. Liss, Inc., New York, pp. 82–90.

Brown, F.H., Sarna-Wojcicki, A.M., Meyer, C.E., Haileab, B., 1992. Correlation of Pliocene and Pleistocene tephra layers between the Turkana Basin of East Africa and the Gulf of Aden. Quaternary International 13/14, 55–67.

Brown, F.H., Haileab, B., McDougall, I., 2006. Sequence of tuffs between the KBS Tuff and the Chari Tuff in the Turkana Basin, Kenya and Ethiopia. Journal of the Geological Society of London 163, 185–204.

Brunet, M., Guy, F., Pilbeam, D.R., Mackaye, H.T., Likius, A., Ahounta, D., Beauvilain, A., Blondel, C., Bocherens, H., Boisserie, J.-R., de Bonis, L., Coppens, Y., Dejax, J., Denys, C., Duringer, P., Eisenmann, V., Fanone, G., Fronty, P., Geraads, D., Lehmann, T., Lihoreau, F., Louchart, A., Mahamat, A., Merceron, G., Mouchelin, G., Otero, O., Campomanes, P.P., Ponce de Leon, M., Rage, J.-C., Sapanet, M., Schuster, M., Sudre, J., Tassy, P., Valentin, X., Vignaud, P., Viriot, L., Zazzo, A., Zollikofer, C., 2002. A new hominid from the upper Miocene of Chad, Central Africa. Nature 418, 145–151.

Buzas, M.A., 1990. Another look at confidence limits for species proportions. Journal of Paleontology 64(5), 842–843.

Campisano, C.J., Behrensmeyer, A.K., Bobe, R., Levin, N., 2004. High resolution paleoenvironmental comparisons between Hadar and Koobi Fora: preliminary results of a combined geological and paleontological approach. PaleoAnthropology A35.

Cande, S.C., Kent, D.V., 1995. Revised calibration of the geomagnetic polarity timescale for the late Cretaceous and Cenozoic. Journal of Geophysical Research 100, 6093–6095.

Cerling, T.E., Harris, J.M., Passey, B.H., 2003. Diets of East African Bovidae based on stable isotope analysis. Journal of Mammalogy 84(2), 456–470.

Coppens, Y., 1975. Evolution des Hominidés et de leur environnement au cours du Plio-Pléistocène dans la basse vallée de l'Omo en Ethiopie. Comptes Rendus de l' Académie des Sciences. Paris, Série D, Sciences Naturelles 281(22), 1693–1696.

Coppens, Y., Howell, F.C., Isaac, G.L., Leakey, R.E. (Eds.), 1976. Earliest Man and Environments in the Lake Rudolf Basin: Stratigraphy, Paleoecology, and Evolution. University of Chicago Press, Chicago.

DeGusta, D., Vrba, E.S., 2003. A method for inferring paleohabitats from the functional morphology of bovid astragali. Journal of Archaeological Science 30, 1009–1022.

de Heinzelin, J. (Ed.), 1983. The Omo Group. Musée Royale de l'Afrique Central, Tervuren.

de Heinzelin, J., Haesaerts, P., 1983a. The Shungura Formation. In: de Heinzelin, J. (Ed.), The Omo Group. Musée Royale de l'Afrique Central, Tervuren, pp. 25–127.

de Heinzelin, J., Haesaerts, P., 1983b. The Usno Formation. In: de Heinzelin, J. (Ed.), The Omo Group. Musée Royale de l'Afrique Central, Tervuren, pp. 129–139.

deMenocal, P.B., 1995. Plio-Pleistocene African Climate. Science 270(5233), 53–59.

deMenocal, P.B., 2004. African climate change and faunal evolution during the Plio-Pleistocene. Earth and Planetary Science Letters 220, 3–24.

Eck, G.G., 2007. The effects of collection strategy and effort on faunal recovery: a case study of the American and French collections from the Shungura Formation, Ethiopia. In: Bobe, R., Alemseged, Z., Behrensmeyer, A.K. (Eds.), Hominin Environments in the East African Pliocene: An Assessment of the Faunal Evidence. Springer, Dordrecht.

Eck, G.G., Jablonski, N.G., Leakey, M.G. (Eds.), 1987. Les Faunes Plio-Pléistocène de la Basse Vallée de l'Omo (Ethiopie): Cercopithecidae de la Formation de Shungura. CNRS, Paris.

Feibel, C.S., 1997. Debating the environmental factors in hominid evolution. GSA Today 7(3), 1–7.

Feibel, C.S., 2003a. Stratigraphy and depositional history of the Lothagam sequence. In: Leakey, M.G., Harris, J.M. (Eds.), Lothagam: The Dawn of Humanity in Eastern Africa. Columbia University Press, New York, pp. 17–29.

Feibel, C.S., 2003b. Stratigraphy and depositional setting of the Pliocene Kanapoi Formation, lower Kerio valley, Kenya. In: Harris, J.M., Leakey, M.G. (Eds.), Geology and Vertebrate Paleontology of the Early Pliocene Site of Kanapoi, Northern Kenya. Natural History Museum of Los Angeles County, Los Angeles, pp. 9–20.

Feibel, C.S., Brown, F.H., McDougall, I., 1989. Stratigraphic context of fossil hominids from the Omo Group deposits: Northern Turkana Basin, Kenya and Ethiopia. American Journal of Physical Anthropology 78, 595–622.

Feibel, C.S., Harris, J.M., Brown, F.H., 1991. Palaeoenvironmental context for the late Neogene of the Turkana Basin. In: Harris, J.M. (Ed.), Koobi Fora Research Project, Volume 3: The fossil ungulates: geology, fossil artiodactyls, and paleoenvironments. Clarendon Press, Oxford, pp. 321–370.

Gagnon, M., Chew, A.E., 2000. Dietary preferences in extant African Bovidae. Journal of Mammalogy 81(2), 490–511.

Gentry, A.W., 1978. Bovidae. In: Maglio, V.J., Cooke, H.B.S. (Eds.), Evolution of African Mammals. Harvard University Press, Cambridge, pp. 540–572.

Gentry, A.W., 1985. Pliocene and Pleistocene Bovidae in Africa. In: Coppens, Y. (Ed.), L'Environnement des Hominidés au Plio-Pléistocène. Masson, Paris, pp. 119–122.

Geraads, D., 1994. Evolution of bovid diversity in the Plio-Pleistocene of Africa. Historical Biology 7, 221–237.

Geraads, D., Coppens, Y., 1995. Évolution des faunes de mammifères dans le Plio-Pléistocène de la basse vallée de l'Omo (Éthiopie): apports de l'analyse factorielle. Comptes Rendus de l' Académie des Sciences. Paris, série II a 320, 625–637.

Gray, B.T., 1980. Environmental Reconstruction of the Hadar Formation (Afar, Ethiopia). Ph.D. Thesis, Case Western Reserve University.

Greenacre, M.J., 1993. Correspondence Analysis in Practice. Academic Press, London.

Greenacre, M.J., Vrba, E.S., 1984. Graphical display and interpretation of antelope census data in African wildlife areas, using correspondence analysis. Ecology 65(3), 984–997.

Haileab, B., Brown, F.H., 1992. Turkana Basin-Middle Awash Valley correlations and the age of the Sagantole and Hadar Formations. Journal of Human Evolution 22, 453–468.

Haileab, B., Brown, F.H., 1994. Tephra correlations between the Gadeb prehistoric site and the Turkana Basin. Journal of Human Evolution 26, 167–173.

Haileab, B., Brown, F.H., McDougall, I., Gathogo, P.N., 2004. Gombe Group basalts and initiation of Pliocene deposition in the Turkana depression, northern Kenya and southern Ethiopia. Geological Magazine 141(1), 41–53.

Haile-Selassie, Y., Suwa, G., White, T.D., 2004. Late Miocene Teeth from Middle Awash, Ethiopia, and Early Hominid Dental Evolution. Science 303(5663), 1503–1505.

Harris, J.M. (Ed.), 1983. Koobi Fora Research Project, Volume 2: The fossil ungulates: Proboscidea, Perissodactyla, and Suidae. Clarendon Press, Oxford.

Harris, J.M. (Ed.), 1991. Koobi Fora Research Project, Volume 3: The fossil ungulates: geology, fossil artiodactyls, and paleoenvironments. Clarendon Press, Oxford.

Harris, J.M., 2003. Bovidae from the Lothagam succession. In: Leakey, M.G., Harris, J.M. (Eds.), Lothagam: The Dawn of Humanity in Eastern Africa. Columbia University Press, New York, pp. 531–579.

Harris, J.M., Leakey, M.G. (Eds.), 2003. Geology and Vertebrate Paleontology of the Early Pliocene Site of Kanapoi, Northern Kenya. Natural History Museum of Los Angeles County, Los Angeles.

Harris, J.M., Brown, F.H., Leakey, M.G., 1988. Stratigraphy and Paleontology of Pliocene and Pleistocene Localities West of Lake Turkana, Kenya. Natural History Museum of Los Angeles County, Los Angeles.

Hayek, L.-A.C., Buzas, M.A., 1997. Surveying Natural Populations. Columbia University Press, New York.

Howell, F.C., 1968. Omo research expedition. Nature 219, 567–572.

Howell, F.C., Haesaerts, P., de Heinzelin, J., 1987. Depositional environments, archeological occurrences and hominids from Members E and F of the Shungura Formation (Omo basin, Ethiopia). Journal of Human Evolution 16, 665–700.

Johanson, D.C., Taieb, M., 1976. Plio-Pleistocene hominid discoveries in Hadar, Ethiopia. Nature 260, 293–297.

Johanson, D.C., Taieb, M., Coppens, Y., 1982. Pliocene hominids from the Hadar Formation, Ethiopia (1973–1977): stratigraphic, chronologic, and paleoenvironmental contexts, with notes on hominid morphology and systematics. American Journal of Physical Anthropology 57(4), 373–402.

Kalb, J.E., Oswald, E.B., Mebrate, A., Tebedge, C., Jolly, C., 1982. Stratigraphy of the Awash Group, Middle Awash Valley, Afar, Ethiopia. Newsletters on Stratigraphy 11, 95–127.

Kappelman, J., Plummer, T., Bishop, L., Duncan, A., Appleton, S., 1997. Bovids as indicators of Plio-Pleistocene paleoenvironments in East Africa. Journal of Human Evolution 32, 229–256.

Kimbel, W.H., Walter, R.C., Johanson, D.C., Reed, K.E., Aronson, J.E., Assefa, Z., Marean, C.W., Eck, G.G., Bobe, R., Hovers, E., Rak, Y., Vondra, C.F., Yemane, T., York, D., Chen, Y., Evenssen, N., Smith, P., 1996. Late Pliocene *Homo* and Oldowan tools from the Hadar Formation (Kada Hadar Member), Ethiopia. Journal of Human Evolution 31, 549–561.

Kingdon, J., 1982a. East African Mammals: An Atlas of Evolution in Africa, Volume III, Part C: Bovids. Academic Press, London.

Kingdon, J., 1982b. East African Mammals: An Atlas of Evolution in Africa, Volume III, Part D: Bovids. University of Chicago Press, Chicago.

Lanphere, M.A., Champion, D.E., Christiansen, R.L., Izett, G.A., Obradovich, J.D., 2002. Revised ages for tuffs of the Yellowstone Plateau volcanic field: Assignment of the Huckleberry Ridge Tuff to a new geomagnetic polarity event. GSA Bulletin 114, 559–568.

Leakey, L.N., 2001. Body weight estimation of Bovidae and Plio-Pleistocene faunal change, Turkana Basin, Kenya. Ph.D. Thesis, University College London.

Leakey, M.G., Harris, J.M. (Eds.), 2003. Lothagam: The Dawn of Humanity in Eastern Africa. Columbia University Press, New York.

Leakey, M.G., Leakey, R.E. (Eds.), 1978. Koobi Fora Research Project, Volume 1: The fossil hominids and an introduction to their context, 1968–1974. Clarendon Press, Oxford.

Leakey, M.G., Feibel, C.S., McDougall, I., Walker, A., 1995. New four-million-year-old hominid species from Kanapoi and Allia Bay, Kenya. Nature 376, 565–571.

Leakey, M.G., Feibel, C.S., McDougall, I., Ward, C.V., Walker, A., 1998. New specimens and confirmation of an early age for Australopithecus anamensis. Nature 393, 62–66.

Leakey, M.G., Spoor, F., Brown, F.H., Gathogo, P.N., Kiarie, C., Leakey, L.N., McDougall, I., 2001. New hominin genus from eastern Africa shows diverse middle Pliocene lineages. Nature 410, 433–440.

Ludwig, J.A., Reynolds, J.F., 1988. Statistical Ecology: A Primer on Methods and Computing. John Wiley & Sons, New York.

Magurran, A.E., 1988. Ecological Diversity and Its Measurement. Princeton University Press, Princeton.

156        R. BOBE ET AL.

McDougall, I., 1985. K-Ar and $^{40}$Ar/$^{39}$Ar dating of the hominid-bearing Pliocene–Pleistocene sequence at Koobi Fora, northern Kenya. Geological Society of America Bulletin 96, 159–175.

McDougall, I., Feibel, C.S., 1999. Numerical age control for the Miocene-Pliocene succession at Lothagam, a hominoid-bearing sequence in the northern Kenya Rift. Journal of the Geological Society of London 156, 731–745.

McDougall, I., Feibel, C.S., 2003. Numerical age control for the Miocene–Pliocene succession at Lothagam, a hominoid-bearing sequence in the northern Kenya Rift. In: Leakey, M.G., Harris, J.M. (Eds.), Lothagam: The Dawn of Humanity in Eastern Africa. Columbia University Press, New York, pp. 43–64.

McDougall, I., Davies, T., Maier, R., Rudowski, R., 1985. Age of the Okote Tuff Complex at Koobi Fora, Kenya. Nature 316, 792–794.

McDougall, I., Brown, F.H., Cerling, T.E., Hillhouse, J.W., 1992. A reappraisal of the geomagnetic polarity time scale to 4 Ma using data from the Turkana Basin, East Africa. Geophysical Research Letters 19, 2349–2352.

McHenry, H.M., Coffing, K., 2000. Australopithecus to Homo: transformations in body and mind. Annual Review of Anthropology 29, 125–146.

Plummer, T.W., Bishop, L.C., 1994. Hominid paleoecology at Olduvai Gorge, Tanzania as indicated by antelope remains. Journal of Human Evolution 27, 47–75.

Potts, R., 1998. Environmental hypotheses of hominin evolution. Yearbook of Physical Anthropology 41, 93–136.

Quade, J., Levin, N., Semaw, S., Stout, D., Renne, P., Rogers, M., Simpson, S., 2004. Paleoenvironments of the earliest stone toolmakers, Gona, Ethiopia. Geological Society of America Bulletin 116(11–12), 1529–1544.

Sarna-Wojcicki, A.M., Meyer, C.E., Roth, P.H., Brown, F.H., 1985. Ages of tuff beds at East African early hominid sites and sediments in the Gulf of Aden. Nature 313, 306–308.

Senut, B., Pickford, M., Gommery, D., Mein, P., Cheboi, K., Coppens, Y., 2001. First hominid from the Miocene (Lukeino Formation, Kenya). Comptes Rendus de l' Académie des Sciences. Paris, Sciences de la Terre et des planetes 332, 137–144.

Shackleton, N.J., 1995. New data on the evolution of Pliocene climatic variability. In: Vrba, E.S., Denton, G.H., Partridge, T.C., Burckle, L.H. (Eds.), Paleoclimate and Evolution, with

Emphasis on Human Origins. Yale University Press, New Haven, pp. 242–248.

Shipman, P., Harris, J.M., 1988. Habitat preference and paleoecology of Australopithecus boisei in Eastern Africa. In: Grine, F.E. (Ed.), Evolutionary History of the "Robust" Australopithecines. Aldine de Gruyter, New York, pp. 343–381.

Spencer, L.M., 1997. Dietary adaptations of Plio-Pleistocene Bovidae: implications for hominid habitat use. Journal of Human Evolution 32(2–3), 201–228.

Sponheimer, M., Reed, K.E., Lee-Thorp, J.A., 1999. Combining isotopic and ecomorphological data to refine bovid paleodietary reconstruction: A case study from the Makapansgat Limeworks hominin locality. Journal of Human Evolution 36(6), 705–718.

Tauxe, L., Deino, A.D., Behrensmeyer, A.K., Potts, R., 1992. Pinning down the Bruhnes/Matuyama and upper Jaramillo boundaries: a reconciliation of orbital and isotopic time scales. Earth and Planetary Science Letters 109, 561–572.

Tiercelin, J.J., 1986. The Pliocene Hadar Formation, Afar depression of Ethiopia. In: Frostick, L.E. (Ed.), Sedimentation in the African Rift. Geological Society Special Publication, pp. 221–240.

Van Couvering, J.A. (Ed.), 1996. The Pleistocene Boundary and the Beginning of the Quaternary. Cambridge University Press, Cambridge.

Vrba, E.S., 1974. Chronological and ecological implications of the fossil Bovidae at the Sterkfontein australopithecine site. Nature 256, 19–23.

Vrba, E.S., 1980. The significance of bovid remains as indicators of environment and predation patterns. In: Behrensmeyer, A.K., Hill, A. (Eds.), Fossils in the Making: Vertebrate Taphonomy and Paleoecology. University of Chicago Press, Chicago, pp. 247–271.

Vrba, E.S., 1987. Ecology in relation to speciation rates: some case histories of Miocene – recent mammal clades. Evolutionary Ecology 1, 283–300.

Vrba, E.S., 1988. Late Pliocene climatic events and hominid evolution. In: Grine, F.E. (Ed.), The evolutionary history of the robust australopithecines. Aldine de Gruytersr, New York, pp. 405–426.

Vrba, E.S., 1995a. The fossil record of African antelopes (Mammalia: Bovidae) in relation to human evolution and paleoclimate. In: Vrba, E.S., Denton, G.H., Partridge, T.C., Burckle, L.H. (Eds.), Paleoclimate and Evolution with Emphasis on Human Origins. Yale University Press, New Haven, pp. 385–424.

Vrba, E.S., 1995b. On the connection between paleoclimate and evolution. In: Vrba, E.S., Denton, G.H.,

Partridge, T.C., Burckle, L.H. (Eds.), Paleoclimate and Evolution with Emphasis on Human Origins. Yale University Press, New Haven, pp. 24–45.

Vrba, E.S., Denton, G.H., Prentice, M.L., 1989. Climatic influences on early hominid behavior. Ossa 14, 127–156.

Walter, R.C., 1994. The age of Lucy and the first family: single-crystal $^{40}$Ar/$^{39}$Ar dating of the Denen Dora and lower Kada Hadar members of the Hadar Formation, Ethiopia. Geology 22, 6–10.

Walter, R.C., Aronson, J.L., 1993. Age and source of the Sidi Hakoma Tuff, Hadar Formation, Ethiopia. Journal of Human Evolution 25, 229–240.

White, T.D., Johanson, D.C., Kimbel, W.H., 1981. *Australopithecus africanus*: its phyletic position reconsidered. South African Journal of Science 77, 445–470.

White, T.D., Suwa, G., Hart, W.K., Walter, R.C., WoldeGabriel, G., de Heinzelin, J., Clark, J.D., Asfaw, B., Vrba, E.S., 1993. New discoveries of *Australopithecus* at Maka in Ethiopia. Nature 366, 261–265.

Wood, B.A., 1991. Koobi Fora Research Project, Volume 4: Hominid cranial remains. Clarendon Press, Oxford.

Wynn, J.G., 2004. Influence of Plio-Pleistocene aridification on human evolution: evidence from paleosols of the Turkana Basin, Kenya. American Journal of Physical Anthropology 123(2), 106–118.

# 7. Comparability of fossil data and its significance for the interpretation of hominin environments

## A case study in the lower Omo Valley, Ethiopia

Z. ALEMSEGED
*Department of Human Evolution*
*Max Planck Institute for Evolutionary Anthropology*
*Deutscher Platz, 6, 04103 Leipzig, Germany*
*zeray@eva.mpg.de*


R. BOBE
*Department of Anthropology*
*The University of Georgia*
*Athens, GA 30602, USA*
*renebobe@uga.edu*


D. GERAADS
*UPR 2147 CNRS-44 rue de l'Amiral Mouchez*
*75014 Paris, France*
*dgeraads@ivry.cnrs.fr*

Keywords: Human evolution, paleoenvironments, Plio-Pleistocene, Shungura Formation, fossil databases

## Abstract

Unraveling the context in which the evolution and diversification of early hominins occurred has become one of the core and highly debated subjects in paleoanthropology. Over the past three decades substantial progress has been made due to the proliferation of fieldwork and a consequently expanding fossil record, and development of new methods of analysis. The present study uses data of fossil mammals from the Shungura Formation of Ethiopia, with specimens collected semi-independently by French and American research teams who worked in the southern and northern parts of the Shungura area respectively. We compare these two samples in terms of collection methods, taxonomy, taphonomy, and local environmental differences. The following results were obtained: (1) No major taphonomic differences were observed between the two collections. The effect of a major taphonomic shift that occurred in the middle of Member G (G-13) is observed in both samples and is caused by the important change in the depositional environment from fluvial to lacustrine conditions. (2) The French team collected more specimens than the American team, in part because it had a larger area of exposures, and it spent two extra seasons in the field. Additionally, the French team collected more large-sized taxa including their postcranial elements, while the American team recovered a restricted set of postcranial bones. In contrast, the American team collected more primates and carnivores than the French team. (3) Despite these differences, comparable taxonomic composition and number of species are observed in both collections. (4) A study of changes in relative abundance in bovid tribes indicates that similar patterns of variation through time are observed in both samples. This is considered to be evidence for the prevalence of generally similar habitats (and habitat change through time) in the north and

*R. Bobe, Z. Alemseged, and A.K. Behrensmeyer (eds.) Hominin Environments in the East African*
*Pliocene: An Assessment of the Faunal Evidence, 159–181.*

south of the Shungura area. (5) However, habitat differences may have occurred locally, as inferred by differences in taxonomic abundances at the species level. For example, the bovid *Menelikia lyrocera* was more common in the southern parts of the Shungura exposures, while *Kobus sigmoidalis* was more common in the north. (6) Finally, the present study underscores the importance of the quality of data in unraveling past environments and patterns of faunal changes through time. Well-controlled and standardized collecting methods and systematic documentation procedures are critical for future fieldwork activities. This will improve the quality of our data, facilitate comparisons across regions, and lead to more robust hypotheses.

## Introduction

Our understanding of the origin, diversification, and evolution of early hominins is tightly linked to our knowledge of the paleoenvironments in which these processes took place. Over the last few decades, increasing attention has been paid to the environmental context of human evolution. New hypotheses have been proposed, and novel approaches developed for paleoenvironmental research. Some of these approaches rely on the hominin fossils themselves. The functional anatomy of postcranial elements may be used to provide information about locomotion and therefore about substrate (Senut, 1980; Senut and Tardieu, 1985; Susman and Stern, 1991; Stern, 2000; Ward, 2002). However, our understanding of the adaptive significance of anatomical characters is limited. The hominin fossil record has also been used to elucidate evolutionary patterns and possible links to climatic and environmental change (e.g., Vrba, 1988), but the rarity and discontinuity of this record make potential links between evolutionary patterns and broader climatic factors highly problematic (e.g., White, 1995; Behrensmeyer et al., 1997). Other approaches used to investigate the paleoenvironments and paleoecology of Plio-Pleistocene hominins in Africa include the analysis of fossil pollen, the study of paleosols and their isotopic composition, and the analysis of vertebrate remains that are commonly encountered in the fossil record. In the study of fossil vertebrates, the most common methods rely on faunal composition and taxonomic abundance data to derive paleoenvironmental information from hominin localities (e.g., Coppens, 1975; Geraads

and Coppens, 1995; Bobe and Eck, 2001; Bobe et al., 2002; Alemseged, 2003; Suwa et al., 2003). Other approaches rely on ecomorphology or on the analysis of ecological community structure. Ecomorphology deals with interpreting fossil remains in terms of functional anatomy and its relationship to environmental conditions (e.g., Kappelman, 1988; Spencer, 1997; Bishop, 1999). These approaches have strengths and weaknesses, and whenever possible they should be combined and cross-checked to evaluate the consistency of their respective signals.

One of the better known attempts to tie human evolution to environmental and climatic changes is the "turnover pulse hypothesis" (Vrba, 1988, 1992, 1995, 1999). This idea posits that most if not all speciation and extinction is due to climatic change: the majority of evolution occurs fairly rapidly and is concentrated during periods of dramatic climatic change causing significant pulses of speciation and/or extinction over time. Turnover pulses are concentrations of first and last appearance data (FADs and LADs) of species' temporal ranges, as shown for bovids (Vrba, 1995), and more recently for cercopithecids (Frost, 2002, 2007) and carnivores (Lewis and Werdelin, 2007). However, calculating the number of FADs and LADs is not as clear-cut as it might appear (e.g., Hill, 1987; White, 1995). In particular, detecting the timing of major biotic changes and potential links to climatic signals derived from other sources, such as marine sediments and stable isotopes (deMenocal, 1995; deMenocal and Bloemendal, 1995; Kennett, 1995; Shackleton, 1995; Denton, 1999), has resulted in discrepancies between different approaches (Vrba, 1988, 1995, 2000; Feibel et al., 1991; Behrensmeyer et al., 1997; Bobe and Eck, 2001; Bobe et al., 2002; Alemseged, 2003). For Vrba (1988, 1995, 2000), the major turnover pulse among bovids occurred in the interval 2.8–2.5 Ma. Other researchers indicate that major faunal changes in East Africa occurred at around 2 Ma (Harris et al., 1988; Feibel et al., 1991). Behrensmeyer et al. (1997) analyzed various mammalian taxa to suggest that the most important episode of faunal change occurred at around 1.8 Ma. Based on their study on the Shungura fauna, Bobe and Eck (2001) and Bobe et al. (2002) indicate that an episode of significant faunal change occurred at about 2.8 Ma and was followed by an interval of stability from 2.7 to 2.5 Ma. Geraads and Coppens (1995) and Alemseged (2003) on the other hand detected major faunal change at around 2.3 Ma in the Shungura sequence. Finally Frost (2002, 2007) showed that major turnover in monkeys happened at around 3.4 and 2.0 Ma, in broad agreement with the results of Lewis and Werdelin (2006) in their analysis African carnivores.

All studies discussed above have contributed considerably to our understanding of the relationships between mammalian evolution and climate change, and to different aspects of paleoenvironment and paleoecology in the Plio-Pleistocene of Africa. Particularly, the results of these studies have shed light on the role played by environment and climate in shaping the evolution of our own family since ca. 6 Ma. Nonetheless, discrepancies among these studies clearly demonstrate the complexity of the issues. One important aspect of these approaches is that they are all susceptible to the quality and comparability of the data, which depends in turn on how well we control biases introduced by taphonomy, collection strategies, and stratigraphic and provenance uncertainties.

In this study we address the issue of data comparability using information from the well-known Shungura Formation in the southwest of Ethiopia. Our main goal is to explore the differences and similarities of two fossil samples that were collected from comparable sedimentary contexts, time interval, and geographic areas, but by two independent paleontological teams. The large data set from the Shungura Formation, where American and French research teams conducted fieldwork during the 1960s and 1970s, is used for the comparative study. Given that both collections come from the same stratigraphic context and sedimentary basin we expect the two research teams to document similar taphonomic, paleoenvironmental, and paleoecological information.

The Shungura sequence is unrivalled by any hominin bearing Plio-Pleistocene site for its continuity, abundance of fossils, and quality of dating methods used. All these added together make this sequence the best candidate to undertake comparisons in terms of taxonomic abundance, species richness, taphonomic and collecting biases. The fact that there are two independent samples of fossil fauna provides an opportunity to evaluate equivalent fossil assemblages of the same geological age and region.

## The Shungura Formation: Background

The Shungura Formation is located in the lower Omo Valley of southwestern Ethiopia, west of the Omo River and north of Lake Turkana (Brown and Heinzelin, 1983) (Figure 1). The composite stratigraphic section of the formation measures nearly 800 m and radiometric ages indicate that it covers the time span from at least 3.6 Ma to about 1 Ma (Feibel et al., 1989) (Figure 2). The sedimentary cycles of the formation are grouped into 12 members (Basal, A, B, C, D, E, F, G, H, J, K, and L), each (except the Basal Member) commencing with a volcanic tuff designated by the same letter. The sequence is typically composed of fluvial sediments, but episodes of lacustrine deposition also occurred, particularly in the Basal Member, upper Member G, and upper Member L (de Heinzelin and Haesaerts, 1983).

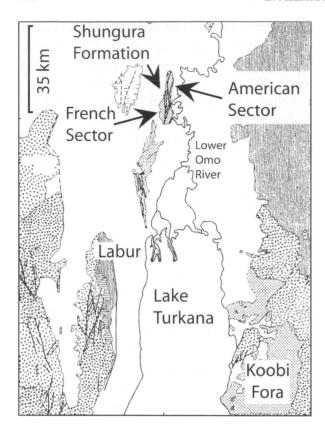

Figure 1. Geographic position and distribution of major rock types in the lower Omo valley and Turkana. (Adapted from de Heinzelin, 1983).

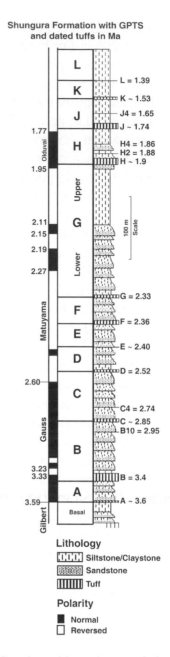

Figure 2. Stratigraphic column of the Shungura Formation (after de Heinzelin, 1983).

The Mission Scientifique de l'Omo, led by C. Arambourg in 1932 and 1933, was the first expedition to conduct systematic paleontological work in the lower Omo Valley (Arambourg and Coppens, 1967; Coppens, 1976). In 1966, the International Omo Research Expedition (IORE) was created under the direction of L.S.B. Leakey, C. Arambourg, and F.C. Howell (Coppens et al., 1976). In 1967, three contingents of the IORE independently explored the sedimentary exposures of the Omo region. The French one, under the direction of C. Arambourg and Y. Coppens, worked principally in the "type area" of the Shungura Formation. The Kenyan contingent, under the direction of R.E.F. Leakey, and the American contingent, under the direction of F.C. Howell, worked farther to the north, the Kenyans in the Kibish and Mursi Formations, and the Americans in the Usno Formation. In 1968,

quitting his research in the Mursi and Kibish Formations, Leakey moved to the Koobi Fora area of northern Kenya. Howell on the other hand arranged with Arambourg to move south and jointly explored the type area of the Shungura Formation. The American contingent ceased work in the lower Omo Valley at the end of the 1974 field season, and the French expedition continued until 1976.

In the nine years of fieldwork between 1967 and 1976 (no fieldwork was conducted by either team in 1975), nearly 50,000 paleontological specimens were collected, 21,858 by the Americans and 27,409 by the French. Most of these were recovered during surface survey, but large paleontological excavations were also carried out, producing 6,692 American and 3,417 French specimens (see Johanson et al., 1976).

About 220 hominin specimens were recovered by the two research teams. The oldest hominin remains were recovered from Member B and the Usno Formation. Most specimens are teeth and span the interval from 3.3 to ca.1.0 Ma. There are 21 hominin specimens between 3.3 and 3.0 Ma, 45 between 3.0 and 2.5 Ma, 145 between 2.5 and 2.0 Ma, 3 between 2.0 and 1.5 Ma, and 3 between 1.5 and 1.0 Ma (Suwa, 1990). Among these specimens a "gracile australopithecine" is recognized in Member B of the Shungura Formation and Usno (Howell and Coppens, 1973; Suwa, 1988). In Member C, *Australopithecus aethiopicus* is identified (Arambourg and Coppens, 1967, 1968; Coppens, 1976; Suwa, 1988, 1990; Suwa et al., 1996). This species is believed to have existed up to lower Member G (Suwa et al., 1996). *Australopithecus boisei* and *Homo* sp. are recognized at the base of Member G; however, the genus *Homo* could have existed during the times of Member E (Howell and Coppens, 1974; Howell et al., 1987; Suwa et al., 1996).

## Comparative Studies

### DEPOSITIONAL ENVIRONMENTS AND TAPHONOMY

As noted above, nearly all of the richly fossiliferous deposits of the Shungura Formation consist of fluvial sediments laid down by a major river similar in size to the modern Omo River. These sediments consist of sands deposited in the river channel, silts deposited on the banks of the river near the channel, and, silty clays deposited during periods of high water more distal to the channel. The fossil specimens were dominated by elements resistant to damage: jaw fragments and teeth, dense postcranial elements, and bovid horn cores. This is evidence for hydraulic sorting, which is not a surprise in a depositional context characterized by a major river system (Alemseged et al., 1996; Bobe and Eck, 2001; Alemseged, 2003). While it is hard to determine to what extent this differential representation of diverse skeletal elements affects the resulting fossil taxonomic composition in the Shungura Formation, it is possible to use differences in relative abundance of skeletal elements between the two collections to evaluate some taphonomic aspects of the two samples. This is done by comparing the relative change of the number of skeletal elements through time in both collections, which allows the assessment of taphonomic conditions in relation to depositional environments. The results show that patterns of variation through time of the number of isolated teeth collected by both teams are very similar (Table 1, Figure 3). It is true that in almost all members the French team collected more teeth, particularly in lower Member G, which reflects the greater total number of specimens collected by this team. However, as one moves from one member to the next, the number of teeth collected fluctuates in the same manner in both samples. A similar pattern is observed when postcranial elements are considered (Table 1, Figure 4). These two observations are probably indicative of similar changes in both areas with respect to taphonomic context. The differences observed, particularly in lower Member G where the French sample is much larger, can be attributed mainly to differences in the number of specimens (NISP) between the two collections.

To make each sample comparable to one another we used ratios of the number of teeth or postcranial elements to the total number of specimens (Table 1, Figures 5 and 6). Isolated teeth constitute a high proportion of specimens

*Table 1. Number of specimens for various skeletal elements (top) and ratios of the numbers of teeth and postcrania vs. the total of skeletal elements (bottom) in the American and French collections*

| American collection | A | B | C | D | E | F | G(L) | G(U) | H | J | K | L | Total |
|---|---|---|---|---|---|---|---|---|---|---|---|---|---|
| Isolated teeth | 29 | 576 | 1607 | 556 | 661 | 725 | 2190 | 113 | 236 | 103 | 159 | 128 | 7,083 |
| Mandibles | 1 | 56 | 220 | 74 | 108 | 67 | 301 | 26 | 37 | 16 | 8 | 20 | 934 |
| Crania | 3 | 87 | 273 | 87 | 178 | 217 | 1002 | 41 | 44 | 21 | 29 | 37 | 2,019 |
| Postcrania | 6 | 168 | 858 | 136 | 310 | 283 | 706 | 254 | 144 | 70 | 62 | 90 | 3,087 |
| Total | 39 | 887 | 2958 | 853 | 1257 | 1292 | 4199 | 434 | 461 | 210 | 258 | 275 | 13,123 |
| **French collection** | | | | | | | | | | | | | |
| Isolated teeth | 233 | 1008 | 1744 | 233 | 1000 | 659 | 5555 | 495 | 309 | 48 | 118 | 342 | 11,744 |
| Mandibles | 11 | 68 | 169 | 30 | 125 | 66 | 903 | 68 | 30 | 5 | 19 | 16 | 1,510 |
| Crania | 6 | 73 | 136 | 42 | 102 | 64 | 1172 | 85 | 32 | 8 | 45 | 17 | 1,782 |
| Postcrania | 84 | 281 | 1034 | 186 | 544 | 253 | 2908 | 772 | 288 | 84 | 229 | 97 | 6,760 |
| Total | 334 | 1430 | 3083 | 491 | 1771 | 1042 | 10538 | 1420 | 659 | 145 | 411 | 472 | 21,796 |
| **Ratios** | | | | | | | | | | | | | |
| **American collection** | A | B | C | D | E | F | G(L) | G(U) | H | J | K | L | Total |
| Isolated teeth | 0.744 | 0.649 | 0.543 | 0.652 | 0.526 | 0.561 | 0.522 | 0.26 | 0.512 | 0.49 | 0.616 | 0.465 | 0.54 |
| Mandibles | 0.026 | 0.063 | 0.074 | 0.087 | 0.086 | 0.052 | 0.072 | 0.06 | 0.08 | 0.076 | 0.031 | 0.073 | 0.071 |
| Crania-maxillae | 0.077 | 0.098 | 0.092 | 0.102 | 0.142 | 0.168 | 0.239 | 0.094 | 0.095 | 0.1 | 0.112 | 0.135 | 0.154 |
| Postcrania | 0.154 | 0.189 | 0.29 | 0.159 | 0.247 | 0.219 | 0.168 | 0.585 | 0.312 | 0.333 | 0.24 | 0.327 | 0.235 |
| **French collection** | | | | | | | | | | | | | |
| Isolated teeth | 0.698 | 0.705 | 0.566 | 0.475 | 0.565 | 0.632 | 0.527 | 0.349 | 0.469 | 0.331 | 0.287 | 0.725 | 0.539 |
| Mandibles | 0.033 | 0.048 | 0.055 | 0.061 | 0.071 | 0.063 | 0.086 | 0.048 | 0.046 | 0.034 | 0.046 | 0.034 | 0.069 |
| Crania-maxillae | 0.018 | 0.051 | 0.044 | 0.086 | 0.058 | 0.061 | 0.111 | 0.06 | 0.049 | 0.055 | 0.109 | 0.036 | 0.082 |
| Postcrania | 0.251 | 0.197 | 0.335 | 0.379 | 0.307 | 0.243 | 0.276 | 0.544 | 0.437 | 0.579 | 0.557 | 0.206 | 0.31 |

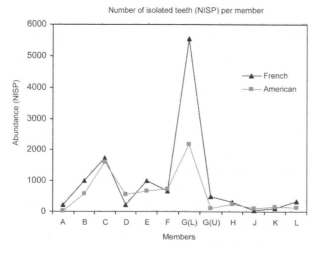

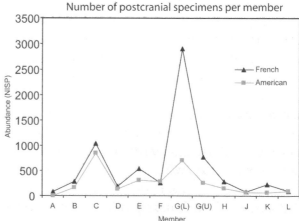

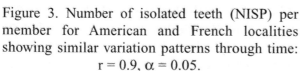

Figure 3. Number of isolated teeth (NISP) per member for American and French localities showing similar variation patterns through time: $r = 0.9$, $\alpha = 0.05$.

Figure 4. Number of postcranial specimens (NISP) per member for American and French localities showing similar variation patterns through time: $r = 0.77$, $\alpha = 0.05$.

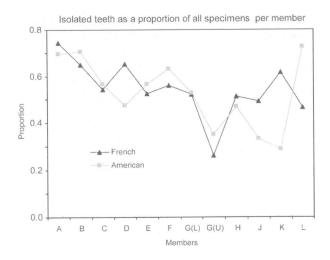

Figure 5. Variation through time of isolated teeth as a proportion of all specimens per member for the American and French collections: r = 0.74, α = 0.05.

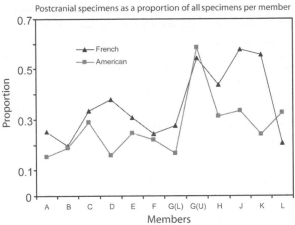

Figure 6. Variation through time of postcranial elements as a proportion of all specimens per member for the American and French collections: r = 0.75, α = 0.05.

in both collections, and show similar patterns of change through time, except for Member D and the younger levels of the sequence (Figure 5). Ratios of both teeth and postcranial elements to the total are significantly correlated when Members J and K are not included. The most striking observation is the drop in the proportion of isolated teeth in the transition from lower Member G to upper Member G observable in both collections. Likewise, the proportion of postcrania shows similar patterns of fluctuation and there is a clear increase of this proportion in the transition from lower G to upper G (Figure 6). Again, we interpret these observations to indicate similarities of taphonomic conditions, at least those related to depositional environments, in both the American and French samples. The relative decrease in teeth and increase in postcrania in the transition from lower to upper Member G is evidence for a transition from a high-energy river system to a low-energy lacustrine system. In other words, teeth, which are more resistant to damage than postcrania, are well represented in both fluvial and lacustrine systems. However, postcranial elements are relatively more common in the lacustrine (lower energy) than in the riverine (higher energy) setting. These results

corroborate conclusions drawn previously from geological data and other sources (Brown and Heinzelin, 1983; Haesaerts et al., 1983).

## ABUNDANCES AT HIGHER TAXONOMIC CATEGORIES

Among the 42,481 specimens that can be identified at the ordinal level, 11 orders of mammals are represented, with artiodactyls clearly predominating, primates having high abundance, and carnivores being uncommon among the mammals of medium to very large body size (Table 2, Figure 7). Out of this total 25,764 and 16,718 come from the French and American localities, respectively. The remaining specimens are identifiable either as Mammalia indet. or belong to nonmammalian taxa, which are not considered in this study. In the combined sample 74% of the total is composed of artiodactyls. However, when this percentage is calculated for the two samples separately, artiodactyls account for 81% of mammals in the French database and only 67% in the American. This shows that there is a relative bias against artiodactyls in the American sample. However, frequency distributions using the Kolmogorov–Smirnov test both at

*Table 2. Numbers of specimens identified at the ordinal level. Kolmogorov–Smirnov's D value of 0.333, for α = 0.05*

|  | American | French | Total |
|---|---|---|---|
| Artiodactyla | 11,053 | 20,731 | 31,784 |
| Primates | 3,632 | 3,052 | 6,684 |
| Proboscidea | 743 | 1,259 | 2,002 |
| Perissodactyla | 390 | 510 | 900 |
| Rodentia | 724 | 96 | 820 |
| Carnivora | 149 | 113 | 262 |
| Chiroptera | 10 | 0 | 10 |
| Insectivora | 10 | 0 | 10 |
| Lagomorpha | 3 | 3 | 6 |
| Hyracoidea | 3 | 0 | 3 |
| Tubulidentata | 1 | 0 | 1 |
| Total | 16,718 | 25,764 | 42,481 |

*Table 3. Number of specimens identified at the family level. Kolmogorov–Smirnov's D value of 0.143, for α = 0.05*

| Macromammals | American | French | Total |
|---|---|---|---|
| Bovidae | 6,295 | 11,007 | 17,302 |
| Hippopotamidae | 2,448 | 5,472 | 7,920 |
| Cercopithecidae | 3,482 | 2,917 | 6,399 |
| Suidae | 1,770 | 3,087 | 4,857 |
| Elephantidae | 517 | 1,062 | 1,579 |
| Giraffidae | 535 | 1,007 | 1,542 |
| Equidae | 332 | 397 | 729 |
| Deinotheriidae | 225 | 196 | 421 |
| Hominidae | 147 | 135 | 282 |
| Rhinocerotidae | 55 | 109 | 164 |
| Felidae | 85 | 42 | 127 |
| Hyaenidae | 16 | 22 | 38 |
| Hystricidae | 15 | 10 | 25 |
| Camelidae | 5 | 16 | 21 |
| Mustelidae | 12 | 3 | 15 |
| Chalicotheriidae | 3 | 4 | 7 |
| Procaviidae | 3 | 0 | 3 |
| Orycteropodidae | 1 | 0 | 1 |
| Total | 15,946 | 25,486 | 41,432 |

ordinal and familial level show that the two samples do not differ significantly (Tables 2 and 3), but abundance comparisons made between the two samples at different taxonomic levels reveal some interesting differences. Comparing numbers of specimens of the first five most common orders (Artiodactyla, Primates, Proboscidea, Perissodactyla, and Carnivora) collected by the two teams shows that the French collected more of every group with the exception of primates and carnivores (Table 2, Figure 7). In other words, in both collections artiodactyls are the most common and carnivores the least, but more primates and carnivores were collected by the American than the French team. This is also reflected when abundance comparison is made at family level. For the first ten most common families (Bovidae, Hippopotamidae, Suidae, Cercopithecidae, Elephantidae, Giraffidae, Equidae, Deinotheriidae, Hominidae, and Rhinocerotidae), there are more specimens in the French collection than there are in the American (Table 3, Figure 8). One major exception is the Cercopithecidae, of which the American team collected more specimens.

It is clear that the intrinsic nature of sediments, where some are more fossiliferous than others, and the size of areas explored by the different teams, have affected the overall difference in the number of specimens collected. However, other factors may explain some of the differences observed above. Among these, differences in collection protocols between the two teams have played a role. The higher number of total mammalian specimens collected by the French can be explained in part by the fact that they included postcranial elements of all taxa, even those of very large mammals such as hippopotamids. Consequently, while most mammalian families constitute comparable percentages in both the French and the American collections, hippopotamids comprise 21% of the French but only 16% of the American collections (Figures 9 and 10). This illustrates that the American team was biased against large mammals, particularly their postcrania (see Eck, 2007). In comparison, cercopithecids differ clearly in their relative abundance in the two collections, constituting 22% in the American but only 12% in the French collection (Figures 9 and 10). The higher number of cercopithecids amassed by the American

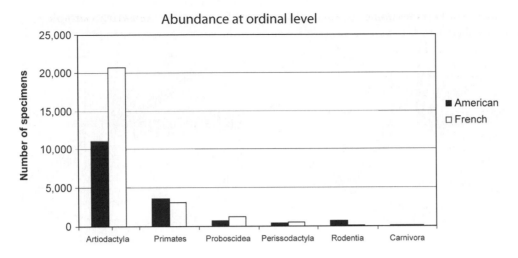

Figure 7. Comparative abundance of the first six most common mammalian orders in the French and American samples.

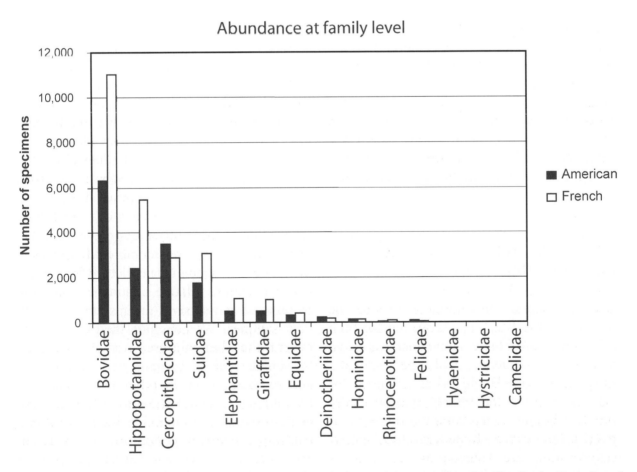

Figure 8. Comparative abundance of the first 14 most common mammalian families in the French and American samples.

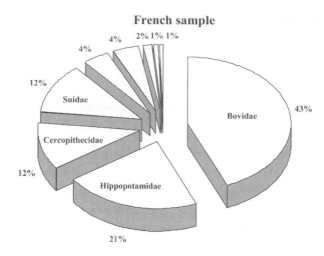

Figure 9. Percentages of the most common mammalian families within the French collection.

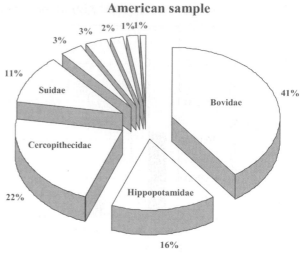

Figure 10. Percentages of the most common mammalian families within the American collection.

contingent is noteworthy considering that the overall French sample exceeds the American sample by almost 10,000 specimens. It is clear that the American team was collecting monkeys with higher intensity than other major taxa, which accounts for the differences observed (see Eck, 2007). Thus, paleoenvironmental interpretations based on abundance of monkeys relative to other major taxa should be considered with caution.

## TAXONOMIC ABUNDANCE: VARIATION THROUGH TIME

Given that similar taphonomic history can be assumed in the Shungura sequence for both collections, taxonomic abundance variation through the sequence could be evaluated using some taxa. Bovid tribes are used for this purpose for the widely accepted reason that they are very common and habitat specific. Four tribes – Reduncini, Alcelaphini, Aepycerotini, and Tragelaphini are considered. We used these four tribes and looked at how their relative abundances vary through the Shungura sequence in the American and French samples.

Results of comparisons between the two collections do not show major differences in relative abundance variation through time from Members B to upper G among the taxa under consideration (Table 4, Figures 11–14). Patterns of variation are almost identical for Reduncini, Tragelaphini, and Alcelaphini and similar fluctuation patterns are observed in both collections for Aepycerotini, even though after Member E American proportions are higher for the latter. These observations indicate that overall, despite the considerable differences in the number of specimens collected by the two research teams, relative abundance variation patterns of major bovid groups are similar in both samples. We conclude therefore that in general the Shungura area was characterized by similar type of habitats both in northern and southern parts. In other words, the two research teams sampled areas that overall were characterized by comparable depositional and paleoenvironmental conditions. As a result both samples document similar patterns of changes through time in terms of taphonomy and taxonomic abundance.

*Table 4. Number of specimens of bovid tribes across the members of the Shungura Formation in the American and French collections*

| French | B | C | D | E | F | G(L) | G(U) |
|---|---|---|---|---|---|---|---|
| Tragelaphini | 41 | 110 | 21 | 69 | 74 | 731 | 66 |
| Bovini | 37 | 50 | 5 | 12 | 20 | 53 | 5 |
| Reduncini | 110 | 76 | 15 | 30 | 68 | 1645 | 198 |
| Hippotragini | 0 | 0 | 0 | 0 | 1 | 6 | 1 |
| Aepycerotini | 69 | 92 | 30 | 29 | 68 | 545 | 124 |
| Alcelaphini | 6 | 11 | 4 | 3 | 26 | 68 | 33 |
| Antilopini | 2 | 0 | 1 | 0 | 0 | 8 | 5 |
| **American** | **B** | **C** | **D** | **E** | **F** | **G(L)** | **G(U)** |
| Tragelaphini | 29 | 178 | 66 | 117 | 87 | 375 | 2 |
| Bovini | 24 | 73 | 15 | 13 | 15 | 38 | 2 |
| Reduncini | 63 | 74 | 30 | 56 | 69 | 684 | 14 |
| Hippotragini | 0 | 3 | 0 | 1 | 0 | 2 | 0 |
| Aepycerotini | 48 | 94 | 56 | 91 | 141 | 461 | 2 |
| Alcelaphini | 6 | 5 | 6 | 9 | 25 | 51 | 1 |
| Antilopini | 1 | 3 | 1 | 0 | 5 | 4 | 0 |

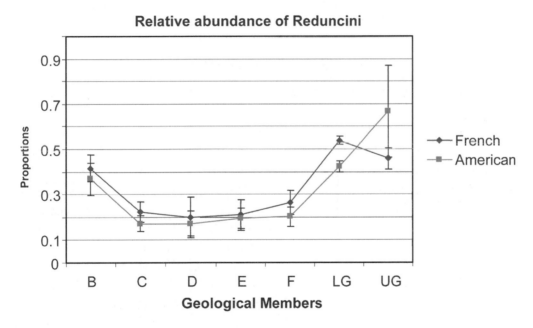

Figure 11. Abundance variation through time of Reduncini in the American and French collections. (95% CI calculations based on Buzas, 1990).

## ARE THERE LOCAL HABITAT DIFFERENCES?

As mentioned above, although the French sample is larger than the American one, both collections are characterized by similar species richness. Excluding micromammals, there are 55 species of mammals in the French collection and 60 species in the American one. This shows that the higher number of specimens in the French collection is caused in part by the inclusion of more postcranial elements rather than sampling more paleohabitats. Likewise, abundances of major bovid tribes are similar in both collections and

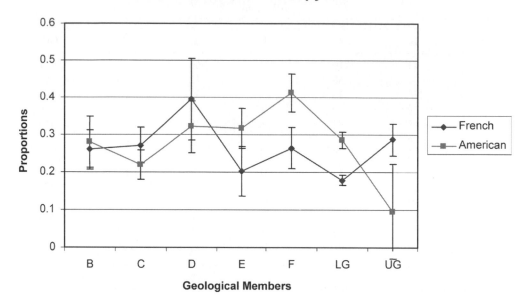

Figure 12. Abundance variation through time of Aepycerotini in the American and French collections (95% CI calculations based on Buzas, 1990).

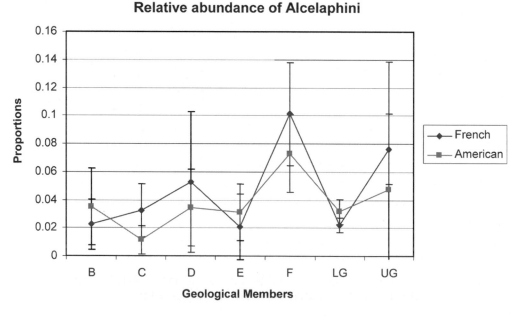

Figure 13. Abundance variation through time of Alcelaphini in the American and French collections (95% CI calculations based on Buzas, 1990).

show a high degree of comparability in terms of taxonomic diversity both in time and space. These observations lead to us to conclude that the overall underlying paleoenvironmental and paleoecological conditions both in the southern and northern parts of Shungura were similar. However, it is possible that there were local habitat differences characterized by higher or lower proportions of various species. A correspondence analysis (CA) was

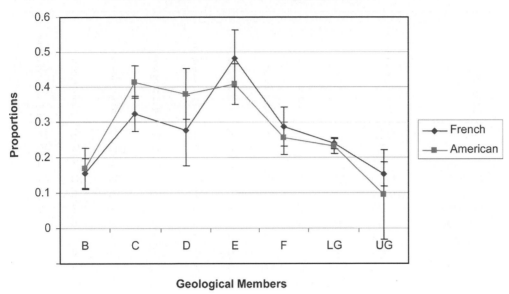

Figure 14. Abundance variation through time of Tragelaphini in the American and French collections (95% CI calculations based on Buzas, 1990).

chosen to look into this question. CA allows projecting rows and columns simultaneously onto a two- or three-dimensional graph, thus allowing us to see which taxa have higher or lower proportions in which localities. CA is a statistical visualization method for picturing the associations between the levels of a two-way contingency table. It is a geometric technique for displaying the rows and columns as points in a low-dimensional space, such that the positions of the row and column points are consistent with their associations in the table. In our case the objective is to explore if localities are particularly characterized by a given taxon or group of taxa. The expectation is that there should not be significant differences among localities of the two teams in terms of species proportions. Based on these assumptions we conducted an analysis at lower taxonomic levels (genera and species) in a restricted time interval, lower Member G (dating from ~2.3 to 2.1 Ma) (see Table 5). This interval was chosen because of the high abundance of fossils from a relatively restricted

time period. Some localities that are within this interval did not yield large enough numbers of fossil and were excluded from the analysis. We also excluded upper Member G because it differs from lower Member G in depositional environments, taphonomic conditions, and abundance of fossils. As stated above, the transition from lower G to upper G is characterized by a shift from fluvial to lacustrine conditions, with a lake expanding to the north from the center of the Turkana Basin. The geographic distribution of the localities considered is presented in Figure 15.

A correspondence analysis on localities from lower Member G using genera and species as variables shows that there is some differentiation between the French and the American localities, implying that there may be local faunal (and perhaps habitat) differences (Figure 16). A similar analysis using only bovid species shows a stronger differentiation between French and American localities, with a higher proportion of *Menelikia lyrocera* in French localities and *Kobus sigmoidalis* in

Table 5. *Abundance of selected mammalian species and genera in lower Member G localities*

| Locality | Stratigraphic level | T. nakuae | T. gaudryi | M. lyrocera | K. sig moidalis | K. ancystrocera | A. shungurae | Kolpochoerus | Notochoerus | Metridiochoerus |
|----------|---------------------|-----------|------------|-------------|-----------------|-----------------|--------------|--------------|-------------|-----------------|
| Omo29 | G(L) | 62 | 38 | 8 | 6 | 2 | 117 | 22 | 36 | 14 |
| Omo47 | G8 | 103 | 25 | 39 | 0 | 7 | 42 | 104 | 24 | 25 |
| Omo48 | G12 | 15 | 1 | 1 | 0 | 4 | 10 | 4 | 1 | 0 |
| Omo75 | G(L) | 130 | 57 | 50 | 27 | 16 | 171 | 105 | 64 | 40 |
| Omo113 | G10–11 | 2 | 0 | 17 | 0 | 11 | 0 | 0 | 0 | 0 |
| Omo210 | G3 | 7 | 1 | 4 | 5 | 0 | 5 | 0 | 3 | 0 |
| Omo309 | G6 | 0 | 0 | 1 | 7 | 0 | 5 | 0 | 0 | 0 |
| Omo308 | G4 | 1 | 2 | 2 | 5 | 0 | 0 | 1 | 0 | 0 |
| Omo310 | G8 | 1 | 0 | 66 | 0 | 1 | 0 | 3 | 0 | 1 |
| Omo323 | G8 | 28 | 2 | 81 | 0 | 5 | 3 | 31 | 4 | 5 |
| SH 1 | G8–9 | 12 | 27 | 6 | 2 | 3 | 22 | 13 | 15 | 0 |
| Omo50 | G(L) | 16 | 3 | 9 | 0 | 4 | 16 | 10 | 9 | 4 |
| Omo311 | G8 | 0 | 0 | 13 | 0 | 0 | 2 | 4 | 0 | 0 |
| L7 | ~G5 | 22 | 38 | 14 | 37 | 0 | 18 | 16 | 12 | 12 |
| L16 | G4 | 11 | 7 | 5 | 32 | 1 | 20 | 6 | 2 | 1 |
| L25 | G13 | 7 | 0 | 5 | 11 | 0 | 40 | 10 | 5 | 0 |
| L35 | G5 | 8 | 5 | 2 | 4 | 1 | 10 | 3 | 1 | 1 |
| L43 | G12 | 2 | 0 | 2 | 5 | 0 | 1 | 0 | 0 | 2 |
| L67 | ~G8 | 19 | 3 | 8 | 36 | 0 | 5 | 0 | 4 | 0 |
| L73 | G12 | 6 | 1 | 2 | 3 | 0 | 15 | 2 | 1 | 2 |
| L74 | G4 | 5 | 0 | 1 | 4 | 0 | 1 | 2 | 4 | 0 |
| L80 | G4 | 5 | 12 | 1 | 10 | 0 | 0 | 5 | 0 | 0 |
| L112 | G7 | 2 | 0 | 5 | 3 | 0 | 0 | 3 | 0 | 0 |
| L627 | G12 | 21 | 21 | 4 | 4 | 6 | 109 | 27 | 6 | 6 |

American localities (Figure 17). These two species of Reduncini might have preferred only part of the Shungura paleolandscape, suggesting possible local habitat differences between the two areas.

Can this pattern be extrapolated over the whole range of the Shungura area? And, is there any consistent differentiation in the distribution of these taxa in the north and south? To answer these questions we considered only the tribe Reduncini (and its species), and examined their distribution in the whole geographic range covered by the two research teams. The goal was to see if there were relative abundance differences among closely related species of this tribe (in this case *Menelikia lyrocera*, *Kobus sigmoidalis*, and *Kobus ancystrocera*) in the

southern and northern parts of the Shungura area. If these differences exist, this would indicate possible local differences in habitat, which the various species of Reduncini would occupy, even though they shared general adaptations to waterlogged environments characteristic of the tribe as a whole.

For this purpose, abundances in each locality were plotted on the maps redrawn from those published at a scale of 1:10,000 by de Heinzelin (1983). For localities to be comparable we used ratios of taxonomic abundances that are represented by different sizes and colors of circles as illustrated in Figure 18. This was done because the term "locality" covers a wide range of collecting units, from a single spot where only one

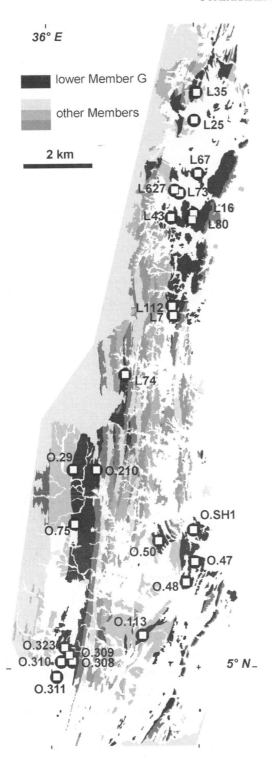

Figure 15. Geographical distribution of American (top) and French (bottom) localities from lower Member G of the Shungura Formation.

fossil specimen was collected (common in the American collection), to huge areas where thousands of fossils were amassed (such as the French Locality Omo-75). For every taxon (species, tribe) and anatomical element (horn cores, teeth, postcrania), we calculated its relative abundance as the ratio of its total number (numerator) to the number of the same element in the immediately higher taxonomic category (denominator). For instance, abundance of Tragelaphini teeth at locality Omo-323 is the ratio (N = Tragelaphini teeth from locality Omo-323)/(N = Bovid teeth from Omo-323), but abundance of *Tragelaphus nakuae* horn cores in the same locality is the ratio (N = *T. nakuae* horn cores from Omo-323)/(N = *Tragelaphus* horn cores from Omo-323). Thus, the size of the circles is proportional to the denominator and its color (lighter or deeper) reflects the abundance of the numerator. In Figure 18 for example *Menelikia* horn cores are compared to the number of Reduncini horn cores. A deeper color means that the *proportion* of *Menelikia* is greater, while the size of the circle, which is proportional to the number of Reduncini horn cores, reflects the *significance* of this proportion at this particular locality. Since for many localities the number of collected specimens is low, the computed ratios are not always significant. For instance, if only one reduncin horn core is found in a locality, the relative abundance of the occurring species will be one, but the small size of the resulting circle will reflect the insignificance of that particular locality. It is important to note that this method allows us to control taphonomic or collecting bias, because the results are not altered by differential preservation of teeth vs. long bone or selective collecting of horn cores vs. teeth.

Obviously, color depth cannot have the same meaning for all taxa, as there are rare and common ones. To estimate the "rarity" or "commonness" of taxa, we could have calculated the average proportion in the whole collection, but this would have given too much weight to the rich

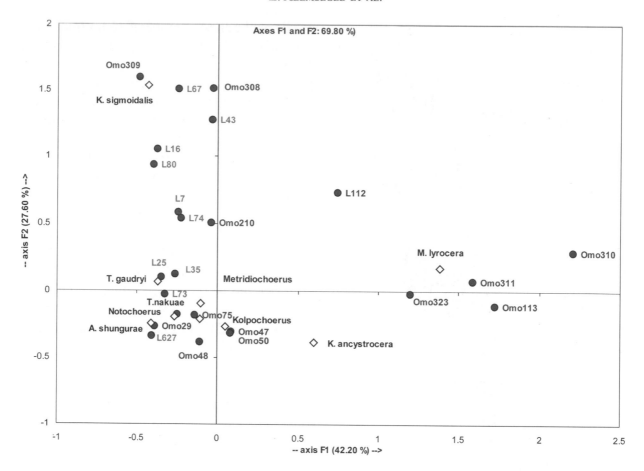

Figure 16. Correspondence analysis showing locality distributions on axes 1 and 2. Each point represents a locality. The American and French localities tend to separate. In this analysis columns represent mammalian taxa and rows represent localities. Omo = French localities, L = American localities.

localities. Alternatively, calculating the mean of ratios (sum of the ratios in each locality, divided by the number of localities) would have given too much weight to less fossiliferous localities. Thus, we compared these ratios to the average ratios in the 71 localities that yielded more than 100 specimens. Circle color reflects this ratio (the higher the proportion, the deeper the color), as follows:

very rare (very light color): less than half the average proportion;
rare (light color): between half and the average proportion;
common (deep color): between the average and twice the average;
very common (very deep color): more than twice the average.

For Reduncini horn cores, these ratios are: Reduncini/Bovidae=0.28; *Menelikia*/Reduncini = 0.24; *Kobus ancystrocera*/Reduncini = 0.07; *Kobus sigmoidalis*/Reduncini = 0.28. However, before analyzing differences at the species level within the Reduncini, we calculated the relative abundance of the tribe itself compared to the total number of bovids in different localities using the same approach. This was done by dividing the number of Reduncini horn cores in every locality considered by the total number of bovid horn cores in that same locality. The ratio indicates that the relative distribution of the tribe Reduncini in the north and south was not even. Figure 19 (column A) shows that we have comparable number of bovids in the north and south as is illustrated by the comparability of the size of the circles, but we have more

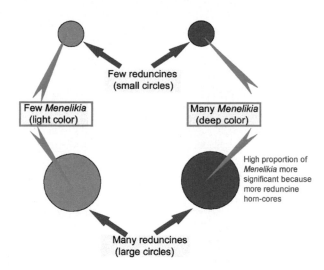

Figure 18. Schematic representation of the relative proportion of a taxon in a locality. The size of the circles is proportional to the denominator and its color (lighter or deeper) reflects the abundance of the numerator. In this example *Menelikia* horn cores are compared to the number of Reduncini horn cores. A deeper color means that the proportion of *Menelikia* is greater, while the size of the circle, which is proportional to the number of Reduncini horn cores, reflects the significance of this proportion at this particular locality.

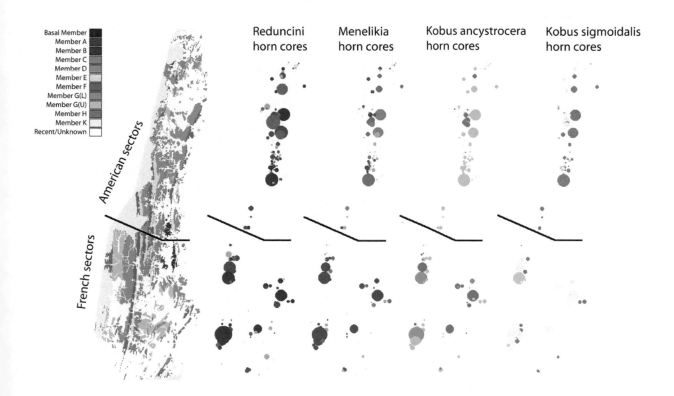

Figure 19. Distribution of Reduncini and its species over the whole range of the Shungura area for the French and American localities in the north and south. Every circle represents a locality. The depth of the color indicates the proportion of the taxa in a particular locality. Note that Reduncini are more common in the south than in the north in general. However, specimens of *Menelikia lyrocera* and *Kobus ancystrocera* are more common in the south, while those of *Kobus sigmoidalis* are more common in the north. These differences are statistically significant, $X^2 = 153$, $P < 0.001$.

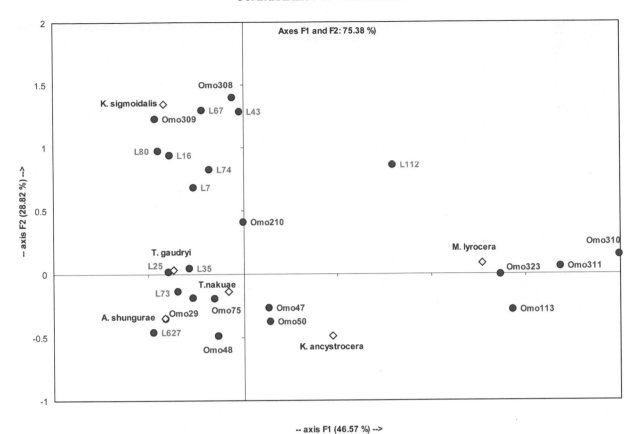

Figure 17. Correspondence analysis showing locality and taxa profiles on axes 1 and 2. Note the separation along the first axis of *Menelikia lyrocera* and *Kobus sigmoidalis*, two species of the tribe Reduncini.

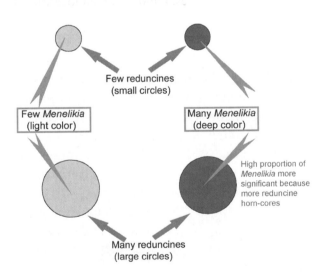

Figure 18. Schematic representation of the relative proportion of a taxon in a locality. The size of the circles is proportional to the denominator and its color (lighter or deeper) reflects the abundance of the numerator. In this example *Menelikia* horn cores are compared to the number of Reduncini horn cores. A deeper color means that the proportion of *Menelikia* is greater, while the size of the circle, which is proportional to the number of Reduncini horn cores, reflects the significance of this proportion at this particular locality.

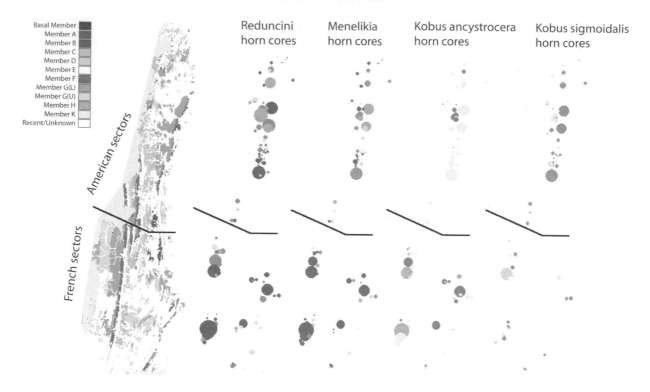

Figure 19. Distribution of Reduncini and its species over the whole range of the Shungura area for the French and American localities in the north and south. Every circle represents a locality. The depth of the color indicates the proportion of the taxa in a particular locality. Note that Reduncini are more common in the south than in the north in general. However, specimens of *Menelikia lyrocera* and *Kobus ancystrocera* are more common in the south, while those of *Kobus sigmoidalis* are more common in the north. These differences are statistically significant, $X^2 = 153$, $P < 0.001$.

deep-color circles in the south pointing to a relatively higher number of Reduncini in the area, as shown also by the table below (322 horn cores on the American collection from units G3–G13, 432 in the French one).

|  | American collection | French collection |
|---|---|---|
| *Kobus ancystrocera* | 28 | 61 |
| *Kobus sigmoidalis* | 193 | 82 |
| *Menelikia lyrocera* | 101 | 289 |

Given the differences in the relative abundance of Reduncini in the north and south, we undertook the same calculation for the different species within this tribe to see if they also show differences in their relative abundance in different geographic secotrs. Horn cores of *Menelikia lyrocera*, *Kobus ancystrocera*, and *Kobus sigmoidalis* are used for this purpose. The result shows that there are more large-size circles in the south than in the north confirming the relative higher abundance of the tribe Reduncini in the south. However, looking at the depth of colors of circles reveals that there are more deep colored circles in the south for *M. lyrocera* and *K. ancestrocera* showing the higher relative abundance of these tow species in the south, i.e. in the French localities. In contrast for *K. sigmoidalis*, there are more deep colored circles in the north. This indicates that specimens of *Menelikia* and

*Kobus ancystrocera* are more common in the south (French localities), while those of *K. sigmoidalis* are more common in the north (American Localities) (Figure 19 column B, C, D) and this difference is significant (X2 = 153, p<0.001). This may provide evidence for local ecological differences within the lower Omo basin during lower Member G times.

The inverse abundance of the two most common reduncine species (*M. lyrocera* and *K. sigmoidalis*) had already been noticed by Gentry (1985) who linked his observation to ecological differences, since he assumed that *Menelikia* was an open-country form. Alemseged (1998) further substantiated this ecological separation, but following Spencer (1997) he assumed that *Menelikia* was adapted to more wooded environments. The preferred kind of habitat of these two extinct species remains to be determined by further ecomorphological studies, but our results provide some evidence that these species may have lived in different habitats, or may have excluded each other. Their abundances were negatively correlated among Shungura localities, and overall *K. sigmoidalis* was more common in the north whereas *M. lyrocera* was so in the south. Further studies should test the possibility of local habitat differences in the lower Omo Basin. Results of such studies could shed light on issues pertaining to habitat preferences of hominin species.

Geraads and Coppens (1995) found that in Member G the American team collected more bovid horn cores relative to teeth than the French team, while the French collected proportionately more teeth. Eck (2007) found the same results and suspects that these differences were introduced because the French team less consistently collected horn cores. However, the same pattern may be produced by the French collecting a greater proportion of teeth, especially fragmentary ones. Even though the American team collected more horn cores than the French team, there is no reason to believe that either team was particularly biased for or against horn cores of any particular taxon. Therefore, it is likely that the differences in taxonomic abundances based on horn cores within each collection reflect the reality on the ground. In this contribution we have shown that both collections are generally comparable in taxonomic composition and patterns of relative abundance, but at finer levels of resolution there are some intriguing differences. Eck's (2007) suspicion that differences in skeletal element abundances between the two teams introduced taxonomic biases within the collections does not distinguish taphonomic from taxonomic biases. We suggest that some of these differences may have environmental causes, but further work in the Shungura Formation would be required to settle the issue of environmental vs. collection factors.

## Summary

The present study addressed issues that relate to data comparability and standardization based on fossil collections from the well-dated Shungura Formation in southwestern Ethiopia. Large numbers of specimens (ca. 50, 000) were collected by French and American research teams working semi-independently in the same stratigraphic sequence and adjacent areas. Here we compared various aspects of the two collections and found important similarities and differences. Taphonomic analyses pertaining to depositional environments show that there are no major differences between the two samples. A major taphonomic shift in the middle of Member G (between units G-13 and G-14) is observed in both samples and is caused by a major change in depositional environments, from fluvial to lacustrine conditions. Effects of this taphonomic shift are expressed in a higher proportion of postcranial elements relative to isolated teeth as depositional environments became more lacustrine. The question of collection bias was addressed by comparing abundances (NISP) of different taxa among the two collections. In this regard, in general the French team collected more specimens than the American team; however, the American team collected more primates and carnivores. In contrast, the French in general recovered more remains of large-sized mammals as well as more postcranial elements of macromammals (other than primates) than the Americans.

Despite these differences, similar taxonomic composition and species richness is documented in both samples. In other words, the types of animals that roamed the paleo-Omo landscape at a given time were found in both areas, in the north and south. In addition a comparative approach that used abundance variation of bovid tribes through time indicated that similar patterns of variation are observed in both samples. This is considered to be additional evidence for the prevalence of generally similar habitats in the northern and southern parts of the Shungura area.

While this is generally true, habitat differences may have occurred locally. Using species of the tribe Reduncini, we were able to demonstrate that some species were more frequent in the southern localities (French) and others in the northern localities (American). In particular, we have shown that *Menelikia lyrocera* was more common in the south while *Kobus sigmoidalis* was so in the north. This means that even if the Shungura area was characterized by generally similar type of habitat in the south and north at any given time, there may have been local ecological differences, indicated by inferred differences in the habitat preference of these species.

## Conclusion

The late Cenozoic fossil record of Africa is growing fast as a result of the proliferation of fieldwork activities in different parts of the continent over the past decades. Major projects are being conducted in Chad, Ethiopia, Kenya, Tanzania, Malawi, South Africa, and Morocco,

to mention some. Several international multidisciplinary research groups undertaking fieldwork in these countries have amassed large numbers of faunal remains including hominins. The faunal collections are useful to understand the paleobiodiversity of a given area and time period. In addition they are one of the best sources of information in exploring the paleoenvironments and paleoecology of our ancestors. Moreover, studies that are carried out to understand the effects of regional and global climatic changes on faunal and hominin evolution require data that are extracted from these collections. In short, data that are recovered in the field remain our primary sources of information in the study of biological, environmental, and ecological evolution of hominins and associated fauna in the Miocene, Pliocene, and Pleistocene. However, research groups tend to work independently following their team-defined approaches, as illustrated by the present study, and there is little or no standardization in the documentation of these fossil collections.

Even though different projects undertake their field activities separately, their research goals are often very similar and the questions they address are strongly linked to each other. While they are explored by separate projects, many paleontological sites are located in the same temporal range and geographical areas; and even sometimes belong to the same sedimentary basin. It is therefore imperative that the various projects coordinate their efforts to maximize the amount of data and information that can be extracted from these irreplaceable resources. For several scientific and nonscientific reasons, it is usually not possible to coordinate different field projects to work together. Yet it is critical that we reach some minimum agreement on how fossil data should be collected in a standardized fashion, so as to establish comparable databases that can subsequently be used for regional and even global understanding of patterns of human evolution.

The present study on comparability of data underscores the importance of the quality of data in unraveling past environments and patterns of change through time. Well-controlled collecting methods and systematic documentation procedures are necessary for the data to be used for these purposes. One way of doing this is to encourage information exchange among different research groups conducting fieldwork in geographically and temporally comparable sites. This can be accomplished in many different ways ranging from informal discussions about current research on specific sites to organized symposia and workshops in which standards and methodologies can be discussed in a comparative fashion. There have been some initiatives over the last few years in this regard that need to be encouraged and expanded. More importantly, a mechanism of data exchange needs to be established among researchers. This will facilitate not only our endeavor towards improving the quality and soundness of databases but also will make our interpretations and hypotheses easier to test and evaluate.

## Acknowledgments

Our gratitude goes to all people who contributed to the collection and curation of the Omo fossils in the field as well as in museums. We thank two anonymous referees and Kaye Reed for providing helpful comments and suggestions.

## References

Alemseged, Z., Geraads, D., Coppens, Y., Guillemot, C., 1996. Taphonomical and paleoenvironmental study of Omo-33, a late Pliocene hominin locality of the lower Omo basin, Ethiopia. Revue de Palébiologie 15, 339–347.

Alemseged, Z., 1998. L'Homininé Omo-323: sa position phylétique et son environnement dans le cadre de l'évolution des communautés de mammifères du Plio-Pléistocène dans la basse vallée de l'Omo (Ethiopie). Ph.D. Dissertation, Museum National d'Histoire Naturelle, Paris.

Alemseged, Z., 2003. An integrated approach to taphonomy and faunal change in the Shungura Formation (Ethiopia) and its implication for hominin evolution. Journal of Human Evolution 44, 451–478.

Arambourg, C., Coppens, Y., 1967. Sur la découverte dans le Pléistocène inférieur de la vallée de l'Omo (Ethiopie) d'une mandibule d'australopithecien. Comptes Rendus de l Academie des Sciences 265, 589–590.

Arambourg, C., Coppens, Y., 1968. Découverte d'un australopithecien nouveau dans les gisements de l'Omo (Ethiopie). South African Journal of Science 64, 58–59.

Behrensmeyer, A.K., Todd, N.E., Potts, R., McBrinn, G.E., 1997. Late Pliocene faunal turnover in the Turkana Basin, Kenya and Ethiopia. Science 278, 1589–1594.

Bishop, L.C., 1999. Reconstructing omnivore paleoecology and habitat preference in the Pliocene and Pleistocene of Africa. In: Bromage, T.G., Schrenk, F. (Eds.), African Biogeography, Climate Change, and Early Hominin Evolution. Oxford University Press, Oxford, pp. 216–225.

Bobe, R., Eck, G.G., 2001. Response of African bovids to Pliocene climatic change. Paleobiology 27, Supplement No. 2, Paleobiology Memoirs 2, 1–47.

Bobe, R., Behrensmeyer, A.K., Chapman, R.E., 2002. Faunal change, environmental variability and late Pliocene hominin evolution. Journal of Human Evolution 42, 475–497.

Brown, F.H., Heinzelin, J., 1983. The lower Omo Basin. Archives of the International Omo Research Expedition. In: de Heinzelin, J. (Ed.), The Omo Group, Annals de Sciences Géologiques. Musée Royal de l'Afrique Centrale, Tervuren, pp. 7–24.

Buzas, M.A., 1990. Another look at confidence limits for species proportions. Journal of Paleontology 64(5), 842–843.

Coppens, Y., 1975. Evolution des Mammifères, de leurs fréquence et de leurs associations, au cours du Plio-Pleistocène dans la basse vallée de l'Omo en Ethiopie. CRAS, Serie D 281, 1571–1574.

Coppens, Y., 1976. Introduction. In: Coppens, Y., Howell, F.C., Isaac, G.Ll., Leakey, R.E. (Eds.), Earliest Man and Environments in the Lake Rudolf Basin. University of Chicago Press, Chicago, pp. 173–176.

de Heinzelin, J., 1983. The Omo group: Archives of the International Omo Research Expedition. In: de Heinzelin, J. (Ed.), Annals des Sciences

180                                          Z. ALEMSEGED ET AL.

Géologiques. Musée Royal de l'Afrique Centrale, Tervuren.

de Heinzelin, J., Haesaerts, P., 1983. The Shungura Formation: Archives of the International Omo Research Expedition. In: de Heinzelin, J. (Ed.), Annales des Sciences Géologiques. Musée Royal de l'Afrique Centrale, Tervuren, 85, 25–127.

deMenocal, P.B., 1995. Plio-Pleistocene African climate. Science 270, 53–59.

deMenocal, P.B., Bloemendal, J., 1995. Plio-Pleistocene climatic variability in subtropical Africa and the paleoenvironment of hominin evolution: a combined data model approach. In: Vrba, E.S., Denton, G.H., Partridge, T.C., Burckle, L.H. (Eds.), Paleoclimate and Evolution. Yale University Press, New Haven, pp. 262–288.

Denton, G.H., 1999. Cenozoic climate change. In: Bromage, T.G., Schrenk, F. (Eds.), African Biogeography, Climate Change, and Human Evolution. Oxford University Press, Oxford, pp. 94–114.

Eck, G.G., 2007. The effects of collection strategy and effort on faunal recovery: a case study of the American and French collections from the Shungura Formation, Ethiopia. In: Bobe, R., Alemseged, Z., Behrensmeyer, A.K. (Eds.), Hominin Environments in the East African Pliocene: An Assessment of the Faunal Evidence. Springer, Dordrecht.

Feibel, C.S., Brown, F., McDougall, I., 1989. Stratigraphic context of fossil hominins from the Omo Group Deposits: Northern Turkana Basin, Kenya and Ethiopia. American Journal of Physical Anthropology 78, 623–632.

Feibel, C.S., Harris, J.M., Brown, F.H., 1991. Paleoenvironmental context for the Late Neogene of the Turkana Basin. In: Harris, J.M. (Ed.), Koobi Fora Research Project, Volume 3, The Fossil Ungulates: Geology, Fossil Artiodactyls, and Paleoenvironments. Clarendon, Oxford, pp. 321–346.

Frost, S.R., 2002. East African cercopithecid fossil record and its relationship to global climatic changes. American Journal of Physical Anthropology Supplement 32, 72–73.

Frost, S.R., 2007. African Pliocene and Pleistocene cercopithecid evolution and global climatic change. In: Bobe, R., Alemseged, Z., Behrensmeyer, A.K. (Eds.), Hominin Environments in the East African Pliocene: an Assessment of the Faunal Evidence. This volume, Springer, Dordrecht.

Gentry, A.W., 1985. The Bovidae of the Omo Group Deposits, Ethiopia. In: Coppens, Y., Howell, F.C. (Eds.), Les faunes Plio-Pléistocenes de la basse

vallée de l'Omo (Ethiopie). Cahiers de paléontologie. travaux de paléontologie est-africaine. CNRS, Paris.

Geraads, D., Coppens, Y., 1995. Evolution des faunes de mammifères dans le Plio-Pléistocène de la basse vallée de l'Omo (Ethiopie): apports de l'analyse factorielle. CRAS. Paris, Serie IIA 320, 625–637.

Harris, J.M., Brown, F.H., Leakey, M.G., 1988. Stratigraphy and paleontology of Pliocene and Pleistocene localities west of Lake Turkana, Kenya. Contributions in Science 399, 1–128.

Haesaerts, P., Stoops, G., Van Vliet-Lanoe, B., 1983. Data on sediments and fossil soils: Archives of the International Omo Research Expedition. In: de Heinzelin, J. (Ed.), Annales des Sciences Géologiques. Musée Royal de l'Afrique Centrale, Tervuren 85, 149–185.

Hill, A., 1987. Causes of perceived faunal change in the later Neogene of East Africa. Journal of Human Evolution 16, 583–596.

Howell, F.C., Coppens, Y., 1973. Deciduous teeth of homininae from the Plio-Pleistocene of the lower Omo Basin, Ethiopia. Journal of Human Evolution 2, 461–472.

Howell, F.C., Coppens, Y., 1974. Inventory of remains of homininae from Plio-Pleistocene formations of the lower Omo Basin, Ethiopia (1967–1972). American Journal of Physical Anthropology 40, 1–16.

Howell, F.C., Haesaerts, P., de Heinzelin, J., 1987. Depositional environments, archaeological occurrences and hominins from Member E and F of the Shungura Formation (Omo Basin, Ethiopia). Journal of Human Evolution 16, 665–700.

Johanson, D.C., Splingaer, M., Boaz, N.T., 1976. Paleontological excavations in the Shungura Formation, lower Omo Basin, 1969–1973. In: Coppens, Y., Howell, F.C., Isaac, G.Ll., Leakey, R.E.F. (Eds.), Earliest Man and Environments in the Lake Rudolf Basin. University of Chicago Press, Chicago, pp. 402–420.

Kappelman, J., 1988. Morphology and locomotor adaptations of the bovid femur in relation to habitat. Journal of Morphology 198, 119–130.

Kennett, J.P., 1995. A review of polar climatic evolution during the Neogene, based on the marine sediment record. In: Vrba, E.S., Denton, G.H., Partridge, T.C., Burckle, L.H. (Eds.), Paleoclimate and Evolution, with Emphasis on Human Origins. Yale University Press, New Haven, pp. 49–64.

Lewis, M.E., Werdelin, L., 2007. Patterns of change in the Plio-Pleistocene carnivorans of eastern Africa: implications for hominin evolution. In: Bobe, R., Alemseged, Z., Behrensmeyer,

A.K. (Eds.), Hominin Environments in the East African Pliocene: an Assessment of the Faunal Evidence. Springer, Dordrecht.

Senut, B., 1980. New data on the humerus and its joints in the Plio-Pleistocene hominins. College of Anthropology 1, 87–93.

Senut, B., Tardieu, C., 1985. Functional aspects of Plio-Pleistocene hominin limb bones: implications for taxonomy and phylogeny. In: Delson, E. (Ed.), Ancestors: The Hard Evidence. Alan R. Liss, New York, 193–201.

Shackleton, N.G., 1995. New data on the evolution of Pliocene climatic variability. In: Vrba, E.S., Denton, G.H., Partridge, T.C., Burckle, L.H. (Eds.), Paleoclimate and Evolution, with Emphasis on Human Origins. Yale University Press, New Haven, pp. 242–248.

Spencer, L.M., 1997. Dietary adaptation of Plio-Pleistocene Bovidae: implications for hominin habitat use. Journal of Human Evolution 32, 201–228.

Stern, J.T., 2000. Climbing to the top: a personal memoir of Australopithecus afarensis. Evolutionary Anthropology 9, 113–133.

Susman, R.L., Stern, J.T., 1991. Locomotor behavior of early hominins: epistemology and fossil evidence. In: Senut, B., Coppens, Y. (Eds.), Origine(s) de la bipedie chez les Hominines. CNRS, Paris, pp. 121–132.

Suwa, G., 1988. Evolution of the "robust" australopithecine in the Omo succession: evidence from mandibular premolar morphology. In: Grine, F. (Ed.), Evolutionary History of the Robust Australopithecine. De Gruyter, New York, pp. 199–222.

Suwa, G., 1990. A comparative analysis of hominin dental remains from the Shungura and Usno Formations, Omo Valley, Ethiopia. Ph.D. Dissertation, University of California, Berkeley.

Suwa, G., White, T.D., Howell, F.C., 1996. Mandibular postcanine dentition from the Shungura Formation, Ethiopia: crown morphology, taxonomic allocation, and Plio-Pleistocene hominin evolution.

American Journal of Physical Anthropology 101, 247–282.

Suwa, G., Nakaya, H., Asfaw, B., Sacgusa, H., Amzage, A., Kono, R.T., Beyene, Y., Katoh, S., 2003. Plio-Pleistocene terrestrial mammal assemblage from Konso, Southern Ethiopia. Journal of Vertebrate Paleontology 23, 901–916.

Vrba, E.S., 1988. Late Pliocene climatic events and hominin evolution. In: Grine, F. (Ed.), Evolutionary History of the Robust Australopithecines. De Gruyter, New York, pp. 405–426.

Vrba, E.S., 1992. Mammals as a key to evolutionary theory. Journal of Mammalogy 73(1), 1–28.

Vrba, E.S., 1995. On the connections between paleoclimate and evolution. In: Vrba, E.S., Denton, G.H., Partridge, T.C., Burckle, L.H. (Eds.), Paleoclimate and Evolution with an Emphasis on Human Origins. Yale University Press, New Haven, pp. 24–45.

Vrba, E.S., 1999. Habitat theory in relation to the evolution of African Neogene biota and hominins. In: Bromage, T.G., Schrenk, F. (Eds.), African Biogeography, Climate Change, and Early Hominin Evolution. Oxford University Press, Oxford, pp. 19–34.

Vrba, E.S., 2000. Major features of Neogene mammalian evolution in Africa. In: Partridge, T.C., Maud, R.R. (Eds.), The Cenozoic of Southern Africa. Oxford University Press, Oxford, pp. 277–304.

Ward, C.V., 2002. Interpreting the posture and locomotion of Australopithecus afarensis: where do we stand? Yearbook of Physical Anthropology, Volume 45, pp. 185–215.

White, T.D., 1995. African Omnivores: global climatic change and Plio-Pleistocene hominins and suids. In: Vrba, E.S., Denton, G.H., Partridge, T.C., Burckle, L.H. (Ed s.), Paleoclimate and Evolution, with Emphasis on Human Origins. Yale University Press, New Haven, pp. 369–384.

# 8. The effects of collection strategy and effort on faunal recovery

A case study of the American and French collections from the Shungura Formation, Ethiopia

G.G. ECK
*Department of Anthropology*
*University of Washington*
*Seattle, Washington, USA*
*ggeck@u.washington.edu*

**Keywords**:   Collection bias, Omo, Pliocene, Pleistocene

## Abstract

Although preserved by sediments that were contemporaneously deposited by the same river and lake system and exposed in contiguous areas, the American and French collections of fossil specimens from the Shungura Formation of southwestern Ethiopia produce differences in specimen counts that are surprisingly large. Some of these differences were caused by well-documented differences in geography and geology of the formation and the history of the research efforts of the two expeditions. Other differences apparently arose because of factors that are less well documented. The following paper briefly describes the well-documented factors leading to differences in specimen counts, including differences in the sizes of areas explored, months of active fieldwork, and numbers of sites excavated for the recovery of macro- and microfaunal remains. Further, it proposes methods for discovering factors that are less well documented, likely related to differences in research strategies and the inherent richness of the sediments explored. And finally, it suggests how the collections might be best used to avoid the effects of biases that they apparently contain.

## Introduction

It is widely recognized that the analysis of counts and relative frequencies of individuals representing taxa provide much richer evidence for environmental interpretation and reconstruction than does analysis of simple presence and absence. Useful analyses of counts and relative frequencies in fossil assemblages, however, require that they represent these quantities in the living communities from which they derive and, as taphonomists regularly warn, the pathway between living individual and recovered fossil is a complex one that is affected by many factors both natural and human.

Using catalogs of fossil specimens collected from the Shungura Formation of southwestern Ethiopia, the following paper explores some of the natural and some of the human factors that affected the collection of these specimens and thus their counts and relative frequencies.

*R. Bobe, Z. Alemseged, and A.K. Behrensmeyer (eds.) Hominin Environments in the East African*
*Pliocene: An Assessment of the Faunal Evidence, 183–215.*

These factors include the intrinsic richness of the sediments, their areas of exposure, and the collection strategies and effort used in the recovery of the specimens. Lessons are drawn as to how these effects might be better understood and the biases that they introduced into the collections minimized. Readers are directed to Alemseged et al. (2006) for additional analysis of this dataset and alternative interpretations of some of the patterns discussed.

## Geographic, Geologic, and Historic Setting

The Shungura Formation is located in the lower Omo Valley of southwestern Ethiopia, west of the Omo River and north of Lake Turkana (Brown and de Heinzelin, 1983) (Figure 1). The composite stratigraphic section of the formation measures nearly 800 m and radiometric ages indicate that it covers the time span from 3.6 to 1.05 Ma (Feibel et al., 1989) (Figure 2). The sedimentary cycles of the formation are grouped into 12 members (Basal, A, B, C, D, E, F, G, H, J, K, and L), each (except the Basal Member) commencing with a volcanic tuff designated by the same letter. The sequence is typically composed of fluvial sediments, but episodes of lacustrine deposition also occurred, particularly in the Basal Member, upper Member G, and upper Member L (de Heinzelin and Haesaerts, 1983).

The Mission Scientifique de l'Omo, led by C. Arambourg in 1932 and 1933, was the first expedition to conduct systematic paleontological work in the lower Omo Valley (Arambourg, 1947; Coppens et al., 1976). Heselon Mukiri, field assistant of L.S.B. Leakey, visited the exposures in the early 1940s and made unprovenanced fossil collections that were placed in the Coryndon Museum (now the National Museums of Kenya). In 1954, F.C. Howell examined these collections during a prolonged visit to eastern and southern Africa and discussed with Leakey the advisability of working in the lower Omo Valley one day. Howell then visited the Turkana Basin, during July of 1959,

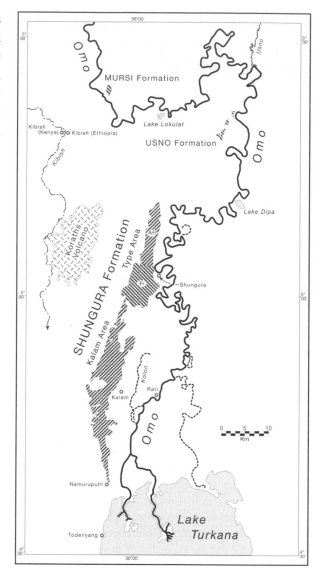

Figure 1. Omo Group formations of the lower Omo Valley.

and collected vertebrate fossils from exposures of the Shungura Formation, which he left in the care of local authorities in southern Ethiopia. During this visit, Howell recognized the protracted sedimentary sequence, the volcanic ash accumulations, and intraformational faulting exposed in the lower Omo Valley. Seven years later, at the urging of the Emperor Haile Selassie, the Ethiopian Government gave permission for a joint international scientific consortium to work there. In 1966, the International Omo Research Expedition (IORE) was created

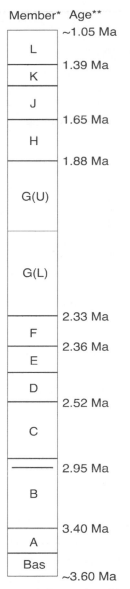

Member* Age**

~1.05 Ma

L

1.39 Ma

K

J

1.65 Ma

H

1.88 Ma

G(U)

G(L)

2.33 Ma

F

2.36 Ma

E

D

2.52 Ma

C

2.95 Ma

B

3.40 Ma

A

Bas

~3.60 Ma

\* Member thicknesses are scaled to stratigraphic thickness.
\*\* Geologic ages are radiometric and from Feibel *et al.* (1989) except the first and the last, which are extrapolations.

Figure 2. Composite stratigraphic section of the Shungura Formation.

under the direction of Leakey, Arambourg, and Howell (Coppens et al., 1976). In 1967, three contingents of the IORE independently explored the sedimentary exposures of the lower Omo Valley. The French one, under the direction of Arambourg and Coppens, worked principally in what came to be known as the "Type Area" of the Shungura Formation. The Kenyan con-

tingent, under the direction of R.E.F. Leakey, and the American contingent, under the direction of Howell, worked further to the north, the Kenyans in the Kibish and Mursi Formations, and the Americans in the Usno Formation. In 1968, disappointed with the size or richness of the Mursi and Kibish Formations, Leakey moved to the eastern shores of Lake Rudolf (now Lake Turkana), which he had observed from the air-contained extensive sedimentary outcrops. For similar reasons, Howell arranged with Coppens to move south and jointly explore the Type Area of the Shungura Formation. It was agreed that the Americans would work the Type Area north of the "watering road," while the French would work south of it. This boundary would later be formalized by de Heinzelin and Haesaerts (1983), who drew the boundaries of Geological Sectors 15/16 and 17 generally along the watering road. Only in the later years of the expedition, mostly after 1972, did both the French and American contingents begin to explore the Kalam Area of the Shungura Formation, located to the southwest of the Type Area. The American contingent ceased work in the lower Omo Valley at the end of the 1974 field season, the French ended their work there in 1976. The two research teams conducted their research separately with little or no coordination of research strategies or collection methods, except for the watering road between them.

In the nine years of fieldwork between 1967 and 1976, neither contingent mounted an expedition in 1975, nearly 50,000 paleontological specimens were collected: 21,858 by the Americans and 27,409 by the French (Figure 3). Most of these were recovered during surface survey, but large paleontological excavations were also carried out, producing 6,692 American and 3,417 French specimens (see, for example, Johanson et al., 1976).

The vast majority of these are of mammals because neither contingent regularly collected lower vertebrates or invertebrates and plant specimens were very rare. Only 10% of the American collection comprises nonmammalian

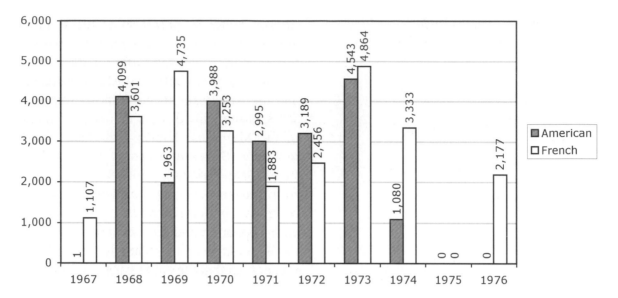

Figure 3. Numbers of specimens collected by each contingent by year.

specimens, while they constitute 6% of the French collection.

Among the 42,481 mammalian specimens that can be identified at the ordinal level, 11 orders are represented, with artiodactyls clearly predominating, primates making a surprisingly strong showing, and carnivores having the smallest numbers among the mammals of medium to very large body size, as is expected (Table 1).

Table 2 presents the numbers of mammalian specimens that have been identified at the family level.

*Table 1. Numbers of specimens identified at the order level*

|  | American | French | Total | Percentage |
|---|---|---|---|---|
| Artiodactyla | 11,053 | 20,731 | 31,784 | 74.819 |
| Primates | 3,632 | 3,052 | 6,684 | 15.734 |
| Proboscidea | 743 | 1,259 | 2,002 | 4.713 |
| Perissodactyla | 390 | 510 | 900 | 2.119 |
| Rodentia | 724 | 96 | 820 | 1.930 |
| Carnivora | 149 | 113 | 262 | 0.617 |
| Chiroptera | 10 | 0 | 10 | 0.024 |
| Insectivora | 10 | 0 | 10 | 0.024 |
| Lagomorpha | 3 | 3 | 6 | 0.014 |
| Hyracoidea | 3 | 0 | 3 | 0.007 |
| Tubulidentata | 1 | 0 | 1 | 0.002 |
| Total | 16,718 | 25,764 | 42,481 |  |

## Factors Affecting the Numbers

### DEPOSITIONAL AND EROSIONAL FACTORS

As noted above, nearly all of the richly fossiliferous deposits of the Shungura Formation consist of fluvial sediments laid down by a major river similar in size to the modern Omo River. These sediments consist of gravels and sands deposited in the river channel, light-colored silts deposited on the banks of the river near the channel, and dark-colored, silty clays deposited during periods of high water more distal to the channel.

Fossilization was clearly most complete in specimens from the channel sands and gravels. These were very well mineralized, hard, and resistant to breakage as they eroded onto the surface. Unfortunately, because they had in most cases undergone multiple episodes of burial and erosion and substantial transport by the river, they were already in a fragmentary state when they were deposited. Predepositional damage to specimens from the channel sands and gravels thus produced a strong bias towards the densest limb bones and limb bone parts, bovid horn cores, jaw

*Table 2. Number of specimens identified at the family level*

| Macromammals | American | French | Total |
|---|---|---|---|
| Bovidae | 6,295 | 11,007 | 17,302 |
| Hippopotamidae | 2,448 | 5,472 | 7,920 |
| Cercopithecoidea | 3,482 | 2,917 | 6,399 |
| Suidae | 1,770 | 3,087 | 4,857 |
| Elephantidae | 517 | 1,062 | 1,579 |
| Giraffidae | 535 | 1,007 | 1,542 |
| Equidae | 332 | 397 | 729 |
| Deinotheriidae | 225 | 196 | 421 |
| Hominidae | 147 | 13 5 | 282 |
| Rhinocerotidae | 55 | 109 | 164 |
| Felidae | 85 | 42 | 127 |
| Hyaenidae | 16 | 22 | 38 |
| Hystricidae | 15 | 10 | 25 |
| Camelidae | 5 | 16 | 21 |
| Mustelidae | 12 | 3 | 15 |
| Chalicotheriidae | 3 | 4 | 7 |
| Procaviidae | 3 | 0 | 3 |
| Orycteropodidae | 1 | 0 | 1 |
| Total | 15,946 | 25,486 | 41,432 |
| **Micromammals** | **American** | **French** | **Total** |
| Muridae | 611 | 60 | 671 |
| Sciuridae | 63 | 7 | 70 |
| Viverridae | 36 | 12 | 48 |
| Cricetidae | 23 | 6 | 29 |
| Soricidae | 10 | 0 | 10 |
| Dipodidae | 3 | 4 | 7 |
| Bathyergidae | 1 | 5 | 6 |
| Thryonomyidae | 5 | 1 | 6 |
| Leporidae | 3 | 3 | 6 |
| Hipposideridae | 5 | 0 | 5 |
| Lorisidae | 3 | 0 | 3 |
| Emballonuridae | 3 | 0 | 3 |
| Pteropodidae | 1 | 0 | 1 |
| Total | 767 | 98 | 865 |

fragments, teeth, and tooth fragments. Complete or partial crania and mandibles were rare and articulated or partially articulated skeletons or limbs were unknown from these sediments.

In contrast, specimens deposited in the near channel silts were less well mineralized and more subject to postdepositional damage as they eroded onto the surface. The erosional surface of silt deposits were thus often littered by a rich array of bone fragments, but only those of very dense foot bones, bovid horn cores, jaw fragments, and teeth were complete enough to identify to taxon and body part. Excavations into the silts carried out by the American contingent showed, however, that they often contained beautifully complete specimens, including complete crania and mandibles and partially articulated skeletons (see Johanson et al., 1976). Postdepositional damage typically occurred as the weathering front in the sediments, usually 10 to 15 cm below the surface, moistened the specimens, reducing them to the fragments noted above, which then dispersed as they eroded onto the surface.

The more distal over-bank deposits of dark-colored, silty clays rarely if ever produced fossil specimens.

It also seems that the teeth, especially isolated teeth, of large mammals suffered greater rates of fragmentation than did those of medium-sized animals. Impressions derived from surface survey suggest that a larger proportion of the teeth of the larger pigs, hippos, rhinos, deinotheres, and elephants occurred as fragments than those of smaller animals. This increased fragmentation might have resulted from the fact that the teeth of large mammals contain a greater proportionate volume of dentine and cementum. Especially their worn teeth, containing proportionately less enamel, may have been more subject to predepositional damage in the high-energy channel deposits. In the lower energy silts, dentine and cementum were poorly mineralized, leading to postdepositional damage. It thus may be that dental specimens of large mammals are underrepresented in the collections by comparison to those of the smaller animals. How large this disproportion might be is not known.

The sample of specimens scattered on the erosional surface of the Shungura was thus clearly biased towards those most resistant to damage: jaw fragments and teeth, dense postcranial elements, and bovid horn cores. It was also probably biased against the teeth of large mammals. Whether these biases differed between the American and French collections areas is not known.

## FACTORS OF TIME AND SPACE

The French collection is substantially larger than the American, as noted above and shown in Figure 3. There are many reasons why this is so, but one is clearly the time spent searching for fossils. Although both contingents had field seasons of roughly the same length (about 10 weeks), the French had expeditions to the Shungura Formation in 1967 and 1976 when the American did not (see Figure 3).

In the Type Area, exposures of Members K and L do not occur in either the American or French areas, nor do exposures of Member J in the American area. Counts of fossils from these members in Table 3 and Figure 4 are of specimens recovered in the Kalam Area and, for reasons that will become clear below, they will not be considered further. The Basal Member is not exposed in the American area of the Type Area and the exposures of Member A are very small, explaining the big differences in numbers between the areas.

Also of importance is the fact that the French area of exposure in the Type Area is substantially larger than the American. Given the same numbers of specimens in the ground, larger areas of exposure will produce larger numbers during surface survey. This difference can be accurately measured on de Heinzelin's

*Table 3. Numbers of specimens found in each member*

|       | American | French |
|-------|----------|--------|
| Ba    | 2        | 34     |
| A     | 60       | 429    |
| B     | 2,118    | 1,637  |
| C     | 3,821    | 3,414  |
| D     | 993      | 540    |
| E     | 3,814    | 2,026  |
| F     | 4,473    | 4,585  |
| G(L)  | 4,714    | 11,220 |
| G(U)  | 503      | 1,549  |
| H     | 499      | 695    |
| J     | 222      | 150    |
| K     | 287      | 447    |
| L     | 361      | 492    |
| Total | 21,867   | 27,218 |

Geological Map of the Shungura Formation (de Heinzelin, 1983). I have done this by scanning the map to produce a digital image, separating the areas of the different members using Photoshop, and measuring the area of each member using NIH Image J. Because the Kalam Area of the Shungura Formation remains largely unmapped, I could make these measurements only for exposures of the members in the Type Area. I have followed de Heinzelin's (1983) convention of dividing Member G into

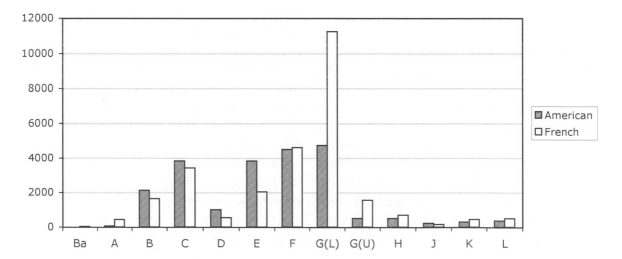

Figure 4.  Numbers of specimens found in each member.

an upper and lower part because lower Member G (Member G(L)) consists mostly of fluvial deposits that are richly fossiliferous, while upper Member G (Member G(U)) is composed mainly of lacustrine deposits that are poorly so. My results can be seen in Table 4 and Figure 5 (note that this approach to data standardization differs from that of Alemseged et al., 2006).

Although the absolute values cited in Table 4 are subject to some uncertainty because of the uncertainty in the scale of de Heinzelin's map, stated as approximately 1:10,000, the relative differences are very accurate because of the precision of the maps. The French area of exposure is substantially larger than the American, but much of the difference lies in Member G(U), which is poorly fossiliferous.

The factors of time and space probably explain much of the difference in the sizes of the American and French collections, but, as will be seen below, other factors play a role as well.

## THE FACTOR OF MULTIPLE SPECIMENS REPRESENTING SINGLE INDIVIDUALS

Fossil individuals can of course be represented in the record by a very large range of specimen counts, ranging from one to perhaps many thousands. Differential fragmentation rates of individuals, leading to widely ranging counts, obscure the more interesting variation in numbers of individual organisms. Given the fluvial environments of deposition, which are dominant in the Shungura Formation, it is likely that most surface specimens, derived from transported remains, came from different individuals and that specimen counts roughly represent individual counts. This was clearly, however, not always the case. During surface survey, one

*Table 4. Areas of exposure of members in the Type Area in km²*

|      | American | French |
|------|----------|--------|
| Ba   | 0.000000 | 0.021607 |
| A    | 0.001545 | 0.078394 |
| B    | 0.256655 | 0.272918 |
| C    | 2.449999 | 1.281457 |
| D    | 1.794332 | 1.048239 |
| E    | 1.447085 | 1.179028 |
| F    | 1.555917 | 1.607318 |
| G(L) | 3.807007 | 4.549295 |
| G(U) | 0.555443 | 4.250264 |
| H    | 0.051012 | 0.588353 |
| J    | 0.000000 | 0.026139 |
| K    | 0.000000 | 0.000000 |
| L    | 0.000000 | 0.000000 |
| Total | 11.918995 | 14.903012 |

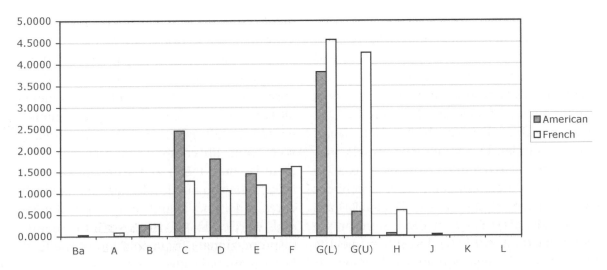

Figure 5. Areas of exposure of members in the Type Area in km².

often encountered patches of specimens that, based on similarity in preservation, individual age, and complementarity of elements, certainly or almost certainly belonged to single individuals. Beginning late in the 1968 field season, the American contingent began to enter into their field catalogs information indicating which specimens comprised single individuals, based on the above mentioned characteristics. This notation was later extended to the specimens collected in earlier 1968 as well. It is thus possible, in the American catalog, to remove multiple specimens representing single individuals, bringing the specimen counts more in line with the ideal individual counts. The French contingent did not regularly keep similar information and, as the French catalog now stands, it is less often possible to know which single specimens likely represent individuals and which ones do so as a group. In the analyses described below, it is assumed that the number of individuals represented by multiple specimens is essentially the same in the two collections and thus that this factor does not cause important differences in numbers. Whether or not this assumption is warranted can only be determined by future work on the French collection.

## FACTORS OF COLLECTION STRATEGY AND EFFORT

The difference in numbers of excavated specimens occurring in the two collections is easily explained by the fact the Americans spent more time at excavation. Various members of the contingent, directed by M. Splingaer, D.C. Johanson, D.D. Dechant-Boaz, N.T. Boaz, H.B. Wesselman, D. Cramer, and myself, over several years carried out 17 excavations at 15 different localities producing 6692 specimens (see Johanson et al., 1976; Dechant-Boaz, 1994 for details). Some of these were only a few m$^2$ in extent, while the largest, in Locality 398, located at the bottom of Member F, covered 178 m$^2$ and produced 2642 speci-

mens. In contrast, the French carried out one large excavation, directed by C. Guillemot, in Locality Omo 33, also located at the bottom of Member F, which produced most of the 3417 specimens recovered from the locality.

Differences in effort and probably luck explain the differences in micromammal recoveries too. Luck played a role in that concentrations of micromammals seem to be very rare in the Shungura Formation and their occurrence is very difficult if not impossible to predict. During the 1970 field season, J.J. Jaeger recovered modest numbers in the French area. In 1972 and 1973, H.B. Wesselman sampled many localities in the American area and recovered most of the micromammals in the American collection. Both used special techniques in excavating and wet washing sediments that are very different from methods used to recover the larger mammalian fossils (see Wesselman, 1984).

Differences in strategy and effort surely affected numbers of specimens collected during surface survey by both contingents, but these are more difficult to disentangle because neither contingent kept detailed records in this regard. From late in the 1968 field season until the middle of the 1972 field season, I directed crews that collected most of the surface specimens recovered from the American area. In late 1968, I decided, given the very large numbers of surface specimens, to collect only a limited set of those found, but to collect all of these. The collected set included: all recognized specimens of primates and carnivores, no matter how small or fragmentary; all complete or relatively complete crania and mandibles, all upper and lower jaw fragments with teeth, all complete or very nearly complete isolated teeth, and all astragali of other mammalian taxa. In addition, all bovid horn cores and all bovid and camelid distal metapodials were collected. Because of the seemingly disproportionate fragmentation of the teeth of large mammals discussed above, I collected teeth of large pigs, rhinos, deinotheres, and elephants that were more fragmentary than those of other mammals other than primates and carnivores. This surface collection strategy was also generally followed by American crews in the years after 1972. The American collection is thus clearly biased, in most cases, towards cranial specimens. Whether or not the French collection is similarly biased, but perhaps in other ways, is less well documented. I think there is a way to discover biases in both collections, as explicated below.

## Differences in Numbers Caused by Differences in Collection Strategy and Effort

Of the major factors controlling specimen numbers—intrinsic richness of the sediments, areas of exposure, and collection strategy and effort—only areas of exposure are accurately enough documented, by de Heinzelin's Geologic Map, to allow numerical corrections for the specimen counts based on this factor. In the discussion below, I will thus investigate specimen densities, the number of specimens divided by the area of exposure of the member, rather than counts. Because the Kalam

Area is essentially unmapped, Members J, K, and L, which principally crop out there, are excluded from further discussion. Similarly, because the Basal Member is not exposed in the American portion of the Type Area and the exposures of Member A are very small, these members too will be excluded. I will thus investigate specimen densities only from Members B through H. Excavated specimens, both micro and macro, have also been excluded because of the known differences in collection effort between the areas or rates of pre- and postdepositional damage between surface and excavated specimens.

Calculation of the densities of specimens of macromammals recovered on the surface of the two parts (American and French) of the Type Area gives the values presented in Figure 6 (counts of specimens used to calculate the densities can be found in the Appendix).

An intriguing pattern is apparent, one in which the highest density alternates between the American and French areas as one moves from Members B to H. Although the absolute values vary greatly from highs in Member B and lows in G(U) and the differences vary greatly as well, with the biggest in Member C and the smallest in G(U), I think that the

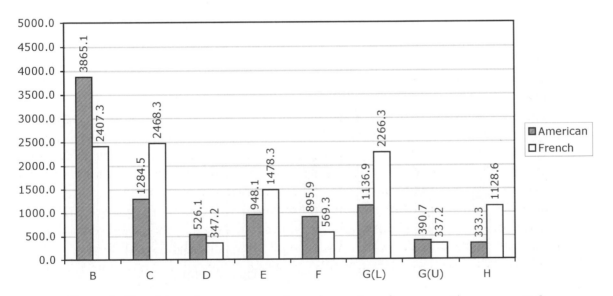

Figure 6. Densities of macromammal specimens (specimen count/exposure area).

important pattern is seen in the simple alternation of highest densities. I argue below that it is caused by intrinsic differences in numbers of fossil specimens preserved in the two areas of each member and by the differential distribution of especially rich surface occurrences of specimens. I further argue that significant differences in numbers caused by differences in collection strategy or effort produce a different pattern that will become apparent below. Because I know, as explained above, or suspect that different families of macromammals experienced different collection intensities, the following analyses are organized by family, beginning with the Hominidae.

## HOMINIDAE

I choose to begin this discussion with the Hominidae because of all the mammals they were certainly the most sought after. Howell had come to the Omo to find hominids, among other things. The French clearly recognized their importance as well. Competition between the two camps in the hominids they found was clear. They were also egged on by Leakey's successes at Koobi Fora. I think that the effort to find hominids was great on both sides and the strategy clear, collect them all.

The densities of hominid specimens are shown in Figure 7. In both the American and French collections, one hominid individual is represented by a large number of fragments, the American specimen, L894–1, with 31 fragments from Member G(U) and the French specimen, Omo 323–896, with 21 fragments from Member G(L). In these cases, the counts for the specimens have been reduced to one in Figure 7. Importantly, the densities of hominid specimens show the alternating pattern seen in the macromammals as a whole. These densities are principally determined by the numbers of isolated teeth, but even the densities of jaw fragments, based on a total of only 15 specimens, show the pattern as seen in Figure 8.

That the hominids, surely the most intensely searched for of all the mammals, share the alternating density pattern with the total macromammal sample, even though they consti-

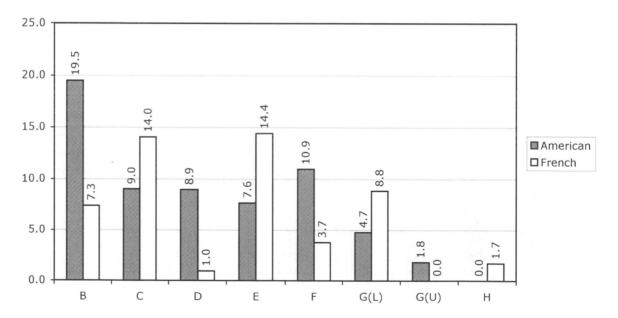

Figure 7. Densities of hominid specimens.

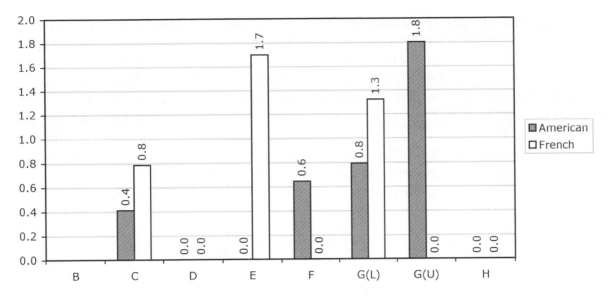

Figure 8. Densities of hominid jaw fragments.

tute only 0.5% of it, suggests to me that the alternating pattern results from factors other than differences in collection strategy and effort between the two contingents.

## CERCOPITHECIDAE

Because of Howell's and my interests in monkeys, they too were collected with great effort,

as noted above. One might expect a strong bias in the counts and densities of monkey specimens towards the American collection. This expectation is partially met in that the differences in monkey densities in Members E and G(L) are not as great as they are for the macromammals as a whole, and they are reversed in Member H, as can be seen in Figure 9. The bias towards the American collection is not as strong, however, as one might have expected.

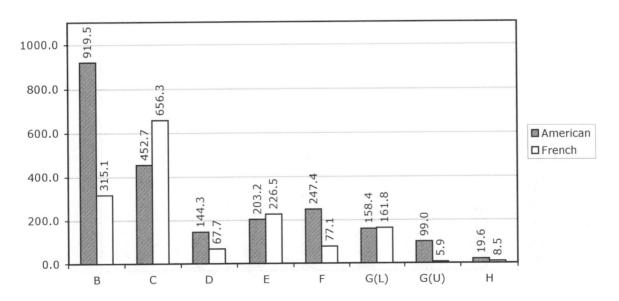

Figure 9. Densities of monkey specimens.

A weak bias towards the American collection is also seen in the densities of all monkey dental specimens, both complete and fragmentary teeth. The differences are small in Member E and the pattern is reversed in Member G(L) and H, as can be seen in Figure 10.

The bias towards the American collection is weaker yet, or perhaps disappears, if one looks only at the densities of complete teeth, remov-ing all fragments from the counts (Figure 11). Only Member H is now out of alternating pattern and its densities are determined by a total of only four specimens. That the French were less compulsive about collecting tooth fragments is also suggested by the fact that 47% of their dental sample is composed of fragments, compared with 55% of the American sample.

The alternating pattern of high densities is again well established in the combined

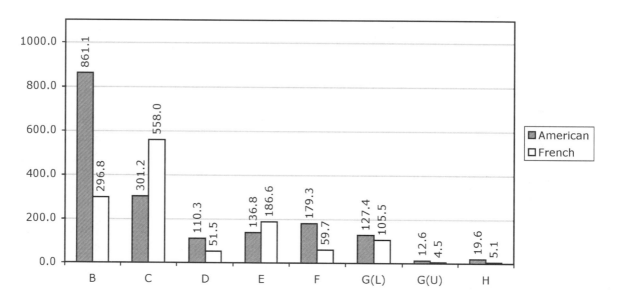

Figure 10. Densities of monkey complete and fragmentary isolated teeth.

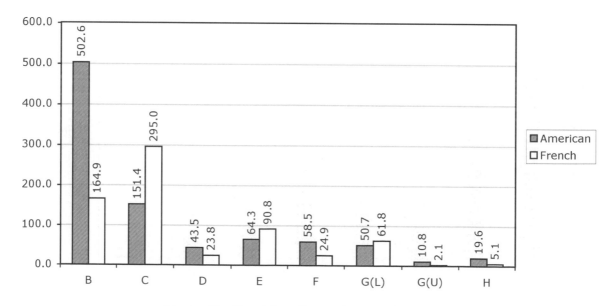

Figure 11. Densities of complete monkey teeth.

densities of complete teeth, jaw fragments, and complete and partial crania and mandibles. Only Member H is reversed, but the densities are determined by only four specimens (Figure 12). It would seem then that when it came to well preserved parts of monkey skulls the French were just as intense in their collecting as the Americans

were. Only tooth fragments are biased in the American favor.

The same cannot be said when in comes to monkey postcrania. Here there is a very clear and strong bias in the American favor. Americans consistently collected more and a wider variety of postcranial elements (Figure 13), collecting more than twice as many as the

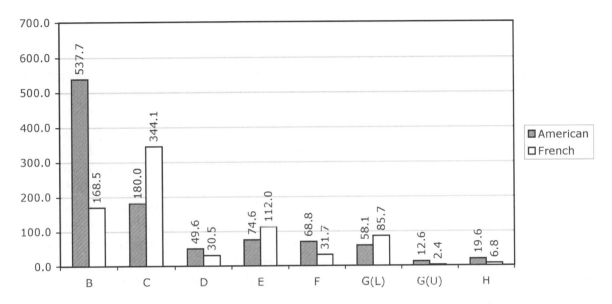

Figure 12. Densities of monkey complete teeth, jaw fragments, and complete and partial crania and mandibles.

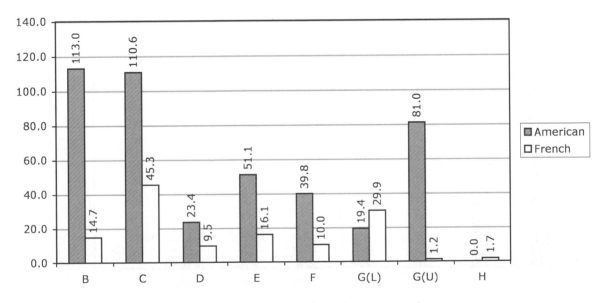

Figure 13. Densities of monkey postcrania.

French did, with average density three times that of the French. Clearly the Americans were much more compulsive than the French when it came to monkey postcrania.

The monkey densities suggest, that when high densities alternate between the two collections as one moves from Member B to Member H, they do so because of differences in intrinsic numbers of specimens within the sediments of the American and French areas. I think it highly unlikely that either the Americans or the French would have left undiscovered or uncollected the large numbers of highly valued specimens required to produce the observed differences in surface densities. The alternating pattern of high densities signals similarity in collection strategy and effort. When, however, densities are clearly biased towards one collection or the other, as they are for monkey postcrania, this signals differences in strategy and effort.

## CARNIVORA

Even as early as 1968, Howell had long harbored a strong interest in the evolution of carnivores and encouraged the American survey teams to intensely search for and collect carnivore specimens of medium to large body size. One might thus expect that the American collection would be biased towards these very rare taxa (a total of only 181 specimens were collected by both teams from the surface of Sectors 1 though 27). The expected bias does not appear though, for, except for Member C, the alternating pattern of densities is seen (Figure 14).

## ELEPHANTIDAE

Elephants are big and thus have the fortune of being easy to find, but the misfortune of being hard to carry and to store. The French had special interests in them because Coppens was writing a dissertation on elephant evolution. The Americans had a somewhat different interest because of their known potential for dating sediments. Specimens of elephant teeth were apparently equally sought by both contingents as can be seen in their alternating densities, being information rich and easy to carry (Figure 15).

The Americans were not as diligent, however, when it came to the heavier and more cumbersome cranial, maxillary, and mandibular specimens. The densities of these have a distinct French bias (Figure 16).

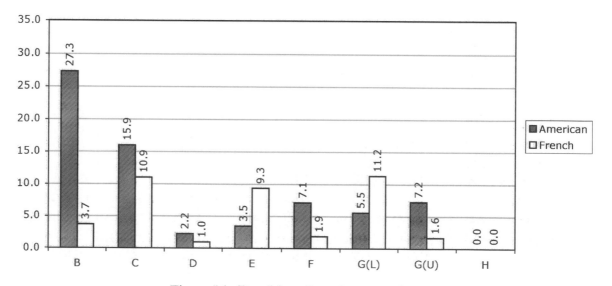

Figure 14. Densities of carnivore specimens.

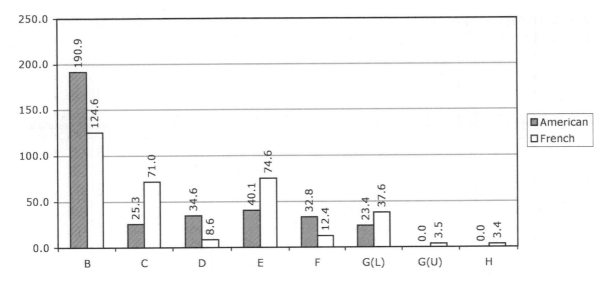

Figure 15. Densities of elephant dental specimens.

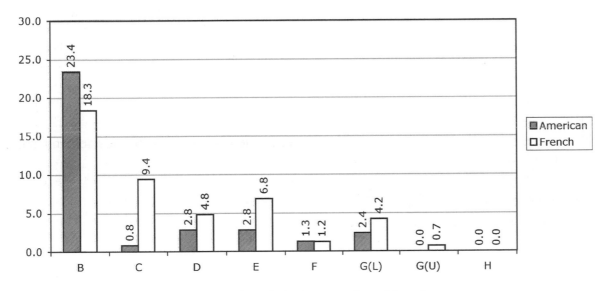

Figure 16. Densities of elephant crania and jaw fragments.

And clearly elephant postcrania were almost entirely a French affair, being important for evolutionary studies, but much less so for dating (Figure 17).

The American bias against postcrania is clearly apparent in the elephant densities. Even astragali did not fare well; only one was collected by the Americans, whereas the French recovered 13.

## HIPPOPOTAMIDAE

Hippos are very common elements in the Shungura fauna and may have suffered because of their ubiquity. In addition, as mentioned above, their worn teeth tended to fragment at high rates. The French were clearly more dedicated to the recovery of hippo specimens in general (Figure 18).

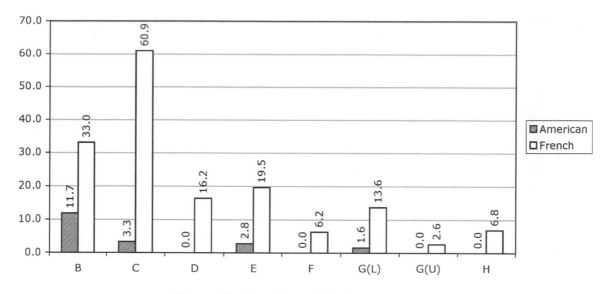

Figure 17. Densities of elephant postcrania.

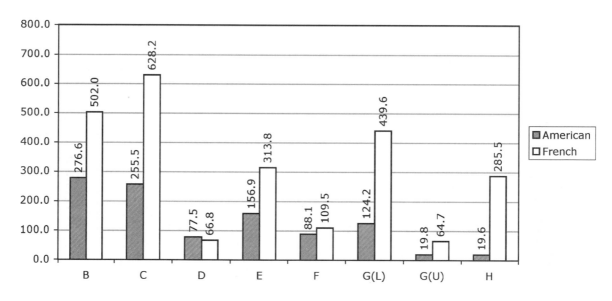

Figure 18. Densities of hippo specimens.

Elements that were part of the American collection protocol, crania, jaw fragments, and complete premolars and molars, however, were apparently collected with equal effort (Figure 19).

As in the case of the elephants, the French more consistently collected a wider array of hippo postcrania than did the Americans (Figure 20), because most of these were not part of the American collection protocol.

The French collected more than twice the number of hippo specimens as the Americans (1685 and 4116, respectively), with much of the difference in numbers accounted for by tooth fragments and postcranial elements.

In contrast, hippo astragali, part of the American collection protocol, were consistently collected. In their case, densities alternate except for Member D, suggesting similarity in collection effort (Figure 21).

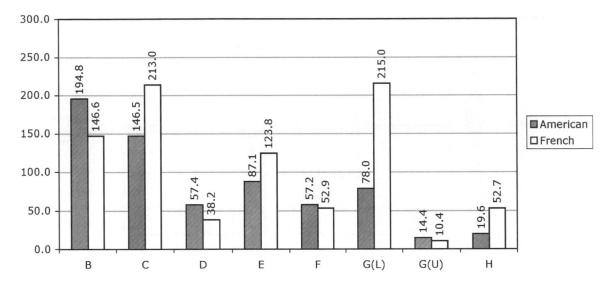

Figure 19. Densities of hippo crania, jaw fragments, and complete Ps and Ms.

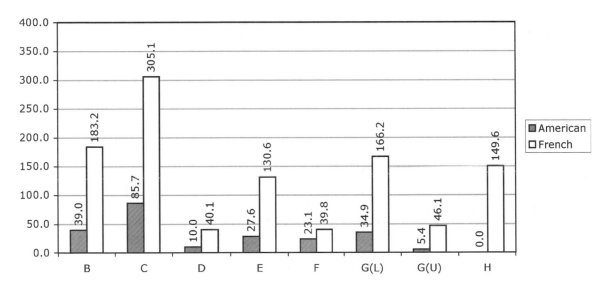

Figure 20. Densities of hippo postcrania.

## SUIDAE

The densities of pig specimens alternate through the members, although weakly so in Members F and G(U), suggesting that both contingents sought them with about equal effort (Figure 22). This pattern for all pig specimens masks, however, interesting collection biases in certain elements.

For example, H. B. S. Cooke had suggested to me that even fragmentary third molars, typically the talons and talonids, might be important in questions of dating. Thus, in an exception to the standard collection protocol, I collected all pig third molars, including fragments consisting of the posterior part of these teeth. The practice led to a clear collection bias in pig third molars except in Members G(U) and H, where American exposures were very small (Figure 23).

In contrast, the densities of complete premolars and molars alternate, except in Members F and G(U), suggesting if anything a slight French bias in collecting (Figure 24).

A similar pattern, with a slight French bias, is seen in the densities of crania and jaw

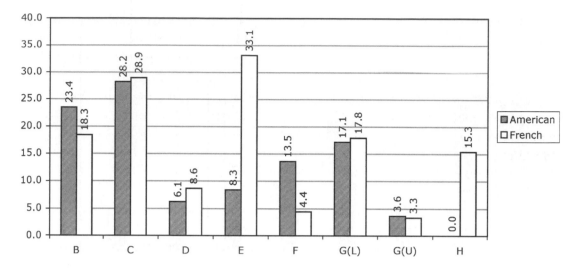

Figure 21. Densities of hippo astragali.

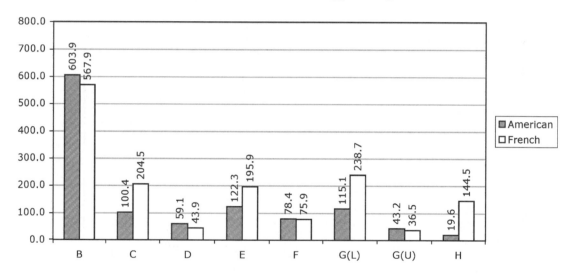

Figure 22. Densities of suid specimens.

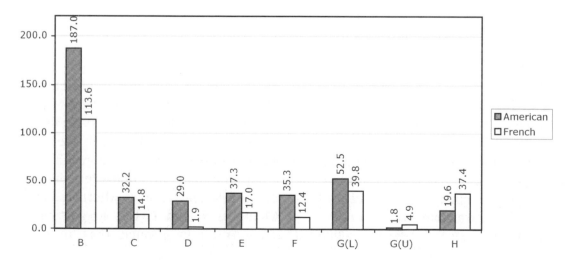

Figure 23. Densities of complete and fragmentary suid third molars.

fragments, but here only Member D breaks the alternating pattern (Figure 25).

As is to be expected there is a strong bias in the favor of the French with regard to densities of suid postcrania (Figure 26). The pattern is even stronger if one removes multiple specimens that constitute single individuals from Member G(U) of the American collections. The corrected value of the density then becomes 7.2 specimens per km². 

The alternating pattern of densities is seen again with regard to suid astragali (Figure 27).

Although over all suid densities suggest essentially equal effort in search and recovery of these specimens, two clear biases are hidden in these numbers, the American bias towards collection of fragmentary third molars and the French bias towards postcrania.

## GIRAFFIDAE

Giraffid specimens also have alternating densities, except in Members G(U) and H where American numbers are very small,

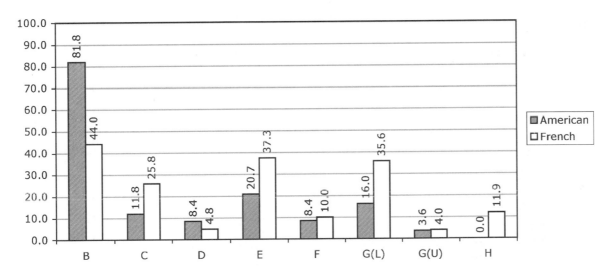

Figure 24.  Densities of complete premolars and molars.

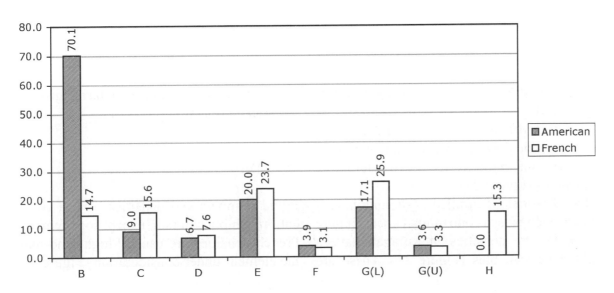

Figure 25.  Densities of suid crania and jaw fragments.

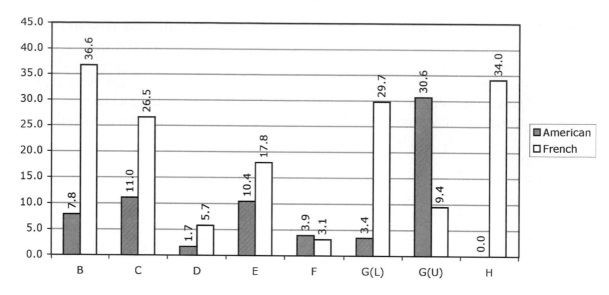

Figure 26. Densities of suid postcrania.

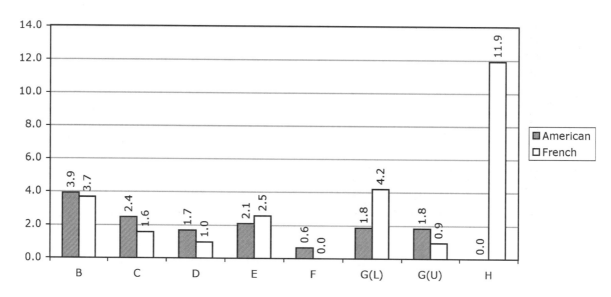

Figure 27. Densities of suid astragali.

suggesting similar search and recovery effort (Figure 28).

Densities of various cranial and dental elements (crania, jaw fragments, and complete premolars and molars) also have alternating frequencies (Figure 29).

The densities of giraffid postcrania have a slight French bias, as one might expect (Figure 30), while the densities of astragali have a slight American bias (Figure 31). These are both expectations one might have, given the American collection protocol.

EQUIDAE

Equid numbers and densities have a consistent French bias, in that, except for Member B, French densities are usually higher no matter the element under consideration (Figure 32).

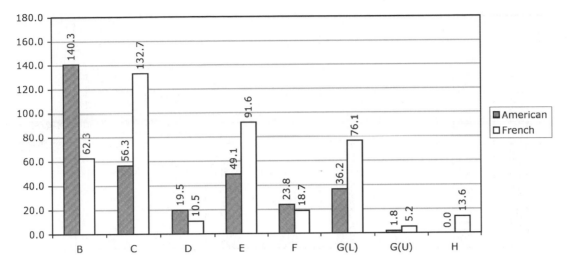

Figure 28. Densities of giraffid specimens.

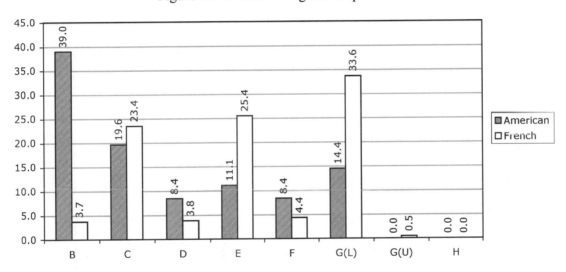

Figure 29. Densities of giraffid crania, jaw fragments, and complete Ps and Ms.

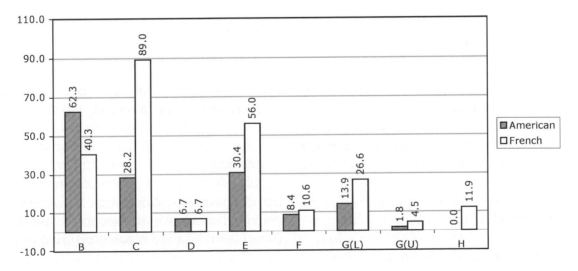

Figure 30. Densities of giraffid postcrania.

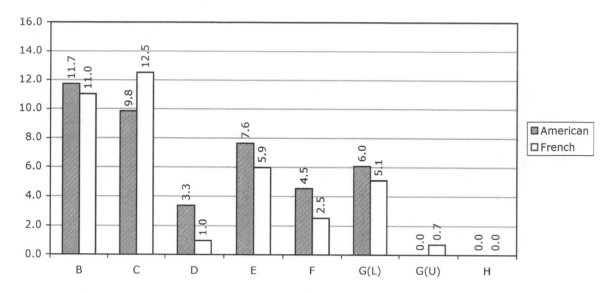

Figure 31. Densities of giraffid astragali.

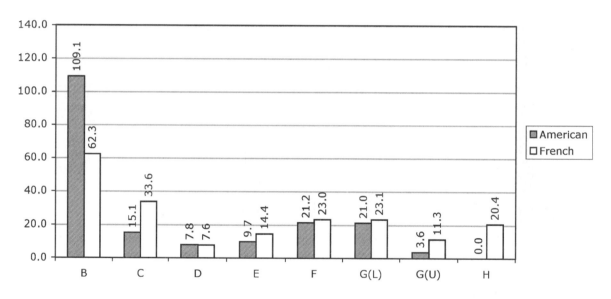

Figure 32. Densities of equid specimens.

This is true whether one is concerned with cranial and dental specimens (Figure 33) or with postcrania (Figure 34), a pattern that is not seen in any other family and one for which I have no ready explanation.

## RHINOCEROTIDAE

Rhinos are rare in the Shungura Formation as large mammals go (see Table 2) and seemingly suffer from high rates of dental fragmentation as discussed above. Thus, contrary to standard collection protocol, the Americans collected all dental fragments, as apparently did the French. Perhaps as a consequence, rhino densities, in contrast to those of equids, show neither an American or French bias (Figure 35). Unusually, the Americans collected more rhino postcranial specimens (nine) than did the French (six) from Sectors 1 through 27, but all of these were collected in 1968 before the standard collection protocol was instituted.

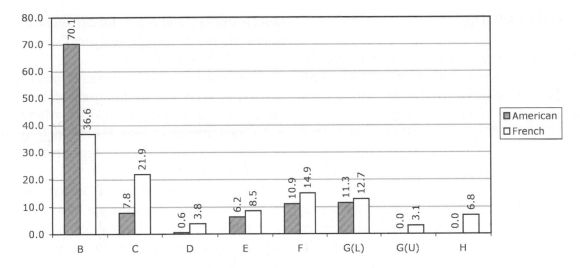

Figure 33. Densities of equid crania, jaw fragments, and complete premolars and molars.

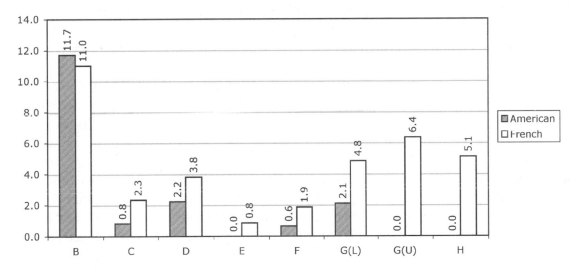

Figure 34. Densities of equid postcranial specimens.

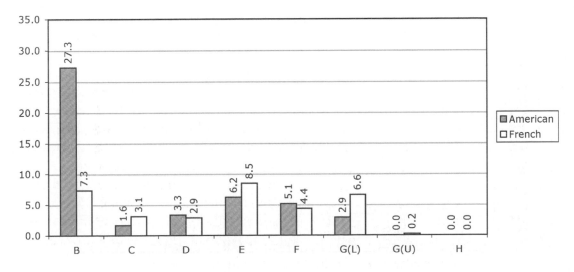

Figure 35. Densities of rhino specimens.

## DEINOTHERIIDAE

Deinothere specimens too are uncommon in the Shungura Formation and, because their teeth contain vast volumes of dentine with a very thick covering of enamel, they are typically found as enamel fragments. Because of the rarity of complete teeth, the Americans decided to pick up even single dental fragments if they seemed to represent single individuals. This appears to have produced a bias towards the American collection with regard to deinothere cranial and dental fragments (Figure 36). In contrast, the bias is clearly towards the French with regard to postcranial elements, for they collected 34 specimens, while the Americans collected none. The American bias for cranial and dental and the French for postcrania clearly play out even in this uncommon taxon.

## BOVIDAE

And finally I come to the Bovidae, by far the most numerous of the macromammalian families, constituting nearly half of all the specimens found (Table 2). When all body elements are considered together, the American and French teams seem to have collected specimens with similar strategy and effort, for densities of specimens alternate through the members, except for Member G(U) (Figure 37).

A similar but somewhat weaker alternating pattern is seen in the densities of complete teeth and jaw fragments, in that, Member F has equal densities and the French higher densities in Member G(U) suggesting a slight French bias (Figure 38).

The collections of bovid postcranial elements also have a slight French bias. Members D and G(U) have higher French densities, whereas both have higher American densities in the alternating pattern (Figure 39).

As might be expected given the American collection protocol, the French bias in astragali densities is weaker than for postcrania taken as a whole, for only Member G(U) breaks the alternating pattern (Figure 40).

Given that the American and French densities of bovid teeth, jaw fragments, and postcrania suggest similar collection strategies and intensities, with, at most, a slight bias towards higher French densities, it comes as some surprise that the collections show a consistent American bias in the densities of crania with horn cores (rare), frontlets (conjoined horn cores), and horn cores with bases (Figure 41) (also noted by Geraads and Coppens, 1995). Only in Member H are

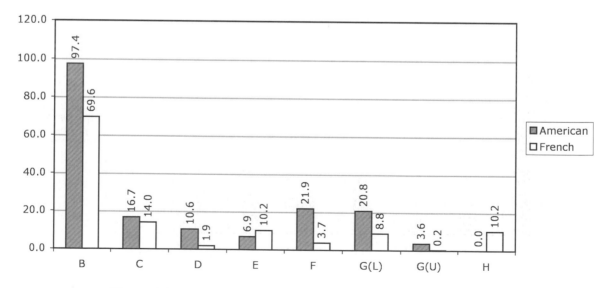

Figure 36. Densities of deinothere cranial and dental specimens.

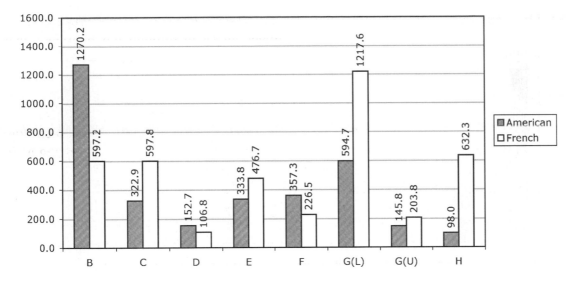

Figure 37. Densities of bovid specimens.

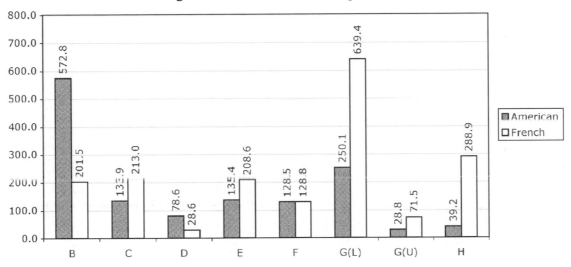

Figure 38. Densities of complete bovid teeth and jaw fragments.

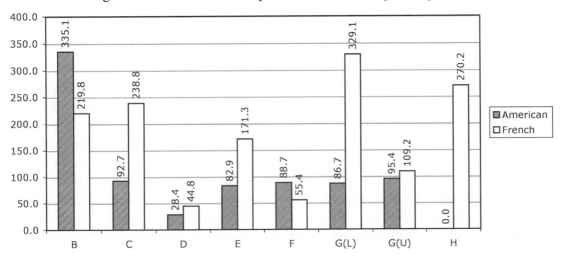

Figure 39. Densities of bovid postcrania.

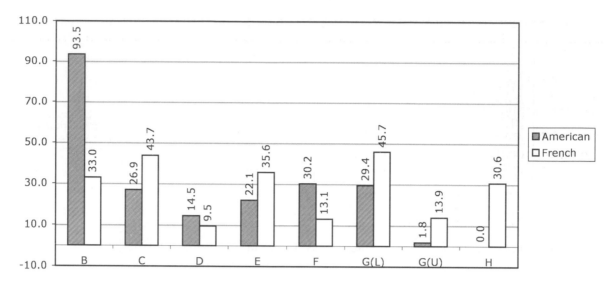

Figure 40. Densities of bovid astragali.

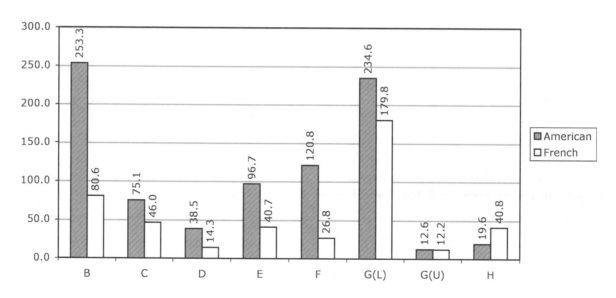

Figure 41. Densities of bovid crania, frontlets, and horn cores.

American densities lower than the French and, as will be remembered, the American area of Member H is very small.

The higher American densities are most puzzling in Member G(L), because in nearly every other taxon and skeletal element discussed above, French densities are highest in this member. Clearly something curious is going on with regard to the densities of bovid horn cores. I suspect that the higher American densities result from collection bias in which the French less consistently collected horn cores, especially fragmentary ones, than did the Americans, a collection bias similar to those documented on both sides in the numerous cases discussed above. If this is the case, then it is possible that the counts of French horn-core specimens and their relative taxonomic abundances are not representative of the numbers actually on the ground, leading to differences in taxonomic abundance between the American and French collection that are apparent, but not real. If my suspicion is correct, then it presents a significant problem given the potency of the Bovidae and their horn cores in the reconstruction of paleoenvironments. In contrast, Alemseged et al. (2006) think that the differences in counts and relative abundances of horn cores result from paleoenvironmental differences between the American and French collection areas. Resolution of this argument will likely result only from recollection of both the American and French areas, especially in Member G(L), using methods specifically designed to recover representative samples.

## Faunal "Hot Spots"

The large differences in densities between the members that are seen in the above analyses were not expected when I began this analysis. Upon further reflection, however, they may result because of fossiliferous "hot spots" that are unevenly distributed between the American and French areas. Members B, C, E, and G(L) show especially large density differences between the areas in Table 4 and Figure 6. Inspection of the distribution of surface specimens across these members suggests that the "hot spot" concept might be the answer.

Two localities in the American area of Member B produced most of the surface specimens from this member, Locality 1 produced 448 and Locality 2 produced 332, totaling 780. The Americans recovered 992 macromammalian surface specimens from the surface of Member B, thus these two localities produced 78.6% of the total. Locality 1 is one of the larger American localities, covering 50,384 m² on de Heinzelin's map. Locality 2 is much smaller covering 5,012 m². The total exposures of Member B in the American area cover 256,655 m². Localities 1 and 2, thus, make up 21.6% of the total area. This analysis is complicated by the fact the French began to collect in what became American Locality 1 in

1967, calling it their Locality Omo 28. When the Americans moved south to the Type Area in 1968, it was agreed that the French could continue to collect in Locality Omo 28 in later years, which they did, especially in 1968. If I add the French specimens from Locality Omo 28 to American Locality 1, its total goes to 1,287 and the total for Member B goes to 1,831. Now the specimens from Localities 1 and 2 comprise 88.4% of the total number from the American area of Member B—good illustrations of what I mean by hot spots. In contrast, in the French area of Member B, the locality with the highest count, Locality Omo 3, produced only 217 specimens, whereas, all the others, some at least half the size of American Locality 1, produced far fewer (see Figure 42).

In contrast, French densities are usually highest in Member C, again probably resulting from a differential distribution of hot spots. French Locality Omo 18 produced 1,243 macromammalian surface specimens from an area of 102,572 m². Locality Omo 40 produced 358 specimens from an area of 88,004 m². The two localities thus produced 50.6% of the specimens collected in the French area of Member C, from only 14.9% of its area. The American area of Member C also contains a hot spot, Locality 32, which produced 396 macromammalian surface specimens from an area of 28,176 m², 12.6% of the specimens from American Member C in only 1.1% of its area. The larger number of specimens from Omo 18 and 40 outweigh, however, the higher density of Locality 32, producing the overall higher densities in the French area.

The French area of Member E typically produces the highest densities of macromammalian surface specimens and contains a single hot spot in this member, Locality Omo 57. This locality produced 39.3% of the total specimens from Member E (685 out of 1,743) in 2.8% of the area (32,492 m² of 1,179,028 m²). The locality with the highest count in the American area of Member E is Locality 146 with 215 specimens.

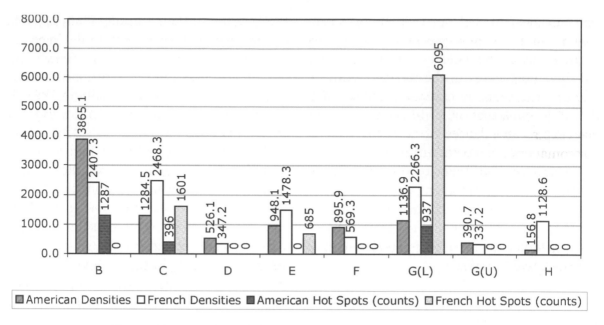

Figure 42. American and French densities and hot spot counts.

The French area also most often produced the highest densities of specimens from Member G(L). Four hot spots warrant discussion. Locality Omo 75 is a huge area (hardly a spot) whose boundaries are only partly marked on de Heinzelin's map. Locality Omo 75 includes three other localities that were established at a later time—Localities Omo 138, 139, and 140. Its area, including the areas of the three newer localities, is estimated to be 374,583 m². The four localities together produced 2,274 macromammalian surface specimens out of the 10,310 specimens recovered from Member G(L) or 22.1%. Locality Omo 75 covers 8.2% of the total area (4,549,295 m²) of the member. Locality Omo 29 is also a very large (208,838 m²) and poorly bounded locality, including four localities that were established at a later time—Localities Omo 222, 223, 231, and 234. Together, the five localities produced 1,156 macromammalian surface specimens or 11.1% of the specimens in 4.6% of the surface area of Member G(L). In contrast, Locality Omo 47 is well defined on de Heinzelin's map and small, covering only 24,315 m². It includes an earlier defined locality, Locality Omo Sh 2–3. The two locali-

ties produced 1,753 macromammalian surface specimens, 17.0% of the total from Member G(L), in 0.5% of it area. Locality Omo 47 is an exemplary hot spot. These three localities— Omo 29, 47, and 75 (and their included localities) produced 50.1% of the specimens from Member G(L) in 13.4% of its area. Locality Omo 323 produced 912 macromammalian surface specimens, so it too probably qualifies as a hot spot, but it was defined and collected in 1976, after de Heinzelin had completed his work in the Shungura Formation, and not included in his maps. I therefore do not know nor can I measure its areal extent.

The American area of Member G(L) contains two hot spots of macromammalian surface specimens, Localities 7 and 627. Locality 7 produced 462 (10.7%) and Locality 627 produced 475 (11.0%) of the 4,328 specimens recovered from the surface of Member G(L). Locality 7 covers 41,170 m² (1.1%) and Locality 627 covers 9,567 m² (0.3%) of the 3,807,007 m² of exposures of the member. Again, although the densities of these two localities are very high, the numbers of specimens they produced are overwhelmed by those of the French hot spots.

Members D, F, G(U), and H appear not to contain hot spots, although the methods I use may not be able to detect them in these members, given their small counts and/or areas.

The occurrence of differentially distributed hot spots in the American and French areas may explain the differences in densities of macromammalian surface specimens from the various members. Why they alternate in density is not known, but may result purely from chance, remembering that the boundary between them, the watering road, was chosen because it was well marked and known to everyone, not because it had any paleontological consequence.

## Lessons to be Learned from This Analysis

Given that the American and French collection areas were contiguous, of roughly the same size, with their centers lying only 9 km apart and that the sediments in the two areas were laid down by the same river at the same time, one might expect that the two collections would be very similar. That this is not the case is clear from the discussion above. It should also be apparent that the causes for the differences from this expectation are varied and complex.

Some of the differences result from well-documented variation in areas of exposure available to the contingents or to differences in effort. For example, the larger French collection, at least in part, results from two more years of field expeditions to the Shungura Formation and to larger areas of exposures. The larger number of excavated specimens and micromammals in the American collection result from more excavation effort over a longer period of time. The common pattern of larger numbers of postcranial elements in the French collection is explained by the fact that the Americans decided not to collect them for most taxa, while the French did. Other differences are less easily explained, however,

principally because collecting methods and records were not formulated or not kept.

One of the more important of these concerns the large differences in apparent specimen density in the areas of the two contingents. Although differences in effort might have produced these differences, I think it unlikely for a number of reasons. First, the pattern of densities of primate taxa, especially hominids but also monkeys, both highly valued, is generally similar to that of the fauna as a whole. One might expect that if effort determined the densities, then those of the most highly valued taxa would differ from those of more general concern. Second, it is nearly universally the case, whatever taxon one considers, that American densities are highest in Member B, for example, while those of the French are highest in Member G(L). I think that systematic differences in effort are unlikely to have produced these differences in densities, given the very different interests that the members of the expeditions held and given that much of the collecting occurred before the stratigraphic structure of the formation was well understood or mapped. I think, therefore, that the alternating pattern of densities results from similar effort being applied to exposures of sediments of inherently different fossil richness. The American and French areas differ in density, in all likelihood, because of the differential distribution of specimen "hot spots." Why they alternate in density may result purely from chance.

How all of this affects taphonomic and taxonomic analyses, especially those based on specimen frequencies, is of some interest. I suggest, for example, that those body elements whose densities show the alternating density pattern, indicating similar collection strategy and effort, are more likely to be representative samples of the exposed sediments than those whose density patterns are biased towards one contingent or the other. These elements then might be best used in comparisons of the collections aimed at determining the primary cause of the density differences (e.g., paleoecological). Similarly, the collections of these specimens might be combined to provide larger samples for use in comparing the faunae from different members to discover faunal differences in time. In contrast, body elements whose density patterns suggest biases principally produced by differences in strategy or effort might be avoided for these types of analyses.

And finally, although I developed implicit rules concerning the collection of specimens, as describe above, and these rules were generally followed by other members of the expedition after they were formulated, the rules were never made part of an explicit research strategy nor was their rationale discussed. In hindsight, I now think that the implicitness of the rules lead to some of the biases in the American collection discussed above; even I did not always follow them. If undocumented biases are not to creep into fossil collections, collection strategies should be explicit, their rationale generally agreed upon, and strict adherence to them rewarded.

## Acknowledgments

I wish to thank René Bobe, Zeresenay Alemseged, Clark Howell, Donald Grayson, and Kay Behrensmeyer for useful discussion and comments.

## References

Alemseged, Z., Bobe, R., Geraads, D., 2007. Comparability of fossil data and its significance for the interpretation of hominin environments: a case study in the lower Omo valley, Ethiopia. In: Bobe, R., Alemseged, Z., Behrensmeyer, A.K. (Eds.), Hominin Environments in the East African Pliocene: An Assessment of the Faunal Evidence. Springer, Dordrecht.

Arambourg, C., 1947. Mission Scientifique de l'Omo 1932–1933, Tome I, Fascicule III, Géologie—Anthropologie. Muséum National d'Histoire Naturelle. Éditions du Muséum, Paris.

Brown, F.H., de Heinzelin J., 1983. The lower Omo Basin. In: de Heinzelin, J. (Ed.), The Omo Group: Archives of the International Omo Research Expedition, Chapter 2, Annals, Serie in 8°, Sciences Geologiques, n° 85. Musée Royal de l'Afrique Centrale, Tervuren, Belgique.

Coppens, Y., Howell, F.C., Isaac, G.Ll., Leakey, R.E.F., 1976. Preface. In: Coppens, Y., Howell, F.C., Isaac, G.Ll., Leakey, R.E.F. (Eds.), Earliest Man and Environments in the Lake Rudolf Basin. University of Chicago Press, Chicago, pp. xvii–xxi.

Dechant-Boaz, D.D., 1994. Taphonomy and the fluvial environment—examples from Pliocene deposits of the Shungura Formation, Omo basin, Ethiopia. In: Corruccini, R.S., Ciochon, R.L. (Eds.), Integrative Paths to the Past, Paleoanthropological Advances in Honor of F. Clark Howell. Prentice Hall, Englewood Cliffs, N.J., pp. 377–414.

de Heinzelin, J., 1983. The Omo Group: Archives of the International Omo Research Expedition, Volume 2, Maps, Annals, Serie in 8°, Sciences Geologiques, n° 85. Musée Royal de l'Afrique Centrale, Tervuren, Belgique.

de Heinzelin, J., Haesaerts P., 1983. The Shungura Formation. In: de Heinzelin, J. (Ed.), The Omo Group: Archives of the International Omo Research Expedition, Chapter 3, Annals, Serie in 8°, Sciences Geologiques, n° 85. Musée Royal de l'Afrique Centrale, Tervuren, Belgique.

Feibel, C.S., Brown, F., McDougall, I., 1989. Stratigraphic context of fossil hominids from the Omo Group Deposits: Northern Turkana Basin, Kenya and Ethiopia. American Journal of Physical Anthropology 78, 595–622.

Geraads, D., Coppens, Y., 1995. Evolution des faunes de mammifères dans le Plio-Pléistocène de la basse vallée de l'Omo (Ethiopie): apports de l'analyse factorielle. Comptes rendus de l' Academie des Sciences. Serie II Fascicule A 320, 625–637.

Johanson, D.C., Splingaer, M., Boaz, N.T., 1976. Paleontological excavations in the Shungura Formation, Lower Omo Basin, 1969–1973. In: Coppens, Y., Howell, F.C., Isaac, G.Ll., Leakey, R.E.F. (Eds.), Earliest Man and Environments in the Lake Rudolf Basin. University of Chicago Press, Chicago, pp. 402–420.

Wesselman, H.B., 1984. The Omo Micromammals. In: Hecht, M.K., Szalay, F.S. (Eds.), Contributions to Vertebrate Evolution, Volume 7. Karger, New York.

# Appendix

*Counts of macromammals specimens from the surface of Sectors 1 to 27 (Type Area) of the Shungura Formation*

| | American Collection | | | | | | | | | | | |
| | Ba | A | B | C | D | E | F | G(L) | G(U) | H | J | Total |
|---|---|---|---|---|---|---|---|---|---|---|---|---|
| Hominidae | 0 | 0 | 5 | 22 | 16 | 11 | 17 | 18 | 1 | 0 | 0 | 90 |
| Jaw fragments | 0 | 0 | 0 | 1 | 0 | 0 | 1 | 3 | 1 | 0 | 0 | 6 |
| Cercopithecidae | 0 | 0 | 236 | 1109 | 259 | 294 | 385 | 603 | 55 | 1 | 0 | 2942 |
| Complete and fragmentary isolated teeth | 0 | 0 | 221 | 738 | 198 | 198 | 279 | 485 | 7 | 1 | 0 | 2127 |
| Complete isolated teeth | 0 | 0 | 129 | 371 | 78 | 93 | 91 | 193 | 6 | 1 | 0 | 962 |
| Complete teeth and skull fragments | 0 | 0 | 138 | 441 | 89 | 108 | 107 | 221 | 7 | 1 | 0 | 1112 |
| Postcrania | 0 | 0 | 29 | 271 | 42 | 74 | 62 | 74 | 45 | 0 | 0 | 597 |
| Carnivora | 0 | 0 | 7 | 39 | 4 | 5 | 11 | 21 | 4 | 0 | 0 | 91 |
| Elephantidae | | | | | | | | | | | | |
| Dental specimens | 0 | 0 | 49 | 62 | 62 | 58 | 51 | 89 | 0 | 0 | 0 | 371 |
| Crania and jaw fragments | 0 | 0 | 6 | 2 | 5 | 4 | 2 | 9 | 0 | 0 | 0 | 28 |
| Postcrania | 0 | 0 | 3 | 8 | 0 | 4 | 0 | 6 | 0 | 0 | 0 | 21 |
| Hippopotamidae | 0 | 0 | 71 | 626 | 139 | 227 | 137 | 473 | 11 | 1 | 0 | 1685 |
| Crania, jaw fragments, complete Ps and Ms | 0 | 0 | 50 | 359 | 103 | 126 | 89 | 297 | 8 | 1 | 0 | 1033 |
| Postcrania | 0 | 0 | 10 | 210 | 18 | 40 | 36 | 133 | 3 | 0 | 0 | 450 |
| Astragali | 0 | 0 | 6 | 69 | 11 | 12 | 21 | 65 | 2 | 0 | 0 | 186 |
| Suidae | 0 | 3 | 155 | 246 | 106 | 177 | 122 | 438 | 24 | 1 | 0 | 1272 |
| Complete and fragmentary M3s | 0 | 0 | 48 | 79 | 52 | 54 | 55 | 200 | 1 | 1 | 0 | 490 |
| Complete Ps, M1s and M2s | 0 | 1 | 21 | 29 | 15 | 30 | 13 | 61 | 2 | 0 | 0 | 172 |
| Crania and jaw fragments | 0 | 0 | 18 | 22 | 12 | 29 | 6 | 65 | 2 | 0 | 0 | 154 |
| Postcrania | 0 | 0 | 2 | 27 | 3 | 15 | 6 | 13 | 17 | 0 | 0 | 83 |
| Astragali | 0 | 0 | 1 | 6 | 3 | 3 | 1 | 7 | 1 | 0 | 0 | 22 |
| Giraffidae | 0 | 1 | 36 | 138 | 35 | 71 | 37 | 138 | 1 | 0 | 0 | 457 |
| Crania, jaw fragments, complete Ps and Ms | 0 | 1 | 10 | 48 | 15 | 16 | 13 | 55 | 0 | 0 | 0 | 158 |
| Postcrania | 0 | 0 | 16 | 69 | 12 | 44 | 13 | 53 | 1 | 0 | 0 | 208 |
| Deinotheriidae | 0 | 0 | 25 | 41 | 19 | 10 | 34 | 79 | 2 | 0 | 0 | 210 |

*(Continued)*

*Counts of macromammals specimens from the surface of Sectors 1 to 27 (Type Area) of the Shungura Formation—cont'd.*

**American Collection**

| | Ba | A | B | C | D | E | F | G(L) | G(U) | H | J | Total |
|---|---|---|---|---|---|---|---|---|---|---|---|---|
| Bovidae | 0 | 0 | | 326 | 791 | 274 | 483 | 556 | 2264 | 81 | 5 | 4780 |
| Jaw fragments and complete teeth | 0 | 0 | | 147 | 328 | 141 | 196 | 200 | 952 | 16 | 2 | 1982 |
| Postcrania | 0 | 0 | | 86 | 227 | 51 | 120 | 138 | 330 | 53 | 0 | 1005 |
| Astragali | 0 | 0 | | 24 | 66 | 26 | 32 | 47 | 112 | 1 | 0 | 308 |
| Crania, frontlets, and horn cores | 0 | 0 | | 65 | 184 | 69 | 140 | 188 | 893 | 7 | 1 | 1547 |

**French Collection**

| | Ba | A | B | C | D | E | F | G(L) | G(U) | H | J | Total |
|---|---|---|---|---|---|---|---|---|---|---|---|---|
| Hominidae | 0 | 1 | 2 | 18 | 1 | 17 | 6 | 40 | 0 | 1 | 0 | 86 |
| Jaw fragments | 0 | 0 | 0 | 1 | 0 | 2 | 0 | 6 | 0 | 0 | 0 | 9 |
| Cercopithecidae | 1 | 19 | 86 | 841 | 71 | 267 | 124 | 736 | 25 | 5 | 0 | 2175 |
| Complete and fragmentary isolated teeth | 1 | 18 | 81 | 715 | 54 | 220 | 96 | 480 | 19 | 3 | 0 | 1687 |
| Complete isolated teeth | 1 | 13 | 45 | 378 | 25 | 107 | 40 | 281 | 9 | 3 | 0 | 902 |
| Complete teeth and skull fragments | 1 | 13 | 46 | 441 | 32 | 132 | 51 | 390 | 10 | 4 | 0 | 1120 |
| Postcrania | 0 | 1 | 4 | 58 | 10 | 19 | 16 | 136 | 5 | 1 | 0 | 250 |
| Carnivora | 0 | 2 | 1 | 14 | 1 | 11 | 3 | 51 | 7 | 0 | 0 | 90 |
| Elephantidae | | | | | | | | | | | | |
| Dental specimens | 4 | 8 | 34 | 91 | 9 | 88 | 20 | 171 | 15 | 2 | 2 | 444 |
| Crania and jaw fragments | 0 | 1 | 5 | 12 | 5 | 8 | 2 | 19 | 3 | 0 | 0 | 55 |
| Postcrania | 2 | 20 | 9 | 78 | 17 | 23 | 10 | 62 | 11 | 4 | 0 | 236 |
| Hippopotamidae | 9 | 100 | 137 | 805 | 70 | 370 | 176 | 2000 | 275 | 168 | 6 | 4116 |
| Crania, jaw fragments, complete Ps and Ms | 0 | 19 | 40 | 273 | 40 | 146 | 85 | 978 | 44 | 31 | 2 | 1658 |
| Postcrania | 9 | 37 | 50 | 391 | 42 | 154 | 64 | 756 | 196 | 88 | 4 | 1791 |
| Astragali | 0 | 3 | 5 | 37 | 9 | 39 | 7 | 81 | 14 | 9 | 1 | 205 |
| Suidae | 0 | 139 | 155 | 262 | 46 | 231 | 122 | 1086 | 155 | 85 | 3 | 2284 |
| Complete and fragmentary M3s | 0 | 10 | 31 | 19 | 2 | 20 | 20 | 181 | 21 | 22 | 1 | 327 |
| Complete Ps, M1s and M2s | 0 | 29 | 12 | 33 | 5 | 44 | 16 | 162 | 17 | 7 | 1 | 326 |
| Crania and jaw fragments | 0 | 1 | 4 | 20 | 8 | 28 | 5 | 118 | 14 | 9 | 0 | 207 |
| Postcrania | 0 | 6 | 10 | 34 | 6 | 21 | 5 | 135 | 40 | 20 | 0 | 277 |
| Astragali | 0 | 2 | 1 | 2 | 1 | 3 | 0 | 19 | 4 | 7 | 0 | 39 |

| | | | | | | | | | | | |
|---|---|---|---|---|---|---|---|---|---|---|---|
| Giraffidae | 0 | 8 | 17 | 170 | 11 | 108 | 30 | 346 | 22 | 8 | 1 | 721 |
| Crania, jaw fragments, complete Ps and Ms | 0 | 0 | 1 | 30 | 4 | 30 | 7 | 153 | 2 | 0 | 0 | 227 |
| Postcrania | 0 | 2 | 11 | 114 | 7 | 66 | 17 | 121 | 19 | 7 | 1 | 365 |
| Astragali | 0 | 2 | 3 | 16 | 1 | 7 | 4 | 23 | 3 | 0 | 0 | 59 |
| Equidae | 1 | 7 | 17 | 43 | 8 | 17 | 37 | 105 | 48 | 12 | 2 | 297 |
| Crania, jaw fragments, complete Ps and Ms | 1 | 4 | 10 | 28 | 4 | 10 | 24 | 58 | 13 | 4 | 2 | 158 |
| Postcrania | 0 | 3 | 3 | 3 | 4 | 1 | 3 | 22 | 27 | 3 | 0 | 69 |
| Rhinocerotidae | 1 | 3 | 2 | 4 | 3 | 10 | 7 | 30 | 1 | 0 | 5 | 66 |
| Deinotheriidae | 2 | 14 | 19 | 18 | 2 | 12 | 6 | 40 | 1 | 6 | 0 | 120 |
| Bovidae | 10 | 41 | 163 | 766 | 112 | 562 | 364 | 5539 | 866 | 372 | 101 | 8896 |
| Jaw fragments and complete teeth | 0 | 15 | 55 | 273 | 30 | 246 | 207 | 2909 | 304 | 170 | 20 | 4229 |
| Postcrania | 3 | 13 | 60 | 306 | 47 | 202 | 89 | 1497 | 464 | 159 | 65 | 2905 |
| Astragali | 0 | 3 | 9 | 56 | 10 | 42 | 21 | 208 | 59 | 18 | 3 | 429 |
| Crania, frontlets, and horn cores | 6 | 4 | 22 | 59 | 15 | 48 | 43 | 818 | 52 | 24 | 6 | 1097 |

# 9. Serengeti micromammals and their implications for Olduvai paleoenvironments

D.N. REED
*Department of Anthropology*
*The University of Texas at Austin*
*Austin, TX 78712, USA*
*reedd@mail.utexas.edu*

**Keywords**:  Hominid paleoecology, taphonomy, correspondence analysis, taxonomic habitat index

## Abstract

Fossil micromammals are widely used as paleoenvironmental indicators in Pliocene hominin fossil localities, and many assemblages are believed to be accumulated by predators such as owls. This chapter examines modern owl-accumulated micromammal assemblages from Serengeti, Tanzania. These modern roost data are used to examine the fidelity of the taxonomic signal and its sensitivity to change across habitats within an ecosystem. The modern data show that the relative abundance of prey taxa in owl-accumulated assemblages varies across habitats in a predictable fashion. This provides a basis for applying the analysis of fossil micromammal assemblages to intra-basin scales using relative abundance as well as biome or regional scales using presence/absence of taxa. Using the modern micromammal assemblages as analogues, the latter part of the chapter explores taphonomic and paleoenvironmental change through Bed I times at Olduvai Gorge.

## Introduction

Olduvai Gorge is significant for preserving fauna at a transitional period in Earth's climatic history between the warmer, more stable Pliocene and the oscillating extremes of the glacial Pleistocene and also for yielding early discoveries of *Australopithecus boisei*/*Paranthropus boisei* and *Homo habilis*. Synchronic studies of hominin behavioral ecology at Olduvai and elsewhere using a landscape approach require detailed paleoenvironmental reconstructions to effectively test land use models (Blumenschine and Masao, 1991; Peters and Blumenschine, 1995; Blumenschine and Peters, 1998), as do diachronic studies of hominin adaptation in response to climate change.

Detailed paleoenvironmental and paleoecological analyses are bolstered by multiple lines of evidence of which the mammalian fauna is a crucial component (Vrba, 1992). Among mammals, the smallest are well suited for paleoenvironmental analysis and for testing evolutionary models (Avery, 1982). Micromammals are speciose and embody a rich array of adaptations ranging from dedicated faunivory to hyper-grazing. They have the potential to provide paleoenvironmental signals at a finer scale than other lines of evidence (e.g., macromammals and palynology)

*R. Bobe, Z. Alemseged, and A.K. Behrensmeyer (eds.) Hominin Environments in the East African*
*Pleistocene: An Assessment of the Faunal Evidence, 217–255.*
© 2007 *Springer.*

and they are more likely to be accumulated independently from the activities of hominins (but see Fernandez-Jalvo et al., 1999). Owls are one of the primary predators of small mammals, and understanding this predator–prey system is vital to micromammal taphonomy and paleoecology.

This chapter re-examines Plio-Pleistocene Olduvai fossil micromammals in light of new data on modern, owl-accumulated assemblages, or coprocoenoses, taken from different parts of the Serengeti ecosystem. With the actualistic data I pursue two fundamental questions. First, does the fauna found in a coprocoenosis match what one would expect for the surrounding habitat? On one hand predators sampling the biocoenosis are constrained to the prey that inhabit that environment. However, predators exert preferences for certain habitats. Barn owls, for example, exhibit numerous morphological and behavioral adaptations for hunting terrestrial prey in open habitats such as grasslands, raising the question of whether woodland or forest fauna will appear at all in a barn owl assemblage (Andrews, 1983; Tchernov, 1992). This question falls under the rubric of accuracy; it asks, "How accurately does a coprocoenosis represent the surrounding habitat?" A second set of questions focuses on precision. In going from one habitat to another at what point are changes in habitat reflected in the fauna? Or phrased another way, how sensitive is the coprocoenosis to changing habitat, both in terms of taxonomic composition (presence or absence) and the relative abundance of taxa?

These questions are fundamental to the analysis of fossil micromammal assemblages, yet they remain, for the most part, unanswered. Before delving into these issues, a general overview of the predator–prey system formed between owls and small mammals is provided. Following this review, I summarize data collection methods and present results from the analysis of modern assemblages. The latter part of the chapter then turns to applying the results to micromammal faunas at Olduvai Gorge.

## Background

For many years, mammalogists and ornithologists have benefited from the hunting and digestive processes of owls (Glue, 1970; Denbow, 2000). Owls routinely regurgitate the remains of consumed prey in a compact bolus of bone wrapped in fur called a pellet. Pellets provide neontologists with a non-invasive way to study the diet of owls and aid paleontologists by concentrating bones at a single spot (Davis, 1959). Owls sample the surrounding faunal community (or biocoenosis) and return to selected roosting spots where they deposit dense concentrations of pellets. At some fossil localities bone densities are so great that the most reasonable explanation is that owls accumulated them. The phenomenon is common to many cave sites and rock shelters (de Graaff, 1960, 1961; Levinson, 1982; Avery, 1987, 1992; Andrews, 1990), but similar dense concentrations are also known from open-air sites (e.g., Fernandez-Jalvo et al., 1998). Further support that owls were accumulating fossil faunas comes from the discovery of fossilized impressions of pellets (Denys, 1987b; Gawne, 1975), and from detailed taphonomic analysis of micromammal bones (Andrews, 1990; Fernandez-Jalvo and Andrews, 1992; Dauphin et al., 1994, 1997; Denys et al., 1997; Fernandez-Jalvo et al., 1998).

Numerous studies have investigated modern owl pellet assemblages, but the current effort is unique in focusing on the aggregate assemblages resulting from the decay of many pellets (coprocoenosis). Generally, such assemblages are deprecated by neontologists focusing on the ecology of owls or micromammals (e.g., see Lyman and Power, 2003). However, for paleobiologists and zooarchaeologists, the coprocoenosis is the appropriate unit of analysis. Time averaging buffers many short-term fluctuations resulting in an assemblage more like the fossil record. There are two approaches to the taphonomic analysis of coprocoenoses. By one route, one can attempt to reverse taphonomic biases; but this route is difficult because so many interacting factors are in play, and at present we know very little about these processes. A more direct route is to study correlations between the coprocoenosis and the ecological parameters of interest. This approach uses modern taphonomic assemblages as analogues or reference assemblages to be compared with fossil assemblages. Similarities between fossil and taphonomic assemblages are assumed to result from similar ecological processes, though the processes themselves are treated as a "black box." Of course these two approaches are not mutually exclusive. One can start by forming analogues, and exploring correlations, while prying open the box to understand the causal mechanisms responsible for the associations between coprocoenoses and the environments from which they were derived. This is the general approach that I adopted, and either route is preferable to simply ignoring taphonomic biases on micromammal assemblages altogether.

## Modern Micromammal Assemblages

This section covers the analysis of nine modern, owl-accumulated micromammal assemblages from the Serengeti region of northern Tanzania. A brief introduction to the study area is presented first, followed by a summary of the collection methodology, including a description of the ecological trends within the ecosystem that produce habitat differences between roosting sites. The basic faunal composition is tabulated, and this is followed by a description of the habitat proclivities of the species as enumerated in niche models developed from the literature on small mammal ecology. The subsequent sections

examine the patterns of faunal composition and abundance between roosting sites and address potential artifacts such as sample size and predator bias.

## SERENGETI ECOSYSTEM STUDY AREA

Field data were gathered between November 1998 and April 2000. The Serengeti ecosystem straddles the Tanzania–Kenya border in East Africa between 34 and 36° E longitude and 1–2° S latitude. Serengeti National Park in Tanzania encompasses an area of 14,763 km$^2$ but the larger ecosystem—defined as the area covered by the wildebeest migration—extends into neighboring Masai Mara National Reserve (1,510 km$^2$) to the north in Kenya, Ngorongoro Conservation Area (8,094 km$^2$) to the southeast, the Loliondo Game Controlled Area (4,000 km$^2$) to the east, Maswa Game Reserve (2,200 km$^2$) to the southwest, and the Ikoronogo and Grumeti Game Controlled Areas (5,000 km$^2$) to the northwest (Sinclair, 1995b). In total, the ecosystem covers an expanse of roughly 24,000 km$^2$ as shown in Figure 1.

## COLLECTION OF MICROMAMMAL SAMPLES

The category, "small mammals" refers to animals weighing less than 5 kg (after Andrews, 1990), and the term "micromammal" refers to a subset weighing less than a few hundred grams. In Africa, rodents (Order Rodentia) and shrews (Order Insectivora) are the most abundant micromammal prey of owls, but elephant shrews (Order Macroscelidea), bats (Order Chiroptera), rabbits and hares (Order Lagomorpha), and small primates (Order Primates) must also be considered as well as juvenile members of some of the larger mammals.

A total of 61 roosting sites were found in and around the Serengeti National Park, of which nine have been analyzed and the

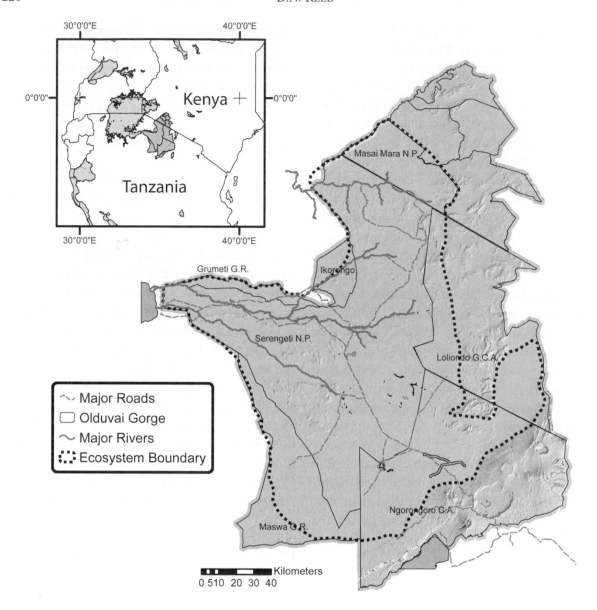

Figure 1. Map of the Serengeti National Park and adjacent protected areas, overlying a shaded area depicting the extent of the Serengeti Ecosystem. The ecosystem is defined as the area covered by the wildebeest (*Connochaetes taurinus*) during their annual migrations. Map inset shows the location of the study area within the East African subregion.

results presented here. Roost locations were recorded on a Garmin XL12 GPS receiver. Table 1 provides summary information on the analyzed roosts. Modern roosting sites were located and identified as such either by direct observation of an owl, or by the presence of pellets and bone detritus from deteriorated pellets. Roosts were located with the help of other researchers and by targeted investigation of rock outcroppings such as escarpments,

inselbergs, and kopjes as well as hollowed trees and woodland thickets.

Many of the roosts had owls in residence. Barn Owls, *Tyto alba*, were identified by their white face disk, dark eyes, orange-buff-colored upper body (dorsal surface) with dark speckles or patches, and white or lightly spotted breasts. Eagle Owls, *Bubo africanus*, have uppers of buff-grey with irregular dark grey or beige patches, yellow eyes, grey breast

Table 1. Summary of the roosts selected for analysis arranged by latitude. Geographical coordinates are given in decimal degrees. An asterisk ( * ) indicates a roost where an owl's identity was confirmed by visual sighting. Identities at other roosts were inferred from pellet morphology, feathers, and the roost type, e.g., cavity or tree crown

| Roost no. | Collection date | Owl species | Latitude | Longitude |
|---|---|---|---|---|
| 44 | September 18, 1999 | *Tyto alba* | −1.64596 | 34.80920 |
| 23 | January 6, 1999 | *Bubo africanus* | −2.36593 | 34.86813 |
| 4 | November 1, 1998 | *Tyto alba*\* | −2.43132 | 34.85326 |
| 12 | December 2, 1998 | *Bubo africanus*\* | −2.43268 | 34.82940 |
| 13 | December 19, 1998 | *Tyto alba* | −2.43625 | 34.95496 |
| 18 | December 30, 1998 | *Bubo africanus* | −2.44666 | 34.98977 |
| 3 | October 26, 1998 | *Tyto alba*\* | −2.47109 | 34.89905 |
| 7 | October 3, 1998 | *Tyto alba*\* | −2.68508 | 34.89518 |
| 24 | January 9, 1999 | *Tyto alba*\* | −2.69849 | 35.06356 |

with bars and "ear" tufts (Zimmerman et al., 1996). In northern Tanzania, these two species are similar in size, with the spotted eagle owl being slightly larger. At roosts where the owls were not present, indirect evidence from feathers, pellet morphology, and the physical structure of the roost, provide a reliable indicator of roost occupation. Barn owls are restricted to roosting in cavities such as the hollowed interiors of trees, or vertical fissures in rock outcroppings. Eagle owls, on the other hand roost in exposed settings such as tree crowns, or on the ground near rocks. These differences and their potential impact on faunal composition are discussed below.

Barn owls and eagle owls were the only birds observed in direct association with collected pellets and coprocoenoses; however, other owls large enough to prey on small mammals are known to occur in the study area, including: Verreaux's Eagle Owl, *Bubo lacteus*; the African Wood Owl, *Strix woodfordii nigricantior*; and the Grass Owl, *Tyto capensis*. The first two are reported to roost in tree crowns, while grass owls prefer to ground roost in wet grasslands (Vernon, 1972; Fry et al., 1988). It is possible that all of these species may have contributed fauna to roosts that have been attributed to *Bubo africanus*, though it is considered unlikely for various reasons. First, *Bubo africanus* is the most abundant, exposed-roosting owl in the

study area based on the observations made during this study. Second, the prey items are all in the size range expected for *Tyto alba affinis* and *Bubo africanus* with no evidence of larger species, such as hedgehogs (*Erinaceus albiventris*) that are preferred prey of *Bubo lacteus* (Fry et al., 1988). For these reasons it is reasonable to presume that cavity roosts contain prey primarily accumulated by barn owls, and that exposed roosts are primarily the work of spotted eagle owls.

Micromammal specimens were iteratively sorted with the aid of printed and digital identification keys (Davis, 1965; Foster and Duff-Mackay, 1966; Coetzee, 1972; Delany, 1975; Rogers and Stanley, 2003). Final taxonomic assignments were made by comparison with collections at the American Museum of Natural History, New York NY (AMNH); Field Museum of Natural History, Chicago IL (FMNH); National Museum of Natural History, Washington DC (NMNH); and the Zoologisches Forschungsinstitut und Museum Alexander Koenig, Bonn, Germany (ZFMK). The assemblages include many complete or partially complete skulls, but most taxa are also represented by isolated teeth. Identification relied primarily on discrete dental characteristics of the molars in order to maintain a consistent pattern of taxonomic assignment across specimens with different preservation. Comparisons were made against all taxa known to occur in the subregion as reported in Davies and Berghe (1994) and Wilson and Reeder (1993). The taxonomic classification used here follows that of Wilson and Reeder (1993). Shrews were identified only by maxillary specimens because the mandibles and lower dentition of some genera cannot be distinguished readily. The analysis is conducted at the generic level as this is the lowest common ranking at which all specimens can be identified accurately and efficiently from discrete diagnostic criteria. An exception is made for bats, which are identified to suborder only, and *Mus* for which two subgenera, *Mus* and *Nannomys*, are readily diagnosed. Statistical analyses were conducted using the R statistical package (Ihaka and Gentleman, 1996; Maindonald and Braun, 2003).

## SERENGETI VEGETATION AND ROOST HABITATS

Roost specific habitat analysis included the area within a 1.5-km radius surrounding each roost. The 1.5-km analysis radius covers approximately 707 ha and is based on ranging behavior of *Tyto alba* in North American and European telemetry studies (Colvin, 1984; Taylor, 1994). No studies have been conducted yet on the ranging behavior of either *Tyto* or *Bubo* species in Africa. Land cover data were compiled from multiple sources as part of a combined project on Serengeti vegetation mapping (Reed, 2003). The principal land cover data are derived from over 800 spot surveys of vegetation throughout the ecosystem. These data were combined with Landsat 7 ETM + satellite imagery to produce a detailed land cover map of the entire Serengeti-Mara ecosystem. Additional vegetation and land cover data derive from published maps and unpublished databases. From these data, details of vegetation cover, precipitation, and topographic heterogeneity could be quantified for the areas surrounding each roosting site.

Generally, the distributions of woody and herbaceous plant cover across the ecosystem follow a pattern resulting from three levels of influence: climate, topography, and disturbances (Pratt and Gwynne, 1977). These factors are largely interdependent and in concert produce ecological gradients and repeated patterns of land form. The most important gradients relevant to land cover include a north-by-northwest rainfall gradient (Figure 2) with the lowest mean annual precipitation (ca. 400 mm) and a more unimodal pattern of annual rainfall found at the heart of the rain shadow just northwest of the Ngorongoro highlands and trending toward higher precipitation (ca. 1200 mm) with a more bimodal pattern in the north (Norton-Griffiths et al., 1975). Much of the southern Serengeti ecosystem is blanketed by natrocarbonatitic ash from Pleistocene and Holocene eruptions of nearby volcanoes (Dawson, 1963; Hay, 1976). The pattern of ashfall from the eruptions followed the prevailing winds, the same factor inducing the rainfall gradient. Thus, there is a compound gradient in precipitation, topographic heterogeneity, soil mineral composition, and soil depth. Local variation in edaphic conditions due to topography, i.e., soil catenas (Milne, 1935; Jager, 1982), augment the compound gradient and influence soil texture, mineral composition, and soil moisture availability over short distances (ca. 10–100 m).

Disturbance factors, such as fire, grazing, browsing, and burrowing, have important local influence on plant species composition and community structure (Bell, 1969, 1982; McNaughton, 1983; Dublin and Douglas-Hamilton, 1987; Dublin et al., 1990; McNaughton and Banyikwa, 1995; Sinclair, 1995a), but seem to be secondary determinants of woody/grass ratios compared with more pervasive climatic factors (e.g., precipitation, temperature, winds) or edaphic factors (including soil moisture availability, mineral composition, texture) (Belsky, 1990, 1995; Coughenour and Ellis, 1993).

The analyzed collections are distributed along the gradient in different land cover zones as shown in Figure 2. Roosts 24 and 7 are located to the south, in the short- to mid-grass

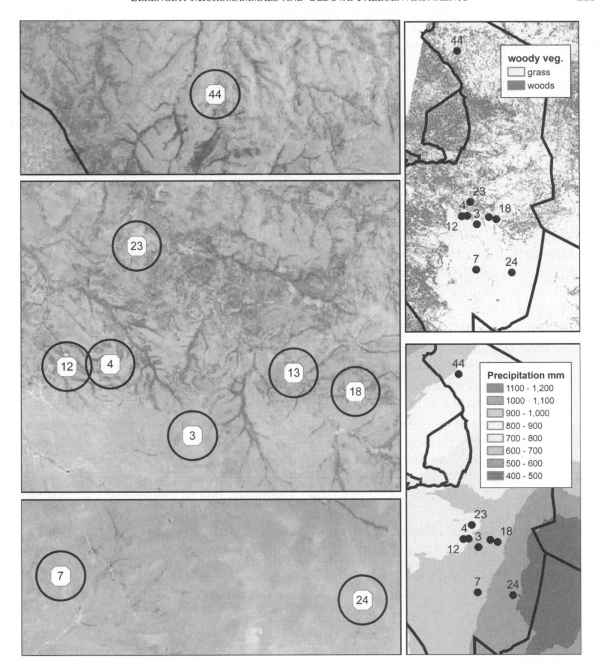

Figure 2. Distribution of analyzed roost sites. The upper right pane shows the roosts against a background map of woody vegetation, and against precipitation in the lower right. Remaining panes show close-ups of roosts outlined by 1.5-km buffer against a Landsat background. The background includes a semi-transparent overlay of the vegetation classification to highlight woody vegetation.

plains. Roost 3 is located in shrubbed grasslands, at the transition from grasslands to woodlands. Roosts 13 and 18 are to the east in the catchment of the Ngare Nanyuki River where deeper soils support woody vegetation, but under relatively low rainfall due to adjacent hills. Two roosts, 4 and 12, are in the vicinity

of the Serengeti Wildlife Research Center and Seronera. This area is a mosaic of grasslands and stands of dense woodland. Further north, roost 23 is firmly established in the shrubbed woodlands. Far to the north, roost 44 lies along the tributaries of the Mara River system in tall, moist grasslands adjacent to dense-canopy, evergreen forest and marsh. Mean annual precipitation at the roosting sites increases from its lowest value of 527 mm at roost 24 in the southern grasslands up to 886 mm at roost 44 in the northern extension (Norton-Griffiths et al., 1975). Table 2 summarizes climatic and land cover attributes for the different roosts.

## FAUNAL DATA

Complete listings of the mammalian fauna and frequencies for each taxon as number of identified specimens (NISP), and minimum number of individuals (MNI) are given in Tables 3 and 4. Since shrews were tabulated only by maxilla, an adjusted NISP value, NISPn, was calculated by dividing the raw NISP values for each taxon by the expected number of elements for that taxon. The number of expected elements for most taxa was three, a skull and two mandibles. Shrews were identified by maxillae only, so their expected value is one. The minimum number of individuals, MNI, was tabulated by counting the number of occurrences of each dental element (e.g., left $M^1$) and taking the maximum value across all elements as the MNI value for that taxon. No attempt was made to separate specimens based on wear stage, but side was taken into account. Generally MNI values were used for most statistical procedures, while NISPn is used for comparisons of relative abundance (Grayson, 1984).

A pie chart showing relative abundances of the major groups across all assemblages is given in Figure 3. The fauna is dominated by small mammals though passerine birds, insects, and reptiles were also observed. Overall, crociduran shrews are the most abundant mammalian taxon (30%). The abundance of *Crocidura* results from the patterning of diversity between generic and species ranks in shrews and rodents. Among the 59 species of soricid shrews in East Africa, 42 belong to the genus *Crocidura*. A different pattern occurs in rodents where 113 species are distributed among 40 genera and none have more than 10 species (Davies and Berghe, 1994). Thus, much of the biodiversity in shrews occurs at the species level, while

Table 2. *Summary of ecological and land cover characteristics at the roosts. Elevation values are given in meters, precipitation in millimeters, slope as percent. Land cover is given in percent pixels for an equal area around each roost. Dashes indicate an absence of that land cover; pluses indicate the land cover is present at less than 0.5%. The final column shows the average across all roosts. Land cover densities are coded as s = sparse, o = open, d = dense, and c = closed*

| | | Roost no. | | | | | | | | | |
| | | 24 | 7 | 3 | 18 | 13 | 12 | 4 | 23 | 44 | Avg. |
|---|---|---|---|---|---|---|---|---|---|---|---|
| Vegetation summary | Codes | | | | | | | | | | |
| Mean elevation | ELEV | 1759 | 1632 | 1583 | 1607 | 1562 | 1509 | 1536 | 1472 | 1465 | 1533 |
| Elevation (s.d.) | SELE | 10 | 8 | 6 | 11 | 11 | 8 | 13 | 23 | 18 | 13 |
| Mean annual precipitation | MAP | 527 | 614 | 679 | 635 | 655 | 708 | 702 | 709 | 886 | 711 |
| Mean percent slope | MPS | 1.5 | 1.3 | 1.1 | 2.5 | 2.8 | 2.1 | 2.5 | 4.6 | 4.9 | 2.9 |
| Percent slope (s.d.) | PSD | 0.6 | 0.5 | 0.4 | 1.0 | 1.3 | 0.8 | 1.1 | 3.5 | 2.2 | 1.5 |
| Percent woody veg. cover | PWV | 0 | 0 | 2 | 26 | 20 | 35 | 33 | 28 | 28 | 25 |
| Percent land cover | | | | | | | | | | | |
| Sparse-open grassland | soG | 13% | 27% | + | 1% | 1% | 8% | 3% | 7% | 1% | 4% |
| Dense-closed grassland | dcG | 75% | 70% | 20% | 40% | 12% | 9% | 13% | 15% | 19% | 25% |
| s-o Bushed grassland | soBG | + | 0% | 21% | 29% | 38% | 39% | 41% | 26% | 43% | 30% |
| d-c Bushed grassland | dcBG | 13% | 3% | 58% | 4% | 29% | 9% | 9% | 23% | 10% | 19% |
| o-d Grassed bushland | odGB | + | + | 2% | 23% | 19% | 27% | 31% | 23% | 26% | 19% |
| d-c Grassed bushland | dcGB | – | – | – | 3% | 1% | 8% | 2% | 5% | 2% | 3% |
| d-c Forest | F | – | – | – | – | – | – | – | – | + | + |

*Table 3. Taxonomic representation presented as the number of identified specimens (NISP). Taxa are grouped by order and the Rodentia are further subdivided into subfamilies*

| Taxa | Roost no. | | | | | | | | | Total NISP |
| | 3 | 4 | 7 | 12 | 13 | 18 | 23 | 24 | 44 | |
|---|---|---|---|---|---|---|---|---|---|---|
| **Order Insectivora** | | | | | | | | | | |
| *Crocidura* | 114 | 91 | 76 | 9 | 66 | 47 | 5 | 75 | 27 | 510 |
| *Suncus* | 8 | 0 | 0 | 0 | 4 | 1 | 0 | 0 | 0 | 13 |
| **Order Macroscelidea** | | | | | | | | | | |
| *Elephantulus* | 0 | 0 | 0 | 0 | 1 | 1 | 4 | 0 | 0 | 6 |
| **Order Chiroptera** | | | | | | | | | | |
| *Microchiroptera* gen. Indet. | 5 | 0 | 0 | 0 | 0 | 4 | 2 | 0 | 0 | 11 |
| **Order Rodentia** | | | | | | | | | | |
| Subfamily Murinae | | | | | | | | | | |
| *Acomys* | 1 | 0 | 0 | 1 | 3 | 2 | 0 | 0 | 2 | 9 |
| *Aethomys* | 0 | 0 | 0 | 0 | 0 | 0 | 0 | 1 | 16 | 17 |
| *Arvicanthis* | 15 | 79 | 23 | 0 | 4 | 11 | 48 | 2 | 68 | 250 |
| *Dasysmys* | 0 | 0 | 0 | 0 | 0 | 0 | 0 | 0 | 12 | 12 |
| *Lemniscomys* | 8 | 8 | 14 | 1 | 4 | 11 | 14 | 5 | 1 | 66 |
| *Mastomys* | 27 | 140 | 8 | 1 | 9 | 6 | 56 | 8 | 34 | 289 |
| *Mus (Mus)* | 69 | 37 | 94 | 3 | 29 | 33 | 15 | 11 | 0 | 291 |
| *Mus (Nannomys)* | 47 | 39 | 9 | 0 | 46 | 61 | 8 | 2 | 4 | 216 |
| *Praomys* | 1 | 0 | 0 | 0 | 0 | 4 | 0 | 0 | 1 | 6 |
| *Thallomys* | 18 | 9 | 0 | 10 | 11 | 10 | 85 | 0 | 0 | 143 |
| *Zelotomys* | 13 | 22 | 11 | 0 | 17 | 1 | 1 | 0 | 1 | 66 |
| Subfamily Cricetomyinae | | | | | | | | | | |
| *Saccostomus* | 18 | 24 | 10 | 1 | 8 | 17 | 36 | 0 | 0 | 114 |
| Subfamily Dendromurinae | | | | | | | | | | |
| *Dendromus* | 67 | 104 | 60 | 3 | 79 | 23 | 7 | 52 | 10 | 405 |
| *Steatomys* | 97 | 137 | 182 | 8 | 216 | 44 | 3 | 309 | 22 | 1018 |
| Subfamily Gerbillinae | | | | | | | | | | |
| *Gerbillus* | 8 | 2 | 142 | 0 | 48 | 3 | 0 | 107 | 0 | 310 |
| *Tatera* | 2 | 15 | 33 | 0 | 13 | 8 | 7 | 223 | 10 | 311 |
| **Total** | **518** | **707** | **662** | **37** | **558** | **287** | **291** | **795** | **208** | **3401** |
| Subtotal Insectivora | 122 | 91 | 76 | 9 | 70 | 48 | 5 | 75 | 27 | 523 |
| Subtotal Rodentia | 391 | 616 | 586 | 28 | 487 | 234 | 280 | 720 | 181 | 2937 |
| Subtotal Murinae | 199 | 334 | 159 | 16 | 123 | 139 | 227 | 29 | 139 | 1365 |
| Subtotal Dendromurinae | 164 | 241 | 242 | 11 | 295 | 67 | 10 | 361 | 32 | 1423 |
| Subtotal Gerbillinae | 10 | 17 | 175 | 0 | 61 | 11 | 7 | 330 | 10 | 621 |

in rodents the diversity is partitioned among genera. At the ordinal level, rodents make up 69% of the assemblage.

Overall, the fauna recovered from the modern assemblages is consistent with a tropical mosaic of grassland and woodland, albeit one in which the larger, more diurnal taxa are underrepresented. A bias toward nocturnal species is expected for an owl-accumulated assemblage. Looking more closely, Tables 3 and 4 indicate differences between roosts. For example, the genus *Thallomys* represents 18 and 28% of the fauna at roosts 12 and 23 respectively but less than 3% elsewhere. Gerbils make up 21 and 35% of the fauna at roosts 7 and 24 but are rare at the other roosts. Before exploring the cause for these differences it is useful to review the habitat proclivities of the different taxa involved. For this purpose, habitat preferences are enumerated into niche models as described below. Following the discussion of niche models comes a more thorough look at the abundance patterns between roosting sites in different habitats using correspondence

*Table 4. Taxonomic representation presented as the minimum number of individuals (MNI). Taxa are grouped by order and the Rodentia are further subdivided into subfamilies*

| Taxa | Roost no. | | | | | | | | | Total MNI |
| --- | --- | --- | --- | --- | --- | --- | --- | --- | --- | --- |
| | 3 | 4 | 7 | 12 | 13 | 18 | 23 | 24 | 44 | |
| **Order Insectivora** | | | | | | | | | | |
| *Crocidura* | 104 | 89 | 76 | 9 | 63 | 40 | 5 | 73 | 27 | 486 |
| *Suncus* | 8 | 0 | 0 | 0 | 4 | 1 | 0 | 0 | 0 | 13 |
| **Order Macroscelidea** | | | | | | | | | | |
| *Elephantulus* | 0 | 0 | 0 | 0 | 1 | 1 | 2 | 0 | 0 | 4 |
| **Order Chiroptera** | | | | | | | | | | |
| *Microchiroptera* gen. Indet. | 4 | 0 | 0 | 0 | 0 | 3 | 1 | 0 | 0 | 8 |
| **Order Rodentia** | | | | | | | | | | |
| Subfamily Murinae | | | | | | | | | | |
| *Acomys* | 1 | 0 | 0 | 1 | 2 | 1 | 0 | 0 | 1 | 6 |
| *Aethomys* | 0 | 0 | 0 | 0 | 0 | 0 | 0 | 1 | 5 | 6 |
| *Arvicanthis* | 5 | 21 | 7 | 0 | 3 | 4 | 14 | 1 | 17 | 72 |
| *Dasysmys* | 0 | 0 | 0 | 0 | 0 | 0 | 0 | 0 | 4 | 4 |
| *Lemniscomys* | 4 | 3 | 7 | 1 | 2 | 5 | 6 | 4 | 1 | 33 |
| *Mastomys* | 9 | 35 | 3 | 1 | 4 | 2 | 21 | 4 | 10 | 89 |
| *Mus (Mus)* | 21 | 14 | 37 | 2 | 11 | 9 | 7 | 5 | 0 | 106 |
| *Mus (Nannomys)* | 20 | 13 | 4 | 0 | 13 | 21 | 3 | 1 | 1 | 76 |
| *Praomys* | 1 | 0 | 0 | 0 | 0 | 2 | 0 | 0 | 1 | 4 |
| *Thallomys* | 7 | 2 | 0 | 5 | 4 | 3 | 26 | 0 | 0 | 47 |
| *Zelotomys* | 3 | 6 | 2 | 0 | 5 | 1 | 1 | 0 | 1 | 19 |
| Subfamily Cricetomyinae | | | | | | | | | | |
| *Saccostomus* | 6 | 7 | 5 | 1 | 2 | 7 | 11 | 0 | 0 | 39 |
| Subfamily Dendromurinae | | | | | | | | | | |
| *Dendromus* | 24 | 33 | 21 | 2 | 26 | 9 | 4 | 17 | 4 | 140 |
| *Steatomys* | 32 | 44 | 61 | 4 | 64 | 14 | 2 | 93 | 7 | 321 |
| Subfamily Gerbillinae | | | | | | | | | | |
| *Gerbillus* | 3 | 1 | 44 | 0 | 15 | 3 | 0 | 29 | 0 | 95 |
| *Tatera* | 1 | 7 | 10 | 0 | 4 | 3 | 3 | 66 | 3 | 97 |
| **Total** | **253** | **275** | **277** | **26** | **223** | **129** | **106** | **294** | **82** | **1665** |
| Subtotal Insectivora | 112 | 89 | 76 | 9 | 67 | 41 | 5 | 73 | 27 | 499 |
| Subtotal Rodentia | 137 | 186 | 201 | 17 | 155 | 84 | 98 | 221 | 55 | 1154 |
| Subtotal Murinae | 71 | 94 | 60 | 10 | 44 | 48 | 78 | 16 | 41 | 462 |
| Subtotal Dendromurinae | 56 | 77 | 82 | 6 | 90 | 23 | 6 | 110 | 11 | 461 |
| Subtotal Gerbillinae | 4 | 8 | 54 | 0 | 19 | 6 | 3 | 95 | 3 | 192 |

analysis. Sampling issues are also considered, with attention given to both sample size effects and the influence of different owl species that accumulate the assemblages. These steps lay the groundwork for the testing and calibration of two common analytical methods, taxonomic ratios, and taxonomic habitat indices (THI). The modern data are examined with each of these techniques and the results compared against the actual habitats surrounding each roost.

## MICROMAMMAL NICHE MODELS

Rather than provide lengthy written description of the taxa, condensed numerical summaries of habitat preference are given in the form of niche models. Niche models, though not termed as such, were developed as a component of the taxonomic habitat index (THI) by Nesbit-Evans et al. (1981), who described an animal's habitat preference using five major tropical habitat types: forest, woodland–bushland,

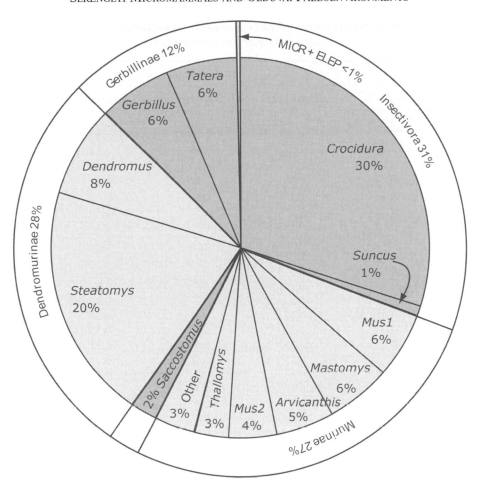

Figure 3. Pie chart of relative abundances (% NISPn) of all mammalian taxa from all roosts combined.

grassland, desert and semi-desert, and wet or swamp habitats. An animal's total habitat use was partitioned into these classes based on published descriptions of the taxon by Meester and Setzer (1971) and Kingdon (1974a, b). For example, Nesbit-Evans et al. (1981, p. 102) score the African elephant, "0.33 forest, 0.33 woodland–bushland, 0.23 grassland, and 0.11 semi-desert." This set of closed-sum numerical weights for a given taxon is what I call a niche model. Averaging the niche model values for all taxa appearing in a habitat gives the THI. A variant of THI uses averages weighted by the taxon abundance (Andrews, 1990).

A shortfall of this method is that the inventors provide no clear guidelines on how to distribute the values in the niche model across the habitat classes. Theoretically a niche model

should be based on probability of a particular habitat (say, Habitat 1) for a given taxon (Taxon A); $p(H_1|T_A)$. This probability could be ascertained by systematic survey of museum collection records, a task that will only become practical as records and field collection notes are digitized. Quantitative assessments of habitat use are often available in the literature, though care should be taken not to confuse habitat indication with habitat association. The former is given by the probability above and is the appropriate niche estimator for paleoenvironmental interpretation. Habitat association asks what is the probability of finding Taxon A, given a particular habitat, $p(T_A|H_1)$. The two are clearly different. A rare and endemic species may be unique to a single habitat, thus $p(H_1|T_A) = 1.00$, but because it occurs

Table 5. Niche models for Serengeti rodents based on data from Fernandez-Jalvo et al. (1998) except those taxa indicated by an asterisk (*)

| | Land cover and ranks | | | | |
|---|---|---|---|---|---|
| | Forest | Woodland | Bushland | Grassland | Semi-Arid |
| **Taxa** | 5 | 4 | 3 | 2 | 1 |
| *Arvicanthis* | 0 | 0 | 0.25 | 0.75 | 0 |
| *Aethomys* | 0.18 | 0.25 | 0.4 | 0.18 | 0 |
| *Mastomys* | 0 | 0.33 | 0.33 | 0.33 | 0 |
| *Mus* | 0.35 | 0.19 | 0.26 | 0.2 | 0 |
| *Oenomys* | 0.5 | 0.5 | 0 | 0 | 0 |
| *Pelomys* | 0 | 0 | 0.5 | 0.5 | 0 |
| *Thallomys* | 0 | 0.5 | 0.5 | 0 | 0 |
| *Grammomys* | 0.4 | 0.35 | 0.2 | 0 | 0.05 |
| *Zelotomys* | 0 | 0 | 0.2 | 0.7 | 0.1 |
| *Gerbillus* | 0 | 0 | 0.2 | 0.2 | 0.6 |
| *Tatera* | 0 | 0 | 0.4 | 0.6 | 0 |
| *Steatomys* | 0.2 | 0.2 | 0.2 | 0.2 | 0.2 |
| *Dendromus* | 0.05 | 0.27 | 0.4 | 0.28 | 0 |
| *Saccostomus* | 0 | 0.33 | 0.33 | 0.33 | 0 |
| *Otomys* | 0 | 0.25 | 0.5 | 0.25 | 0 |
| *Xerus* | 0 | 0.33 | 0.66 | 0 | 0 |
| *Heterocephalus* | 0 | 0.2 | 0.3 | 0.4 | 0.1 |
| *Acomys** | 0 | 0.17 | 0.17 | 0.17 | 0.5 |
| *Dasysmys** | 0 | 0 | 0.2 | 0.8 | 0 |
| *Lemniscomys** | 0 | 0.1 | 0.1 | 0.5 | 0.3 |
| *Praomys** | 0.8 | 0.2 | 0 | 0 | 0 |

in low abundance the probability of finding that species in Habitat 1, may be as low as 1 in 100, i.e., $p(H_1|T_A) = 0.01$. The trapping records associated with most museum specimens provides a first approximation of habitat preferences, but due to predator preferences and other taphonomic factors it is better to calculate the niche model from coprocoenoses, such as those presented here.

Nesbit-Evans et al. (1981) focused on large mammals, but recently niche models were compiled for Olduvai fossil micromammals by Fernandez-Jalvo et al. (1998). Their niche models are reproduced here in Table 5. These authors use a slightly different set of habitat categories than did Nesbit-Evans et al. There are still five classes: forest, woodland, bushland, grassland, and semi-arid. The Aquatic-swamp category has been dropped and the woodland–bushland category split. The Olduvai fossil microfauna overlaps greatly with that of the modern Serengeti coprocoenoses, leaving only four genera that were not covered

by Fernandez-Jalvo et al. (1998) and for which new niche models were needed: *Acomys*, *Dasymys*, *Lemniscomys*, and *Praomys*.

To build the new niche models and validate the models developed by Fernandez-Jalvo et al. (1998), autecological summaries were compiled for all Serengeti genera based on Kingdon (1974a, b); and verified against other published descriptions (Vesey-Fitzgerald, 1966; Delany, 1972, 1986; Hubbard, 1972; Andrews et al., 1975; Avery, 1982). A summary is given in Table 6. Bats are excluded because they are very rare in owl assemblages and are not as intimately associated with land cover as are their non-volant cousins. Shrews are listed in the table, but they too are excluded from analyses because either too little is known about their autecology in East Africa (as is the case for *Suncus*) or there are too many species within the genus for it to be informative (*Crocidura*). The niche index values that appear in Table 6 are derived from the niche models for each taxon ($t_i$) as

$$Niche\ index\,(t_i\,) = \sum_{j=1}^{5} (R_j \bullet W_j) \qquad (1)$$

where $R_j$ is the habitat rank (1 for semi-arid, 2 for grassland, etc.) and $W_j$ is the weighting for the taxon in the habitat. For example, *Arvicanthis* has a value of 2.25 = (5 × 0) + (4 × 0) + (3 × 0.25) + (2 × 0.75) + (1 × 0). The niche index is used to give a simple ranking of habitat preferences with larger values indicating preference for more moist/closed habitats. Comparing the niche index against the habitat summary in the adjacent column of Table 6 serves as a check of the niche models.

Generally the summary of habits and habitat preferences given in Table 6 agree with the niche models developed by Fernandez-Jalvo et al. (1998) listed in Table 5. Gerbils (*Gerbillus* and *Tatera*) are characterized in the literature as arid adapted based on geographic distribution and habitat use within that distribution. Thus, they are expected to have a low niche index and heavy weighting on semi-arid and grassland habitats in the niche model, which they do. Similarly, most of the other taxa are ordered sensibly, indicating that the niche models proposed by Fernandez-Jalvo et al. (1998) are reasonable. However, there are a few incongruencies.

*Dasymys* has a low niche index value because it favors grassland habitats; however, it is specific to wetlands and marshes. Similarly, *Arvicanthis* favors moist grasslands or woodlands with a moist-grass understory. Their position in part reflects the limitations of a simple linear index that combines dry and wet grasslands.

*Heterocephalus* is particularly indicative of arid environments (Kingdon, 1974a, b), yet it ranks rather high on the niche index due to partial weighting in woodland environments. Temperature and soil characteristics are probably more critical than woody vegetation cover or moisture, and again this taxon may differ on an independent axis of variation that is distorted by the linear niche scale.

Table 6. *Body size and habitat summaries derived from Kingdon (1974) and others (see text). The niche index is a summary of habitat preference and ranges from open/xeric (1) to closed/mesic (5)*

| Taxa | Approx. body mass (g) | Habitat | Niche index |
|---|---|---|---|
| *Acomys* | 23 | Dry sav.—rocky | 1.86 |
| *Arvicanthis* | 78 | Grassl/dry–moist sav. | 2.25 |
| *Aethomys* | 100 | Dry–moist savanna | 3.03 |
| *Dasysmys** | 103 | Marsh/moist grassland | 2.2 |
| *Grammomys* | 42 | Dry–moist sav./sec. growth | 4.05 |
| *Lemniscomys** | 55 | Grassl./dry–moist sav. | 1.9 |
| *Mastomys* | 50 | Dry–moist savanna | 2.64 |
| *Mus (Mus)* | 10 | Dry savanna | 3.15 |
| *Mus (Nannomys)* | 12 | Dry–moist savanna | 3.15 |
| *Oenomys* | 90 | Sec. growth | 4.5 |
| *Pelomys* | 68 | Grass/sec. growth | 2.5 |
| *Praomys* | 35 | Sec. growth/forest | 3.8 |
| *Thallomys* | 68 | Dry–moist savanna | 3 |
| *Zelotomys* | 60 | Dry–moist savanna | 2.1 |
| *Otomys* | 157 | Grassl/sec. growth | 3 |
| *Saccostomus* | 63 | Dry savanna | 2.64 |
| *Dendromus* | 12 | Grassl./dry–moist sav. | 2.77 |
| *Steatomys* | 37 | Grassl./dry savanna | 2.6 |
| *Gerbillus* | 38 | Grassland | 1.6 |
| *Tatera* | 128 | Grassl./dry savanna | 2.4 |
| *Xerus* | 622 | Grassl./dry savanna | 3.3 |
| *Heterocephalus* | 55 | Semi-arid | 2.6 |

For *Steatomys*, Fernandez-Jalvo et al. (1998) propose a catholic habitat distribution with equal rankings in every habitat class. However, Kingdon (1974b) attributes this species to more xeric environments, and even at the most mesic edge of their distribution they favor moist woodland or "savanna" but not forest (Genest-Villard, 1979).

A revised nich model, making use of the comments above is beyond the scope of this chapter. For comparability I retain the niche models employed by Fernandez-Jalvo et al. (1999) but with the caveats described above.

## PATTERNS IN THE DISTRIBUTION AND ABUNDANCE OF MICROMAMMALS BETWEEN ROOSTS

The niche models give *a priori* expectations as to where certain taxa should be most abundant. This section examines patterns of relative abundance between roosts to see if these expectations are manifested. The first question to address is whether relative abundance is significantly different from one roost to the next on a taxon-by-taxon basis. Here, a statistical test of independence using the Pearson's $\chi^2$ (often called a chi-square test), examines whether changes in relative abundance of a taxon are significantly greater than would be expected by chance alone. The statistic examines a contingency table of abundance values (in this case MNI) and tests whether the observed values differ from what one would expect given the size of the sample. Or put another way, it tests whether the abundance of the taxon is independent of the roost examined. The proportion of a taxon at any one roost is tested against its global proportion across all roosts. This approach mitigates the effect of sample size differences between roosts. The Pearson statistic is compared against a chi-square probability distribution (Sokal and Rohlf, 1995). Significant values indicate the taxon is strongly associated with at least one roost, and not randomly distributed across the roosts. Table 7 lists the results. Most taxa are not randomly distributed; however bats, *Elephantulus*, *Acomys*, *Lemniscomys*, *Praomys*, *Zelotomys*, *Dendromus* could not be distinguished from chance in their distribution across roosting sites and these taxa are thus excluded from subsequent analysis. The remaining 13 taxa were found to differ significantly from expected abundances.

The pattern of faunal associations is demonstrated on a per-roost basis using correspondence analysis (CA). Correspondence analysis is a multivariate ordination technique that maps the 13 dimensions of variability (one dimension for each taxon found significant in Table 7) onto a two-dimensional space. The analysis examines covariance between each of the dimensions and attempts to preserve the spatial relationships between the points (Greenacre and Vrba, 1984; Johnson and Wichern, 2002). The technique uses a chi-square distance measure and is thus amenable to count and frequency data among samples occurring along an ecological gradient (McCune and Grace, 2002). The technique also allows roosts to be mapped in the same space with the taxa so one can easily visualize both sets of variables. Taxon points that are closer together in the CA plot tend to appear together at the same roosts.

A contingency table of NISPn values is used for the CA to best represent the relative abundance of taxa (Figure 4). However, one obtains similar results using MNI, %MNI, or %NISPn. The first axis of variation splits roosts along a general gradient of dry/open to the left and more wet/closed roosts to the right. Taxa such as *Gerbillus*, *Steatomys*, and *Tatera* appear to the left, near grassland roosts 24 and 7. Shrubbed or partially wooded habitats in the dryer zones at roosts 3, 13, 18 are near the origin. Those roosts with the greatest woody cover and higher precipitation (4, 12, and 23) are to the right of the origin and associated with *Saccostomus* and *Thallomys*.

Table 7. Each row shows the results for a test of the independence of the taxon across all roosts, significant results indicate that the abundance of the taxon is not random from one roost to another. The critical value of 23.774 corresponds to an experiment-wide alpha of 0.5 adjusted for 20 unplanned comparisons; alpha = 0.05/20 = 0.0025, with eight degrees of freedom. NS = not Significant, * significant at alpha > 0.001, ** significant at alpha < 0.001

| Taxa | Chi-square | Probability | Result |
|---|---|---|---|
| Crocidura | 53.8 | 0.00010 | ** |
| Suncus | 29.8 | 0.00020 | * |
| Elephantulus | 16.8 | 0.03180 | NS |
| Microchiroptera gen. indet. | 21.8 | 0.00540 | NS |
| Acomys | 16.3 | 0.03770 | NS |
| Aethomys | 79.8 | 0.00010 | ** |
| Arvicanthis | 104.1 | 0.00010 | ** |
| Dasymys | 77.4 | 0.00010 | ** |
| Lemniscomys | 14.2 | 0.07750 | NS |
| Mastomys | 111.2 | 0.00010 | ** |
| Mus (Mus) | 42.3 | 0.00010 | ** |
| Mus (Nannomys) | 70.3 | 0.00010 | ** |
| Praomys | 15.7 | 0.04740 | NS |
| Thallomys | 231.9 | 0.00010 | ** |
| Zelotomys | 9.4 | 0.31280 | NS |
| Saccostomus | 47.0 | 0.00010 | ** |
| Dendromus | 15.6 | 0.04900 | NS |
| Steatomys | 84.8 | 0.00010 | ** |
| Gerbillus | 103.1 | 0.00010 | ** |
| Tatera | 183.1 | 0.00010 | ** |

The taxa *Dasymys* and *Aethomys* are unique (or nearly so) to roost 44. *Dasymys* occurs only at this roost. *Aethomys* is most abundant here but a single specimen was also observed at roost 24. The uniqueness of roost 44 is most strongly expressed on Axis 2, indicating that it differs in its own way from roosts 24 and 23. *Mastomys* and *Arvicanthis* fall between roosts 23 and 44. These taxa share identical patterns of association. The remaining taxa are clustered about the origin.

taxon) and sample size (total NISP) for all taxa using both Pearson's and Kendall's rank correlation tests. NISP is appropriate here because the tests examine rank order of sample size, the size of the samples themselves does not affect the significance of the result. Given the large number of tests (13), a conservative significance level should be set at an alpha of 0.05/13 = 0.0038. At this level none of the taxa show patterns of relative abundance that are correlated with sample size.

## SAMPLE SIZE EFFECTS

Grayson (1984) notes that faunal abundances may covary with sample sizes. The exact causes are not consistent across data sets but it is often the case that small samples have biased relative abundance values. As a simple precaution, tests were made on the correlation between relative abundance (%NISP for each

## PREDATOR EFFECTS

As mentioned earlier, the pattern of roost occupation shows a consistent segregation between two owl species. The barn owls, *Tyto alba affinis*, were commonly observed inside cavities, such as small vertical fissures of granitic rock outcroppings (kopjes) or the hollowed interior of baobab trees, *Adansonia digitata*. Alternately,

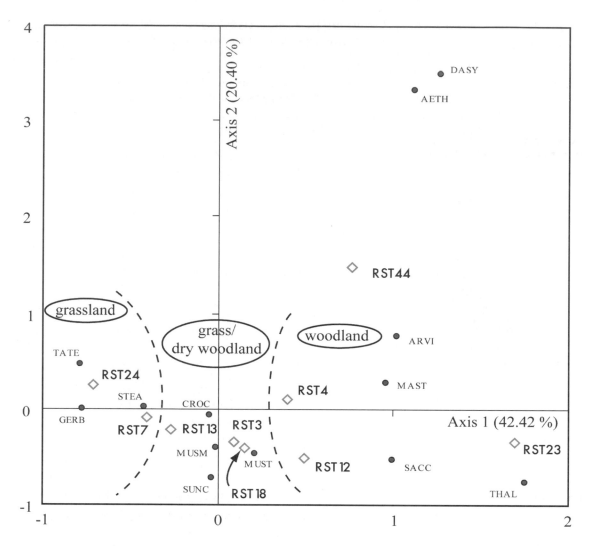

Figure 4. Correspondence analysis based on NISPn, Axes 1 and 2. The first two axes explain 62.8% of the inertia. Diamonds indicate roosts, and circles indicate taxa. Taxon codes are from the first four letters of the taxon name.

spotted eagle owls, *Bubo africanus* were found in exposed circumstances such as the sparse crowns of *Acacia* trees or roosting on the ground near rocks (Reed, 2003). These observations agree with other published reports on the roosting habits of these species (Brain, 1981; Demeter, 1982; Taylor, 1994). If the predators have very different prey preferences or hunting habits then strong differences are expected between assemblages even when they are accumulated in the same habitat. However, an analysis of the prey composition in regard to roost type (and hence predator species), when habitat is constrained, showed identical faunal composition (presence/absence)

between cavity and exposed roosts, but differences in the relative abundance of some taxa (Reed, 2005). Significant differences occurred in the relative abundance of *Lemniscomys*, and a strong shift in the rank abundance of *Thallomys*. In both cases these taxa were more common at exposed (eagle owl) roost than at cavity roosts. Additionally, exposed roosts demonstrated a significantly greater mean prey body mass than the cavity roosts. The prevalence of the arboreal, *Thallomys*, at exposed roosts is consistent with opportunistic predation by a tree-roosting eagle owl, and both *Thallomys* and *Lemniscomys* are larger taxa that may be preferred by the larger predator, the eagle owl. However, theses differences are insufficient to explain the patterns of faunal composition seen between roosts in different habitats.

## SUMMARY OF FAUNAL ABUNDANCE PATTERNS

The CA plot gives a graphical representation of the relationships between species and roosts. The overall pattern of faunal distribution is in agreement with the niche models developed from the literature and presented in Tables 5 and 6. The dry, open grassland roosts (7 and 24) have the most significant associations with *Gerbillus*, the most arid-adapted taxon with the lowest niche index value (1.6). Both roosts also share strong negative associations with *Mastomys*, *Mus* (*Nannomys*), and *Thallomys*, all of which have high niche index values (2.64, 3.15, and 3 respectively). At the other extreme the mesic woodland roosts have different constellations of fauna, but all composed of species with more mesic habits such as *Thallomys* (niche index = 3), *Saccostomus* (n. i. = 2.64), *Mastomys* (n. i. = 2.64). Arid adapted fauna such as *Gerbillus* are negatively associated.

Roosts 3 and 44 do not neatly fit this pattern. Roost 3 is a grassland roost but at higher precipitation than 7 or 24 and right on the edge of the transition to woodlands. This roost is dominated by *Crocidura* and has very negative associations with the arid grassland fauna. Absent more detailed data on the habits of crociduran shrews this result remains enigmatic.

Roost 44 shows a negative association with dry grassland taxa consistent with its position at the wettest portion of the precipitation gradient. Semi-aquatic taxa such as *Dasymys* appear here as is expected since tributaries of the Mara River provide a perennial source of water. Likewise moist grassland taxa such as *Arvicanthis* are present beside other mesic taxa such as *Mastomys* and *Aethomys*. However, arboreal taxa, such as *Thallomys*, are conspicuously absent. Two explanations deserve consideration. Predator bias may explain the absence in part. *Thallomys* is more common at eagle owl roosts as was noted in the section Predator Effects. Roost 44 is a cavity roost and thus presumably the work of a barn owl. Another factor may be forest type. *Thallomys* prefers drier *Acacia* or *Brachystegia* woodland (Linzey and Kesner, 1997). The vegetation at roost 44 includes grasslands that abut dense riverine forests composed of broadleaf evergreen species (e.g., *Diospyros*, *Drypetes*, *Teclea*) and not *Acacia* (Herlocker, 1974). Thus, *Thallomys* may be absent from this habitat altogether, and not just absent from the owl pellets.

Returning to the questions that motivated this research the following points are emphasized: (1) The two species of owl examined here are capable of taking similar prey taxa, but with subtle biases in the relative abundances of prey items that may stem from prey size, with eagle owls preferring slightly larger prey. An alternate explanation is that roosting habits bias prey preferences. For example, the arboreal *Acacia* tree rat, *Thallomys*, may be opportunistically preferred by *Bubo africanus* since this owl roosts in tree crowns. (2) Over the subtle habitat transitions considered in this study, the relative abundance of the accumulated fauna differs significantly and predictably from roost to roost in different habitats. These differences agree with habitat predictions based on trapping studies and thus time-averaged, owl-accumulated taphonomic assemblages are good indicators of environments within a radius of 1.5 km surrounding the site of accumulation.

### Calibration of Paleoecological Methods

Identifying the faunal differences between habitats as was the focus in the previous section remains one step removed from a paleoenvironmental reconstruction. Two techniques used to integrate data into composite paleoenvironmental reconstructions are reviewed below, taxonomic ratios and the taxonomic habitat index. These methods are applied to the modern owl assemblages and the results compared against the existing vegetation.

## TAXONOMIC RATIOS

The use of ratios of indicator taxa is a popular method for interpreting the past. For example, Vrba (1980, 1985, 1995) has used the proportion of Alcelaphini and Antilopini bovid tribes as an indicator of open habitats. Similarly, the ratio of Gerbillinae to Murinae (both sub-families in Rodentia, Muridae) has been proposed as an indicator of open habitats (Jaeger, 1976; Dauphin et al., 1994; Fernandez-Jalvo et al., 1998). Figure 5 shows the relationship between the Gerbillinae:Murinae (G:M) ratio and the percent woody vegetation surrounding the Serengeti roosting sites. A trend is evident toward decreasing proportions of gerbils as the roost environment becomes more wooded (Spearman rank correlation test, *rho* = −0.0773, *p* = 0.015). The relationship is largely driven by the abundance of *Gerbillus*,

which accords with the niche model for this taxon. Similar patterns emerge from ratios of Soricidae:Murinae and Dendromurinae: Murinae (Table 8). The consistent direction of the relationship implies that murines tend to predominate in mesic woody vegetation. Roost 3 again appears as an outlier to the pattern, with a greater number of murines (especially *Mus* spp.) than would be expected for a grassland roost. As mentioned previously, roost 3 is a marginal grassland and may have been more wooded in the recent past. With this one exception, the G:M ratio performs well using the modern data.

## TAXONOMIC HABITAT INDEX

The taxonomic habitat index is a method for combining the niche models of each taxon into an overall picture of the habitat preferred by its constituents. The pooled value given to each habitat class is called the taxonomic habitat index (THI) and is defined by the following relationship:

$$THI_j = \frac{\sum_{i=1}^{i} w_{ij}}{i} \quad (2)$$

where $w_{ij}$ is the habitat indication of the $i$th taxon for the $j$th habitat taken from the niche model. The analysis works from a contingency table of niche models with taxa as row headings and the different habitats as column headings (Table 9). The niche model proportions each species across habitats as was shown in Table 5.

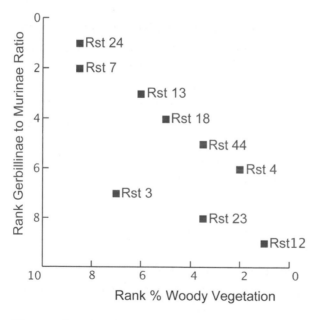

Figure 5. Bivariate rank plot of Gerbillinae: Murinae ratio against roost percent woody vegetation.

Table 8. Spearman rank correlation, rho, for three taxonomic ratios (rows), each tested against percent woody vegetation. Alpha values of 0.005 adjusted for three unplanned tests; 0.05/3 = 0.015

| Ratio | Rho | P | Result |
|-------|------|------|--------|
| G:M | −0.773 | 0.015 | * |
| D:M | −0.681 | 0.044 | NS |
| S:M | −0.580 | 0.102 | NS |

Table 9. *Example niche model contingency table*

|         | Habitat 1   | Habitat 2   | ...  | Habitat $j$ |
|---------|-------------|-------------|------|-------------|
| Taxon 1 | $w_{11}$    | $w_{12}$    |      | $w_{1j}$    |
| Taxon 2 | $w_{21}$    |             |      |             |
| ...     |             |             |      |             |
| Taxon $i$ | $w_{i1}$  |             |      | $w_{ij}$    |

The values in the table are then summed by habitat type (i.e., down the columns) for those species present in the assemblage and divided by the total number of species to produce a habitat index. The results indicate which habitats are most strongly represented by the species in the assemblages (Nesbit-Evans et al., 1981; Andrews, 1990). A histogram charting the THIs for all habitats is called a habitat spectrum (van Couvering, 1980).

The quality of results from a THI analysis depends on several factors: the niche models for the taxa, the taxonomic rank that is used (species niche models are more specific than those for genera, family, etc.), and the assumption of taxonomic uniformity (i.e., transferred ecology). This assumption states that modern representatives are suitable analogues for the fossil taxa. Generally this assumption becomes less tenable with the older the fossil assemblage, and also depending on the evolutionary history of the lineages involved. Has the lineage experienced a recent radiation with the appearance of new species? If so, the assumption of transferred ecology is probably not as strong as for a lineage that has been morphologically stable.

Another factor to consider is how the taxa should be weighted with regard to their relative abundance. Equal weighting ignores relative abundance and assumes that each taxon is equally informative about the habitat. Weighting taxa by relative abundance assumes that the most abundant are best adapted to the surrounding habitat and should have a greater influence in the analysis. As a test, THI is applied to the modern coprocoenoses using both assumptions.

Taxonomic habitat spectra using equal weighting are shown in Figure 6. Given the broad niche breadth of most rodents, habitat spectra will usually have all habitat classes represented in at least some small amount, hence the presence of a habitat in the spectrum is not necessarily indicative of that in reality. This is clear, for example, at roost 24 where forest, woodland, and bushland habitats are indicated in the spectrum even though they are absent (or present in very small quantities) at that roost. Forest appears in the spectrum for every roost, but this landcover type is only present in significant quantities at roost 44.

The THI spectrum succeeds in returning indices in their proper rank order. For example, roost 24 yields grassland (36%) > bushland (~26%) > woodland (~15%) > semi-arid (~12%) > forest (~9%) in that order. This agrees with the remote-sensing habitat analysis that indicated 88% grasslands, and 12% bushed grasslands. The woody component at roost 24 is mostly low shrubs, with small numbers of trees at the kopjes themselves. Some bare ground is present both as part of the rocky kopjes and at salt flats. There is no forest.

This same rank order is returned by the THI spectra for each roost, except roost 12. From the perspective of accuracy this is appropriate as grassland is the dominant land cover category at all roosts. The only inaccurate result occurs at roost 12 where woody vegetation is over-represented in the analysis. However, at roost 12 the top three habitat indices are all very close to one another (26.1, 26.6, and 25.1), so the erroneous result may be due to that roost's small sample size.

The results are less consistent when comparing one roost to the next. Given the absence of woody vegetation from the area around roost 7 and 24, it is expected that these roosts should have the highest grassland and semi-arid values and the lowest forest values. Encouragingly, the four roosts at the drier/open end of the spectrum have the greatest values for the semi-arid

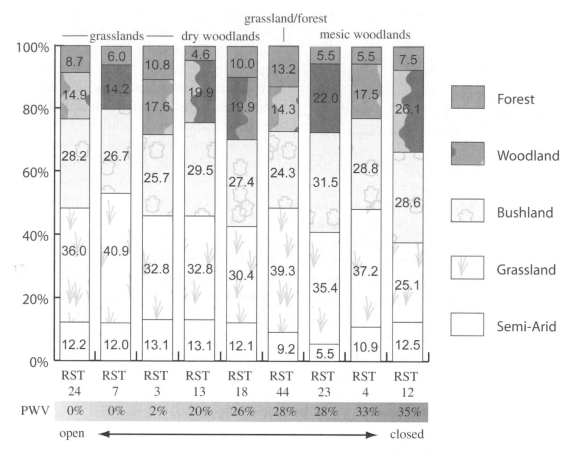

Figure 6. Taxonomic habitat spectra based on Serengeti taphonomic assemblages. Each bar shows cumulative percentages for each habitat index at the roost. Roosts are arranged from left to right in ascending order of percent woody vegetation cover. Each taxon is equally weighted in the analysis (abundance is not incorporated). General habitat descriptions are given across the top of the chart.

class; however, roost 12 has a value that is higher than expected. Similarly, for the grassland category roosts 7 and 24 have high values as expected but roosts 44 and 4 are also unexpectedly high. Combining the grassland and semi-arid categories, a generous observer may find a decreasing trend, but it is clearly disrupted by roosts 44 and 4.

The bushland category is very similar across all roosts while woodlands follow an increasing trend to the right as would be expected. Roost 44 has a small woodland component consistent with the actual habitat there (grasslands bordering forest). Furthermore, roost 44 has the largest value for forest. However, some of the other more closed roosts have very

small forest contributions. The fact that these values are small is not surprising given that true forests are not present in abundance, but it is surprising to find higher values at the grassland roosts.

Weighting by taxon abundance (NISPn) tends to amplify the results. THI values from the weighted analysis are illustrated in Figure 7. Grasslands remain the dominant vegetation type at roosts 24, 7, and 44, which accords with the actual habitat at these roosts, and roosts 24 and 7 retain high values for the semi-arid class, whereas this value drops off in the other woodland roosts. However, the forest class nearly disappears from roost 44. Woodland becomes the first ranking habitat class at all the remaining roosts.

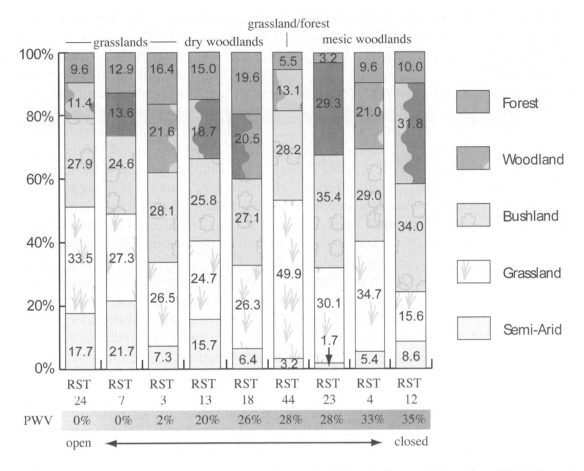

Figure 7. Taxonomic habitat spectra based on taxonomic habitat indices weighted by NISPn. Roosts are given in ascending order of woody vegetation starting with the least wooded roost on the left.

## SUMMARY OF ANALYTICAL METHODS

Two common methods of micromammal faunal analysis are tested against modern assemblages. The taxonomic ratio of Gerbillinae to Murinae performs well. It returns rank order results consistent with the habitats observed around each roost. The dry, grassland roosts have G:M ratios greater than 1 with the exception of roost 3. This roost sits right on the border between grasslands and woodlands and it may at times in the past have been more wooded. Arid-adapted, Egyptian gerbils of the genus *Gerbillus* play a prominent role in determining this ratio, especially when found in association with other burrowing rodents such as *Steatomys* and *Tatera*. *Tatera* also occur in

woodlands that have a grass understory, making this genus a less reliable indicator of open grassland environments. *Steatomys*, though modeled as a catholic species, occurs predominantly in open grassland environments in Serengeti. All three species are burrowing, an adaptation that is crucial for predator avoidance in habitats with sparse vegetation cover. Burrowing is also an effective strategy against frequent fires. In the more mesic and wooded environments, the denser understory provides shelter for many of the murine species. Murine species such as *Dasymys* and *Arvicanthis* also filled the semi-aquatic edaphic grassland niche as evidenced at roost 44. Thus, non-burrowing murines come to replace burrowing gerbils and dendromurines in wet and water margin environments. In the middle are burrowing rodents, such as *Arvicanthis* and *Lemniscomys* that also rely on runways through the herbage to avoid predation.

Results from the THI analysis were less robust and differed depending on whether the analysis used equal weighting or weighting by taxon abundance. Equal weighting returned results for which the dominant THI matched the prevailing habitat at the roost (grasslands in all cases) except for roost 12, where bushland was the dominant THI class. With equal weighting woodlands (areas with >20% woody canopy cover) have lower THI values for grassland and semi-arid classes compared to grasslands and generally higher values for forest and bushland classes. However, the differences are relatively small and there is some overlap that makes interpretation difficult. Weighting by abundance amplifies the differences but also produces results that appear less accurate, as for example, the small value of forest at roost 44.

Fernandez-Jalvo et al. (1998) compare THI values of fossil assemblages against results from modern faunas. The Serengeti data used here indicate that such comparisons are probably inappropriate because variability within a park or census area may be as great as the variability between them. For example, the three grassland areas (roost numbers 24, 7, 3) show a range of variability greater than that shown between grassland and woodland areas. The general trends in the data are encouraging, but the technique will need further refinement before confidence can be placed in the results.

## Olduvai Paleoenvironments

Olduvai Gorge lies on the southeastern edge of the Serengeti (35.3483° E, 2.9881° S), where the plains slope down to meet the foot of the volcanic highlands formed by the Ngorongoro caldera and neighboring volcanoes. This topographic depression forms a shallow basin now dissected by a seasonally flowing river that runs along the bottom of the gorge and flows

from two alkaline lakes further to the west (Lake Masek and Lake Ndutu). At its eastern edge the Gorge empties into the Ol'Balbal swamp. This swamp is also fed by freshwater streams coursing down from the volcanic highlands to the southeast (Figure 8).

Archaeological and paleontological sites are found throughout the gorge, including the FLK Bed I sequences at the crux between the main gorge and the side gorge. Exposures in the eastern portion of the gorge represent alternating fluvial and lacustrine sequences formed by fluctuating paleolake margins (Hay, 1976; Denys et al., 1996; Ashley and Driese, 2000). This region has been reconstructed as alkaline mudflats grading into moist grasslands and spring fed marshes along a volcanic piedmont between paleolake Olduvai and the adjacent highland to the southeast (Hay, 1976; Deocampo et al., 2002; Blumenschine et al., 2003). Archeological sites at FLK have yielded rich Plio-Pleistocene faunas including both large and small mammals (Butler and Greenwood, 1976; Jaeger, 1976; Gentry, 1978a, b). The oldest FLK sites occur in Middle and Upper Bed I deposits and span a time interval of approximately 50,000 years between Tuff IB at $1.798 \pm 0.014$ Ma and Tuff IF at $1.749 \pm 0.007$ Ma (Walter et al., 1991). Middle Bed I is represented at FLK North–North by three levels below Tuffs IC and ID (FLKNN1–3). Level FLKNN1 has few faunal remains but is considered contemporaneous with the "Zinjanthropus" floor at FLK 1 level 22, hereafter referred to as FLK-Zinj (Leakey et al., 1971). Above Tuff ID the Upper Bed I deposits include six levels at site FLK North, (FLKN1–6) extending up to Tuff IF.

## THE PLIO-PLEISTOCENE RODENT FAUNA AT OLDUVAI GORGE

Lavocat (1965) gave a brief description of the micromammals recovered from FLK. More detailed taxonomic treatments followed for the elephant shrews (Butler and Greenwood, 1976) and rodents (Jaeger, 1976; Denys, 1990,

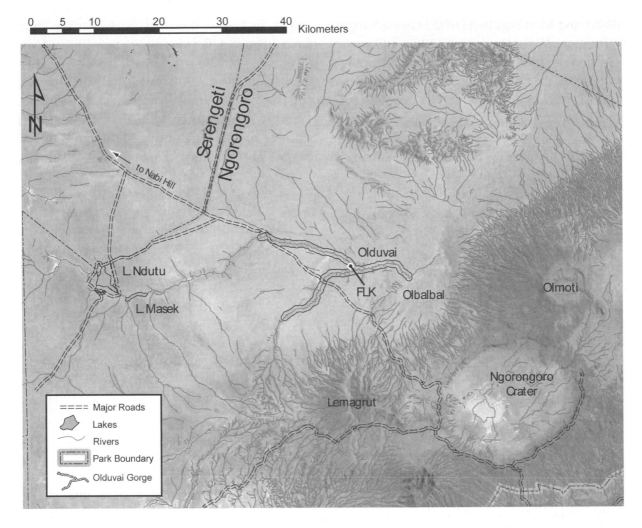

Figure 8. Satellite image background with map overlay showing the location of Olduvai Gorge and site FLK.

1999; Denys and Tranier, 1992. Table 10 lists the taxa occurring in the Bed I FLK sites at Olduvai Gorge.

The genera represented in the fossil Bed I assemblages overlap considerably with the taxa known from the modern Serengeti eco-system and surrounding region. Of the 17 fossil rodent genera recovered at Olduvai, all but two, *Heterocephalus* and *Oenomys*, have been noted in the modern Serengeti ecosystem and Ngorongoro highlands either through owl pellet or trapping studies. The two extra-limital taxa are known from surrounding regions. The modern distribution of *Heterocephalus* includes the arid regions of Somalia, Ethiopia, and Kenya, while *Oenomys* is found in forest habitats

around Lake Victoria. Three genera: *Pelomys*, *Otomys*, *Grammomys*, have been recorded in the Serengeti through trapping studies or pre-vious owl-pellet studies (Swynnerton, 1958; Andrews, 1983) but did not appear in the owl pellet assemblages analyzed for this project (Table 10). The consistency between fossil and modern assemblages implies a relatively sta-ble metacommunity structure for East African rodents through the Pleistocene.

In East Africa, roughly contemporaneous Late Pliocene rodent faunas are known from the Peninj group of deposits at West Lake Natron (Denys, 1987a), the Koobi Fora Formation at East Turkana, between the KBS and Okote Tuffs (ca. 1.6 Ma) (Black, 1984; Black and Krishtalka,

1986), and Members D–H of the Omo Shungura Formation (Wesselman, 1982, 1984, 1995).

Clues to the origins and biogeographical history of these Late Pliocene rodent faunas can be found in older Pliocene deposits such as the Laetolil and Ndolanya Beds at Laetoli (Denys, 1987b) and the Ibol Member from the Manonga Valley in Tanzania (Winkler, 1997). In Kenya there are assemblages from the Chemeron Formation in the Tugen Hills (Winkler, 2002), the Mio-Pliocenc deposits at Lothagam (Winkler, 2002), and as yet undescribed material from Kanapoi (Winkler, 1998). Further north along the rift valley are sizeable assemblages from Members B and C of the Omo Shungura Formation (Wesselman,

Table 10. *Abundance (MNI) of Olduvai fossil rodents from FLK sites modified from Fernandez-Jalvo et al. (1998). Taxa are organized by niche index with the more arid-adapted taxa to the left. Presence or absence of taxa in modern roosts and trapping studies is indicated below each taxon. Asterisks next to taxon names indicate the taxon shows significant change in abundance between levels. The upper frame lists MNI values and the lower frame percent MNI*

| | Niche index > | 1.60 | 2.10 | 2.25 | 2.40 | 2.50 | 2.60 | 2.97 | 2.97 | 3.00 | 3.00 | 3.09 | 3.30 | 3.46 | 3.50 | 3.69 | 4.05 | 4.50 | |
|---|---|---|---|---|---|---|---|---|---|---|---|---|---|---|---|---|---|---|---|
| | Stratigraphic Level | *Gerbillus** | *Zelotomys* | *Arvicanthis* | *Tatera* | *Pelomys* | *Heterocephalus** | *Mastomys** | *Saccostomus* | *Steatomys* | *Otomys** | *Dendromus* | *Xerus** | *Aethomys* | *Thallomys** | *Mus* | *Grammomys* | *Oenomys** | Total MNI |
| | Trapping | + | + | + | + | + | − | + | + | + | + | + | + | + | + | + | + | − | |
| | Owl Assemb. | + | + | + | + | − | − | + | + | + | − | + | + | + | + | + | − | − | |
| | **Absolute abundance (MNI)** | | | | | | | | | | | | | | | | | | |
| Upper Bed I | FLKN1 | 21 | 10 | 0 | 10 | 0 | 0 | 2 | 3 | 16 | 27 | 10 | 0 | 3 | 0 | 3 | 0 | 0 | 105 |
| | FLKN2 | 12 | 8 | 0 | 11 | 0 | 0 | 1 | 10 | 13 | 17 | 2 | 0 | 8 | 2 | 2 | 0 | 0 | 86 |
| | FLKN3 | 6 | 3 | 0 | 14 | 0 | 2 | 0 | 9 | 8 | 19 | 2 | 5 | 10 | 1 | 0 | 0 | 0 | 79 |
| | *Taphonomic shift* | | | | | | | | | | | | | | | | | | |
| | FL KN4 | 16 | 3 | 1 | 7 | 1 | 11 | 9 | 4 | 9 | 16 | 1 | 0 | 11 | 1 | 1 | 0 | 0 | 91 |
| | FLKN5 | 8 | 2 | 0 | 3 | 1 | 8 | 11 | 1 | 4 | 8 | 1 | 0 | 3 | 0 | 1 | 0 | 0 | 51 |
| | FLKN6 | 2 | 1 | 0 | 2 | 0 | 0 | 0 | 2 | 4 | 2 | 1 | 0 | 2 | 0 | 1 | 0 | 0 | 17 |
| | *Tuffs IC and ID* | | | | | | | | | | | | | | | | | | |
| Middle Bed I | FLK-Zinj | 0 | 3 | 2 | 3 | 1 | 0 | 2 | 2 | 3 | 0 | 1 | 0 | 16 | 11 | 1 | 0 | 1 | 46 |
| | FLKNN2 | 0 | 3 | 1 | 9 | 2 | 0 | 2 | 0 | 4 | 1 | 0 | 0 | 3 | 1 | 1 | 0 | 1 | 28 |
| | FLKNN3 | 0 | 2 | 2 | 3 | 0 | 0 | 1 | 0 | 9 | 15 | 0 | 0 | 4 | 1 | 1 | 1 | 4 | 43 |
| | **Percent abundance** | | | | | | | | | | | | | | | | | | |
| Upper Bed I | FLKN1 | 20 | 10 | − | 10 | − | − | 2 | 3 | 15 | **26** | 10 | − | 3 | − | 3 | − | − | 100 |
| | FLKN2 | 14 | 9 | − | 13 | − | − | 1 | 12 | 15 | 20 | 2 | − | 9 | 2 | 2 | − | − | 100 |
| | FLKN3 | 8 | 4 | − | 18 | − | 3 | − | 11 | 10 | **24** | 3 | 6 | 13 | 1 | − | − | − | 100 |
| | *Taphonomic shift* | | | | | | | | | | | | | | | | | | |
| | FLKN4 | 18 | 3 | 1 | 8 | 1 | 12 | 10 | 4 | 10 | 18 | 1 | − | 12 | 1 | 1 | − | − | 100 |
| | FLKN5 | 16 | 4 | − | 6 | 2 | 16 | 22 | 2 | 8 | 16 | 2 | − | 6 | − | 2 | − | − | 100 |
| | FLKN6 | 12 | 6 | − | 12 | − | − | − | 12 | **24** | 12 | 6 | − | 12 | − | 6 | − | − | 100 |
| | *Tuffs IC and ID* | | | | | | | | | | | | | | | | | | |
| Middle Bed I | FLK-Zinj | − | 7 | 4 | 7 | 2 | − | 4 | 4 | 7 | − | 2 | − | **35** | 24 | 2 | − | 2 | 100 |
| | FLKNN2 | − | 11 | 4 | 32 | 7 | − | 7 | − | 14 | 4 | − | − | 11 | 4 | 4 | − | 4 | 100 |
| | FLKNN3 | − | 5 | 5 | 7 | − | − | 2 | − | **21** | **35** | − | − | 9 | 2 | 2 | 2 | 9 | 100 |

1982, 1984, 1995) and the Sidi Hakoma, Denen Dora, and Kada Hadar Members at Hadar, Ethiopia (Sabatier, 1982).

Taking a broader geographic perspective, one can compare the East African faunas to large assemblages from South African Pliocene localities including, Langebaanweg (Pocock, 1987; Denys, 1990), Makapansgat (Pocock, 1987), Kromdraai (de Graaff, 1961; Pocock, 1987), Taung (McKee, 1993), Sterkfontein and Swartkrans (Pocock, 1987; Avery, 1995, 998, 2001). Two other micro-mammal sites are the Plio-Pleistocene fissure fillings at Humpata, Angola (Pickford et al., 1994) and the Lusso Beds in the Democratic Republic of Congo (formerly Zaire) (Boaz et al., 1992). Fossil micromammals also occur at numerous North African Mio-Pliocene localities, but these have greater affinities to European and circum-Mediterranean mammal communities than to those in eastern and southern Africa (Geraads, 1998).

In a review of the biogeography of East African rodents, Denys (1999) demonstrates that the Plio-Pleistocene and modern rodent faunas from Olduvai and Serengeti are typical of the Somali–Masai vegetation biome. This biome extends from northern Tanzania up through Kenya and the horn of Africa. Rift valley tectonic activity and volcanism increased paleobiodiversity in the region during the Pliocene by creating a mosaic of habitats ranging from arid environments along the valley floor to montane forests on the slopes of newly formed volcanoes. Transverse faults along the rift created mountain chains that segmented the rift valley into isolated basins, simultaneously restricting the movements and terrestrial animals and elevating the potential for fracturing (i.e., vicariance) of species' geographic ranges (Denys et al., 1986). East African rodent diversity was also influenced by migration from southern Africa. Shortly after the start of the Pliocene, ca. 4–3 Ma, southern Africa develops a differentiated Zambezian fauna that comes to influence the southern parts of eastern Africa. For example, *Zelotomys* and *Otomys* are prevalent South African Pliocene taxa that appear in East Africa at Laetoli and Olduvai but are not recorded as far north as the Omo or Hadar (Denys, 1999).

## PREDATOR INFLUENCE ON DIACHRONIC CHANGES IN OLDUVAI BED I RODENT FAUNAS

The general consensus among early studies of the Olduvai faunas was that the micromammals associated with Upper Bed I (especially FLKN level 1) represented a more xeric-adapted community than those associated with Middle Bed I (FLKNN level 3-2 and FLK-Zinj). Butler and Greenwood (1976) note that xeric-adapted macroscelideans, such as *Elephantulus*, become increasingly more abundant through Upper Bed I times. They observe that "a marked change takes place in the insectivore fauna between FLK NNI and FLK NI... This must imply a change of environment, and the most likely change would be a reduction in rainfall" (Butler and Greenwood, 1976, p. 48). However, Andrews (1983) proposed that some of these faunal shifts might reflect changes in the predators accumulating the assemblages rather than real environmental change. For example, the abundance of *Gerbillus* in the Upper Bed I deposits may be an artifact of eagle owls preferring gerbils over murines. Accounting for the bias led Andrews to conclude that the fauna at FLKN1–2 are "indicative of a wooded habitat that was perhaps closer to the denser and wetter woodlands of the northwestern part of the Serengeti ecosystem rather than to any of the habitats in the immediate vicinity of Olduvai Gorge today" (p. 84). Subsequent work on taphonomic processes affecting micromammal assemblages (Andrews, 1990) eventually led to a thorough investigation the taphonomy of Olduvai microfauna by Fernandez-Jalvo et al. (1998). One of their important contributions is the detailed analysis of bone breakage and surface modification. Their findings are summarized in the "accumulator" and "modification" columns of Table 11. From this analysis the authors identify three intervals, each with different taphonomic biases. The oldest, Middle Bed I, assemblages at FLKNN and FLK-Zinj exhibit less breakage and less surface etching of the bones (Table 11). These patterns are consistent with owls, and the barn owl specifically in the case of FLKNN2. There is then a switch to a more destructive pattern of breakage and surface modification that the authors identify as a mammalian carnivore perhaps in combination with owls. This pattern maintains for levels 4–6 at FLKN. Further up the sequence (FLKN1–3), the assemblages exhibit moderate breakage consistent with an eagle owl.

One of the most important differences between Middle and Upper Bed I faunas is the absence of *Gerbillus* from the older members and its appearance in Upper Bed I. Is the appearance of *Gerbillus* in Middle Bed I the results of environmental change or a change in predator? This taxon has a very low niche index value (see Table 5) and is one of the best indicators or open and semi-arid habitats. Its abundance influences other types of analyses such as the ratio of Gerbillinae to Murinae, and THI

method discussed previously. Thus, taphonomic biases that affect *Gerbillus* can have a strong impact on any analysis. Fernandez-Jalvo et al. (1998, p. 166) argue that predator selectivity rather than environment "may produce changes in species composition... between FLKNN and FLKN." They argue (ibid, p. 166) Middle Bed I (especially FLKNN2) was the work of a non-destructive accumulator such as the barn owl, *Tyto alba*, that "may favor murines against gerbils, as seen in modern assemblages (Laurie, 1971; Andrews, 1990)." Thus, they conclude that the absence of gerbils from the Middle Bed I assemblages is an artifact of barn owls selecting against gerbils. However, the modern Serengeti roost data does not appear to support this assertion. Gerbils are abundant in the drier grassland roosts of the current study (roosts 7 and 24). Both these roosts are very likely the work of barn owls. In both instances, barn owls were found as current occupants and both are "cavity" roosts, the type favored by barn owls to the near exclusion of eagle owls (Reed, 2003, 2005). At roost 7 in the current study *Gerbillus* was the third most abundant taxon (ca. 17% NISPn) behind the shrew *Crocidura* and the dendromurine *Steatomys*. The one previous pellet study in Serengeti by Laurie (1971) does not

*Table 11. Stratigraphic summary of the Middle and Upper Bed I deposits including taphonomic interpretation for the microfauna. Accumulator and modification columns are taken from Fernandez-Jalvo et al. (1998)*

|  | Level | Accumulator | Modification |
|---|---|---|---|
| Upper Bed I | Tuff IF | 1.749 Ma | |
| | FLKN1 | Bubo leakeyae | Intermediate |
| | FLKN2 | Bubo lacteus | Low |
| | FLKN3 | Bubo lacteus | Low |
| | FLKN4 | Mammal + B. lacteus | Extreme + low |
| | FLKN5 | mammalain carnivore | Extreme |
| | FLKN6 | Unknown | Unknown |
| | Tuff ID | 1.764 Ma | |
| | Tuff iC | 1.761 Ma | |
| Bed I | FLK-Zinj | Bubo? | Intermediate |
| | FLKNN2 | Tyto alba | Very low |
| | FLKNN3 | Owl | Low |
| | Tuff IB | 1.798 Ma | |

report any *Gerbillus*. The discrepancy may be due to Laurie's emphasis on fresh pellets, which are sensitive to short-term population dynamics of the prey species. The current study relies on larger, time-averaged assemblages of decayed pellets that should provide a more robust representation of the biocoenosis. Also, pellet studies from western Africa, South Africa, and Israel indicate that barn owls take gerbils, including *Gerbillus*, as prey, and when available they are captured in abundance (Vernon, 1972; Rekasi and Hovel, 1997; Pokines and Peterhans, 1998; Ba et al., 2000). The modern data do not indicate a substantial bias against gerbils by barn owls as proposed by Fernandez-Jalvo et al. (1998). Rather, the available data support Andrews's (1990) summary of barn owl diet; "muri[nes] such as *Praomys* [are] the most common [prey] in Africa, replaced by *gerbils in drier regions*" (p. 179 emphasis added).

The abundance patterns in *Gerbillus* raise the issue of whether differences across levels at Olduvai are statistically significant and meaningful. Sample sizes of identifiable elements for the Olduvai micromammal assemblages range from very small (17 and 28 upper first molar specimens at FLKN6 and FLKNN2 respectively) to moderate (105 at FLKN1). With small, unequal samples it is possible that chance plays a large role in the shifting abundance values. Tests of independence using the Pearson $\chi^2$ statistic reveal that 8 of the 17 taxa show significant differences across stratigraphic levels (probability threshold is adjusted for an experiment wide error rate of 0.05 with 17 unplanned comparisons; $p < 0.003$, $df = 8$). These taxa are highlighted in Table 10 and shown at right in Figure 9. *Gerbillus* is among the eight taxa with significant changes in relative abundance across levels at FLK but *Tatera* is not (though the $\chi^2$ value for *Tatera* is nearly significant). Any argument incorporating the changing abundance between levels should emphasize these eight taxa.

Figure 9 shows trends in the Gerbillinae: Murinae ratio using the entire fauna (solid line) and just the eight significant taxa (dashed line). Of the eight significant taxa, three make a first or last appearance at Tuff IC/D. The xeric-adapted taxa *Gerbillus* and *Heterocephalus* appear above Tuff IC/D while the mesic-adapted genus *Oenomys* disappears. No similar pattern appears in the interval between FLKN4–6 and FLKN1–3 as suggested by Fernandez-Jalvo et al. (1998). The ground squirrel, *Xerus* is the only taxon to appear at this transition and the environmental implications of this are not clear as the genus includes species that range from open to wooded environments.

Much of the prior taphonomic research into micromammal assemblages has focused on identifying different predators. The results of this study indicate that it is perhaps time to focus on the significance imposed by different predators once identified. The results from the modern Serengeti data indicate that barn owls do not exhibit a strong bias against gerbils and that at least some predators, such as barn owls and spotted eagle owls, have very similar trophic habits and limitations such that they may be treated as isotaphonomic under certain analyses (Reed, 2005).

## PALEOENVIRONMENT

Given the degree of faunal overlap between the fossil and extant assemblages, the modern results may serve as a model or starting point for evaluating the environmental signal indicated by the fossil faunas. Figure 10 shows a correspondence analysis using the intersection of taxa found in the modern and fossil assemblages. The plot illustrates associations between fossil and modern micromammal assemblages based on the abundance patterns of shared taxa. The first axis positions the more open, xeric, grassland roosts (7 and 24) to the left, grading into the more mesic, closed roosts to the right. The second axis separates modern from fossil roosts, but there is some overlap. FLK-Zinj is located furthest to the upper right, indicating a mesic environment most like that

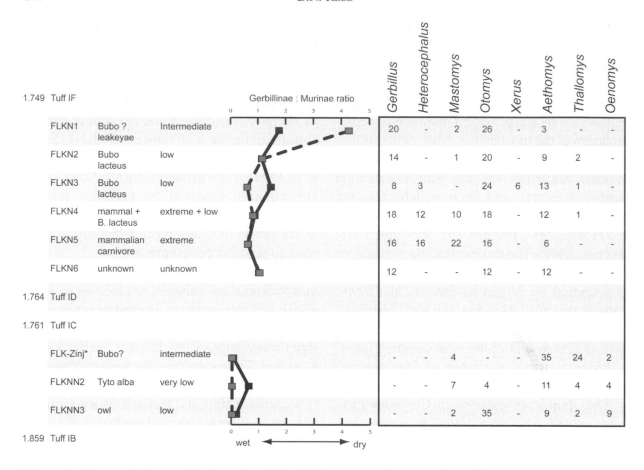

Figure 9. Temporal changes in the ratio of Gerbillinae to Murinae as reflected in the entire fauna (solid line). The right pane shows percent MNI of those eight taxa with significant interlevel differences in abundance (dashed line).

of modern roost 44. Its position is influenced largely by *Aethomys* and *Thallomys*. Both are *Acacia*/*Brachystegia* shrubland/woodland species. These taxa are found in the modern assemblages but not together in high abundance. The FLK-Zinj assemblage is consistent with a woodland environment, or moist savanna with evergreen gallery forest. Roost 44 occurs at higher rainfall with tall grasslands grading into relic evergreen, broadleaf forests. Extending this analogy to the paleo-Olduvai basin would produce a lake margin habitat hosting both dense and open woodland habitats that included some component of grass understory as habitat for *Tatera* and *Steatomys*.

At the other extreme, FLKN1 falls to the left along Axis 1. It also has the lowest position on Axis 2 indicating that this level has

the greatest affinity to the modern analogues. FLKN1 is most closely positioned to roost 24, the driest and most xeric of the modern analogues. The association is based on high abundances of two xeric-adapted taxa, *Gerbillus* and *Steatomys*. Both species shelter below ground, and *Steatomys* especially is an active burrower whose presence indicates soft, well-drained soils. However, the FLKN1 assemblages differ from all the modern assemblages in having *Otomys* as the dominant taxon. Modern species of *Otomys* are grazers that tolerate a broad variety of habitats including thicket and secondary growth to marshes and montane grasslands. Any interpretation of the paleoenvironment at FLKN1 must reconcile *Otomys* with the xeric affinities of the remaining taxa.

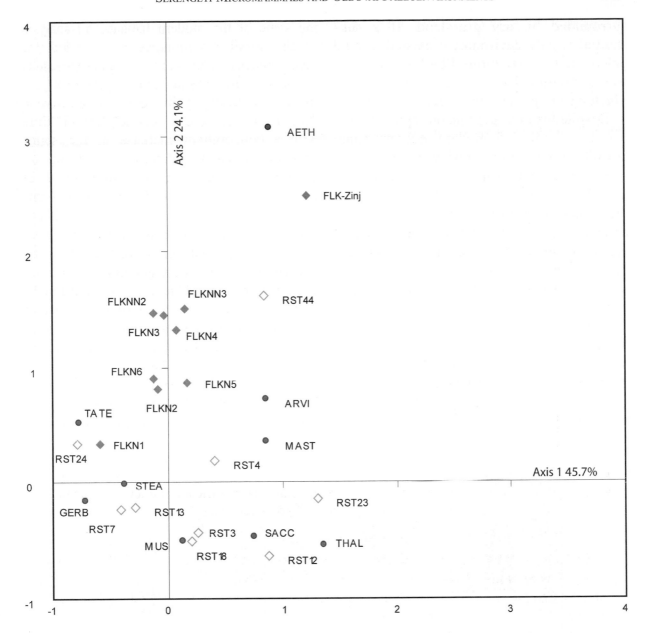

Figure 10. Correspondence analysis based on modern faunas (using NISPn abundance values). Fossil assemblages are included in the plot as ancillary data; fossil faunas were not used to calculate the topography of the plot. The position of the fossil assemblages is based only on those taxa that overlap with the modern assemblages. Modern data are shown as open diamonds, the taxa as closed circles, and the fossil assemblages as closed diamonds.

*Otomys* is present throughout the Bed I sequence. Its persistence may be associated with a stable moist grassland or marsh along the paleolake margin. It could also indicate lake margin or riparian bushland. However, none of the other taxa at FLKN1 give an indication of woody vegetation. Two closed habitat taxa, *Aethomys* and *Mus*, are in low abundance, but similar low abundances of these taxa may be found in the modern grassland roosts (e.g., *Aethomys* and *Mus* both appear at roost 24). With the available data, it is not possible to rule out freshwater moist grasslands fed by montane streams,

surrounded by dry grassland. If a substantial woody environment existed around
paleolake Olduvai during FLKN1 times, one
would predict *Thallomys*, *Aethomys*, and
*Mastomys* in greater abundance.

Despite the overlap in taxon representation
between fossil and modern micromammal
assemblages, the correspondence analysis also
serves to illustrate that the fossil assemblages
are all more like each other than they are to
the modern assemblages and vice versa. It is
certainly possible that the fossil assemblages
are not analogous to any modern assemblage.
The uniqueness of fossil Olduvai communities demonstrates that it is necessary both to
expand the range of modern analogues and at
the same time find more generalized factors
for comparison than taxonomy, such as ecomorphology (Alexander, 1988; Damuth, 1992
or ecological structure analysis.

Although the THI analysis of the extant
roost samples had difficulty differentiat-

ing some of the modern habitats, an analysis
of the fossil assemblages is provided for
completeness. Figures 11 and 12 show the results
calculated under assumptions of even weighting and weighting by abundance, respectively.
Proportions of forest decline through time
with a concomitant increase in the semi-
arid component. Bushland and grassland are
the dominant habitat classes in every level
except FLK-Zinj where the woodland component ranks second behind the bushland
component when computed with abundance
weighting. The pattern of results from the
THI analysis follows the general drying trend
already described. Under an assumption of
even weighting, there is a marked increase in
the abundance of the semi-arid class between
Middle and Upper Bed I assemblages, and not
a very marked change between FLKN4–5–6
and FLKN1–2–3 as reported by Fernandez-
Jalvo et al. (1998). The same pattern appears
when the analysis is run with abundance

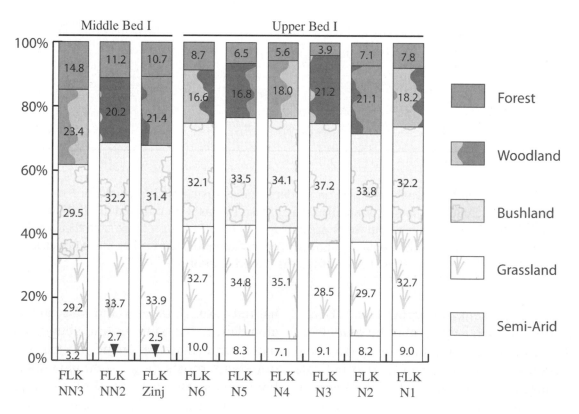

Figure 11. Taxonomic habitat spectra of Olduvai Bed I micromammal assemblages calculated with even weightings for all taxa.

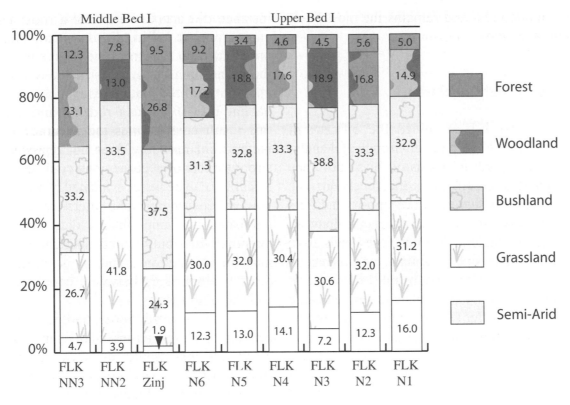

Figure 12. Taxonomic habitat spectra of Olduvai Bed I micromammal assemblages weighted by taxon abundance.

weighting, but in this analysis a change in the semi-arid category is more pronounced between FLKN4 and FLKN3. The same THI analysis was conducted by Fernandez-Jalvo et al. (1998) using the same data set. Their results show a similar pattern although forest values are much greater (see their Figure 8).

In summary, an analysis of the micromammals that incorporates modern data along with taphonomic analysis supports the initial interpretation that Middle Bed I represents a more wooded and mesic environment. This is followed by a transition to a drier phase in Upper Bed I. At this transition *Gerbillus* appears in the assemblages, *Thallomys* is lost or present at low abundance and *Grammomys* drops out as well, although the sample size for *Grammomys* is too small to say this definitively. This faunal transition occurs beside a change in the accumulating agent, but with current data we may reject the hypothesis that the low abundance

of *Gerbillus* in Middle Bed I (FLKNN and FLK-Zinj) was predator induced. The shift in the dominant predator may still influence the assemblages, but the effects need to be investigated with regard to specific methods of analysis. The ratio of Gerbillinae to Murinae increases throughout the Middle Bed I assemblages until they attain a community structure and composition similar to modern grassland assemblages, but with the addition of *Otomys* in high abundance. The best null hypothesis for the FLKN1 assemblage is that it represents a grassland environment with little woody vegetation cover but with some component of moist grassland or wetlands. This is a testable hypothesis. Should continued excavation in the FLKN1 level produce woodland species such as *Thallomys*, *Aethomys*, or forest species such as *Grammomys* the hypothesis should be rejected. True wetlands should also provide habitat for *Pelomys* or *Dasymys*. Without these taxa, moist grassland remains the most likely landscape type.

## Summary and Conclusion

This chapter has examined the efficacy of micromammals for paleoenvironmental analysis with regard to predator bias, accuracy of habitat representation, and precision. The results indicate that faunal composition differs significantly one roost to the next along an ecological gradient despite biases owls have for hunting in open habitats, and a related study (Reed, 2005) demonstrates that at least two owl species (barn owls and spotted eagle owls) do not differ significantly in the proportions of most prey taxa that they take.

The prey composition and relative abundances noted in the modern Serengeti data appear to accurately represent the habitats that are present at the roost. The grassed plains are characterized by gerbils (especially *Gerbillus*), and the dendromurine *Steatomys* concurrent with the absence or very low abundance of arboreal or semi-arboreal taxa such as *Thallomys*. A reciprocal pattern occurs for *Acacia* woodland roosts, which are characterized by the absence or low abundance of *Gerbillus*, and the presence of *Thallomys* and *Mastomys*. Woodlands also provided habitat for moist grassland taxa such as *Arvicanthis*. Tall grasslands and evergreen forest occur in the northern extension where precipitation exceeds 800 mm annually and there are perennial water sources. Here suitable habitats exist for *Aethomys* and *Dasymys*. Ratios of Gerbillinae to Murinae were significantly correlated to percent woody vegetation cover along the ecological gradient and THI values calculated from previously devised niche models were consistent with the modern observed habitats.

Dominant prey taxa are those one would expect based on independent trapping studies reported in the ecology literature, and in no

instance did a taxon occur at a roost where an appropriate habitat for that taxon was not present within 1.5 km of the roost site. Given the broad niche tolerances of many rodents, and that limited data are available on the habitat use of African rodents and shrews, this result corroborates the accuracy of the method but probably does not represent the strongest test of accuracy one could perform. Accuracy can be better evaluated by improving how we document niche tolerance, collating published and unpublished data on micromammal distribution and abundances from museum collections, and by simultaneous, direct comparison between coprocoenoses and trapping results.

The nine analyzed roosts have largely overlapping taxonomic representation, and a simple taxonomic list would diagnose habitats at some roosts, but would overlook a great deal of meaningful ecological information. Using the relative abundance of taxa, it is possible to distinguish subtle differences in habitats over distances of tens of kilometers. This study demonstrates the potential of micromammal assemblages for depicting habitats within biomes at finer spatial scales.

Applying the general model to the Olduvai microfauna gives results that differ from the paleoenvironmental interpretations proposed by Fernandez-Jalvo et al. (1998) and corroborate earlier interpretations of the micromammals made by Jaeger (1976) and Butler and Greenwood (1976). Whereas Fernandez-Jalvo et al. (1998) argue that the faunal transition between Middle and Upper Bed I (specifically between FLKNN/Zinj and FLKN4–5–6) is the result of biases in the accumulating agent, this is not borne out by the modern data that indicate that barn owls hunting in open and arid habitats produce assemblages rich in *Gerbillus*. This finding does not rule out the possibility of a faunal transition during Upper Bed I (between FLKN4–5–6 and FLKN1–2–3) as they propose. The Gerbillinae to Murinae ratio as well as the THI analysis both indicate a general drying trend over this interval. This pattern corroborates results from other, independent lines of evidence (Kappelman, 1984). Applying Vrba's (1980, 1985, 1995) Alcelaphini + Antilopini criteria (AAC) shows increases in the proportional representation of these arid-adapted bovid tribes through the sequence regardless of whether counts are done by minimum number of individuals (MNI) or numbers of individual specimens (NISP) (Gentry, 1978a, b; Kappelman, 1984; Potts, 1988; Plummer and Bishop, 1994) (Figure 13). One discrepancy appears at the FLK-Zinj locality. Among the bovids, FLK-Zinj has an AAC proportion intermediate to the FLKNN levels and the FLKN levels, but

among the micromammal assemblages this is the most wooded and mesic. The contrast may be due to differences in the scale of applicability between micro- and macrofauna, with the former applying to smaller and more local environments, while the bovids include fauna sampled from the region outside the immediate vicinity of paleolake Olduvai.

Using the data provided from modern Serengeti coprocoenoses, the relative, diachronic patterning at Olduvai is well established through multiple lines of evidence. More contentious, is determining the absolute paleoenvironments that occurred at various levels in Bed I, especially the maximum degree of aridity that is represented at FLKN1. The correspondence analysis shows low similarity

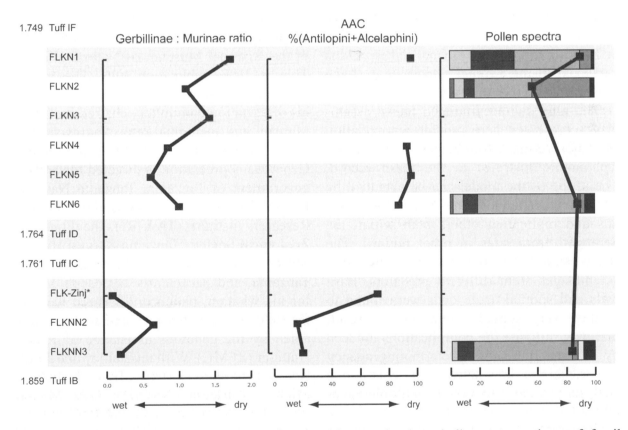

Figure 13. Paleoenvironmental summary showing changes in three indices: proportions of fossil Gerbillinae to Murinae; percent abundance of arid adapted bovid tribes; and pollen spectra with points separating non-arboreal pollen (to left) from arboreal pollen (to right). Sources: Fernandez-Jalvo et al. (1998), Kappelman (1984), and Bonnefille (1984) for micromammals, macromammals, and pollen respectively.

between the mesic Middle Bed I localities and the modern Serengeti assemblages. The gallery forest habitat at roost 44 is the closest modern analogue, but the separation between modern roosts and fossil assemblages indicates the model is incomplete. Drawing upon the modern natural history of the genera found in Middle Bed I, one can infer a more forested and wooded environment based on the proportion of gerbils to murines and on the specific murine taxa that are present, such as the forest- and woodland-dwelling taxa *Oenomys* and *Thallomys* respectively. In contrast, the most arid-adapted assemblage found at FLKN1 shares taxa and a pattern of abundance similar to the drier, grassland roosts in the modern Serengeti, but with the addition of *Otomys* as the most abundant taxon. A plausible explanation for this pattern is a paleolake Olduvai surrounded by dry grassland and with a moist grassland or marsh habitat nearby. Small lakes of this sort exist in the Serengeti today (e.g., Lake Magadi, 70 km NE of Olduvai).

The nine owl-accumulated faunal assemblages presented here provide a foundation for interpreting assemblages with similar taphonomic histories in the fossil record. The scope of the model can be extended by adding owl taphocoenoses from novel habitats and replicating others both within the Serengeti from sites in other regions. The scope can also be extended by incorporating assemblages from different predators. Barn owls and spotted eagle owls were found to produce very similar assemblages, a result that may mitigate the complications induced by different predators. Determining agency will remain a key activity in taphonomic analysis but the discovery of multiple agencies need not mean that assemblages are incomparable. Furthermore, the availability of multiple lines of evidence is proving to be invaluable for paleoecological analysis and paleoenvironmental reconstructions. Simple linear summaries of relative diachronic change such as the Alcelaphini and Antilopini criteria (AAC), or the Gerbillinae:Murinae ratio provide a simple means for comparing results from independent data sets. We lack similar, simple indices for synchronic habitat descriptions, which makes pooling data for habitat reconstruction more challenging than it is for inferring habitat change.

## Acknowledgments

This research is based on portions of the author's Ph.D. Dissertation (Interdisciplinary Program in Anthropological Sciences, Stony Brook University), and was supported by grants from the Wenner–Gren Foundation, the National Science Foundation (98-13706), the Burghard–Turner Fellowship at Stony Brook University, the Ford Foundation, and the Evolution of Terrestrial Ecosystems Program of the National Museum of Natural History. This is ETE publication #96. The author gratefully acknowledges the comments of his dissertation committee, chaired by Curtis Marean, and including Peter Andrews, John Fleagle, Charles Janson, and Elizabeth Stone. Generous appreciation is directed toward the government of Tanzania, Tanzania National Parks (TANAPA), the Tanzanian Wildlife Research Institute (TAWIRI), the Frankfurt Zoological Society, Tanzania Natural History Museum (Arusha) for their support in Tanzania, and particularly to Joseph Masoy for his wisdom, patience, and hard work in the field. Museum research and analysis were aided by the generous assistance of Rainer Hutterer (ZFMK), William Stanley and Julian Kerbis-Petterhans (FMNH), Helen Kafka and Michael Carleton (NMNH), Guy Musser (AMNH). The author is also grateful to three reviewers for helpful critique and comment on the manuscript. Credit for the quality of this work rests with those mentioned above; any mistakes or inaccuracies are the failing of the author alone.

# References

Alexander, R., 1988. The scope and aims of functional and ecological morphology. Netherlands Journal of Zoology, 38, 3–22.

Andrews, P., 1983. Small mammal faunal diversity at Olduvai Gorge, Tanzania. In: Clutton-Brock, J., Grigson, C. (Eds.), Animals and Archacology: 1 Hunters and their Prey. BAR, Oxford.

Andrews, P., 1990. Owls, Caves and Fossils. University of Chicago Press, Chicago.

Andrews, P., Groves, C.P., Horne, J.F.M., 1975. Ecology of the lower Tana River flood plain (Kenya). Journal of the East African Natural History Society of the National Museum 151, 1–31.

Ashley, G., Driese S., 2000. Paleopedology and paleohydrology of a volcaniclastic paleosol interval: implications for early Pleistocene stratigraphy and paleoclimate record, Olduvai Gorge, Tanzania. Journal of Sedimentary Research 70(5), 1065–1080.

Avery, D.M., 1982. Micromammals as paleoenvironmental indicators and an interpretation of the late quaternary in the southern cape province, South Africa. Annals of the South African Museum 85, 183–374.

Avery, D.M., 1987. Late Pleistocene coastal environment of the southern Cape Province of South Africa: micromammals from Klasies River Mouth. Journal of Archaeological Science 405–421.

Avery, D.M., 1992. The environment of early modern humans at Border Cave, South Africa: micromammalian evidence. Palaeogeography, Palaeoclimatology, Palaeoecology 91, 71–87.

Avery, D.M., 1995. Southern savannas and Pleistocene hominid adaptations: The micromammalian perspective. In: Vrba, E., Denton, G.H., Partridge, T.C., Burckle, L.H. (Eds.), Paleoclimate and Evolution, with Emphasis on Human Origins. Yale University Press, New Haven, pp. 459–478.

Avery, D.M., 1998. An assessment of the Lower Pleistocene micromammalian fauna from Swartkrans Members 1–3, Gauteng, South Africa. Geobios 31(3), 393–414.

Avery, D.M., 2001. The Plio-Pleistocene vegetation and climate of Sterkfontein and Swartkrans, South Africa, based on micromammals. Journal of Human Evolution 41, 113–132.

Ba, K., Granjon, L., Hutterer, R., Duplantier, J., 2000. Les micromammifères du Djoudj (delta du Sénégal) par l'analyse du régime alimentaire de la chouette effraie, *Tyto alba*. Bonner Zoologische Beitrage 49, 31–38.

Bell, R.H.V., 1969. The use of the herb layer by grazing ungulates in the Serengeti. In: Watson A. (Ed.), Animal Populations in Relation to Their Food Resources. Blackwell Publishers, Edinburgh, pp. 111–128.

Bell, R.H.V., 1982. The effect of soil nutrient availability on community structure in African ecosystems. In Huntley, B.J. and Walker, B.H. (eds.) Ecology of Tropical Savannas, Springer, pp. 193–216.

Belsky, A.J., 1990. Tree/grass ratios in East African savannas: a comparison of existing models. Journal of Biogeography 17, 483–489.

Belsky, A.J., 1995. Spatial and temporal landscape patterns in arid and semi-arid African savannas. In: Hansson L., Fahrig L., Merriam G. (Eds.), Mosaic Lanscapes and Ecological Processes. Chapman and Hall, London, pp. 31–56.

Black, C.C., 1984. Systematics and Paleoecology of Small Vertebrates from the Plio-Pleistocene Deposits East of Lake Turkana, Kenya. Research reports—National Geographic Society, Washington, D.C., pp. 111–113.

Black, C., Krishtalka, L., 1986. Rodents, bats and insectivores from the Plio-Pleistocene sediments to the east of Lake Turkana, Kenya. National History Museum of Los Angeles County Contributions in Science 372, 1–15.

Blumenschine, R.J., Masao, F.T., 1991. Living sites at Olduvai Gorge, Tanzania? Preliminary landscape archaeology results in the basal Bed II lake margin zone. Journal of Human Evolution 21, 451–462.

Blumenschine, R.J., Peters, C.R., 1998. Archaeological predictions for hominid land-use in the paleo-Olduvai Basin, Tanzania, during lowermost Bed II times. Journal of Human Evolution 34, 565–607.

Blumenschine, R., Peters, C.R., Masao, F.T., Clarke, R.J., Deino, A.L., Hay, R.L., Swisher, C.C., Stanistreet, I.G., Ashley, G.M., McHenry, L.J., Sikes, N.E., van der Merwe, N.J., Tactikos, J.C., Cushing, A.E., Deocampo, D.M., Njau, J.K., Ebert, J.I., 2003. Late Pliocene *Homo* and hominid land use from western Olduvai Gorge, Tanzania. Science 345(6), 565–607.

Boaz, N.T., Bernor, R.L., Brooks, A.S., Cooke, H.B.S., de Heinzelin, J., Dechamps, R., Delson, E., Gentry, A., Harris, J., Meylan, P., Pavlakis, P., Sanders, W., Stewart, K., Vernieers, J., Williamson, P., Winkler, A., 1992. A new evaluation of the significance of the Late Neogene Lusso Beds, Upper Semliki Valey, Zaire. Journal of Human Evolution 22, 505–517.

Bonnefille, R., 1984. Palynological research at Olduvai Gorge. National Geographic Society Research Report, 17, 227–243.

Brain, C.K., 1981. Hunters or the Hunted? An Introduction to African Cave Taphonomy. University of Chicago Press, Chicago.

Butler, P.M., Greenwood, M., 1976. Elephant-shrews (Macroscelididae) from Olduvai and Makapansgat. In: Savage, R.J.G., Coryndon, S.C. (Eds.), Fossil Vertebrates of Africa. Academic Press, London, pp. 1–55.

Coetzee, C.G., 1972. The identification of southern African small mammal remains in owl pellets. Cimbebasia 2(4), 54–62.

Colvin, B.A., 1984. Barn Owl Foraging Behaviour and Secondary Poisoning Hazard from Rodenticide Use on Farms. Bowling Green State University, Bowling Green.

Coughenour, M.B., Ellis, J., 1993. Landscape and climatic control of woody vegetation in a dry tropical ecosystem: Turkana District, Kenya. Journal of Biogeography 22, 107–122.

Danuth, J., 1992, Taxon-free characterization of animal communities. In Behrensmeyer, A.K., Damuth, J.D., DiMichele, W.A.; Potts, R.; Sues, H.; Wing, S.L. (eds.) Terrestrial Ecosystems Through Time; evolutionary Paleoecology of terrestrial plants and animals, University of Chicago Press, pp. 183–204.

Dauphin, Y., Kowalski, C., Denys, C., 1994. Assemblage data and bone and teeth modifications as an aid to paleoenvironmental interpretations of the open-air Pleistocene site of Tighenif (Algeria). Quaternary Research 42, 340–349.

Dauphin, Y., Kowalski, C., Denys, C., 1997. Analysis of accumulations of rodent remains: role of the chemical composition of skeletal elements. Neues Jahrbuch fur Geologie und Palaontologie, Ahbandlungen 203(3), 295–315.

Davies, G., Berghe, E.V. (Eds.), 1994. Check-list of the Mammals of East Africa. East Africa Natural History Society, Nairobi.

Davis, D.H.S., 1959. The barn owl's contribution to ecology and palaeoecology. Ostrich Supplement 3, 144–153.

Davis, D.H.S., 1965. Classification problems with the African Muridae. Zoologica Africana 1, 121–145.

Dawson, J., 1963. Carbonotitic volcanic ashes in northern Tanganyika. Bulletin Volcanologique 27, 1–11.

de Graaff, G., 1960. A preliminary investigation of the mammalian microfauna in Pleistocene deposits of caves in the Transvaal system. Palaeontologia Africana 7, 59–117.

de Graaff, G., 1961. On the fossil mammalian micro-fauna collected at Kromdraai by Draper in 1895. South African Journal of Science 57, 259–260.

Delany, M.J., 1972. The ecology of small rodents in tropical Africa. Mammal Review 2, 1–42.

Delany, M.J., 1975. The Rodents of Uganda. British Museum (Natural History), London.

Delany, M.J., 1986. Ecology of small rodents in Africa. Mammal Review 16, 1–41.

Demeter, A., 1982. Prey of the spotted eagle-owl bubo africanus in the Awash National Park, Ethiopia. Bonner Zoologische Beitrage 33, 283–292.

Denbow, D.M., 2000. Chapter 12. Gastrointestinal Anatomy and Physiology. Sturkie's Avian Physiology, fifth edn. Academic Press, New York, pp. 299–325.

Denys, C., 1987a. Micromammals from the West Natron Pleistocene deposits (Tanzania). Sciences Géologiques Bulletin 40, 185–201.

Denys, C., 1987b. Rodentia and Lagomorpha 6.1: Fossil rodents (other than Pedetidae) from Laetoli. In: Leakey, M.D., Harris, J.M. (Eds.), Laetoli: a Pliocene Site in Tanzania. Oxford University Press, London, pp. 118–170.

Denys, C., 1990a. Deux nouvelles espèces d'Aethomys (Rodentia, Muridae) à Langebaanweg (Pliocène, Afrique du Sud): Implications phylogénétiques et paléoécologiques. Annals de Paléontologie 76, 41–69.

Denys, C., 1990b. First occurrence of Xerus cf. inauris (Rodentia, Scriuridae) at Olduvai Bed I (Lower Pleistocene, Tanzania). Palaontologische Zeitschrift. Z64, 359–365.

Denys, C., 1999. Of mice and men. In: Bromage, T., Shrenk, F. (Eds.), African Biogeography, Climate Change, and Early Hominid Evolution. Oxford University Press, Oxford.

Denys, C., Tranier, M., 1992. Présence d'Aethomys (Mammalia, Rodentia, Muridae) au Tchad et analyse morphometrique preliminaire du complexe A. hindei. Mammalia 56(4), 625–656.

Denys, C., Chorowicz, J., Tiercelin, J.J., 1986. Tectonic and environmental control on rodent diversity in the Plio-Pleistocene sediments of the African Rift System. In: Frostick, L.E. (Ed.), Sedimentation in the African Rifts. Geological Society Special Publication, pp. 363–372.

Denys, C., Williams, C., Dauphin, Y., Andrews, P., Fernandez-Jalvo, Y., 1996. Diagenetical changes in Pleistocene small mammal bones from Olduvai Bed I. Palaeogeography, Palaeoclimatology, Palaeoecology 126, 121–134.

Denys, C., Andrews, P., Dauphin, Y., Williams, T., Fernandez-Jalvo, Y., 1997. Towards a site classification: comparison of stratigraphic, taphonomic and diagenetic patterns and processes. Bulletin de la société géologique de France 168, 751–757.

Deocampo, D.M., Blumenschine, R., Ashley, G.M., 2002. Wetland diagenesis and traces of early hominids, Olduvai Gorge, Tanzania. Quaternary Research 57, 271–281.

Dublin, H.T., Douglas-Hamilton, I., 1987. Status and trends of elephants in the Serengeti-Mara ecosystem. African Journal of Ecology 25, 19–23.

Dublin, H.T., Sinclair, A.R.E., McGlade, J., 1990. Elephants and fire as causes of multiple stable states in the Serengeti-Mara woodlands. Journal of Animal Ecology 59, 1147–1164.

Fernandez-Jalvo, Y., Andrews, P., 1992. Small mammal taphonomy of Gran Dolina, Atapuerca (Burgos), Spain. Journal of Archaeological Science 19, 407–428.

Fernandez-Jalvo, Y., Denys, C., Andrews, P., Williams, T., Dauphin, Y., Humphrey, L., 1998. Taphonomy and palaeoecology of Olduvai Bed-I (Pleistocene, Tanzania). Journal of Human Evolution 34(2), 137–172.

Fernandez-Jalvo, Y., Andrews, P., Denys, C., 1999. Cut marks on small mammals at Olduvai. Journal of Human Evolution 36, 587–589.

Foster, J.B., Duff-Mackay, A., 1966. Keys to the genera of Insectivora, Chiroptera and Rodentia of East Africa. Journal of the East African Natural History Society 15(3), 189–204.

Fry, C.H., Kieth, S., Urban, E.K. (Eds.), 1988. Birds of Africa. Academic Press, London.

Gawne, C.E., 1975. Rodents from the Zia sand, Miocene of New Mexico. American Museum Novitates 2586, 1–25.

Genest-Villard, H., 1979. Écologie de *Steatomys opimus* Pousargues, 1894 (Rongeurs), Dendromuridés en Afrique Centrale. Mammalia 43(3). 275–294.

Gentry, A., 1978a. Fossil Bovidae (Mammalia) of Olduvai Gorge, Tanzania. Part II. Bulletin of the British Museum (Natural History) Geological Series 30, 1–83.

Gentry, A., 1978b. Fossil Bovidae (Mammalia) of Olduvai Gorge, Tanzania. Parts I. Bulletin of the British Museum (Natural History) Geological Series 29, 289–446.

Geraads, D., 1998. Biogeography of circum-Mediterranean Miocene–Pliocene rodents; a revision using factor analysis and parsimony analysis of endemicity. Palaeogeography, Palaeoclimatology, Palaeoecology 137, 273–288.

Glue, D.E., 1970. Avian predator pellet analysis and the mammalogist. Mammal Review 1, 53–62.

Grayson, D.K., 1984. Quantitative Zooarchaeology: Topics in the Analysis of Archaeological Faunas. Academic Press, New York.

Greenacre, M.J., Vrba, E.S., 1984. Graphical display and interpretation of antelope census data in African wildlife areas, using correspondence analysis. Ecology 65(3), 984–997.

Hay, R.L., 1976. The Geology of the Olduvai Gorge. University of California Press, Berkeley, CA.

Herlocker, D., 1974. Woody Vegetation of the Serengeti National Park. Texas A&M University, College Station, Texas.

Hubbard, C.A., 1972. Observations on the life histories and behaviour of some small rodents from Tanzania. Zoologica Africana 7, 419–449.

Ihaka, T., Gentleman, R., 1996. R: a language for data analysis and graphics. Journal of Computational and Graphical Statistics 5(3), 299–314.

Jaeger, J.-J., 1976. Les rongeurs (Mammalia, Rodentia) du Pleistocène inférieur d'Olduvai Bed I (Tanzanie). Ière Partie. In: Savage, R.J.G., Coryndon, S.C. (Eds.), Fossil Vertebrates of Africa. Academic Press, New York, pp. 58–120.

Jager, T., 1982. Soils of the Serengeti Woodlands, Tanzania. Agricultural University, Wageningen, the Netherlands.

Johnson, R., Wichern, D., 2002. Applied Multivariate Statistical Analysis. Prentice Hall, Upper Saddle River, New Jersey.

Kappelman, J., 1984. Plio-Pleistocene of Bed I and Lower Bed II, Olduvai Gorge, Tanzania. Palaeogeography, Palaeoclimatology, Palaeocology 48, 171–196.

Kingdon, J., 1974a. East African Mammals: Volume II A. Chicago University Press, Chicago.

Laurie, W.A., 1971. The food of the Barn Owl in the Serengeti National Park, Tanzania. Journal of the East African Natural History Society 28, 1–4.

Lavocat, R., 1965. Rodents. In: Leakey, L.S.B. (Ed.), Olduvai Gorge 1951–1961, 1: Fauna and Background. Cambridge University Press, Cambridge, pp. 17–18.

Leakey, M.D., Clarke, R.J., Leakey, L.S.B., 1971. New hominid skull from Bed I, Olduvai Gorge, Tanzania. Nature 232, 308–312.

Levinson, M., 1982. Taphonomy of microvertebrates— from owl pellets to cave breccia. Annals of the Transvaal Museum 33(6), 115–121.

Linzey, A.V., Kesner, M.H., 1997. Small mammals of a woodland-savannah ecosystem in Zimbabwe I. Density and habitat occupancy patterns. Journal of the Zoological Society, London 243, 137–152.

Lyman, R.L., Power, E., 2003. Quantification and sampling of faunal remains in owl pellets. Journal of Taphonomy 1, 3–14.

Maindonald, J., Braun, J., 2003. Data Analysis and Graphics Using R. Cambridge University Press, Cambridge.

McCune, B., Grace, J.B., 2002. Analysis of Ecological Communities. MjM, Gleneden Beach, OR.

McKee, J., 1993. The faunal age of the Taung hominid deposits. Journal of Human Evolution 25, 363–376.

McNaughton, S.J., 1983. Serengeti grassland ecology: the role of composite environmental factors and contingency in community organization. Ecological Monographs 53(3), 291–320.

McNaughton, M.M., Banyikwa, F., 1995. Plant communities and herbivory. In: Sinclair, A.R.E., Arcese P. (Eds.), Serenget II: Dynamics, Management and Conservation of an Ecosystem. Chicago University Press, Chicago, pp. 49–70.

Meester, J., Setzer, H.W. (Eds.), 1971. The Mammals of Africa: An Identification Manual. Smithsonian Institution Press, Washington, DC.

Milne, G., 1935. Some suggested units of classification and mapping, particularly for East African soils. Soil Research 4, 183–198.

Nesbit-Evans, E.M., van Couvering, J.H., Andrews, P., 1981. Palaeoecology of Miocene sites in Western Kenya. Journal of Human Evolution 10, 35–48.

Norton-Griffiths, M., Herlocker, D., Pennycuick, L., 1975. The patterns of rainfall in the Serengeti Ecosystem, Tanzania. East African Wildlife Journal 13, 347–374.

Peters, C.R., Blumenschine, R.J., 1995. Landscape perspectives on possible land use patterns for early Pleistocene hominids in the Olduvai Basin, Tanzania. Journal of Human Evolution 29, 321–362.

Pickford, M., Mein, P., Senut, B., 1994. Fossiliferous Neogene karst fillings in Angola, Botswana and Namibia. South African Journal of Science 90, 227–230.

Plummer, T.W., Bishop, L.C., 1994. Hominid paleoecology at Olduvai Gorge, Tanzania as indicated by antelope remains. Journal of Human Evolution 27, 47–75.

Pocock, T.N., 1987. Plio-Pleistocene fossil mammalian microfauna of southern Africa—a preliminary report including descriptions of two new fossil Muroid genera (Mammalia: Rodentia). Palaeontologia Africana. 26, 25–31.

Pokines, J.T., Peterhans, J.C.K., 1998. Barn owl (*Tyto alba*) taphonomy in the Negev desert, Israel. Israel Journal of Zoology 44(1), 19–27.

Potts, R., 1988. Early Hominid Activities at Olduvai. Aldine, Hawthorne, NY.

Pratt, D.J., Gwynne, M.D., 1977. Rangeland Management and Ecology in East Africa. Hooder and Stoughton, London.

Reed, D.N., 2003. Micromammal Paleoccology: Past and Present Relationships between East African Small Mammals and Their Habitats. Stony Brook University, Stony Brook, NY.

Reed, D.N., 2005. Taphonomic implications of roosting behavior and trophic habits in two species of African owl. Journal of Archaeological Science 32, 1669–1676.

Relethford, J.H., Harpending, H.C., 1995. Ancient differences in population size can mimic a recent African origin of modern humans. American Journal of Physical Anthropology Supplement 20, 180.

Rogers, M., Stanley, W., 2003. Tanzania Mammal Key. Field Museum of Natural History, Chicago, http//www.fieldmuseum.org/tanzania/.

Sabatier, M., 1979. Les rongeurs fossiles de la formation de Hadar et leur intérêt paléoécologique. Bulletin de la Société de Géologie, France 21(3), 309–311.

Sabatier, M., 1982. Les rongeurs du site Pliocene a hominides de Hadar (Ethiopie). Palaeovertebrata, 12, 1–56.

Sinclair, A.R.E., 1995a. Equilibria in plant–herbivore interactions. In: Sinclair, A.R.E., Arcese, P. (Eds.), Serengeti II: Dynamics, Management and Conservation of an Ecosystem. University of Chicago Press, Chicago, pp. 91–113.

Sinclair, A.R.E., 1995b. Serengeti past and present. In: Sinclair, A.R.E., Arcese, P. (Eds.), Serengeti II: Dynamics, Management, and Conservation of an Ecosystem. University of Chicago Press, Chicago, pp. 3–30.

Sokal, R.R., Rohlf, F.J., 1995. Biometry: The Principles and Practice of Statistics in Biological Research, third edn. W. H. Freeman and Company, New York.

Swynnerton, G., 1958. Fauna of the Serengeti National Park. Mammalia 22, 435–450.

Taylor, I., 1994. Barn Owls. Cambridge University Press, Cambridge.

Tchernov, E., 1992. Eurasian–African biotic exchanges through the levantine corridor during the neogene and quaternary. Courier Forschunginstitut Senckenberg 153, 103–123.

van Couvering, J.A., 1980. Community evolution in East Africa. In: Behrensmeyer, A.K., Hill, A.P. (Eds.), Fossils in the Making. Chicago University Press, Chicago, pp. 272–298.

Vernon, C.J., 1972. An analysis of owl pellets collected in southern Africa. Ostrich 43, 109–124.

Vesey-Fitzgerald, D.F., 1966. The habits and habitats of small rodents in the Congo River catchment region of Zambia and Tanzania. Zoololica Afrricana 2, 111–122.

Vrba, E.S., 1980. The significance of bovid remains as indicators of environment and predation patterns. In: Behrensmeyer, A., Hill, A.P. (Eds.), Fossils in the Making. University of Chicago Press, Chicago, pp. 247–271.

Vrba, E.S., 1985. Ecological and adaptive changes associated with early hominid evolution. In: Delson, E. (Ed.), Ancestors: The Hard Evidence. Alan R. Liss, New York, pp. 63–71.

Vrba, E.S., 1992. Mammals as a key to evolutionary theory. Journal of Mammalogy 73, 1–28.

Vrba, E.S., 1995. Fossil record of African antelopes (Mammalia, Bovidae) in relation to human evolution and paleoclimate. In: Vrba, E.S., Denton, G.H., Partridge, T.C., Burckle, L.H. (Eds.), Paleoclimate and Evolution with Emphasis on Human Origins. Yale University Press, New Haven, pp. 385–424.

Walter, R.C., Manega, P.C., Hay, R.L., Drake, R.E., Curtis, G.H., 1991. Laser-fusion $^{40}$Ar/$^{39}$Ar dating of Bed I, Olduvai Gorge, Tanzania. Nature 354, 145–149.

Wesselman, H.B., 1982. Pliocene Micromammals from the Lower Omo Valley, Ethiopia: Systematics and Paleoecology. University of California, Berkley.

Wesselman, H.B., 1984. The Omo Micromammals: Systematics and Paleoecology of Early Man Sites from Ethiopia. S. Karger, Basel.

Wesselman, H.B., 1995. Of mice and almost-men: Regional paleoecology and human evolution in the Turkana Basin. In: Vrba, E.S., Denton, G.H., Partridge, T.C., Burckle, L.H. (Eds.), Paleoclimate and Evolution with Emphasis on Human Origins. Yale University Press, New Haven, pp. 356–368.

Wilson, D., Reeder, D.M. (Eds.), 1993. Mammal Species of the World. Smithsonian Institution Press, Washington, DC.

Winkler, A., 1997. Systematics, paleobiogeography and paleoenvironmental significance of rodents from the Ibole Member, Manonga Valley, Tanzania. In: Harrison, T. (Ed.), Neogene Paleontology of the Manonga Valley, Tanzania: A Window into the Evolutionary History of East Africa. Plenum Press, New York, pp. 311–332.

Winkler, A., 1998. New small mammal discoveries from the early Pliocene at Kanapoi, West Turkana, Kenya. Journal of Vertebrate Paleontology 18(Supplement), 87.

Winkler, A., 2002. Neogene paleobiogeography and East African Paleoenvironments: contributions from the Tugen Hills rodents and lagomorphs. Journal of Human Evolution 42, 237–256.

Zimmerman, D., Turner, D.A., Pearson, D.J., Willis, I., Pratt, H.D., 1996. Birds of Kenya and Northern Tanzania. A&C Black, London.

# 10. Taphonomy and paleoecological context of the Upper Laetolil Beds (Localities 8 and 9), Laetoli in northern Tanzania

C. MUSIBA
*Department of Anthropology*
*University of Colorado at Denver and Health Sciences Center*
*Denver, CO 80207, USA*
*charles.musiba@cudenver.edu*


C. MAGORI
*Department of Anatomy and Histology*
*Bugando University College of Health Sciences*
*Mwanza, Tanzania*
*cmagori@buchs.org*


M. STOLLER
*Department of Ecology and Evolution*
*The University of Chicago*
*Chicago, IL 60637, USA*
*mstoller@uchicago.edu*


T. STEIN
*Department of Anatomy and Cell Biology*
*University of Michigan*
*Ann Arbor, MI 48109-0608, USA*
*tastein@umich.edu*


S. BRANTING
*Center for Ancient Middle Eastern Landscapes (CAMEL), The Oriental Institute*
*The University of Chicago*
*Chicago, IL 60637, USA*
*branting@uchicago.edu*


M. VOGT
*GIS Lab, Center for Environmental Restoration Systems*
*Argonne National Laboratory*
*Argonne, IL 60439, USA*
*mvogt@anl.gov*

*R. Bobe, Z. Alemseged, and A.K. Behrensmeyer (eds.) Hominin Environments in the East African*
*Pliocene: An Assessment of the Faunal Evidence, 257–278.*
© 2007 *Springer.*

R. TUTTLE
*Department of Anthropology*
*The University of Chicago*
*Chicago, IL 60637, USA*
*r-tuttle@uchicago.edu*


B. HALLGRÍMSSON
*Department of Cell Biology and Anatomy*
*University of Calgary*
*Calgary, Alberta, Canada T2N 1N4*
*bhallgri@ucalgary.ca*


S. KILLINDO
*Archaeology Unit*
*The University of Dar es Salaam*
*Dar es Salaam, Tanzania*
*skillindo@yahoo.uk.com*


F. MIZAMBWA
*Department of Antiquities*
*Ministry of Tourism and Natural Resources*
*Dar es Salaam, Tanzania*
*oldupai@africaonline.co.tz*


F. NDUNGURU
*Department of Antiquities*
*Ministry of Tourism and Natural Resources*
*Dar es Salaam, Tanzania*
*oldupai@africaonline.co.tz*


A. MABULLA
*Archaeology Unit*
*The University of Dar es Salaam*
*Dar es Salaam, Tanzania*
*aumab@udsm.ac.tz*


**Keywords:**    Pliocene, stratigraphy, paleoecology


## Abstract

The Upper Laetolil Beds at Laetoli in northern Tanzania contain abundant fossil mammalian remains that may help elucidate Pliocene environments and enhance our understanding of the morphological and behavioral adaptations of the Laetoli hominins. The Laetoli vertebrate fossil fauna is also of great interest because its taxonomic composition, especially in the family Bovidae, differs from that of other East African faunas of comparable age. However, the taphonomic history and paleoecological context of the Upper Laetolil fossil faunal assemblage is not fully understood. Furthermore, the depositional environment of the Upper Laetolil Beds was initially associated with a dry, savanna-like environment characterized by grassland, shrubs, and isolated trees. However, revised stratigraphy and taphonomy at Localities 8 and 9 indicate an existence of more complex depositional environments than those previously described. Fossil faunal remains from the Upper Laetolil Beds are represented by highly variable numbers of skeletal parts dominated by heavily fractured and/or modified distal, proximal, and midshaft fragments of varying sizes, and many isolated teeth. Unlike marine fossil deposits, which furnish thick sections with rich fossil accumulations of invertebrates, the Laetoli deposits are most commonly composed of low-density accumulations with terrestrial faunal assemblages that differ significantly from modern counterparts. Observed changes in faunal composition at Laetoli especially between the Lower and Upper Units have been noted, but their cause has not been established. We present a detailed taphonomic and stratigraphic analysis of fossil faunal assemblages from the Upper Laetolil Beds at Localities 8 and 9 in northern Tanzania

## Introduction

The taphonomic history of the Laetoli paleoanthropological site (S3°13 , E35°13 ) in northern Tanzania provides a basis for the formulation of ecological and stratigraphic interpretations. Laetoli, located on the western flank of the Ngorongoro Volcanic Highlands within the Serengeti Plains, consists of fossiliferous sediments that span from 4.3 Ma to 120 Ka. This site is unusual in that its paleobiota is indicative of an upland environmental setting without major rivers or lakes. The Upper Laetolil Beds in particular provide a rich and diverse fossil faunal assemblage, which includes hominins, hominoids, carnivores, rodents, and several bovid species that have been used in ecological interpretations of the ancient landscape.

Paleoecological interpretations of Pliocene Laetoli have been problematic and controversial (Harrison, 2005; Su and Harrison, 2006). For example, Andrews (1989) disputed Leakey et al.'s (1987) suggestion that the Pliocene Laetoli environments were similar to those of the modern Serengeti Plains. Furthermore, Hill's (1987, 1994) analysis of bone weathering from a small sample of Laetoli fossils raised questions about the nature of Laetoli depositional environments. Also, recent paleoecological interpretations of the Upper Laetolil Beds based on the functional morphology of bovid hind limbs

as ecological indicators at Localities 8 and 9 suggested a complex environmental setting for the Upper Laetolil Beds (Musiba, 1999; Musiba and Magori, 2005).

In this paper, we present a new taphonomic history of fossil faunal assemblages from Laetoli in northern Tanzania (Figure 1) and its implications for paleoecological interpretations of the Upper Laetolil Beds at Localities 8 and 9. We focus particularly on the fossil faunal assemblage between the dated horizons of Tuff 8 (3.46 ± 0.12 Ma) and Tuff 1 (≤3.76 ± 0.03) (Drake and Curtis, 1987; Hay, 1987; Manega, 1993) exposed at Localities 8 and 9 (Figure 1).

## Previous Studies

Past generalized ecological studies of Pliocene Laetoli characterized this site as an open savanna grassland similar to the present-day Serengeti Plains (Hay, 1980; Harris, 1985; Bonnefille and Riollet, 1987). Nevertheless, the rich and diverse faunal assemblage of the Upper Laetolil Beds suggests a much more complex environmental setting (Musiba, 1999; Harrison, 2005; Musiba and Magori, 2005). Most of the faunal assemblage, particularly the bovids, rodents, and primates, are diagnostic of specific environmental settings that are no longer present within the Serengeti region. Evidence from fossil pollen collected near the bottom of the

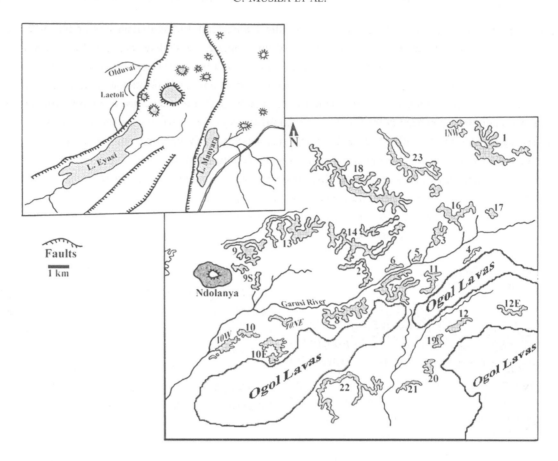

Figure 1. Map of Laetoli showing the paleontological localities with Laetolil Beds.

upper part of the Laetolil Beds, at Tuff 1 and below Tuff 7, suggests vegetational similarities between the Pliocene and the modern Serengeti Plains (Bonnefille and Riollet, 1987). However, it has been suggested that the pollen data are diagnostically and stratigraphically very problematic (Andrews, 1989). For instance, pollen samples used in earlier interpretations derive from sediment traps on termite mounds, and the collected samples contained a high proportion of non-descriptive woodland species that are currently not present within the Serengeti. Bonnefille and Riollet's (1987) palynological spectra of the Laetoli flora revealed high proportions of unknown pollen. This indicates the possible existence of mosaic environments very different from those of the modern Serengeti landscape.

Furthermore, some of the fossil faunal remains and other geologic evidence from the Upper Laetolil Beds are also partly inconsistent with the analogy of the modern Serengeti environments. For example, the Hippotragini are not as abundant at Laetoli as they are in the modern Serengeti (Gentry, 1981, 1987). According to Gentry (1981), "As far as bovids are concerned Laetoli is a strange locality … its faunal assemblage is less similar to present day Serengeti fauna …" The Pliocene bovid community from Laetoli is characterized by a low number of species with many incomplete or uncertain identifications. For example, *Aepyceros* and Reduncini are rarely represented in the Laetolil Beds. According to Gentry (1981) the absence of these bovids combined with the abundance of *Madoqua* indicates a dry-country fauna. However, the distribution of *Madoqua* species in the fossil assemblage within the Laetolil Beds also differs from their modern counterparts and many have no modern representatives (Gentry, 1981, 1987). Furthermore, the presence of Tragelaphini, Cephalophini, and two indeterminate Bovini species at some localities suggests that part of Pliocene Laetoli was characterized by wooded habitats (Gentry, 1987; Bobe et al., 2002).

Additionally, the presence of two species of Chelonia (*Geochelone laetoliensis* and *Geochelone brachygularis*) in the assemblage also indicates that Laetoli Pliocene environments might have contained patches of arid grassland (Meylan and Auffenberg, 1987). However, a modern counterpart of the subgenus *Aldabrachelys*, which is represented by a gigantic tortoise currently seen in the islands of the Pacific and Indian Ocean, inhabits coastal woodland or forest (Meylan and Auffenberg, 1987). In addition, the presence of *Python sebae* suggests the existence of a body of water within the Laetoli area (Meylan and Auffenberg, 1987). Fossil gastropods recovered at Laetoli, however, do not conform to continuous woodland forest or open grassland cover (Verdcourt, 1987; Pickford, pers. communication). However, fossil gastropods from Laetoli, which are represented by Achatinidae (*Achatina zanzibarica*), *Burtoa nilotica*, *Limicolaria martensiana*, *Subulona pseudinvoluta*, and *Trochonanina* sp. are mainly inhabitants of open woodland, coastal forest, and upland forest (Verdcourt, 1987). Modern counterparts of the Laetoli gastropods require moist environments with precipitation ranging from 750 to 2000 mm (Verdcourt, 1987).

The conflicting evidence from previous studies presented here suggests a broad spectrum of environmental settings at Laetoli during the Pliocene and may reflect possible taphonomic biases (spatial mixing or environmental condensation) or may indicate a complex mosaic of environments (Andrews, 1989; Harrison, 2005; Musiba and Magori, 2005). An alternative explanation for the contradictory evidence would be sampling biases, and misinterpretation of the almost indistinguishable strati-graphic events that characterize the sequences in question (Gifford, 1981). Andrews (1989) for example has suggested that sampling biases (mixing by collectors of material from different beds) might be directly responsible for the poor environmental resolution in previous studies of the Laetolil Beds.

The presence of cercopithecines and colobines in the Upper Laetolil Beds (Leakey and Delson, 1987) is also indicative of mosaic environments capable of supporting various primate groups (Kappelman, 1991). In modern African environments it is very rare that more than two closely related cercopithecoid species would occupy the same ecological niche without risking competition and other selective pressures. Of 15 species of rodents represented at Laetoli, *Saccostomus* (a dry savanna inhabitant), *Thallomys* (dependent on *Acacia* woodlands), and *Heterocephalus* (semi-arid inhabitant) dominate the Upper Laetolil fossil assemblage (Denys, 1987). These rodent taxa, if considered as a single community, would support the open-country savanna model at Laetoli 3.5 Ma. However, such a community of several species of rodents occupying a wide range of ecological niches rarely exists in modern savanna ecosystems (Wesselman, 1985). When individually treated, the rodents offer conflicting ecological information (Denys, 1987). The only exception is the naked mole rat (*Heterocephalus*), which can withstand extreme temperatures (Kingdon, 1974, 1997). Therefore, species diversity for rodents is too high for a single community from a very dry savanna habitat (Andrews, 1989).

Additionally, insectivores, which are sparsely represented at Laetoli, contradict the savanna-like model. Only a single species of a giant elephant shrew (*Rhynchocyon pliocaenicus*) has been identified within the Upper Laetolil Beds. This species is somewhat smaller than its living congener (Butler, 1987), which is habitat-specific and requires either gallery forest, woodland and/or closed tree, or bush cover with abundant leaf litter (Corbet and Hanks, 1968).

Evidently, past paleoecological interpretations of the Upper Laetolil Beds present various taphonomic problems including sampling, sediment mixing, and preservation biases. It is however unmistakable that the interpretation of Laetoli paleoenvironments as being drier than today relied exclusively upon the taxonomic distribution of fossil faunal remains and the problematic pollen evidence collected at various localities within the Laetoli area. The fossil faunal assemblages and stratigraphic resolution used in the reconstruction of Laetoli paleoecology represent multiple events of stratigraphically indistinguishable units of well-mixed horizons (Musiba and Magori, 2005). The existence of burrowing animals and termitary mounds within the Laetolil Beds are clearly good indicators of bioturbation and sediment mixing.

## Laetolil Beds in Paleoecological Context

The Laetolil sediments, deposited on a basement rock of basalt origin, also provide important information relevant to the discussion of Laetoli depositional environments and are further reviewed in detail here. The Upper and Lower Laetolil Beds were entirely deposited on land, on the crest and flanks of a broadly uplifted dome overlying the Precambrian bedrock in the Eyasi Plateau. The Laetolil Beds occur in a series of shallow outcrops with many discontinuous exposures spreading about $1600 \, km^2$ to the south and west of Lemagruti, and to the northwest of Lakes Masek and Ndutu (Hay, 1987; Manega, 1993). The Laetolil Beds preserve a unique type of fossil record of hominin footprints and animal trackways that have been dated to 3.5 Ma and provide a snapshot of past environments at Laetoli.

The stratigraphic and taphonomic context of the Upper Laetolil Beds particularly at Localities 8 and 9 is indicative of numerous taphofacies including bioturbation, sediment mixing, and overprinting of short episodes of depositional

environments, especially the last 60 m within the unit (Figure 2). A generalized description of the columnar section of Plio-Pleistocene Laetoli sediments by Hay (1987) indicates that lithologically the area consists of deposits characterized by lava flows, tuffs, and clay stones. The deposits are mainly of nepheline-phonolite, melilitite-carbonatite in composition and/or eolian tuff in origin (Hay, 1978). Because of the varying degrees of exposures and weathering of the Laetolil Beds from one locality to another, the lithologic description provided below is based on observations made at Localities 8, 9, and 9S. Detailed lithologic information has been provided by Manega (1993) and Hay (1987).

The lower unit of the Laetolil Beds consists of graded water-worked tuffs, Lapilli Tuffs, and conglomerates (in the upper 30 m) that are chemically easily identifiable. The unit consists of mudflow deposits, eolian- and water-worked tuffs with channel fillings from the eroded Ogol Lavas, and few layers of conglomerates and breccia (Hay, 1987; Manega, 1993). The topmost part of this unit, however, is about 75% reworked tuff of eolian origin with numerous thin water-worked tuff layers that are 45–60 m thick, indicating the existence of a substantial amount of water in the area during and after their deposition. The remainder of the Upper Laetolil Beds is composed of approximately 20% air-fall volcanic ash. One to two percent of the upper unit consists of easily distinguishable stream-reworked tuffs (Hay, 1987). The water-worked tuffs within the unit are generally composed of fine- to coarse-grained, moderate- to well-rounded, and highly indurated tuffs. They are well sorted with thin laminae that vary in thickness. The water-worked tuffs are also dominated by clastic deposits, which compose about 90% or less of the entire unit. This sedimentary evidence points toward a set of complex depositional environments.

At Locality 8 for example, the exposed Upper Laetolil Unit exhibits a 120°–210° SW strike and a 5° to 10° SW dip with a two-joint system

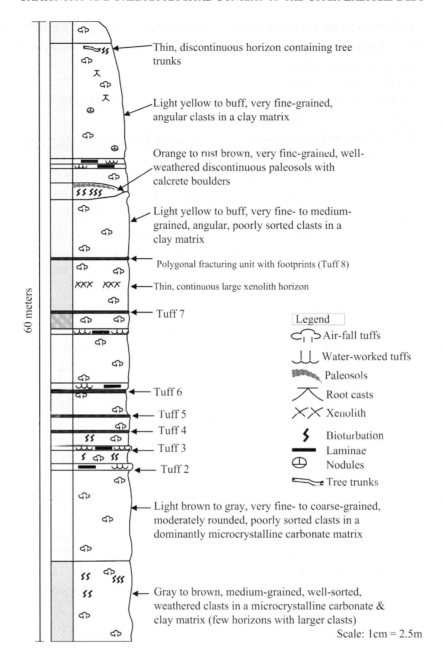

Figure 2. Detailed stratigraphic profile of the Upper Laetolil Beds.

(vertically and horizontally) filled with calcite material of varying size and width ranging from 1 to 60 cm thick. The calcite-filled joints can easily be distinguished within the exposed Upper Laetolil Beds at Localities 8 and 9. The thickness of the two joints at Locality 8 tends to increase as one moves northward towards and beyond the footprint site G. An open joint of about 40 cm thick characterized by a 210° SW and 195° NE striking system with a NE/SW dipping trend was recorded northeast of the footprint site. Similarly at Locality 9, southwest of Locality 8 the same trend was noted. The sediments here tend to be dominated by air- and water-fall tuffs that are distinctively laminated. The laminae are closely interbedded and vary in thickness (12–15 cm). Therefore, they indicate the possible existence of a substantial amount of water that may well have supported a variety of flora and fauna.

Exposures about 3 m thick located northeast of the footprint site G are composed of sediments that reveal evidence of intensive bioturbation. The sediments in this area consist of deposits that are downgraded, heavily worn, reworked, and loosely packed. These deposits consist of laminated layers with medium- to fine-grain sands. About 150 m southeast of site G, the Upper Laetolil exposures bear a 300° SW strike with a 6° SW magnitude dip. The exposures here consist of a sorted fine to medium topmost layer about 60 cm thick, subdivided into sublayers of 15- to 45-cm thick loosely packed tuffs.

Additionally, stream-worked tuffs occur in several places and they are also very common above Tuff 6, where most of the deposits are lenticular and cross-bedded with channel filling that ranges from 0.15 to 1.2 m deep. These deposits grade from finer- to coarser-sorted sediments consisting of xenoliths, airfall, and eolian tuffs. Furthermore, Manega (1993) noted that some sediment mixing occurs above Tuff 7 where stream-worked deposits 45 to 60 cm thick consist of material from younger deposits.

Root casts and nodules are also very common throughout the Upper Laetolil Beds. Most of the root casts are grayish in color, cylindrical in shape, and range in diameters from 1 to 13 cm (Figure 2). The infilling of the root casts consists of calcite material (cryptocrystalline) with rare inclusions of volcanic tuffs. The nodules, which are divided into three types, range in diameter from 1 to 45 cm and consist of: (a) microcrystalline calcite nodules that are circular and uniform in composition, (b) well-undulated tuff nodules that are also circular to oval and uniform in composition, and (c) small nodules with massive rings of tuffs that are separated by thin layers of calcite. From a stratigraphic and lithologic standpoint the Upper Laetolil Beds contain very distinctive sediments with depositional histories that are very important in paleoecological interpretations. This is particularly the case with the water-worked tuffs described above.

## Material and Methods

The data for this study originated from two sources: an excavation at Locality 9 (n = 483) and a systematic surface collection at Localities 8 and 9 (n=812; Figure 3). The fossils were collected using a 100% surface collection method established at Laetoli by Ndessokia (1990). The collection included cranial and postcranial fragments, with few complete and *in situ* bone remains.

At Locality 9, a grid system consisting of three surface collection sections and one trench was established with a north/south bearing datum line. The three surface collection areas (Sections A, B, and C) were established on a 10 × 10 m grid system along the datum line. All exposed bones on each grid were first flagged (Figure 4) and numbered before their orientation relative to the datum line was established and recorded using a Brunton compass. Also the distance from one specimen to the nearest was recorded before the specimens were tagged and bagged. All recovered material, regardless of size, was tagged and recorded. These data were used to generate Rose diagrams of bone orientation. Later on, the sediments within the grid system were swept and screened on a 10 × 10 mm sieve. No floatation recovery technique was used in order to recover fragments that were smaller than 10 mm thick due to lack of sufficient water within the area. The contemporary Laetoli landscape is very dry during the dry season and nearest water sources, which are also used by the nomadic Maasai pastoralists, are located about 20 to 35 km away from Localities 8 and 9. Furthermore, the Ngarusi (Garusi) River, which is a main drainage within the Laetoli area, is seasonal and usually dries out during the dry season.

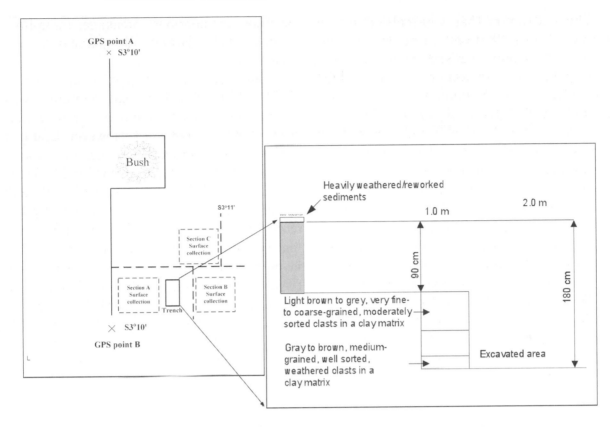

Figure 3. Excavation plan and surface collection grid at Locality 9.

Figures 4. Photo showing flagged and mapped surface finds at Locality 9.

The excavation that was carried out at Locality 9 in 1998 produced numerous complete small mammalian specimens including four pellets of articulated rodents, which are currently under study by Julian Kerbis (at the Field Museum of Natural History in Chicago) and will be published separately in the near future. Small rodents, particularly Gerbillinae and Murinae, dominate most of the burrowing animal assemblages that have been reported from the Upper Laetolil Beds at several localities (Davies, 1987; Denys, 1987). The excavation was conducted in order to establish a control sample to assess the taphonomic biases associated with surface collections at Localities 8 and 9. Taphonomic processes identified and recorded from the assemblage include the following: bone distribution pattern, disarticulation, pitting,

puncture, crenulations, breakage, trampling, etching, root marking, desquamation, and weathering.

Additionally, a small portion of Leakey's (1976–1979) fossil faunal remains was included in the analysis of bone surface modification in our 1998 study. However, Leakey's (1976–1979) field team collected only specimens that were large enough to identify. Small and unidentifiable specimens were collected, counted, and left in piles on the site by her team (Table 1; Figure 5). This method of surface collection produced a sampling bias that favored only robust and easily identifiable bones. Contrary to Leakey's collection strategy, our team systematically recorded all exposed bones in all transects within the established datum line. All bones were tagged, counted, identified, and measured (in terms of length

*Table 1. Leakey's table of vertebrate and invertebrate faunal specimens collected and listed from the Laetolil Beds*

| Taxa | NISP[1] for collected fragments | NISP for fragments left on site | Total |
|---|---|---|---|
| Gastropoda | 316 | 2 | 318 |
| Slug (mantles) | 171 | – | 171 |
| Chelonia (complete) | 8 | – | 8 |
| Chelonia (scutes/bones) | 6 | 132 | 138 |
| Reptilia (various) | 16 | – | 16 |
| Struthionidae (eggshell fragments) | 57 | – | 57 |
| Aves indet. | 31 | – | 31 |
| Soricidae | 8 | – | 8 |
| Lorisidae | 5 | – | 5 |
| Cercopithecidae | 87 | – | 87 |
| Hominidae | 25 | – | 25 |
| Rodentia | 331 | – | 331 |
| Leporidae | 1959 | 902 | 2861 |
| Pedetidae | 191 | 3 | 194 |
| Carnivora | 261 | – | 261 |
| Deinotheriidae | 23 | – | 23 |
| Elephantidae | 97 | 131 | 228 |
| Orycteropodidae | 11 | – | 11 |
| Equidae | 171 | 94 | 265 |
| Chalicotheriidae | 5 | – | 5 |
| Rhinocerotidae | 267 | 377 | 644 |
| Suidae | 234 | 90 | 324 |
| Giraffidae | 456 | 352 | 808 |
| Sivatheriidae | 57 | 1 | 58 |
| Bovidae | 1,684 | 1,597 | 3,281 |
| Totals | 6,420 | 3,738 | 10,158 |

[1] Number of Identified specimens.

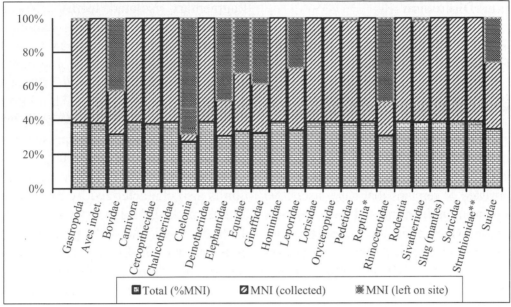

* Various    ** including eggshells

Figure 5 Taxonomic distribution of fossil faunal remains within the Laetolil Beds (including data by Leakey et al., 1987).

and thickness). Only unidentifiable bones that were <1cm long, which could not be identified and used for any taphonomic analyses, were collected but not used in the analyses.

We used Leakey's 1976–1979 Laetoli sample to establish surface bone modification. From that sample, two categories of bone surface modification were established using a scope with a 10× magnification. The two categories we established are physical damage (trampling, breakage on distal ends of long bones such as metapodials, gnawing on distal and proximal ends of long bones, etching, root marks, and boring on shafts of long bones) and chemical alterations (bone surface weathering and exfoliation).

Bone surface modification, particularly chemical alteration, of the Laetoli fossils from Localities 8 and 9 was recorded following the classification system of Andrews and Cook (1985). This system distinguishes five categories of modification: little, moderate, intermediate, great, and extreme. Furthermore,

the system takes into account other biogenic and physical modifiers of bone surfaces in archaeological records. Only three categories from Andrews' (1990) method were used in this study: little (L), moderate (M), and extreme (E) to produce a taphonomic history of bone surface modification of fossils from Localities 8 and 9. Because the differences between "great" and "extreme" as well as "moderate" and "intermediate" are difficult to distinguish, the two categories were not used for this study. Additionally, taphonomic variables such as exfoliation, secondary striations and/or scratch marks, and fragmentation were also recorded. Other bone surface modifications such as polishing or abrasion, striations, trampling, cracking, superficial weathering, grooving, pitting, boring, and color variations were also recorded. Behrensmeyer's (1978) method of bone-weathering classification was also used to score weathering patterns on the fossils from the Upper Laetolil Beds.

## Results and Discussion

### FIELD-BASED OBSERVATION OF BONE SURFACE MODIFICATION

Splitting and cracking, which are very common in the Laetoli faunal assemblage, signal time-averaging events. During the 1998 field season at Locality 9, we noticed that newly eroded fossil bones shortly after the rain season in the Serengeti Plains consisted of remains that have distinctive breakage pattern on the distal and proximal ends of long bones (Figure 6). An actualistic pilot study conducted during the 1998 field season revealed that fragments of densely eroding bones, which have partially been exposed, greatly suffer from animal trampling. Broken fragments are easily transported both horizontally and vertically while the remaining, partially buried pieces, are impacted into the ground through compaction and trampling. Trampling is thus responsible for both vertical and horizontal transportation within the Laetoli landscape particularly on dense small-to medium-sized bones.

### GENERAL PATTERNS OF BONE ASSEMBLAGE FROM LOCALITIES 8 AND 9

The Laetoli fossil fauna from Locality 9 consists of highly variable numbers of skeletal parts, of which 20% are tali, 18% dentition (including fragmentary isolated teeth), 13% calcanei, 12% metapodials, about 8% femora, and 6% vertebrac (Figure 7A). Locality 8 however is characterized by highly fragmented long bones, which make up about 71% of the total collection (Table 2; Figure 7B). Furthermore, the assemblage is characterized by highly fragmented bones with varying surface modifications including cracking, sharp breakage, and splitting of proximal and distal ends. The collection is also characterized by 32.5% weathered, 25.2% cracked, 15.1% exfoliated, 8.7% pitted, 4.1% trampled, and 4.5% polished or rounded bones (Table 3; Figure 8). Likewise, bone surface modifications on teeth and horn cores are common and are usually characterized by cracking and splitting.

### BONE DISTRIBUTION PATTERN

Overall long bone fragments, particularly midshafts, dominate skeletal representation in the fossil assemblage from Localities 8 and 9. Indeterminate shaft fragments (25%), mostly trampled, weathered, and exfoliated, are very common in the assemblage. At Locality 8, the assemblage consists of 54.8% complete tali, 17.7% distal and proximal radii, 12.8% distal and proximal metapodials, 8% complete calcanei, 3% distal and proximal ends of femora, 1.6% scaphoids, and 1.6% pelvis fragments.

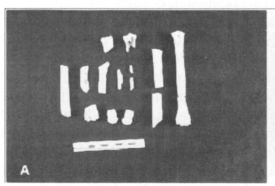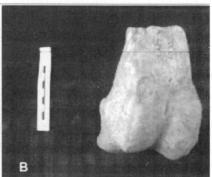

Figure 6. Breakage pattern of long bones at Localities 8 and 9, Upper Laetolil Beds.

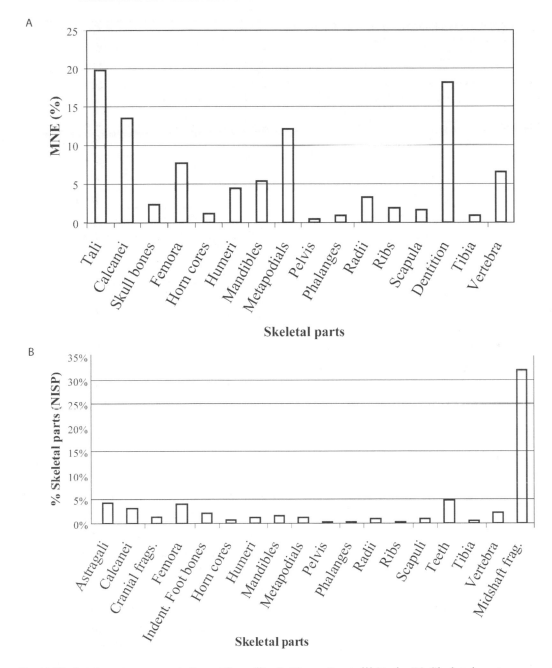

Figure 7. A) Skeletal part representation at Locality 9, Upper Laetolil Beds. B) Skeletal part representation of fossil faunal remains (100% surface collection) at Locality 8, Upper Laetolil Beds.

Scattered long bone and flat bone fragments occur, some *in situ*, and constitute about 95% of the assemblage at Localities 8 and 9 with a NE–SW orientation at Locality 9, and NW–SE bearings at Locality 8 (Figure 9). Isolated teeth and dental fragments of medium- to large-sized bovids (tentatively identified as *Alcelaphus* and *Redunca*) are also common, as are mandibular fragments of small to large ungulates (Gentry, 1987). Small rodents, carnivores, and burrowing animals are also present at Localities 8 and 9. Burrowing animals provide the most complete skeletal remains at Laetoli.

Table 2. Skeletal parts representation from fossil faunal assemblage at Locality 8 (surface collection, n = 812)

| Skeletal parts | NISP |
|---|---|
| Astragali | 35 |
| Calcanei | 25 |
| Cranial fragments | 10 |
| Femora | 33 |
| Indent. tarsal bones | 17 |
| Horn cores | 5 |
| Humeri | 9 |
| Mandibles | 13 |
| Metapodials | 9 |
| Pelvis | 2 |
| Phalanges | 2 |
| Radii | 7 |
| Ribs | 2 |
| Scapulae | 7 |
| Teeth | 39 |
| Tibiae | 4 |
| Vertebrae | 18 |
| Midshaft fragments | 263 |
| Indet. long bones | 312 |
| Total | 812 |

Table 3. Frequencies of taphonomic processes observed on the fossil faunal assemblage from Upper Laetolil fossils (n = 483) at Localities 8 and 9

| Taphonomic process | Total frequency |
|---|---|
| Boring | 12 |
| Cracking | 122 |
| Desquamation | 73 |
| Gnawing | 6 |
| Pitting | 42 |
| Polishing | 17 |
| Root-marks | 9 |
| Rounding | 22 |
| Trampling | 20 |
| Staining | 3 |
| Weathering | 157 |
| Total | 483 |

The dispersal pattern of fossil faunal remains at Laetoli, especially between Tuffs 6, 7, and 8, indicates that the assemblage was time averaged. However, intense volcanism within the Laetoli area during the Pliocene caused short episodes of deposition that resulted in rapid burial, entrapment, and fossilization of small burrowing mammals. Such events, which have been documented

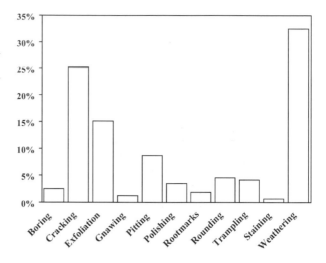

Figure 8. Observed patterns of bone surface modification of the Upper Laetolil fossil faunal assemblage at Locality 9.

by Hay (1987), Drake and Curtis (1987), and Manega (1993), tend to create high fidelity fossil faunal assemblages. For example, rapid sediment deposition in the Upper Laetolil Beds (above Tuff 6) might have caused a maximum amount of spatiotemporal resolution because the rate of bone destruction we observed is lower than that recorded by Leakey and her coworkers in 1987. Fossils recovered above Tuff 6 show no signs of chemical and physical alteration, and they are less fragmented. A similar phenomenon was also observed at various localities within the Laetolil Beds by Leakey and her coworkers in 1987.

Peculiarly, very few immature individual bones (unfused) were recorded for the fossil assemblages from Localities 8 and 9 (<10%). Furthermore, the assemblage is of attritional nature, comprising 95% of bones from mature and 5% from immature individuals of highly diverse species of bovids, leporids, and giraffids. Remains of equids, suids, lagomorphs, rodents, and carnivores are few at Localities 8 and 9.

SURFACE MODIFICATION

Broad contiguous shallow scrap marks on proximal and distal ends of long bones were observed on the fossil assemblage from Localities 8 and

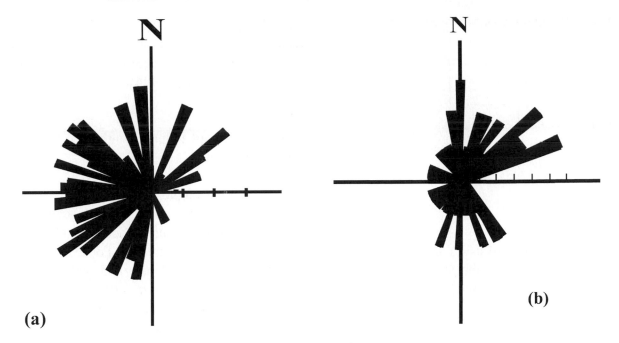

Figure 9. Bone orientation (surface collection) from (left) transects A and B and (right) transect C at Locality 9 (Upper Lactolil Beds).

9 by our team. Such bone surface modifications are normally associated with porcupine bone accumulations (Brain, 1981). However, porcupine species are not common in the Upper Laetolil Beds. Therefore, it is difficult to associate porcupines alone with the surface modification of the fossil assemblages. However, insectivores and carnivores are fairly well represented in the Laetoli fossil fauna (Table 1). The precise identity of the gnawing agent(s) for the Upper Laetolil Beds assemblage is not definitively known, but the gnawing marks on the bones resemble those of carnivorous animals. Perhaps a detailed study of gnawing marks on the Laetoli fossil assemblages will help elucidate some of the taphonomic agents responsible for surface modification and accumulations on the paleolandscape.

It is apparent that damage on long bones from the assemblage indicates that a predator or scavenger with a powerful masticatory apparatus must have applied extensive pressure on both proximal and distal ends of these long bones after they were disarticulated. In modern savanna environments this pattern of bone damage is associated with hyena scavenging activities (Blumenschine,

1989; Hill, 1979, 1980; Selvaggio and Wilder, 2001). Although Laetoli hominid remains are part of the assemblage, their presence does not suggest that they were agents of bone accumulation at Localities 8 and 9. So far there is no conclusive evidence to support such hominid activities at Laetoli. No cut marks on fossil bones from these localities have so far been documented.

## WEATHERING PATTERN

The Laetoli fossil assemblage is dominated by heavily to moderately weathered bone fragments with polishing (abrasive) signatures that indicate prolonged periods of surface exposure. Prolonged exposure also resulted in exfoliation and rounding particularly on the proximal and distal ends of long bones. Exfoliation usually occurs in alkaline environments and has been reported at Olduvai Gorge (Fernández-Jalvo et al., 1998). Following Behrensmeyer (1978) and Tappen (1994), four stages of bone weathering have been recorded at Laetoli. Most of the surface collection is heavily affected by stage IV of bone weathering

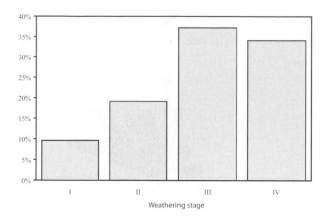

Figure 10. Observed stages of weathering patterns on fossil faunal assemblage from Upper Laetolil Beds (Localities 8 and 9).

(Figure 10). While remains from the excavated trench at Locality 9 show weathering stages II and III, most of the surface collection at this locality is characterized by stages III and IV. This weathering pattern is a good indication of prolonged exposure and time-averaging, especially when combined with cracking and superficial striation.

## BIOTURBATION AND SEDIMENT MIXING

Significant signs of bioturbation were observed at Localities 8 and 9 and have been reported by Hay (1987). Extensive burrowing by termites and land snails is very common and frequent in the Upper Laetolil Beds (Verdcourt, 1987). Burrowing as a taphonomic process has an impact on bone distribution particularly on small, dense, and compact skeletal remains (Behrensmeyer, 1983; Behrensmeyer and Kidwell, 1985). Burrowing and bioturbation generate mixed assemblages. Bioturbation by earthworms, burrowing animals, and termites results in massive reworked sediments that are loosely packed. They can easily allow denser skeletal parts from small mammals (rodents and insectivores) to settle down as loosely packed material within the sediments, thus creating highly mixed assemblages. This kind of mixing process was previously reported at other upland archaeological sites (Martin, 1982; Grant, 1983; Stein, 1983; Armour-Chelu and Andrews, 1994).

## Taphonomy and Paleoecological Interpretations

Stratigraphic, geologic, and biological evidence from Localities 8 and 9 indicate that varying depositional conditions occurred at Laetoli around 3.5 million years ago. For example, the presence of fine laminae and volcaniclastic sediment deposits rich in clay indicates wet climatic conditions, whereas the occurrence of a thin layer of calcite within the Laetolil Beds indicates extreme dry conditions sometime during the middle Pliocene. Also the data derived from different taphofacies and lithofacies in Localities 8 and 9 do not entirely support the suggested savanna-like paleoecological models. For example, rich and diverse floral and faunal assemblages that are diagnostic of specific environmental conditions pose a challenge in constructing assemblage fidelity and stratigraphic acuity (Kidwell and Behrensmeyer, 1988; Behrensmeyer, 1991; Behrensmeyer and Hook, 1992; Behrensmeyer and Chapman, 1993).

The Upper Laetolil Beds, made up of thick layers of air-fall tuff deposited in at least six series of volcanic eruptions from a single source at Lemagruti, represent a time-averaged sequence characterized by weathered sediments and mud cracks produced by extended periods of exposure. The presence of intensely weathered paleosols within this layer suggests that Laetoli experienced highly variable climatic cycles characterized by extremely wet and dry conditions. The weathering of the paleosols reflects substantial precipitation, while bioturbation indicates that the environment may have been highly productive to sustain a wide range of insects. Bioturbation in the Upper Laetolil Beds implies that leaf litter of light to heavy vegetation cover may have characterized the environment. This is further supported by the presence of tree stumps, root casts, and water-worked tuffs of varying degrees. Termitary impressions also support this scenario, while the presence of tree trunks confirms the existence of woodlands within the Laetoli area.

Sedimentological, lithological, and fossil evidence at Localities 8 and 9 indicate that during the Pliocene Laetoli experienced moist and more productive environments than today. The frequency of runoff tuffs with their massive thickness and the high density of termitary tunneling in the Upper Laetolil Beds suggest that sufficient water was available to rework the tuffs over prolonged periods. The presence of exfoliation and root marks, overlapped by striations and cracks on bones, indicates that the assemblage experienced periods of exposure long enough to alter their surface structure. This exposure, however, may not be used to indicate the existence of dry or open country habitats at Laetoli during the Pliocene; these periods of exposure might represent short episodes of extremely variable temperatures followed by wet and cooling phases.

Sedimentation processes in the Laetolil Beds produced stratified sequences of varying degrees of time-averaging (Drake and Curtis, 1987; Hay, 1987). This is consistent with geological observations by other workers. Several episodes of overprinted depositional environments in the footprint tuffs are well documented through sediment mixing by Hay (1987, 1980) and Manega (1993). Furthermore, stratigraphic descriptions by Manega (1993) also provide signs of massive erosion, sediment mixing, reworking, and eventually redeposition that would mask various time-averaged hiatuses.

Although small mammals are good climatic indicators, their representation in the Laetoli fossil assemblages seems to be obscured because they were heavily affected by many taphonomic processes including weathering, fragmentation, predation, and both biological and physical mixing (Hay, 1987; Hill, 1987). Transportation and mixing by biological and geophysical agents obscured the specific composition of small-mammalian remains in the fossil assemblages. This conclusion is supported by evidence from fossilized owl-pellets of regurgitated small mammals indicating the occurrence of long-distance bone transportation within the Laetoli area.

Unfortunately, slow-rate accumulations tend to smooth out short-term shifts in ecological conditions. Changes in faunal compositions at Laetoli have been noted (Harris, 1985; Harrison, 2005; Gentry, 1987), especially between the Lower and Upper Units as well as between the Upper Laetolil and the Ndolanya Beds (Kovarovic et al., 2002). Yet, the cause of these changes has not been fully established. It is unclear whether these faunal changes through time were caused by tectonic, taphonomic, or climatic factors.

Consequently, repeated depositions of volcanic tuffs followed by prolonged periods of precipitation resulted in varying degrees of sedimentary weathering and bone surface modification within the Laetolil Beds. The varying degrees of sedimentary weathering, especially in the Upper Laetolil Beds, may represent several climatic cycles with significant changes that might have taken place during the Pliocene. These changes are also reflected in the distribution of fossil assemblages. For example, the Upper Unit of the Laetolil Beds is richer in fossil remains than the Lower Unit, with the highest fossil faunal concentration occurring just above and below Tuff 6. However, the fauna is composed of highly fragmented long and flat bones. The assemblage includes fragmented mandibles, isolated teeth, fragmented cranial bones, and occasionally articulated small burrowing mammals.

The geologic and stratigraphic information of the Upper Laetoli Beds at Localities 8 and 9 supports a mosaic environmental setting characterized by grassland and galleries of woodland for Pliocene Laetoli. In addition, studies based on functional related morphological characters of bovid femoral heads and metapodials by Musiba (1999) also favor the model of mosaic environments at Laetoli, especially at Localities 8 and 9. Therefore, hominins from the Upper Laetoli Beds, though not so abundant, were part of a mosaic environment. The Upper Laetoli Beds preserve a record of prolonged changes in the environmental settings with a continuous presence of hominins and other mammals. The Laetoli mosaic environment and its varying relief may have been very crucial in preserving the ichnofossil record that documents evidence of hominin footprints and animal trackways during the Pliocene. Its complex geology and stratigraphy makes Laetoli an important site for detailed paleoanthropological studies in the future, especially of the ecological settings and the affinities of the Laetoli footprint makers.

## Conclusion

The fossil faunal assemblage from Laetoli is dominated by heavily fragmented long bones, particularly on the distal and proximal ends, and midshafts (Table 2). Few complete bones were recovered at Localities 8 and 9 in the Upper Laetolil Beds during our fieldwork. Bone modifications on this assemblage include boring, tunneling, exfoliation, cracking, and weathering (Table 3). Trampling and bone weathering (stages II–IV) were the most dominant processes affecting the fossil assemblage at Localities 8 and 9. However, small mammalian bones were much more affected than bones from medium to large mammals. Rounding and polishing due to water transportation on proximal and distal ends of long bones is common in the Upper Laetolil fossil assemblage. Exfoliation tends to affect both exposed and *in situ* bones by staining them to appear as if they were charred. This process, which is very common at Laetoli, is attributed to mineral leaching from the sediments into bone matrix. In general, bone modifications of the faunal assemblage point towards the presence of streams or channels.

Taphonomic observations and stratigraphic interpretations of the fossil assemblage from Upper Laetolil Beds at Localities 8 and 9 are summarized below. Localities 8 and 9 consist of highly fragmented bone assemblages that point toward multiple depositional events during the Pliocene. Taphonomically, these assemblages are characteristic of active mass accumulation that may have been created by carnivores, rodents, and in some cases predatory birds.

Bone distribution is sparsely scattered and heavily dominated by compact, round, and dense bones as well as long bones. Tali (20%) and radii (18%) dominate the assemblage as the most common skeletal parts. Calcanei and metapodials represent about 14% of the total skeletal parts in the fossil assemblage. Scaphoids and pelvis fragments are least represented in the assemblage, while isolated teeth, which are highly fragmented, make up only about 18%. The highly fragmented teeth are also heavily exfoliated, probably as a result of secondary mineralization after reexposure.

The disarticulation pattern is also nonspecific, with the exception of occasional trap specimens from burrows. The Upper Laetolil faunal assemblage from Localities 8 and 9 reflects several phases of cyclic environmental conditions dominated by phases of low-energy regimes. In addition, the assemblage indicates a high degree of predation and scavenging activities as reflected in bone-breakage patterns, gnawing, and tooth marks. In general, the order of disarticulation of the assemblage influenced other processes such as bone surface modification, transportation, and spatial patterning of bones on the landscape. Of particular interest in regard to disarticulation pattern is chemical and physical weathering, which is indicative of a prolonged surface exposure.

High frequencies of bone weathering (stages III–IV), pitting, and burrowing indicate that the disarticulated bones were not rapidly buried after death. There were prolonged periods of carcass exposure, followed by rapid decay and bone disarticulation and weathering. The prolonged carcass exposure is reflected through bone surface modifications and damages on proximal and distal ends of long bones. Other factors that had considerable effect on the assemblage are insect boring and bioturbation. Animal trampling, rodent gnawing, grooving, root marking, and bacterial and algal staining also suggest that the Laetoli death assemblages suffered prolonged exposure.

Taphonomically, the Upper Laetolil Beds are characterized by a low density of time-averaged fossil accumulations. The beds record a fairly stable sedimentation rate (Drake and Curtis, 1987), but are also characterized by brief episodes of rapid deposition from volcanic eruptions from nearby sources at Lemagruti (Hay, 1987; Manega, 1993). Based on surface bone modifications and other observed taphonomic modifications, the Upper Laetolil faunal assemblage was partially accumulated by carnivores before fossilization. Thereafter the assemblage was exposed and reworked through physical and chemical processes. The assemblage was further modified and vertically and horizontally dispersed via bioturbation, trampling, and compaction.

Overall, classification of the Laetoli fossil bovid assemblage by Musiba and Magori (2005) and Musiba (1999) in a discriminant function analysis of fossil bovid limb morphologies as eco-indicators points towards mosaic-like environments. Using discriminant function analysis and PNN (Probabilistic Neural Network protocol in MatLab) statistics for bovid limb morphologies, Musiba (1999) placed the analyzed bovid limb elements in four habitat types: open, intermediate, woodland, and forest. This interpretation of the Laetoli Pliocene environment is consistent with the geological and stratigraphic information provided in this study for Localities 8 and 9 within the Upper Laetoli Beds.

Fossil bovid remains from the Upper Laetolil Beds at Localities 8 and 9 provide an interesting picture of their evolutionary and behavioral ecology. Gentry (1987) acknowledged the peculiarity of the taxonomic composition of the Laetoli bovids. These fossils differ in their overall morphology and composition from their extant sister groups (Gentry, 1981). The Laetoli fossil bovids present a mosaic of femoral morphologies with implications ranging from open country habitat to woodland or woodland/forest mix. Their morphologies reflect the locomotor behaviors favored by the Laetoli bovids, which range from *Simatherium kohllarseni*—a large-size Bovini that may have favored an intermediate habitat type—to species of Cephalophini (rarely found as fossils), which may have preferred forested or wooded habitats at Laetoli. This broad spectrum of habitats during the Pliocene makes Laetoli very different from other East African paleoanthropological sites (such as Koobi Fora in Kenya and Hadar in Ethiopia) of comparable age, in that its faunal composition was highly diverse.

Furthermore, Laetoli fossil Neotragini, a tribe that is abundant in the Upper Laetolil Beds, may be represented by *Madoqua avifluminis* and probably a *Raphicerus* species (Gentry, 1987). Their extant counterparts tend to favor rather open country habitats. Nevertheless, the appearance of Neotragini in the Upper Laetolil Beds as an abundant taxon would be consistent with open country habitats characterized by galleries of mixed vegetation of light to dense woodland. But *Madoqua* species (extant dik-diks), which are mainly browsers, prefer light to heavily wooded habitats consisting of *Acacia* trees, bush, and thorn scrub, and tend to avoid open grasslands (Kingdon, 1982).

Previously proposed paleoenvironmental models for Pliocene Laetoli placed it in open country grassland environments similar to the present-day Serengeti (Leakey et al.,

1987). However, Laetoli's complex faunal remains and its geology and stratigraphy as discussed in this paper point towards a mosaic environment. For example, *Madoqua* is the most common bovid taxon in the Upper Laetoli Beds (Gentry, 1987) and its modern counterparts prefer light cover habitats. Additionally, the existence of *Tragelaphus* sp., *Simatherium kohllarseni*, *Cephalophus* sp., *Hippotragus* sp., and sp. indet. aff. *Pelea* in the fossil assemblage is also in contrast with the open country or savanna grassland models. Furthermore, viverrids, *Herpestes*, and several mollusk species within the Upper Laetolil Beds also point towards a mosaic type of ecological settings. The geologic and stratigraphic evidence further contradicts the savanna grassland interpretation of Laetoli's Pliocene environments, especially at Localities 8 and 9.

This study represents a small fraction of the entire Laetoli site. Therefore, the results presented here may not be used to generate an overall interpretation of the greater Laetoli paleoanthropological site ($200\,km^2$) without taking into account the size of the area and the variable topography that existed during the Pliocene. Therefore, systematic and uniform studies of the entire site in the future are recommended in order to model overall Pliocene environments for the entire site. We believe that the taphonomic history of Laetoli is much more complex than previously considered in that it is characterized by episodes of highly variable time-averaged bone accumulations. Some of these accumulations between Tuffs 6, 7, and 8 occurred over a time span of approximately 300,000 years and are separated by multiple depositional events.

## Acknowledgments

We would like to thank Drs. Ronald Singer (deceased), and Susan Kidwell (University of Chicago), Chapurukha Kusimba, Bruce Patterson, Julian Kerbis (FMNH), and Roger (ANL), for their criticism, comments, and suggestions. We are indebted to Drs. Paul Manega, Pelaji Kyauka, and E.B. Chausi for their invaluable time and contribution that shaped this study.

We also thank our Tanzanian colleagues, Godfrey Ole Moita, O.S. Kileo, Samuel Lauwo, Cosmas Dawi, and Lazaro Mariki, for comments, field logistics, and field assistance during our 1996–1998 fieldwork at Laetoli. We are very grateful to the Tanzanian Ministry of Tourism and Natural Resources, the Antiquities Department, the Ngorongoro Conservation Area Authority, and the Commission for Science and Technology (COSTECH) for granting us permission to conduct our research at Laetoli. We are very grateful to Dr. Y. Kohi, H.M. Nguli, and Donatius Kamamba for issuing the COSTECH research permits and the Antiquities excavation license.

We extend our thanks to the Lutheran Bishop of Dodoma, Rev. Dr. Peter L. Mwamasika and Dr. Abel T. Nkini for their encouragement and negotiations for our research permits in Tanzania. Furthermore, we extend our special thanks to the anonymous reviewers and the editors of this volume for their tireless work. This project was funded by the Wenner Gren Foundation (Gr. 5980), the Dahlberg Memorial Fund, the Hinds Fund, the University of Chicago's Division of Social Sciences Century Scholarship, the Connecticut State University System, the National Geographic Society, and a generous grant from Dr. Melissa K. Stoller.

## References

Alemseged, Z., 2003. An integrated approach to taphonomy and faunal change in the Shungura Formation (Ethiopia) and its implication for hominid evolution. Journal of Human Evolution 44, 461–478.

Andrews, P., 1989. Paleoecology of Laetoli. Journal of Human Evolution 18, 173–181.

Andrews, P., Cook, J., 1985. Natural modifications to bones in a temperate setting. Man 20, 675–691.

Andrews, P., 1990. Owls, Caues and Fossils, British Museum (Natural History) London.

Armour-Chelu, M., Andrews, P., 1994. Some effects of bioturbation by earthworms (Oligochaeta) on archaeological sites. Journal of Archaeological Science 21, 433–443.

Behrensmeyer, A.K., 1978. Taphonomic and ecological information from bone weathering. Palaeobiology 4, 150–162.

Behrensmeyer, A.K., 1983. Patterns of natural bone distribution on recent land surfaces: implications for archaeological site formation. In: Clutton-Brock, J., Grigson, C. (Eds.), Animal and Archaeology 1: Hunters and Their Prey. B.A.R. International Series 163, Oxford, pp. 93–106.

Behrensmeyer, A.K., 1991. Terrestrial vertebrate accumulations. In: Allison, P., Briggs, D.E.G. (Eds.), Taphonomy: Releasing the Data Locked in the Fossil Record. Plenum, New York, pp. 291–335.

Behrensmeyer, A.K., Chapman, R.E., 1993. Models and simulations of time averaging in terrestrial vertebrate accumulations. In: Kidwell, S.M., Behrensmeyer, A.K. (Eds.), Taphonomic Approaches to Time Resolution in Fossil Assemblages. University of Tennessee, Knoxville, pp. 125–149.

Behrensmeyer, A.K., Hook, R.W., 1992. Paleoenvironmental contexts and taphonomic modes. In: Behrensmeyer, A.K., Damuth, J.D., DiMichele, W.A., Potts, R., Sues, H.D., Wing, S.L. (Eds.), Terrestrial Ecosystems Through Time. University of Chicago Press, Chicago, pp. 15–136.

Behrensmeyer, A.K., Kidwell, S.M., 1985. Taphonomy's contributions to paleobiology. Palaeobiology 11(1), 105–119.

Blumenschine, R.J., 1989. Landscape taphonomic model of the scale of prehistoric scavenging oppurtunities. Journal of Human Evolution 18(4), 345–371.

Bobe, R., Behrensmeyer, A.K., Chapman, R.E., 2002. Faunal change, environmental variability and late Pliocene hominin evolution. Journal of Human Evolution 42, 475–497.

Bonnefille, R., Riollet, G., 1987. Palynological spectra from the Upper Laetolil Beds. In: Leakey, M.D., Harris, J.M. (Eds.), Laetoli: A Pliocene Site in Northern Tanzania. Clarendon Press, Oxford, pp. 52–61.

Brain, C.K., 1981. The Hunters or the Hunted? An Introduction to African Cave Taphonomy. University of Chicago Press, Chicago.

Butler, P.M., 1987. Fossil insectivores from Laetoli. In: Leakey, M.D., Harris, J.M. (Eds.), Laetoli: A Pliocene Site in Northern Tanzania. Clarendon Press, Oxford, pp. 85–87.

Corbet, G.B., Hanks, J., 1968. A revision of the elephant-shrews, family Macroscelididae. Bulletin of the British Museum of Natural History (Zoology) 16, 47–111.

Davies, C., 1987. Fossil Pedetidae (Rodentia) from Laetoli. In: Leakey, M.D., Harris, J.M. (Eds.), Laetoli: A Pliocene Site in Northern Tanzania. Clarendon Press, Oxford, pp. 171–189.

Denys, C., 1987. Fossil rodents (other than Pedetidae) from Laetoli. In: Leakey, M.D., Harris, J.M. (Eds.), Laetoli: A Pliocene Site in Northern Tanzania. Clarendon Press, Oxford, pp. 118–170.

Drake, R., Curtis, G.H., 1987. Geochronology of the Laetoli fossil localities. In: Leakey, M.D., Harris, J.M. (Eds.), Laetoli: A Pliocene Site in Northern Tanzania. Clarendon Press, Oxford, pp. 48–52.

Fernández-Jalvo, Y., Denys, C., Andrews, P., Williams, T., Dauphin, Y., Humphrey, L., 1998. Taphonomy and paleoecology of Olduvai Bed-I (Pleistocene, Tanzania). Journal of Human Evolution 34, 137–172.

Gentry, A.W., 1981. Pliocene and Pleistocene Bovidae in Africa. In: Beden, M., et al. (Eds.), L'Environnement des Hominidés au Plio-Pléistocène. Fondation Singer Polignac, Paris, pp. 119–122.

Gentry, A.W., 1987. Pliocene Bovidae from Laetoli. In: Leakey, M.D., Harris, J.M. (Eds.), Laetoli: A Pliocene Site in Northern Tanzania. Clarendon Press, Oxford, pp. 378–408.

Gifford, D.P., 1981. Taphonomy and paleoecology: a critical review of archaeology's sister disciplines. Advances in Archaeological. Methods and Theory 4, 365–438.

Grant, J.D., 1983. The activities of earthworms and the fates of seeds. In: Satchell, J.E. (Ed.), Earthworm Ecology. Chapman and Hall, London, pp. 107–122.

Harris, J.M., 1985. Age and Paleoecology of the Upper Laetolil Beds, Laetoli, Tanzania. In: Delson, E. (Ed.), Ancestors: The Hard Evidence. Alan Riss, New York, pp. 76–81.

Harrison, T., 1992. A reassessment of the taxonomic and phylogenetic affinities of the fossil catarrhines from Fort-Ternan, Kenya. Primates 33(4), 501–522.

Harrison, T., 2005. Fossil bird eggs from the Pliocene of Laetoli, Tanzania: their taxonomic and paleoecological relationships. Journal of African Earth Sciences 41, 289–302.

Hay, R.L., 1978. Melilitite-Carbonatite tuffs in the Laetoli Beds of Tanzania. Contributions to Mineralogy and Petrology 67, 357–367.

Hay, R.L., 1980. Paleoenvironment of the Laetolil Beds, northern Tanzania. In: Rapp, G., Vondra, C.F. (Eds.), Hominid Sites: Their Geologic Settings. AAAS Selected Symposium 63. Boulder, Westview Press, pp. 7–24.

Hill, A., 1979. Disarticulation and scattering of mammal skeletons. Paleobiology 5, 261–274.

Hill, A., 1980. Early post-mortem damage to the remains of some contemporary east African mammals. In: Behrensmeyer, A.K., Hill, A. (Eds.), Fossil in the Making. University of Chicago Press, Chicago, pp. 131–152.

Hill, A., 1987. Damage to some fossil bones from Laetoli. In: Leakey, M.D., Harris, J.M. (Eds.), Laetoli: A Pliocene Site in Northern Tanzania. Clarendon Press, Oxford, pp. 543–545.

Hill, A., 1994. Early hominid behavioral ecology: a personal postscript. Journal of Human Evolution 27, 312–328.

Kappelman, J., 1991. The paleoenvironment of Kenyapithecus at Fort Ternan. Journal of Human Evolution 20, 95–129.

Kidwell, S.M., Behrensmeyer, A.K., 1988. Overview: ecological and evolutionary implications of taphonomic processes. Paleogeography, Paleoclimatology, Paleoecology 63, 1–13.

Kingdon, J. 1974. East African Manmals. An Atlas of Evolution in Africa. Volume 11. Part B. (Hares and Rodents) Academic Press, London.

Kingdon, J., 1982. East African Mammals. An Atlas of Evolution in Africa. Vol. III, Parts C and D (Bovids). Academic Press, London.

Kingdon, J., 1997. The Kingdon Field Guide to African Mammals. Academic Press, London.

Kovarovic, K., Andrews, P., Aiello, L., 2002. The Paleoecology of Upper Ndolanya Beds at Laetoli, Tanzania. Journal of Human Evolution 43, 395–418.

Leakey, M.D., Beden, M., Guérin, C., Renders, E.M., Sondaar, P., 1987. Animal prints and trails. In: Leakey, M.D., Harris, J.M. (Eds.), Laetoli: A Pliocene Site in Northern Tanzania. Clarendon Press, Oxford, pp. 451–489.

Leaky, M.D., Harris, J.M. 1987. Lactoli: A Pliocene site in Northern Tanzania Clarendon Press, oxford.

Manega, P.C., 1993. Geochronology, geochemistry and isotopic study of the Plio-Pleistocene hominid sites and the Ngorongoro volcanic highland in northern Tanzania. Ph.D. thesis, University of Colorado, Boulder, Colorado.

Martin, N.A., 1982. The interaction between organic matter in soil and the burrowing activity of three species of earthworms (Oligochaeta: Lumbricidae). Pedobiologia 24, 185–190.

Meylan, P.A., Auffenberg, W., 1987. The chelonians from the Laetolil Beds. In: Leakey, M.D., Harris, J.M. (Eds.), Laetoli: A Pliocene Site in Northern Tanzania. Clarendon Press, Oxford, pp. 62–78.

Musiba, C.M., 1999. Laetoli Pliocene Paleoecology: a reanalysis via morphological and behavioral approaches. Ph.D. thesis, The University of Chicago, Illinois.

Musiba, C.M., Magori, C.C., 2005. Laetoli Pliocene paleoecology: predictive behavioral ecology model based on functional morphology and sediment proxy data. In: Bertran B.B. Mapunda and Paul Msemcia (Eds.), Salvaging Tanzania's Cultural Heritage, Dar Es Salaam University Press, pp. 137–157.

Ndessokia, P.N.S., 1990. The Mammalian Fauna and Archaeology of the Ndolanya and Olpiro Beds, Laetoli, Tanzania. Ph.D. thesis, University of California, Berkeley.

Selvaggio, M., Wilder, J., 2001. Identifying the involvement of multiple Carnivore Taxa with archaeological assemblages. Journal of Archaeological Science 28, 465–470.

Stein, J.K., 1983. Earthworm activity: a source of potential disturbance of archaeological sediments. American Antiquity 48, 277–289.

Su, D., Harrison, T., 2006. The paleoecology of the Upper Laetolil Beds at Laetoli: a reconsideration of the large mammal evidence. In: Bobe, R., Alemseged, Z., Behrensmeyer, A.K. (Eds.), Hominin Environments in the East African Pliocene: An Assessment of the Faunal Evidence. Springer, Dordrecht.

Tappen, M., 1994. Bone weathering in the tropical rain forest. Journal of Archaeological Science 21, 667–673.

Verdcourt, B., 1987. Mollusca from the Laetolil and Upper Ndolanya Beds. In: Leakey, M.D., Harris, J.M. (Eds.), Laetoli: A Pliocene Site in Northern Tanzania. Oxford, Clarendon Press, pp. 438–450.

Vrba, E.S., 1985. Environment and evolution: alternative causes of the temporal distribution of evolutionary events. South African Journal of Science 81, 229–236.

Wesselman, H.B. 1985. Fossil micromamonalls asindicates of climate Change about 2.4 myr ago in the Orno Valley, Ethiopia. South African Jornal of Science, 81: 260–261.

# 11. The paleoecology of the Upper Laetolil Beds at Laetoli

## A reconsideration of the large mammal evidence

D.F. SU
*Human Evolution Research Center*
*Department of Integrative Biology*
*University of California, Berkeley*
*Berkeley, California 94720, USA*
*denisefu@calmail.berkeley.edu*


T. HARRISON
*Department of Anthropology*
*Center for the Study of Human Origins*
*New York University*
*25 Waverly Place*
*New York, NY 10003, USA*
*terry.harrison@nyu.edu*

**Keywords:** Ecological diversity analysis, ecovariables

## Abstract

Laetoli, in northern Tanzania, is one of the most important paleontological and paleoanthropological sites in Africa. Apart from Hadar, it has yielded the largest sample of specimens attributable to the mid-Pliocene hominin, *Australopithecus afarensis*, including the type specimen. As such, it is important to explore the paleoenvironment at Laetoli, especially the different habitat types that may have been exploited by *A. afarensis*. Previous interpretations of the paleoecology at Laetoli have led to quite different conclusions. Initially, the paleoenvironment was reconstructed as an arid to semi-arid grassland with scattered bush and tree cover, and patches of acacia woodland, similar to the modern-day local setting. However, some aspects of the fauna do indicate that the range of habitats may have included more dense bush cover and more extensive tracts of woodland than seen in the region today. The main objective of this paper is to re-examine this issue by more thoroughly documenting the paleoecological setting by conducting a more detailed and comprehensive comparative analysis of the mammalian fauna. To this end, the ecovariable structure of the mammalian fauna at Laetoli is compared to other Plio-Pleistocene hominin-bearing fossil localities and modern faunal communities from different habitats, including forest, woodland, open woodland, bushland, shrubland, grassland, and desert. Principal components analysis (PCA) and bivariate analyses of predictor ecovariables were conducted. An important finding was the general distinctiveness of fossil assemblages, including Laetoli, from modern communities. Terrestrial mammals were found to have the greatest impact on the uniqueness of fossil communities, with fossil assemblages having very high proportions of terrestrial mammals when compared to modern communities. Furthermore, the high frequency of grazers and terrestrial mammals, combined with the low occurrence of arboreal and frugivorous mammals at Laetoli, indicates affinities with modern mammalian communities living in grassland, savanna, and open woodland settings. Taking into account

*R. Bobe, Z. Alemseged, and A.K. Behrensmeyer (eds.) Hominin Environments in the East African Pliocene: An Assessment of the Faunal Evidence, 279–313.*

the results of this study, and the presence of indicator species, we reconstruct the paleoecology of the Upper Laetolil Beds as a mosaic habitat dominated by grassland and shrubland, with areas of open- to medium-cover woodlands, as well as some closed woodland and possibly gallery forest along seasonal river courses.

## Introduction

Laetoli, in northern Tanzania, is renowned as one of the most important paleontological and paleoanthropological sites in Africa. Apart from Hadar, it has yielded the largest sample of specimens attributable to the mid-Pliocene hominin, *Australopithecus afarensis*, including the type specimen, as well as the remarkable discovery of well-preserved trails of hominin footprints (Leakey et al., 1976; White, 1977, 1980a, b, 1981, 1985, 1989; Leakey and Hay, 1979, 1982; Leakey, 1981, 1987a, b, c; Tuttle, 1985, 1987, 1990; Robbins, 1987; White and Suwa, 1987; Tuttle et al., 1991, 1992). An intensive program of research directed by Mary Leakey from 1974 to 1982 laid the foundation for a greater understanding of the geology, geochronology, and paleontology of Laetoli (see papers in Leakey and Harris, 1987). More recent fieldwork at Laetoli has included further excavations and collections (Kyauka and Ndessokia, 1990; Ndessokia, 1990; Kaiser et al., 1995), refinements in the geochronology (Ndessokia, 1990; Manega, 1993), efforts to conserve the fossil footprints (Anonymous, 1995; Agnew et al., 1996), and research on the paleoecology and taphonomy (Musiba, 1999; Kovarovic et al., 2002). In 1998, one of us (TH) began directing renewed paleontological and geological investigations at Laetoli, a project that is currently ongoing. A major aim of this renewed work at Laetoli is to provide a better understanding of the paleoecology, especially the types of habitats that could potentially have been exploited by *A. afarensis*.

Previous interpretations of the paleoecology at Laetoli have led to quite different conclusions. Initially, the paleoecology was reconstructed as an arid to semi-arid grassland with scattered bush and tree cover, and patches of acacia woodland, similar to the modern-day local setting (e.g., Hay, 1981, 1987; Bonnefille and Riollet, 1987; Gentry, 1987; Harris, 1987a; Leakey, 1987a; Meylan, 1987; Watson, 1987). The major lines of evidence that supported this conclusion were derived from analyses and interpretations of the geology, palynology, and vertebrate paleontology. These include: (1) extensive wind transportation of sand-sized ash particles, indicating poor vegetation coverage on land surfaces (Hay, 1987); (2) caliche paleosols with ash particles cemented by phillipsite, most likely formed under alkaline environments favored by semi-arid to arid conditions, at least seasonally (Hay, 1987); (3) the Footprint Tuff (the lower part of Tuff 7) directly overlies a tuffaceous layer containing small roots, interpreted as grass rootlets, while the regularity of the contact between these tuffs implies that the land surface on which the Footprint Tuff was deposited was largely barren of grass, and possibly heavily grazed (Hay, 1987); (4) the base of the Footprint Tuff contains a layer rich in fossil twigs and leaf impressions that resemble modern species of *Acacia* (Hay, 1987); (5) fossil pollen assemblages indicate an arid savanna vegetation characterized by a high diversity of herbaceous plants, dominated by grasses, and with a sparse tree cover, possibly associated with a warmer and drier climate than today (Bonnefille and Riollet, 1987); (6) the snake and avian fauna is typically associated with savanna, bushland, and open woodland habitats (Meylan, 1987; Watson, 1987); (7) several genera of rodents (i.e., *Saccostomus*, *Xerus*, *Thallomys*, and *Pedetes*) as well the leporid, *Serengetilagus*, are indicative of dry grassland–savanna habitats (Denys, 1985, 1987; Davies, 1987a, b); (8) the abundance and taxonomic diversity of herpestids indicate savanna to open woodland conditions (Petter, 1987); (9) the taxonomic composition of large carnivores at Laetoli resembles the community structure seen today in African savannas (Barry, 1987); (10) the predominance of Alcelaphini, Antilopini, and Neotragini in the bovid fauna implies that non-woodland habitats were present (Gentry, 1987); and (11) the occurrence of the large, hypsodont *Notochoerus euilus* as the dominant suid, and the absence or rarity of *Nyanzachoerus kanamensis*, which is otherwise quite common at contemporary East African localities, suggests relative dry conditions (Harris, 1987a, b). It is reasonable to conclude from these various lines of evidence that grassland, savanna, and open woodland habitats were an important component of the ecological setting at Laetoli during the mid-Pliocene.

However, some aspects of the fauna do indicate that the range of habitats may have included more dense bush cover and more extensive tracts of woodland than seen in the region today (Butler, 1987; Harris, 1987a; Meylan, 1987; Petter, 1987; Verdcourt, 1987). Andrews (1989), using ecological diversity analysis, argues that the fauna reflects a more heavily wooded environment than previously recognized, and that some unusual properties of the mammalian community structure are best accounted for either by mixing of faunas from different ecologies or by habitat changes through the sequence. Subsequent studies of the mammalian fauna by Reed (1997) and Musiba (1999), and of stable carbon isotopes (Cerling, 1992), have provided additional support for a greater representation of wooded habitats at Laetoli. For example, Reed (1997) showed that closed woodlands are indicated by the high taxonomic diversity of arboreal and frugivorous mammals present at Laetoli. There are certainly good ecological indicators in the mammalian fauna that suggest that bushland and woodland habitats were a significant component at Laetoli. Among the bovids, Tragelaphini and

Cephalophini are typically associated with wooded habitats, while *Madoqua*, which is remarkably common at Laetoli, prefers bush and thorn scrub (Kingdon, 1974c, 1997; Gentry, 1987). The species diversity and abundance of giraffids (belonging to three species and representing more than 16% of all artiodactyl specimens from Laetoli) implies a woodland setting that supports a guild of large browsers not represented in contemporary faunas. The suid, *Potamochoerus*, has a strong preference for forest and woodland habitats (Kingdon, 1974e, 1997; Harris, 1987b). Several species of primates are known from Laetoli, including the bushbaby, *Galago sadimanensis*, and at least three species of cercopithecids, *Parapapio ado*, *Paracolobus* sp., and a colobine monkey somewhat larger in size than the extant *Colobus*, and possibly a larger papionin[1] (Leakey and Delson, 1987; Walker, 1987). The diversity of the primate community is suggestive of closed woodland or forest, at least along river courses. Although cercopithecids occupy a range of habitats from grassland to forest today, they do require stands of trees or rocky outcrops as sleeping sites. Reconstructions of the landscape and topography at Laetoli indicate a gently undulating terrain, with no rock outcrops, implying that larger trees would have been important sites of refuge for cercopithecid primates. The postcranial remains attributed to *Paracolobus* and *Parapapio* also imply a significant component of arboreality. Of the small mammals,

---

[1] Ndessokia (1990) lists *Theropithecus darti* in his faunal list of the Upper Laetolil Beds. However, no information is given on the provenience or nature of the find(s) on which this record is based, and we have been unable to locate the original material. No specimens of this taxon have been recovered from the Upper Laetolil Beds by Leakey's or Harrison's teams, so we are inclined to discount this record. However, Mary Leakey did recover *Theropithecus* sp. from the Ngaloba Beds, and it is possible that the record refers to material from this younger horizon.

the occurrence of the bush squirrel, *Paraxerus*, and the giant elephant shrew, *Rhynchocyon*, implies closed woodlands with dense undergrowth and substantial leaf litter (Kingdon, 1974a, b, c, 1997; Butler, 1987; Denys, 1987). The avian community, including at least one small species of francolin, a larger francolin, a guinea fowl, as well as ostriches, implies that the paleoecology at Laetoli was most likely open woodland, bushland, savanna, or grassland (Watson, 1987; Harrison and Msuya, 2005; Harrison, 2005). However, in habitats where grassland predominates, francolins and guinea fowl require low brush and thickets for escape and refuge, as well as trees in which to roost at night. They prefer mosaic ecotonal habitats offering open feeding areas with good visibility, but with dense vegetation cover and patches of woodland nearby (Dörgeloh, 2000; Harrison, 2005). The terrestrial gastropod community at Laetoli includes *Subulona* and *Euonyma* that are found today primarily in evergreen forest (Verdcourt, 1987). Urocyclid slugs are extremely common and ubiquitous, and although they do occur today in dry open woodland and savanna habitats, leaf litter and fallen trees are a necessary requirement as sites for feeding and aestivation. Finally, the greater proportion of Afro-Montane elements in the palynological spectrum compared with the modern pollen rain (Bonnefille and Riollet, 1987), and the density and diversity of macrobotanical remains (such as twigs, leaves, and seeds), indicate that wooded and forested habitats were a more important component of the paleoecology in the Pliocene than they are in the region today.

The balance of evidence implies that the previous emphasis on the predominance of grassland and savanna habitats at Laetoli may have been somewhat overstated. Although one can be confident that grassland and savanna were an important component of the ecological setting, the totality of the faunal and floral evidence suggests that a mosaic of habitats was available, with a greater representation of open and closed woodland than is seen today in the vicinity of Laetoli. Nevertheless, it still remains to be established just what would be the closest modern analog to the paleoecological setting at Laetoli. With a more detailed and comprehensive analysis of the fauna it might be possible to develop a more nuanced interpretation of the paleoecology at Laetoli, one that entails a broader comparison with modern and Pliocene faunas from Africa. This is one of the main objectives of the current study.

However, before the Laetoli fauna can be compared in this way, an initial inference needs to be tested. One possible alternative explanation for the conclusion that Laetoli represents a heterogeneous mosaic of grassland, savanna, and woodland habitats is that the various ecological signals are derived from a composite fauna from different localities that span the entire stratigraphic sequence of the Upper Laetolil Beds. Rather than a complex mosaic of habitats occurring uniformly throughout the sequence, the structure of the fauna might reflect distinct differences in the patterning of vegetation in space and time, of which the composite 'time-averaged' fauna merely offers an ecological palimpsest. To test this hypothesis, we analyze the faunas from the different collecting localities and stratigraphic zones separately to see if there are any significant difference in the faunas in space and time. If there are observed differences, then the paleoecology of each locality and/or stratigraphic zone will need to be reconstructed separately and the general paleoecology of Laetoli reconsidered in light of these findings.

Given these considerations, this paper attempts to answer two critical questions about the paleoecology of Laetoli: (1) Are there significant differences in the composition of the faunas at Laetoli that reflect geographical differences in the local ecology or changes in ecology through time? (2) If the fauna indicates temporal and/or geographical heterogeneity or uniformity, how does this impact on reconstructions of the overall paleoecology at Laetoli? In order to answer these questions we first assess the nature and the degree of differences between the faunas from different localities and stratigraphic zones at Laetoli, then we attempt a reanalysis of the paleoecology based on a more detailed and comprehensive comparative study of the mammalian fauna using ecological diversity analysis.

## Geological Context

The stratigraphy and geochronology of Laetoli have been well documented (Kent, 1941; Pickering, 1964; Hay, 1976, 1978, 1987; Drake and Curtis, 1979, 1987; Hay and Leakey, 1982; see Figure 1). Fossil vertebrates have been recovered from throughout the sedimentary sequence, but the most productive units are the Laetolil and Ndolanya Beds (Hay, 1987; Figure 1). The Laetolil Beds rest unconformably on the Precambrian Basement, and are divided into two lithological units – the upper and lower units. The lower unit consists primarily of aeolian tuffs interbedded with air-fall and water-worked tuffs (Leakey et al., 1976; Hay, 1987). It is dated radiometrically from 3.8 Ma to older than 4.32 Ma (Drake and Curtis, 1987), although based on estimated sedimentation rates at Laetoli, the bottom of the sequence could be as old as 4.6 Ma.

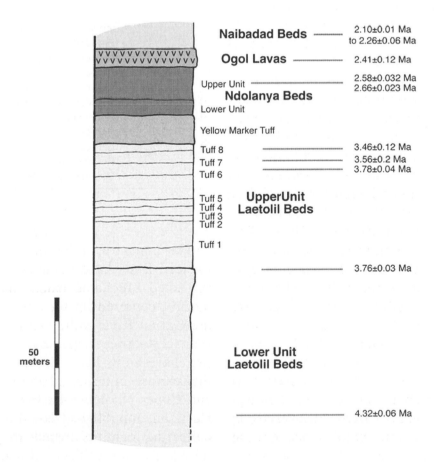

Figure 1. General stratigraphic column of Laetoli (after Hay, 1987; Drake and Curtis, 1987; Ndessokia, 1990; Manega, 1993).

A small fauna has been recovered from the Lower Laetolil Beds (Harris, 1987; Harrison et al., in preparation), but no hominin fossils have yet been found. The Upper Laetolil Beds, from which all of the *Australopithecus afarensis* specimens have been recovered, have been radiometrically dated to ~3.5 to 3.8 Ma (Drake and Curtis, 1987; Figure 1). The sediments consist largely of aeolian tuffs, but also contain a series of air-fall tuffs and some water-worked tuffs (Hay, 1987). Eight of the air-fall tuffs, identified on the basis of their lithology and mineralogical composition, have been identified as marker tuffs (Hay, 1987). These can be used to sub-divide the fauna from the Upper Laetolil Beds into a series of narrow temporal zones (Hay, 1987). Renewed fieldwork at Laetoli since 1998 has allowed a more refined appreciation of the stratigraphical provenience of the fossils at each of the localities at Laetoli, and these data are presented in Table 1.

The Ndolanya Beds consist of a series of tuffs and calcretes, which are subdivided into upper and lower units (Hay, 1987). The lower unit is chiefly clay-rich deposits, with some massive vitric tuffs and limestones. Root markings are common, but no fossil vertebrates have been found. The upper unit is comprised mainly of aeolian- and water-worked tuffs (Hay, 1987). This unit is highly fossiliferous, with a diverse vertebrate fauna, including *Paranthropus aethiopicus* (Harrison, 2002). The fauna from the Upper Ndolanya Beds is consistent with an age of ~2.5 to 2.7 Ma (Harris and White, 1979; Beden, 1987; Gentry, 1987; Hooijer, 1987; Harris, 1987b), and radiometric dates of 2.58 to 2.66 Ma have been reported (Ndessokia, 1990; Manega, 1993). The Ndolanya Beds are overlain by a series of lavas, the Ogol lavas, with an average K-Ar date of 2.41 Ma (Drake and Curtis, 1987; Hay, 1987).

*Table 1. Fossiliferous horizons of the Upper Laetolil Beds at all Laetoli localities included in this study*

| Locality | Fossiliferous horizons |
|----------|------------------------|
| 1 | Between Tuff 6 and Yellow Marker Tuff |
| 2S | Between Tuffs 5 and 7 |
| 2W | Between Tuffs 5 and 7 |
| 3 | Between Tuffs 7 and 8 |
| | Between 4 and 6 |
| 4 | Between Tuffs 6 and 8 |
| 5 | Between Tuffs 3 and 5 |
| 6 | Between Tuffs 5 and 7 |
| 7 | Between Tuffs 5 and 8 |
| 8 | Between Tuffs 5 and 7 |
| | Between Tuff 7 to above Tuff 8 |
| 9 | Between Tuffs 5 and 7 |
| | Between Tuffs 7 and 8 |
| 9S | Between Tuff 2 and below Tuff 1 |
| 10 | Between Tuffs 1 and 3 |
| 10E | Between Tuffs 5 and 7 |
| | Between Tuffs 7 and 8 |
| 10W | Between Tuffs 1 and 3 |
| 11 | Between Tuffs 7 and 8 |
| 12 | Between Tuffs 5 and 7 |
| 12E | Between Tuffs 5 and 7 |
| 13 | Between Tuffs 5 and 8 |
| | Between Tuffs 3 and 5 |
| 15 | Between Tuffs 6 and 7 |
| 16 | Between Tuff 7 to just above Tuff 8 |
| 17 | Between Tuff 7 and Yellow Marker Tuff |
| 19 | Between Tuffs 5 and 8 |
| 20 | Between Tuffs 6 and 8 |
| 21 | Between Tuffs 5 and 7 |
| 22 | Between Tuffs 5 and 7 |
| | Between Tuffs 2 and 5 |
| 22E | Between Tuffs 5 and 7 |

The majority of fossils from the Upper Laetolil Beds have come from a relatively restricted area at Laetoli, covering about 80 km$^2$, centered on the eastern reaches of the Garusi River valley (Figure 2). Thirty-four collecting localities and sub-localities have now been delimited at Laetoli that expose outcrops of the Upper Laetolil and Upper Ndolanya Beds (Leakey, 1987a; Harrison, unpublished data; Figure 2). These collecting localities, which are quite limited in size (no larger than 1 km$^2$), are used as the basic geographical unit in this study.

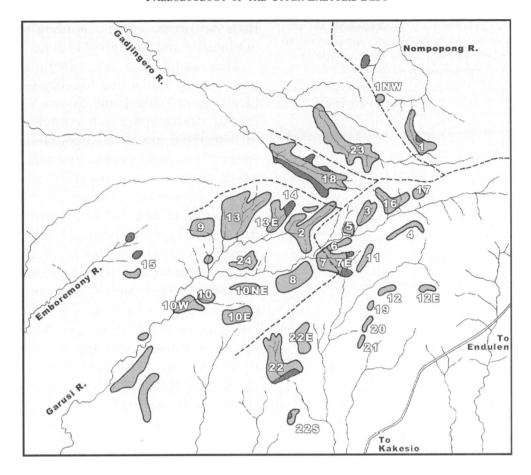

Figure 2. Map of the area of Laetoli and the positions of the localities. The shaded areas are fossiliferous exposures of the Upper Laetolil Beds (light grey) and the Upper Ndolanya Beds (dark grey) (after Leakey, 1987a; Harrison, unpublished data).

## Materials

### MATERIALS FROM THE UPPER LAETOLIL BEDS

The material studied includes fossil mammals from the Upper Laetolil Beds recovered by Mary Leakey from 1974 to 1981 and by Terry Harrison from 1998 to 2001. The Upper Laetolil fauna is represented by 71 mammalian species. For the analyses conducted in this study, taxa with an estimated weight of under 500 g were excluded due to the relative rarity of micromammals at Laetoli. The exclusion of these taxa resulted in a total of 57 mammalian species, which were compiled from a total of 14,575 individual specimens (Mary Leakey Collection: 8,952; Terry Harrison Collection: 5,623). The number of species recovered from each locality is provided in Table 2. Almost all of the material was recovered by surface collection after the specimens had weathered out of the sediments. Previous attempts at excavation have proved unrewarding (see Leakey, 1987a) because fossils are mostly preserved as isolated and fragmentary specimens that were scattered across the paleoland surface rather than in high-density concentrations. Determination of precise stratigraphic provenience (i.e., depth of horizon below a certain marker tuff) is not possible in most cases. Nevertheless, by

Table 2. *The number of species in each locality and sub-locality analyzed in this study. Taxa with estimated weights of less than 500g are not included. There is a total of 57 (>500g) species present at Laetoli*

| Locality | Number of species |
|----------|-------------------|
| 1 | 28 |
| 2S | 19 |
| 2W | 21 |
| 3 | 37 |
| 4 | 22 |
| 5 | 34 |
| 6 | 30 |
| 7 | 35 |
| 8 | 32 |
| 9 | 29 |
| 9S | 28 |
| 10 | 35 |
| 10E | 38 |
| 10W | 33 |
| 11 | 32 |
| 12 | 22 |
| 12E | 16 |
| 13 | 22 |
| 13E | 10 |
| 13 "Snake Gully" | 15 |
| 15 | 15 |
| 16 | 29 |
| 17 | 16 |
| 19 | 8 |
| 20 | 15 |
| 21 | 30 |
| 22 | 28 |
| 22E | 8 |

careful observation of the provenience of *in situ* bones it has been possible to reconstruct the stratigraphic units that have produced the majority of fossils at each locality. In many cases, fossils come from several horizons, so the collected assemblages may contain fossils derived from strata that span one or more marker tuffs. Although this limitation does not allow us to sub-divide the samples according to horizons separated by consecutive marker tuffs, it is possible to divide the faunas according to five stratigraphic zones (i.e., above Tuff 8, between Tuff 7 and 8, between Tuff 5 and 7, between Tuff 3 and 5, below Tuff 3). Specimens from the Upper Laetolil

Beds have been sorted according to collecting locality and stratigraphic zone.

The faunal data are examined in two ways: (1) by collecting locality (see list of localities in Table 1 and Figure 2), regardless of stratigraphic unit, which allows us to determine any local geographic differences; (2) by stratigraphic zone (see Table 1), which permits an assessment of whether or not ecological changes occur during the course of the 300 Kyr represented by the Upper Laetolil Beds. Due to the low number of specimens found above Tuff 8, they were combined with specimens found between Tuffs 7 and 8, resulting in four main divisions of the stratigraphic zones. Using these data, faunas can be compared in a three dimensional spatial-stratigraphic framework to determine the nature of any geographical heterogeneity or temporal change in the ecology.

## MATERIALS FROM COMPARATIVE LOCALITIES

### Modern Localities
Ecological diversity data for modern faunal communities are compared with those from Laetoli in order to assess which modern-day communities are most similar and, therefore, most likely to have a comparable ecology. The modern faunal communities employed in this study can be categorized into seven main habitat types: forest, closed woodland, bushland, open woodland, shrubland, grassland, and desert (Table 3). Definition and categorization of the modern communities follows those of Reed (1996, 1998) and faunal lists are taken from published literature (Swynnerton, 1958; Ansell, 1960, 1978; Lamprey, 1963; Child, 1964; Vesey-FitzGerald, 1964; Rahm, 1966; Sheppe and Osborne, 1971; Smithers, 1971, 1983; Rautenbach, 1978a, b; Behrensmeyer et al., 1979; Happold, 1987; Skinner and Smithers, 1990; Coe et al., 1999).

*Table 3. Modern African localities and vegetation types*

| Locality | Vegetation |
|---|---|
| Congo Rainforest | Forest |
| E. of River Niger | Forest |
| W. of River Niger | Forest |
| E. of River Cross | Forest |
| Zambia Lowland Forest | Forest |
| Zambia Montane Forest | Forest |
| Kilimanjaro | Closed Woodland |
| Guinea Woodland | Closed Woodland |
| Serengeti Bush | Bushland |
| Rukwa Valley | Bushland |
| Mkomazi Game Reserve | Bushland |
| Kafue National Park | Open Woodland |
| Southern Savanna Woodland | Open Woodland |
| Okavango | Open Woodland |
| Botswana Northwest | Open Woodland |
| Sudan Savanna | Open Woodland |
| Southwest Arid | Shrubland |
| Kalahari | Shrubland |
| Kalahari Thornveld | Shrubland |
| Sahel Savanna | Shrubland |
| Chobe National Park | Shrubland |
| Amboseli National Park | Shrubland |
| Tarangire National Park | Shrubland |
| Makgadikgadi Pan | Grassland |
| Serengeti Plains | Grassland |
| SS Grassland | Grassland |
| Namib Desert | Desert |

From: Swynnerton, 1958; Ansell, 1960, 1978; Lamprey, 1963; Child, 1964; Vesey-FitzGerald, 1964; Rahm, 1966; Sheppe and Osborne, 1971; Smithers, 1971, 1983; Rautenbach, 1978; Behrensmeyer et al., 1979; Happold, 1987; Reed, 1996, 1998; Coe et al., 1999.

## Fossil Localities

Data from African Plio-Pleistocene hominin localities of similar age to the Upper Laetolil Beds are also included in order to situate Laetoli in a broader comparative context, and to determine the diversity of habitats that were available to hominins during the Pliocene. The faunal list for each site was compiled from the literature (Gray, 1980; Harris, 1987a; Feibel et al., 1991; Leakey et al., 1995; Reed, 1996; Leakey and Harris, 2003). Current interpretations of their paleoecology are presented in Appendix 1.

## Methods

### ECOLOGICAL DIVERSITY ANALYSIS

Ecological diversity analysis, first applied to the fossil record by Andrews et al. (1979), enables comparisons between the ecological attributes (i.e., body size, feeding habits, and locomotor type) of fossil and extant communities across time and geographic regions without regard to taxonomic affinity. Differences between communities in their ecological diversity reflect differences in habitat. It has been shown, for example, that ecological diversity patterns are similar for similar habitats, regardless of species composition, e.g., tropical rainforest communities in Asia and South America are similar even though quite different taxa are represented (Andrews et al., 1979). This is a valuable method for interpreting the paleoecology of fossil communities because it is based on general ecological principles, rather than inference through closely related modern taxa (Andrews et al., 1979; Reed, 1997). Another advantage of this approach is that preservational and taphonomic biases, which are inherent in specimen counts of fossil assemblages, have less impact on the species represented in the community, especially if small mammals are excluded from the analysis (Andrews et al., 1979; Kovarovic et al., 2002). Due to the relative rarity of micromammals at Laetoli, it is highly unlikely that small mammal taxa are well represented in the Laetoli fossil assemblage. As a result, all taxa with an estimated body weight of less than 500 g were excluded from these analyses.

Trophic and locomotor variables in ecological diversity studies are ideally assigned as a result of ecomorphological studies (such as Kay, 1984; Van Valkenburgh, 1985, 1988, 1990; Janis, 1988, 1990; Damuth, 1990; Spencer, 1995; Kappelman et al., 1997). Once these data are compiled, the total spectrum between communities (e.g., Andrews et al., 1979; Andrews, 1989) or the abundance of each ecological variable can be compared (e.g., Reed, 1997). The trophic and locomotor ecovariables used in this study follow those developed by Reed (1996) (Table 4). There are nine trophic categories: browsers, grazers, fresh grass grazers, mixed feeders, root and tuber feeders, carnivores (includes carnivores that consume bone and carnivores that eat insects), insectivores, frugivores (includes frugivorous mammals that have significant insects and leaves in their diets), and omnivores; and five locomotor ecovariables: aquatic, fossorial, arboreal, terrestrial, and terrestrial/arboreal. In some cases, certain ecovariables are combined for a more robust dataset (i.e., total carnivory and total frugivory) (see Table 4). In this analysis, inferred locomotor and dietary behaviors for Laetoli fossil mammals are taken directly from the literature, including ecomorphologic and isotopic studies (see papers in Leakey and Harris, 1987; also Bishop, 1995, 1999; Spencer, 1995; Reed, 1996; Cerling et al., 1999, 2003b; Sponheimer et al., 1999; Harris and Cerling, 2002; Kovarovic et al., 2002) (Table 5). Locomotor and dietary ecovariables for fauna from comparative fossil and modern communities are taken from published papers (Shortridge, 1934; Maberly, 1950; Blamey

and Jackson, 1956; Ansell, 1960, 1978; Player and Feely, 1960; Mitchell and Uys, 1961; Eloff, 1964; Grafton, 1965; Mitchell et al., 1965; Bothma, 1966; Wilson, 1966; Goddard, 1968; Kummer, 1968; Schaller, 1968; Pienaar, 1969; Tinley, 1969; Owen, 1970; Jungius, 1971; Milstein, 1971; Smithers, 1971, 1983; Child et al., 1972; Grobler and Wilson, 1972; Owen-Smith, 1973; Dunbar and Dunbar, 1974; Kingdon, 1974a–g, 1997; Williamson, 1975; Joubert, 1976; Melton, 1976; Skinner et al., 1976; Sinclair, 1977; Stuart, 1977; Davidge, 1978; Post, 1978; Rasmussen, 1978; Dieckmann, 1980; Skinner et al., 1980; Sharman, 1981; Sauer et al., 1982; Depew, 1983; Novellie, 1983; Ferreira and Bigalke, 1987; Norton et al., 1987; Barton, 1989; Marean, 1989; Gaynor, 1994; Oates, 1994; Bishop, 1995, 1999; Lewis, 1995; Spencer, 1995; Bronikowski and Altmann, 1996; Reed, 1996; Cerling et al., 1999, 2003b; Sponheimer et al., 1999, 2003; Gagnon and Chew, 2000; Fashing, 2001; Werdelin and Lewis, 2001; Avenant and Nel, 2002; Dankwa-Wiredu and Euler, 2002; Harris and Cerling, 2002; Hill and Dunbar, 2002; Kovarovic et al., 2002).

Once ecovariables were assigned, the frequency of each ecovariable was calculated. Before any statistical tests were run, the arcsine transformation was performed on the frequency data in order to normalize the distribution (Zar, 1999). This is because percentages form a binomial, rather than normal, distribution and the deviation from normality is great for small or large percentages (Zar, 1999). A modified chi-square test (Zar, 1999) was used to assess the statistical significance of each ecovariable frequency between collecting localities and stratigraphic zones. For this analysis, only the fossils collected by Harrison were used, because of the greater precision in recording the stratigraphic provenience of the material. In addition, the proportions of each ecovariable from Laetoli were compared with those from other fossil sites and modern communities. In order to do this, principal components analysis

*Table 4. Ecovariable categories used in this study (following Reed, 1996)*

| Code | Locomotor adaptations | Code | Trophic adaptations |
|------|----------------------|------|---------------------|
| T | Terrestrial | G | Grazer |
| T-A | Terrestrial–Arboreal | FG | Fresh Grass Grazer |
| A | Arboreal | B | Browser |
| AQ | Aquatic | MF | Mixed Feeder |
| F | Fossorial | Fg | Frugivore |
| | | F-I | Frugivore–Insect |
| | | FL | Fruit and Leaves |
| | | C | Carnivore |
| | | C-B | Carnivore–Bone |
| | | C-I | Carnivore–Insect |
| | | I | Insectivore |
| | | O | Omnivore |
| | | RT | Root and Tuber |
| | | TC | C + C-B + C-I |
| | | TF | F + F-I + F-L |

*Table 5. List of fossil mammals from the Upper Laetolil Beds (updated from Harris, 1987) and their locomotor and trophic adaptations (see text for references). This list is subject to revision pending further taxonomic studies*

| | Locomotor | Trophic | | Locomotor | Trophic |
|---|---|---|---|---|---|
| **Artiodactyla** | | | *Megantereon* sp. | T-A | C |
| Bovidae | | | *Homotherium* sp. | T | C |
| aff. *Pelea*, sp. indet. | T | G | *Dinofelis* sp. | T-A | C |
| Alcelaphini, large sp. | T | G | *Leo* cf. *pardus* | T | C |
| *Parmularius pandatus* | T | G | *Leo* sp. | T | C |
| *Gazella janenschi* | T | B | *Felis*, large sp. | T-A | C |
| *Simatherium kohllarseni* | T | MF | *Felis*, medium sp. | T-A | C |
| *Brabovus nanincisus* | T | B | *Felis*, small sp. | T-A | C |
| Cephalophini sp. indet. | T | B | Felidae gen. indet. | – | – |
| Hippotragini sp. | T | G | Canidae | | |
| *Praedamalis deturi* | T | G | *?Megacyon* sp. | T | C |
| *Madoqua avifluminis* | T | B | aff. *Canis brevirostris* | T | C-I |
| *Raphicerus* sp. | T | B | *Vulpes* sp. | T | C |
| Reduncini, sp. indet. | T | G | cf. *Otocyon* sp. | T | I |
| *Tragelaphus* sp. | T | B | Canidae gen. indet. | – | – |
| Giraffidae | | | Hyaenidae | | |
| *Giraffa stillei* | T | B | *Crocuta* sp. nov. | T | C-B |
| *Giraffa* cf. *jumae* | T | B | Hyaenidae, *incertae sedis* | T | C-B |
| *Sivatherium* cf. *maurusium* | T | B | Herpestidae | | |
| Suidae | | | *Herpestes (Herpestes) ichneumon* | T | C-I |
| *Notochoerus euilus* | T | G | *Herpestes (Galerella)* | | |
| *Potamochoerus porcus* | T | O | *palaeoserengetensis* | T | C-I |
| **Perissodactyla** | | | *\*Helogale palaeogracilis* | T | C-I |
| Rhinocerotidae | | | *\*Helogale* sp. | T | C-I |
| *Ceratotherium praecox* | T | G | *Mungos dietrichi* | T | C |
| *Diceros bicornis* | T | B | Viverridae | | |
| Ancylotheriidae | | | *Viverra leakeyi* | T | O |
| *Ancylotherium hennigi* | T | B | Mustelidae | | |
| Equidae | | | *\*Propoecilogale bolti* | T | C |
| *Eurygnathohippus* sp. | T | G | *Mellivora capensis* | T | C-I |
| **Proboscidea** | | | **Rodentia** | | |
| Elephantidae | | | Sciuridae | | |
| *Loxodonta exoptata* | T | G | *\*Xerus* cf. *janenschi* | T | F |
| *?Stegodon* sp. | T | – | *\*Paraxerus* sp. Indet. | T-A | F |
| Deinotheriidae | | | *\*Sciuridae gen. et sp. nov.* | – | – |
| *Deinotherium bozasi* | T | B | Cricetidae | | |
| **Tubulidentata** | | | *\*Gerbillinae gen. indet.* | F | B |
| Orycteropodidae | | | *\*Tatera* cf. *inclusa* | T | B |
| *Orycteropus* sp. | F | I | *\*Dendromus* sp. indet. | T-A | G |
| **Primates** | | | *\*Steatomys* sp. indet. | F | B |
| Cercopithecidae | | | *\*Saccostomus major* | F | B |
| cf. *Paracolobus* sp. | T-A | FL | Muridae | | |
| *Parapapio ado* | T-A | FL | *\*Thallomys laetolilensis* | A | B |
| cf. *Papio* sp. | T | FL | *\*Mastomys cinereus* | T | B |
| Colobinae sp. indet. | A | FL | Hystricidae | | |
| Hominidae | | | *Hystrix leakeyi* | T | R |
| *Australopithecus afarensis* | T | O | *Hystrix* cf. *makapanensis* | T | R |
| Galagidae | | | *Xenohystrix crassidens* | T | R |
| *Galago sadimanensis* | A | FI | Bathyergidae | | |
| **Insectivora** | | | *\*Heterocephalus quenstedti* | F | R |
| *Rhynchocyon pliocaenicus* | T | I | Pedetidae | | |
| **Carnivora** | | | *Pedetes laetoliensis* | F | G |
| Felidae | | | **Lagomorpha** | | |
| | | | *Serengetilagus praecapensis* | T | G |

*Species less than 500 g, excluded from analyses.

(PCA) was conducted using STASTICA 6.0, and predictor ecovariables (i.e., arboreality, terrestriality, frugivory, grazing) were used in bivariate plots (Reed, 1996, 1997).

FAUNAL SIMILARITY

In addition, a faunal similarity index was used to provide a measure of how similar the Upper Laetolil fauna is to those from other African Plio-Pleistocene sites (see Appendix 1). Several different faunal similarity indices have been devised (e.g., Simpson, 1960; Nakaya, 1994; Reed, 1996), but the most widely used is Simpson's index (Simpson, 1960). The formula is as follows:

*Simpson's Index* = $C/N_1$
$C$ = number of taxa in common for both faunas
$N_1$ = total number of taxa of the smaller fauna

In this study, the unit of analysis is the species. Indices are transformed into percentages by multiplying $C/N_1$ by 100. Faunal similarity indices are generally used to detect provincial or temporal relationships (Flynn, 1986), rather than habitat similarities, but given that the sites included in this study are regionally and chronologically constrained, it is likely that a significant component of any observed differences is likely to reflect ecological distinctions (Van Couvering and Van Couvering, 1976).

**Results and Discussion**

COMPARISONS OF THE UPPER LAETOLIL FAUNAS FROM DIFFERENT LOCALITIES AND STRATIGRAPHIC ZONES

The faunas from different stratigraphic zones and collecting localities of the Upper Laetolil Beds presented in Table 1 were compared in order to discern whether there was any evidence of temporal or spatial heterogeneity.

Time averaging of faunas over the course of the more than 300 Kyr represented by the Upper Laetolil sequence could produce a composite fauna that reflects a mixture of different habitats. If this is the case, it might account for the high species diversity and unusual composition of the large mammal community. It would also impact on comparisons with present-day mammalian communities, and make it difficult to ascertain the paleoecology of Laetoli based on its closest modern analogs. To test for habitat heterogeneity, the ecological diversity at each of the collecting localities and stratigraphic zones was compared. A modified chi-square test (Zar, 1999) was conducted on the relative proportions of the ecovariables (Table 6).

The results show that there are no significant differences in ecological diversity between the different localities or stratigraphic zones (Table 7). This implies that the composition of

Table 6. *Percentages of locomotor and trophic ecovariables for the large mammalian fauna of the Upper Laetolil Beds*

| | Upper Laetolil (%) |
|------|------|
| T | 79.6 |
| T-A | 9.3 |
| A | 3.7 |
| F | 7.4 |
| AQ | 0.0 |
| G | 22.2 |
| FG | 0.0 |
| B | 22.2 |
| MF | 1.9 |
| Fg | 0.0 |
| FI | 1.9 |
| FL | 7.4 |
| TF | 9.3 |
| C | 16.7 |
| C-I | 7.4 |
| C-B | 5.6 |
| RT | 5.6 |
| O | 5.6 |
| I | 3.7 |

Abbreviations: T = Terrestrial; T-A = Terrestrial–Arboreal; A = Arboreal; F = Fossorial; AQ = Aquatic; G = Grazer; FG = Fresh Grass Grazer; B = Browser; MF = Mixed Feeder; Fg = Frugivore; F-I = Frugivore –Insects; FL = Fruit and Leaves; TF = Total Frugivory; C = Carnivore; C-I = Carnivore–Insects; C-B = Carnivore–Bone; RT = Root and Tuber; O = Omnivore; I = Insectivore.

*Table 7. Significance results for the comparisons of the Upper Laetolil faunas from different localities and stratigraphic zones, using a modified chi-square test (Zar, 1999) where significance is set at $X^2 = 7.815$ (p < 0.05). NS = Not significant*

| Locomotor adaptations | | | Trophic adaptations | | |
|---|---|---|---|---|---|
| | $X^2$ | Significance | | $X^2$ | Significance |
| T | 0.404016 | NS | G | 1.368645 | NS |
| T-A | 0.37787 | NS | FG | – | – |
| A | 0.421882 | NS | B | 0.092018 | NS |
| AQ | – | – | MF | 0.118129 | NS |
| F | 0.365414 | NS | Fg | – | – |
| | | | FI | 3.1933 | NS |
| | | | FL | 1.009128 | NS |
| | | | C | 0.395363 | NS |
| | | | C-B | 0.695276 | NS |
| | | | C-I | 3.1933 | NS |
| | | | I | 0.787727 | NS |
| | | | O | 0.910169 | NS |
| | | | RT | 0.169224 | NS |

the mammalian community in terms of ecovariables was essentially identical throughout the entire Upper Laetolil sequence, and that the general ecological structure remained uniform throughout this time. This is an important finding, because it demonstrates that the large mammal community remained remarkably stable over an extended period of time, regardless of regional and local perturbations in the ecosystem. We can infer from the geology, for example, that periodic inundations of carbonatite ash from the volcano Sadiman were catastrophic events that would have dramatically affected the local vegetation, and in all probability had severe consequences on the local mammalian community. Heavy ash falls would have blanketed the paleoland surface, burying and killing the herbaceous vegetation, and within a short time, the ashes would have formed well-cemented tuffaceous limestones, killing standing trees, preventing root penetration by germinating seeds, and impeding the long-term regeneration of trees and woody shrubs. During these periods, extensive areas of dry grassland, with few or no trees, would

have dominated Laetoli. However, since there are no indications of specialized grassland communities associated with any of the faunal assemblages from the Upper Laetolil horizons, these periods of disruption in the ecosystem were apparently relatively short term (probably on the order of centuries), and grasslands were apparently quickly replaced by the climax vegetation. The uniformity of the mammalian faunal community from Laetoli implies that fossils are almost exclusively preserved in paleosols deposited during periods when the ecosystem was dominated by heavy vegetation, while the very short periods with grassland had relatively low sedimentation rates and produced few vertebrate fossils. The conclusion that can be drawn from these findings is that the general ecosystem at the time of the deposition of the Upper Laetolil Beds was a mosaic of different habitat types (i.e., not a mixture of time-averaged habitats), and one that remained remarkably stable over time, despite the influences of volcanic inundations that probably had only a localized and relatively short-term impact.

For the purposes of this study, given the spatial and geographical uniformity of the faunas from the Upper Laetolil Beds, the composite fauna derived from the entire sequence can now be used to reconstruct the paleoecology of Laetoli. The paleoecological relationships of the Upper Laetolil Beds will be deduced from ecological diversity analyses and faunal similarity indices.

## ECOLOGICAL DIVERSITY ANALYSIS

Ecological diversity data from the Upper Laetolil Beds, and comparative data from other Plio-Pleistocene African sites and modern faunas, were analyzed using principal components analysis (PCA). When PCA was performed, 19 factors were extracted, although PC 1 and 2 accounted for 32.8% and 17.7% of the total variance (Table 8). A bivariate plot of

*Table 8. Eigenvalue and the percentage of total variance for the first six principal components*

| PC | Eigenvalue | % Total variance |
|----|------------|------------------|
| 1 | 6.238171 | 32.8 |
| 2 | 3.363629 | 17.7 |
| 3 | 2.489357 | 13.1 |
| 4 | 1.637822 | 8.6 |
| 5 | 1.280560 | 6.7 |
| 6 | 1.060219 | 5.6 |

the first two factors shows three groupings – modern forest communities, modern non-forest communities, and fossil communities (Figure 3). It is interesting to note, in this regard, that, with the exception of forest communities, all other modern habitat types represented in tropical Africa cluster closely together and are

not easily differentiated, except for Serengeti Plain and Savanna Grassland, which fall within the range of fossil sites (we will return to this point later in the discussion). This may indicate that either the ecovariables or the multivariate methods of analysis used in this study are too coarse to readily distinguish between non-forest habitat types.

There are three possible explanations for the distinctiveness of fossil assemblages, including Laetoli, compared to all modern large mammal communities: (1) *Fossil sites have no modern analogs.* Since faunal communities change and evolve through time, it should not be unexpected to find that the structure of communities in the Pliocene is somewhat different from that of modern-day communities. For instance, Andrews and Humphrey

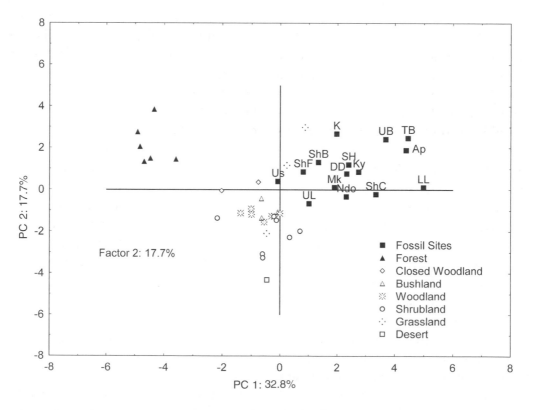

Figure 3. Results of a principal components analysis (PCA). This is a projection of modern and fossil localities on the factor plane (PC 1 × PC 2). There are generally three distinct groupings – modern forest, modern non-forest, and fossil localities. Fossil locality abbreviations: UL = Upper Laetolil Beds; LL = Lower Laetolil Beds; Ndo = Upper Ndolanya Beds; K = Kanapoi; Ap = Apak; Ky = Kaiyumung; SH = Sidi Hakoma; DD = Denen Dora; ShB = Shungura B; ShC = Shungura C; ShF = Shungura F; Us = Usno; TB = Tulu Bor; UB = Upper Burgi; Mk = Makapansgat Member 3.

(1999) hypothesize that many Plio-Pleistocene faunal communities have no equivalence to those found in present-day habitats. Although the general ecological attributes associated with mammalian community structure remain broadly similar through time and space, and there are evidently close taxonomic similarities between the modern fauna and flora to those from the Pliocene of East Africa, this need not necessarily imply that the vegetational and faunal communities were constituted in the exactly same way to produce assemblages comparable to those seen today. If this is the case, then fossil localities cluster together because they share aspects of their ecology that are not found in any modern-day large mammal communities. (2) *There is inherent bias in the fossil record.* Since fossil communities include only a fraction of the taxa represented in the original communities, there is an inherent sampling bias that may affect the outcome of comparative analyses (even if small mammals, which are evidently underrepresented taxonomically at most fossil sites, are excluded; see Dodson, 1973; Korth, 1979; Andrews and Nesbit Evans, 1983; Andrews, 1990; Fernandez-Jalvo and Andrews, 1992; Fernandez-Jalvo, 1995, 1996; Hoffman, 1988). Other than the original sampling bias, there is also the issue of recovery bias. For example, small fossil specimens are more easily destroyed after exposure or overlooked by collectors. If this is the case, then the fossil localities cluster together because certain taxa are uniformly absent or under-represented in the fossil record. (3) *Ecomorphological analyses may not accurately reflect the range of habitat preferences of fossil taxa.* There is an inherent asymmetry in the manner in which ecovariables are assigned to fossil and extant taxa in ecological diversity analyses. Trophic and locomotor categories of modern species are based on direct behavioral observations, while the behavioral categories of fossil taxa are based on inferences of function derived from the morphol-

ogy of preserved anatomical parts. While such inferences can generally be expected to yield equivalent results, there might be a lack of precise correspondence between modern and fossil data that affects the outcome of ecomorphological analyses of fossil communities. For example, it is conceivable that some taxa might be coded incorrectly for habitat type if they show specializations for a particular behavior, even though it may represent a relatively minor component of their overall repertoire (i.e., semi-terrestrial monkeys that spend most of their time in trees, or mixed feeders that include a large component of fruits in their diet). If this is the case, the results of ecomorphological analyses of fossil assemblages would tend to exaggerate the terrestrial, cursorial, and folivorous components of a community. A possible means of circumventing this problem would be to use ecomorphology to assign ecovariables to both fossil and modern taxa. Even so, we suspect that this bias is not profound enough to account for the major differences seen between the fossil and modern large mammal communities, but it may be a contributing factor.

A detailed examination of the large mammal fauna from the Upper Laetolil Beds and modern communities provides insights into whether or not the PCA results are due to differences in community structures in the past, taphonomic biases against certain taxa, or a lack of correspondence between ecomorphological data from modern and fossil communities. Identification of the individual ecovariables that drive the distinctions in the PCA helps to isolate the critical factors that differentiate fossil and modern communities. For PC 1, the highest contribution comes from terrestrial mammals. Direct comparisons of the faunal lists show that fossil communities have a higher proportion of terrestrial mammals. For example, terrestrial mammals in modern communities account for 33.3% to 88.8% of the total large mammal fauna, whereas at fossil sites the proportion of terrestrial mammals

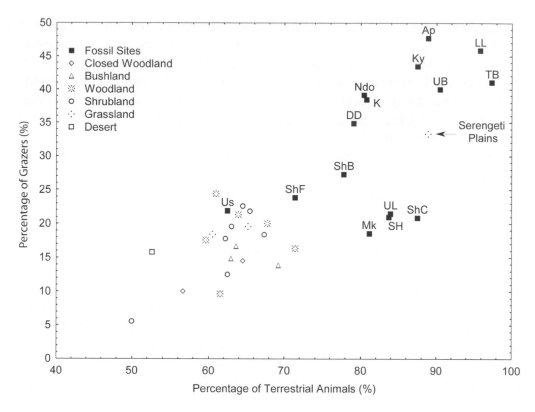

Figure 4. Bivariate plot of the community percentages of terrestrial and grazing mammals from African Plio-Pleistocene localities and modern communities. Note the separation of fossil and modern communities and the placement of the Serengeti Plains (with arrow). The latter is the only modern faunal community to fall within the range of fossil localities. Fossil locality abbreviations: UL = Upper Laetolil Beds; LL = Lower Laetolil Beds; Ndo = Upper Ndolanya Beds; K = Kanapoi; Ap = Apak; Ky = Kaiyumung; SH = Sidi Hakoma; DD = Denen Dora; ShB = Shungura B; ShC = Shungura C; ShF = Shungura F; Us = Usno; TB = Tulu Bor; UB = Upper Burgi; Mk = Makapansgat Member 3.

is generally much higher, ranging from 62.5% to 97.4% (Figure 4).

A closer examination of the faunal lists reveals that the large mammal fauna from the Upper Laetolil Beds differs from modern communities primarily in the relative proportions of carnivores and artiodactyls. A higher proportion of carnivore taxa in modern habitats are non-terrestrial when compared with Upper Laetolil. Non-terrestrial taxa include those that exhibit significant arboreal, terrestrial–arboreal, aquatic, and fossorial locomotor behaviors. Only 26.3% of the carnivore species found in the Upper Laetolil Beds are non-terrestrial, compared to modern communities, which have 37.5% to 75.0% (with

the exception of the Serengeti Plains, 11.1%). Most non-terrestrial carnivores are felids, mustelids, and viverrids. Almost all non-terrestrial carnivores from Upper Laetolil are felids (with the exception of *Mellivora*), and with nine species identified, it is comparable in diversity to many modern communities; thus, it is unlikely that felids are an underrepresented component of the fauna. Mustelids and viverrids, however, are much more impoverished in species number when compared to modern communities, and all of the taxa are classified as terrestrial, except for *Mellivora*. In modern communities, the small carnivore guild is often the most numerous in terms of species numbers and typically includes a large number of non-terrestrial taxa (see Kingdon, 1974d, 1997).

To characterize the relative representation of non-terrestrial carnivores at Laetoli, the Kruskal–Wallis non-parametric significance test was conducted. It showed that there was a statistically significant difference between the proportions of non-terrestrial carnivores of modern and fossil communities ($p = 0.0003$). A whisker plot with 0.95 confidence intervals illustrates the separation of the two sets of communities (Figure 5). The abundance of non-terrestrial carnivores is a significant contributor to the distinctiveness of fossil and modern communities. The relatively low proportion of non-terrestrial carnivores in fossil faunas may be due to ecological differences in community structure between fossil and modern carnivore guilds. The large carnivores from the Plio-Pleistocene of Africa differ from extant communities in having a greater number of species and in exhibiting a different suite of behaviors (Lewis, 1995). It is likely that this taxonomic and paleobiological distinction can also be applied to the small carnivores. Alternatively, the low proportion of non-terrestrial carnivores could possibly be attributed to taphonomic factors. Since many non-terrestrial species are small in size with quite distinctive habitat preferences, preservational or collecting biases may impact on their observed taxonomic diversity. Most likely, the disparity in the proportion of non-terrestrial carnivore species in fossil communities is due to a combination of these factors.

The difference in the proportions of grazing mammals between modern and Upper Laetolil communities can be attributed to artiodactyls, specifically bovids. The proportion of artiodactyl grazers in modern communities

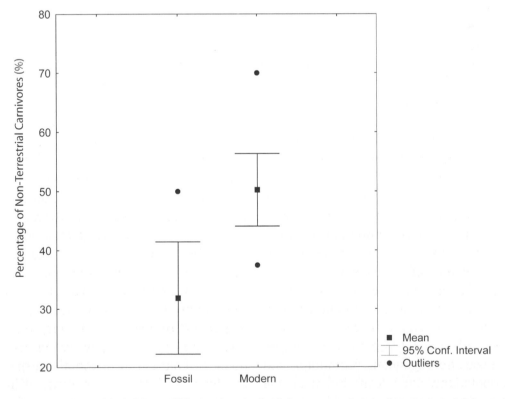

Figure 5. Whisker plot with 0.95 confidence interval of the proportions of non-terrestrial carnivores in fossil and modern communities. There is a clear separation between the fossil and modern communities. The difference is statistically significant (Kruskal–Wallis: $p = 0.0003$).

is relatively low, ranging from 0% to 33.3% (except for Serengeti Plains with 71.4%), compared with 38.9% in Upper Laetolil. The lower frequency of artiodactyl grazers is due to the fact that modern species are classified more often as non-grazers, such as browsers, mixed feeders, and fresh grass grazers, whereas the Laetoli artiodactyls are mostly classified as grazers. As discussed above, ecovariables of modern species are based on direct behavioral observations, while ecovariables of fossil species are based on inferences of function derived from the morphology of preserved elements. This may impact on the accuracy of ecovariable assignment for fossil species. Alternatively, the differences in the proportions of grazers may be due to an ecological difference between modern and fossil communities, such that there is no modern equivalent to the Upper Laetolil fauna. The chance that artiodactyl grazers would have been selectively preserved in the Upper Laetolil Beds compared to browsers or mixed feeders is unlikely, so taphonomic biases can be discounted.

However, when the Kruskal–Wallis non-parametric significance test was conducted on the proportion of grazers in modern and fossil faunal communities, it was found that there was no statistically significant difference between them ($p = 0.3068$). A whisker plot with 0.95 confidence intervals shows overlapping ranges for fossil and extant communities (Figure 6). This suggests that the abundance of grazing artiodactyls may not be an important factor in the separation of fossil and modern communities in a principal components analysis.

To further examine the separation of modern and fossil communities, we conducted an ecological diversity analysis on artiodactyls only. It is instructive to focus on a group with relatively homogeneous locomotor and trophic adaptations, and see its effect on the distribution of the communities. Artiodactyls are suited for this because they are a large and diverse group, and they are usually the most numerous taxon in a community. A principal components analysis (PCA) conducted on the artiodactyls extracted nine factors. PC 1 and 2 accounted for 37.4% and 28.2% of the total variance, respectively (Table 9). A bivariate plot of the first two principal components shows two distinct groupings – modern forest communities versus all non-forest communities, both modern and fossil (Figure 7). Once again, there is no distinction between modern non-forest communities, and they cannot be readily distinguished from the fossil communities. Even though fossil communities have generally higher proportions of artiodactyl grazers compared to modern communities, they cluster together in this PCA, indicating that grazing is not an important factor in distinguishing the fossil and modern communities. The results of this analysis confirm those of the Kruskal–Wallis significance test and whisker plot. While grazers contribute greatly to the variance seen in PC 1 of the principal components analysis, they do not play a key role in distinguishing fossil localities from extant non-forest localities.

As mentioned above, Serengeti Plains (SP) and Savanna Grassland (SG) are the only two modern communities that fall within the range of fossil communities in the PCA (Figure 3). Their positions on the factor plane projection appear to be the result of a combination of factors – the relatively high frequency of terrestrial animals (SP: 88.9%; SG: 60.5%) and the lack or relatively low proportions of arboreal animals (SP: 0.0%; SG: 0.0%), browsers (SP: 5.6%; SG: 2.6%), fruit and leaf eaters (SP: 0.0%; SG: 5.3%), and omnivores (SP: 11.1%; SG: 2.6%). These ecovariables contribute importantly to PC 1 and 2. However, the displacement of Serengeti Plains from modern communities is mostly driven by its high frequency of terrestrial mammals, surpassed only by those of fossil communities. While other modern communities have similar frequencies for individual ecovariables, they do not exhibit the same combination of frequencies. It is

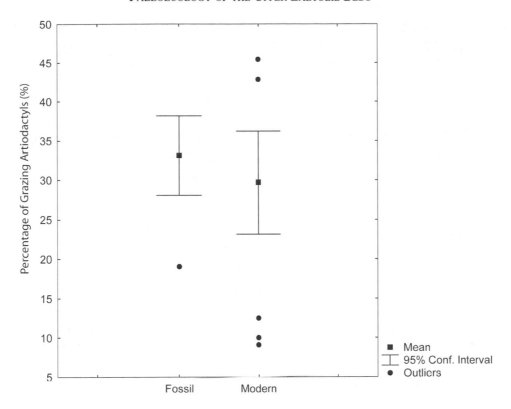

Figure 6. Whisker plot with 0.95 confidence interval of the proportions of grazing artiodactyls in fossil and modern communities. There is no separation between the fossil and modern communities. The difference is not statistically significant (Kruskal–Wallis: $p = 0.3068$).

Table 9. *Eigenvalues and percentage of total variance for the first six principal components from the artiodactyl dataset*

| PC | Eigenvalue | % of Total variance |
|----|-----------|---------------------|
| 1 | 3.366111 | 37.4 |
| 2 | 2.533995 | 28.1 |
| 3 | 1.225692 | 13.6 |
| 4 | 0.799037 | 8.9 |
| 5 | 0.493679 | 5.5 |
| 6 | 0.310580 | 3.5 |

noteworthy that Serengeti Plains is the only modern fauna to fall within the range of fossil sites in a bivariate plot of terrestrial versus grazing animals due to its high proportion of terrestrial mammals (see Figure 4).

The unique position of the Serengeti Plains among modern communities may relate to the limited diversity and uniformity of its vegetation, which results in a relatively impoverished and ecologically specialized large mammal fauna. Among the comparative faunas used in the study, that from the Serengeti Plains represents a special case because its faunal list is limited to mammals observed in the open grassland habitat of the Serengeti National Park (Swynnerton, 1958). Other faunal lists are derived from the entire area of the national park or game reserve in question, which usually includes multiple habitat types. The faunal list from the Serengeti Plains comprises only 18 species from 8 families and 4 orders (Swynnerton, 1958). Of these, 6 species are medium- to large-bodied bovids and 9 species are carnivores that prey on them (resulting in an extremely low proportion of non-terrestrial carnivores). When the Serengeti Plains faunal list is extended to include all of the Serengeti National Park, it no longer clusters with the

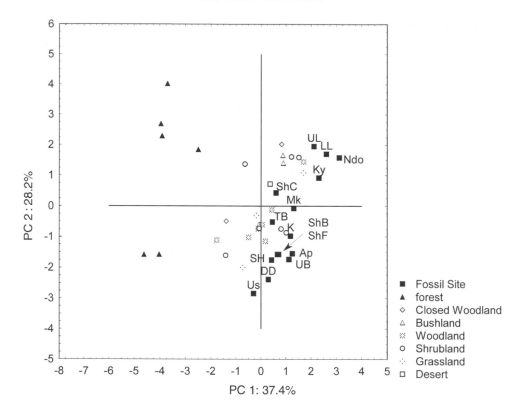

Figure 7. Results of a principal components analysis (PCA) of artiodactyls. This is a projection of modern and fossil localities on the factor plane (Factor 1 × Factor 2). There are two distinct groupings – modern forest and modern and fossil non-forest localities. Fossil locality abbreviations: UL = Upper Laetolil Beds; LL = Lower Laetolil Beds; Ndo = Upper Ndolanya Beds; K = Kanapoi; Ap = Apak; Ky = Kaiyumung; SH = Sidi Hakoma; DD = Denen Dora; ShB = Shungura B; ShC = Shungura C; ShF = Shungura F; Us = Usno; TB = Tulu Bor; UB = Upper Burgi; Mk = Makapansgat Member 3.

fossil localities. Given the balance of evidence, it is unlikely that the faunas from fossil sites are from such constrained habitat types as the Serengeti Plains or that they are representative of communities from homogeneous grassland habitats, but the results of the PCA may imply that extensive grasslands were an important part of the paleolandscape of Africa during the Plio-Pleistocene.

Following Reed (1996, 1997, 1998), predictor ecovariables were used in bivariate plots to examine the habitat types to which the Upper Laetolil Beds may be most similar. Predictor ecovariables are those that are most useful in discriminating habitat types, such as terrestriality, arboreality, frugivory, and grazing. Modern forest communities are

excluded from the bivariate analyses because their community structure is readily distinguishable from all other habitat types. In this case, total frugivory (TF) is used as the fruit-eating category because it encompasses all mammal species with significant proportions of fruit in their diet. The fauna from the Upper Laetolil Beds is characterized by the following distinctive properties: (1) a relative low occurrence of fruit-eating (9.3%) and arboreal (3.7%) mammals; (2) a high frequency of terrestrial mammals (79.6%); and (3) grazers are the most common mammals (22.2%) (Table 6). The Upper Laetolil fauna clusters with those from modern open woodland habitats. This is especially clear when the predictor ecovariables of frugivory and arboreality are used (Figures 8 and 9). Modern open woodland faunas are characterized by relatively low proportions of arboreal (1.6% to 4.4%) and frugivorous (6.6% to 13.5%) taxa and relatively high frequency of terrestrial (59.7% to 71.4%) and grazing (9.6% to 24.4%) mammals. However, there is overlap between faunas from open woodland and shrubland habitats in terms of their ecovariable structure. Among fossil localities, Upper Laetolil is generally grouped with Makapansgat and Sidi Hakoma, both of which have been reconstructed as having mosaic habitats. Makapansgat is considered to have been woodland with some bushland and grassland (Dart, 1952; Wells and Cooke, 1956; Vrba, 1980; Reed, 1996). Sidi Hakoma is reconstructed as having bushland to for-

ested habitats with areas of open grassland (Gray, 1980; Bonnefille et al., 1987, 2004).

The results of the bivariate analyses presented here contradict those presented by Reed (1997), who reconstructed Upper Laetolil paleoecology as being closed to medium density woodlands. This difference is accounted for by Reed's use of a different dataset and ecovariable coding for certain taxa. Reed (1997) utilized a selective list of mammalian taxa from Localities 1 and 7 only, in order to better constrain the temporal and geographical range of the fauna to be analyzed. However, since there are no significant differences in the community structure of the faunas from the entire Upper Laetolil Beds, regardless of their stratigraphic zone or collecting locality, we have been able to use the

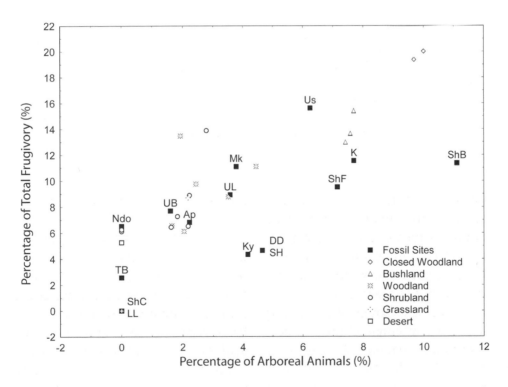

Figure 8. Bivariate plot of the percentages of arboreal and frugivorous mammals from African Plio-Pleistocene localities and modern communities. Upper Laetolil clusters with modern woodland communities, although there is overlap between the shrubland and woodland habitats. Fossil locality abbreviations: UL = Upper Laetolil Beds; LL = Lower Laetolil Beds; Ndo = Upper Ndolanya Beds; K = Kanapoi; Ap = Apak; Ky = Kaiyumung; SH = Sidi Hakoma; DD = Denen Dora; ShB = Shungura B; ShC = Shungura C; ShF = Shungura F; Us = Usno; TB = Tulu Bor; UB = Upper Burgi; Mk = Makapansgat Member 3.

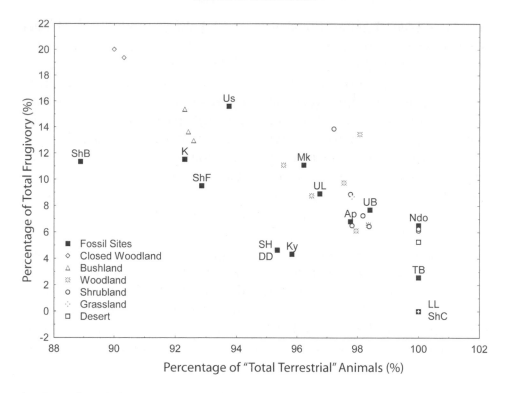

Figure 9. Bivariate plot of the percentages of "total terrestrial" (TT) and frugivorous mammals from African Plio-Pleistocene localities and modern communities. Upper Laetolil clusters with modern woodland communities, although there is overlap between the shrubland and woodland habitats. Fossil locality abbreviations: UL = Upper Laetolil Beds; LL = Lower Laetolil Beds; Ndo = Upper Ndolanya Beds; K = Kanapoi; Ap = Apak; Ky = Kaiyumung; SH = Sidi Hakoma; DD = Denen Dora; ShB = Shungura B; ShC = Shungura C; ShF = Shungura F; Us = Usno; TB = Tulu Bor; UB = Upper Burgi; Mk = Makapansgat Member 3.

larger composite fauna in our analysis. As a consequence, while the number of arboreal and frugivorous species remained the same in both analyses, the number of taxa in other ecovariable categories increased, thereby decreasing the overall percentage of arboreal and frugivorous species. Obviously, sampling of taxa is an important consideration in ecological diversity analyses, and as long as the fauna can be shown to be relatively uniform in space and time, the most inclusive faunal list is preferable, and likely to yield the most accurate inference about paleoecology. In this case, Upper Laetolil can be shown to be most similar to modern mammalian communities that live in medium to open woodlands, rather than in closed woodlands.

## FAUNAL SIMILARITY

Simpson's Similarity Index was used to calculate the similarity of the mammalian fauna from the Upper Laetolil Beds to those from other Pliocene localities (Table 10). In this case, all mammals identified to the species level, regardless of size, were included in the analysis. The results show that the mammal fauna from the Upper Laetolil Beds was most similar to the faunas from the Lower Laetolil Beds (82%) and Upper Ndolanya Beds (59%). This demonstrates that the faunas from Laetoli, regardless of their age, resemble each other more closely in their taxonomic composition than do penecontemporaneous faunas from other regions of Africa (Table 10). For example,

*Table 10. Faunal similarity matrix of African Plio-Pleistocene sites*

|      | UL  | LL  | Ndo | A   | KM  | TB  | UB  | ShB | ShC | ShF | U   | SH  | DD  | K   | Mk3 |
|------|-----|-----|-----|-----|-----|-----|-----|-----|-----|-----|-----|-----|-----|-----|-----|
| UL   | 100 | 82  | 59  | 36  | 50  | 28  | 18  | 13  | 28  | 19  | 50  | 35  | 29  | 38  | 18  |
| LL   | 82  | 100 | 73  | 50  | 33  | 43  | 29  | 14  | 14  | 14  | 21  | 29  | 43  | 31  | 21  |
| Ndo  | 59  | 73  | 100 | 27  | 33  | 30  | 30  | 15  | 16  | 15  | 43  | 25  | 33  | 31  | 15  |
| A    | 36  | 50  | 27  | 100 | 55  | 27  | 14  | 9   | 9   | 9   | 14  | 25  | 32  | 54  | 9   |
| KM   | 50  | 33  | 33  | 55  | 100 | 50  | 17  | 25  | 25  | 25  | 17  | 33  | 42  | 25  | 8   |
| TB   | 28  | 43  | 30  | 27  | 50  | 100 | 48  | 24  | 28  | 24  | 43  | 45  | 42  | 46  | 7   |
| UB   | 18  | 29  | 30  | 14  | 17  | 48  | 100 | 31  | 40  | 34  | 36  | 50  | 38  | 31  | 15  |
| ShB  | 13  | 14  | 15  | 9   | 25  | 24  | 31  | 100 | 76  | 48  | 28  | 35  | 33  | 23  | 18  |
| ShC  | 28  | 14  | 16  | 9   | 25  | 28  | 40  | 76  | 100 | 69  | 34  | 25  | 25  | 15  | 20  |
| ShF  | 19  | 14  | 15  | 9   | 25  | 24  | 34  | 48  | 69  | 100 | 31  | 25  | 25  | 15  | 19  |
| U    | 50  | 21  | 43  | 14  | 17  | 43  | 36  | 28  | 34  | 31  | 100 | 36  | 36  | 15  | 43  |
| SH   | 35  | 29  | 25  | 25  | 33  | 45  | 50  | 35  | 25  | 25  | 36  | 100 | 73  | 31  | 35  |
| DD   | 29  | 43  | 33  | 32  | 42  | 42  | 38  | 33  | 25  | 25  | 36  | 73  | 100 | 54  | 17  |
| K    | 38  | 31  | 31  | 54  | 25  | 46  | 31  | 23  | 15  | 15  | 15  | 31  | 54  | 100 | 8   |
| Mk3  | 18  | 21  | 15  | 9   | 8   | 7   | 15  | 18  | 20  | 19  | 43  | 35  | 17  | 8   | 100 |

Abbreviations: UL = Upper Laetolil Beds; LL = Lower Laetolil Beds; Ndo = Ndolanya Beds; A = Apak; KM = Kaiyumung; TB = Tulu Bor; UB = Upper Burgi; ShB = Shungura B; ShC = Shungura C; ShF = Shungura F; U = Usno; SH = Sidi Hakoma; DD = Denen Dora; K = Kanapoi; Mk3 = Makapansgat.

the faunas from the Upper Laetolil Beds and Sidi Hakoma Member at Hadar (~3.4 Ma), which are similar in age, only share 35% of their fauna.

Other Plio-Pleistocene fossil localities exhibit a similar relationship, in which geographical location is more important in determining faunal similarity than age (Table 10). The Apak and Kaiyumung Members of the Nachukui Formation at Lothagam are most similar to each other (55%) relative to other Plio-Pleistocene sites, even though the reconstructed habitats of these two members are quite different (see overview of Plio-Pleistocene sites). Kanapoi, which is comparable in age and geographical close to the Apak Member at Lothagam shares 54% of its fauna with the Apak Member. The mammalian faunas of the Shungura Formation are most similar to each other, ranging from 48% (Member B and Member F) to 69% (Member C and Member F) to 76% (Member B and Member C), even though there is evidence of changing paleoenvironmental conditions throughout the formation. The hypothesized paleohabitats of the Shungura Formation exhibited gradual

aridification and opening up of habitats in the sequence (Eck, 1976; Gentry, 1976; Bonnefille, 1983; Bonnefille and DeChamps, 1983; Eck and Jablonski, 1985; Wesselman, 1985; Reed, 1997). Finally, the Sidi Hakoma and Denen Dora Members at Hadar share 73% of their fauna. They are more similar faunally to each other than to other fossil localities, even though there was a change in vegetation through the sequence from deciduous and evergreen forest or bushland (lower Sidi Hakoma) to woodland (upper Sidi Hakoma) to wet and dry grassland (Denen Dora) (Bonnefille et al., 1987, 2004).

Clearly, local ecosystems have the potential to remain relatively stable in terms of faunal composition over long periods of time. This suggests that local and regional environmental and ecological conditions exert more influence over the composition of faunas than do large-scale ecological and climatic changes through time. The reconstructed paleoenvironment of the Upper Laetolil Beds, Lower Laetolil Beds, and Upper Ndolanya Beds are inferred to be distinct, but the general taxonomic composition of the fauna retains its overall integrity.

This implies that there were critical aspects of the Laetoli ecosystem that remain stable over time that buffer the mammalian community from faunal turnover. It has been theorized that intrinsic species constraints, such as genetics, development, and behavior, may be more important to species survival than ecological factors of the environment, but that, within the selective environment, local environmental conditions, including climate, geology, flora, and other fauna, may have more effect on a species than regional climate and geology, which in turn, has greater influence than global conditions (McKee, 1999). Thus, as long as a species is not highly specialized and restricted to a narrow set of ecological conditions, it would not be drastically affected by local environmental alterations caused by changes in regional or global conditions.

## Conclusion: Paleoecology of Laetoli Reconsidered

There has been no clear consensus on the paleoecology of Laetoli over the last 15 years, especially for the Upper Laetolil Beds. It has been reconstructed from arid to semi-arid grassland (see papers in Leakey and Harris, 1987) to dense woodland (Reed, 1997). The goal of this paper was to develop a better understanding of the Upper Laetolil paleoecology, based on a more detailed and comprehensive analysis of the large mammal fauna. The major findings of this paper are summarized below.

Time averaging was a significant unresolved issue from previous paleoenvironmental reconstructions of the Upper Laetolil Beds. Given the 300 Kyr time span of the Upper Laetolil Beds, it is conceivable that the fauna represents a composite assemblage that reflects a mixture of different habitats. In order to determine the existence of heterogeneity in the Laetoli large mammal community, the ecological diversity at each of the collecting localities and stratigraphic zones was compared. The results show that there were no statistically significant differences in ecological diversity among the different localities or stratigraphic zones, which allowed for the use of a combined fauna from the entire sequence to reconstruct Upper Laetolil paleoecology. Moreover, this also implied that the general ecological structure throughout the Upper Laetolil sequence remained relatively stable and that the general ecosystem throughout the time of deposition was a mosaic of different habitat types, rather than a mixture of time-averaged habitats.

Ecological diversity data from Upper Laetolil were compared with other African Plio-Pleistocene fossil localities, as well as modern communities, using principal components analysis (PCA). An important finding was the general distinctiveness of fossil assemblages, including Laetoli, from modern communities. Predictor ecovariables (i.e., terrestriality, arboreality, frugivory, grazing) were used in bivariate plots to examine the factors that contributed to the distinctiveness of fossil communities. Terrestrial mammals were found to have the greatest impact on the uniqueness of fossil communities. Fossil assemblages had very high proportions of terrestrial mammals, as well as grazers, when compared to modern communities. The high frequency of terrestrial mammals in the Upper Laetolil Beds was apparently determined mainly by the under-representation of non-terrestrial small carnivores, such as mustelids and viverrids. The over-representation of grazing artiodactyls, particularly bovids, mostly accounted for the high proportion of grazers in the Upper Laetolil Beds. However, the result of a non-parametric Kruskal–Wallis significance test showed that the difference between the proportions of extant and fossil non-terrestrial carnivores was statistically significant, while the abundance of fossil and modern artiodactyl grazers was not significantly different.

Three possible explanations were proposed to account for the distinctiveness of the Upper Laetolil Beds from modern large mammal communities: (1) fossil sites have no modern analogs; (2) there is inherent bias in the fossil record such that certain taxa are absent or under-represented in the fossil record; and (3) ecomorphological analyses may not accurately reflect the range of habitat preferences of fossil taxa.

A key finding of this study is that the Laetoli fauna remained remarkably stable over a long period of time. It was found that there were no statistically significant differences in the ecological diversity among Upper Laetolil localities and stratigraphic zones. This implies that the ecological structure remained relatively uniform throughout the Upper Laetolil sequence, for a period of about 300 Kyr, regardless of regional or local environmental disturbances and changes. Furthermore, faunal similarity measures provide evidence that a certain degree of taxonomic stability extended to the Lower Laetolil Beds and to the Upper Ndolanya Beds, which covers a period of more than 1.6 Myr. This does not imply that the Laetoli faunas did not change over the course of this period as a consequence of regional and global climatic shifts or as a result of community turnover, but shows, relative to other contemporary faunas in East Africa, that the Laetoli mammalian fauna maintained long-term taxonomic affinities that distinguish it regionally, regardless of age. Clearly, fundamental aspects of the Laetoli ecosystem remained stable over time, which buffered the mammalian community from dramatic episodes of taxonomic turnover.

A major goal of this study was to approach the paleoenvironment of the Upper Laetolil Beds from the perspective of a more detailed and comprehensive comparative analysis of the mammalian fauna. Results from predictor ecovariables indicate that Upper Laetolil was unlikely to have been predominantly a closed woodland or forested habitat, since these have a higher proportion of arboreal or semi-arboreal mammals with browsing or frugivorous adaptations. Instead, the high frequency of

grazers and terrestrial mammals, combined with the low occurrence of arboreal and frugivorous mammals, indicates affinities with modern mammalian communities living in grassland, savanna, and open woodland settings. Overall, the Upper Laetolil large mammal fauna exhibits characteristics that most closely approximate modern open woodland communities. Taking into account the results of this study, and the presence of indicator species, we reconstruct the paleoecology of the Upper Laetolil Beds as a mosaic habitat comprising of open woodland, grassland, and shrubland, as well as closed woodland along seasonal river courses. Evidence from the composition and distribution of tuffs, suggests that this climax vegetation was periodically disrupted and replaced for brief periods by extensive tracts of grassland following episodes of volcanic activity. Although these inundations of volcanic ash would presumably have had a profound effect on the local vegetation and mammalian community, the remarkable homogeneity of the Upper Laetolil fauna throughout the stratigraphic sequence suggests that the mammalian community was rapidly reconstituted in its entirety once the climax vegetation re-established itself.

**Acknowledgments**

We would like to thank the following individuals for their participation in the expeditions to Laetoli that contributed to the recovery of the material discussed here: P. Abwalo, P. Andrews, E. Baker, M. Bamford, R. Chami, M. Duru, C. Feibel, S. Hixson, J. Kingston, K. Kovarovic, A. Kweka, M. Lilombero, M. Mbago, K. McNulty, C. Msuya, S. Odunga, C. Robinson, W. Sanders, and L. Scott. We thank the Tanzania Commission for Science and Technology and the Unit of Antiquities in Dar es Salaam for permission to conduct research in Tanzania. Special thanks go to N. Kayombo (Director General), P. Msemwa (Director),

F. Mangalu (Director), A. Kweka and the staff at the National Museums of Tanzania in Dar es Salaam and Arusha for their support and assistance. The Government of Kenya and National Museums of Kenya are thanked for permission to study the collections in Nairobi. We would also like to acknowledge R. Bobe and the three reviewers (P. Andrews, M. Tappen, C. Musiba) for their invaluable comments on this paper. We are grateful to the editors of this volume (K. Behrensmeyer, R. Bobe, and Z. Alemseged) for inviting us to contribute to this volume. Fieldwork at Laetoli was supported by grants from National Geographic Society, the Leakey Foundation, and NSF (BCS-9903434 and BCS-0309513). Data collection and analysis were supported by a grant from NSF (BCS-0216683).

## References

Agnew, N., Demas, M., Leakey, M.D., 1996. The Laetoli footprints. Science 271, 1651–1652.

Alemseged, Z., 2003. An integrated approach to taphonomy and faunal change in the Shungura Formation (Ethiopia) and its implication for hominid evolution. Journal of Human Evolution 44, 451–478.

Andrews, P., 1989. Palaeoecology of Laetoli. Journal of Human Evolution 18, 173–181.

Andrews, P., 1990. Small mammal taphonomy. In: Lindsey, E.H., Fahlbusch, V., Mein, P. (Eds.), European Neogene Mammal Chronology. Plenum Press, New York, pp. 487–494.

Andrews, P., Humphrey, L., 1999. African Miocene environments and the transition to early Hominins. In: Bromage, T.G., Schrenk, F. (Eds.), African Biogeography, Climate Change, and Human Evolution. Oxford University Press, Oxford, pp. 282–300.

Andrews, P., Nesbit Evans, E.M., 1983. Small mammal bone accumulations produced by mammalian carnivores. Paleobiology 9, 289–307.

Andrews, P., Lord, J.M., Nesbit Evans, E.M., 1979. Patterns of ecological diversity in fossil and modern mammalian faunas. Biological Journal of the Linnean Society 11, 177–205.

Anonymous, 1995. Saving our roots from roots. Science 267, 171.

Ansell, W.F.H., 1960. Mammals of Northern Rhodesia. Government Printer, Lusaka.

Ansell, W.F.H., 1978. The Mammals of Zambia. National Park and Wildlife Service, Chilanga, Zambia.

Aronson, J.L., Taieb, M., 1981. Geology and paleogeography of the Hadar hominid site, Ethiopia. In: Rapp, G., Vondra, C.F. (Eds.), Hominid Sites: Their Geologic Settings. Westview Press, Boulder, pp. 165–195.

Avenant, N.L., Nel, J.A., 2002. Among habitat variation in prey availability and use by caracal Felis caracal. Mammalian Biology 67, 18–33.

Barry, J.C., 1987. Large carnivores (Canidae, Hyaenidae, Felidae) from Laetoli. In: Leakey, M.D., Harris, J.M. (Eds.), Laetoli: A Pliocene Site in Northern Tanzania. Clarendon Press, Oxford, pp. 235–258.

Barton, R.A., 1989. Foraging strategies, diet and composition in olive baboons. Ph.D. Thesis, University of St. Andrews, Fife, Scotland.

Beden, M., 1987. Fossil Elephantidae from Laetoli. In: Leakey, M.D., Harris, J.M. (Eds.), Laetoli: A Pliocene Site in Northern Tanzania. Clarendon Press, Oxford, pp. 259–294.

Behrensmeyer, A.K., Western, D., Dechant Boaz, D.E., 1979. New perspectives in vertebrate paleoecology from a recent bone assemblage. Paleobiology 5, 12–21.

Bishop, L., 1995. Reconstructing omnivore paleoecology and habitat preference in the Pliocene and Pleistocene of Africa. Paper for a symposium on African Biogeography, Climate Change, and Early Hominid Evolution. Wenner-Gren Foundation for Anthropological Research: An International Symposium.

Bishop, L., 1999. Suid paleoecology and habitat preferences at African Pliocene and Pleistocene Hominid localities. In: Bromage, T.G., Schrenk, F. (Eds.), African Biogeography, Climate Change, and Human Evolution. Oxford University Press, Oxford, pp. 216–225.

Blamey, A.H., Jackson, W.T., 1956. The elusive little piti. African Wildlife 10, 295–299.

Boaz, N., 1977. Paleoecology of Plio-Pleistocene Hominidae of the Lower Omo Basin. Ph.D. Dissertation, University of California, Oakland, California.

Bobe, R., Behrensmeyer, A.K., Chapman, R., 2002. Faunal change, environmental variability and late Pliocene hominin evolution. Journal of Human Evolution 42, 475–497.

Bonnefille, R., 1983. Evidence for a cooler and drier climate in the Ethiopian uplands toward 2.3 Myr ago. Nature 303, 487–491.

Bonnefille, R., 1986a. Palaeoenvironmental implications of a pollen assemblage from the Koobi For a formation (East Rudolph), Kenya. Nature 264, 403–407.

Bonnefille, R., 1986b. Palynological evidence for an important change in the vegetation of the Omo Basin between 2.5 and 2 million years. In: Coppens, Y., Howell, F.C., Isaac, G.L., Leakey, R.E. (Eds.), Earliest Man and Environments in the Lake Rudolf Basin. University of Chicago Press, Chicago, pp. 421–431.

Bonnefille, R., DeChamps, R., 1983. Data on Fossil Flora. In: de Heinzelin, J. (Ed.), The Omo Group: Archives of the International Omo Research Expedition, Annales, S. 8, Sciences Geologiques, pp. 191–207. Musée de l'Afrique Centrale, Tervuren.

Bonnefille, R., Riollet, G., 1987. Palynological spectra from the Upper Laetolil Beds. In: Leakey, M.D., Harris, J.M. (Eds.), Laetoli: A Pliocene Site in Northern Tanzania. Clarendon Press, Oxford, pp. 52–61.

Bonnefille, R. Vincens, A., Buchet, G., 1987. Palynology and Paleoenvironment of a Pliocene Hominid Site (2.9–3.3 My) at Hadar, Ethiopia. Palaeogeography, Palaeoclimatology, Palaeoecology 60, 249–281.

Bonncfille, R., Potts, R., Chalié, F., Jolly, D., Peyron, O. 2004. High-resolution vegetation and climate change associated with Pliocene *Australopithecus afarensis*. Proceedings of the National Academy of Sciences 101, 12125–12129.

Bothma, J. du P., 1966. Notes on the stomach contents of certain Carnivora (Mammalia) from the Kalahari Gemsbok Park. Koedoe 9, 37–39.

Bronikowski, A.M., Altmann, J., 1996. Foraging in a variable environment: weather patterns and the behavioural ecology of baboons.. Behavioral Ecology and Sociobiology 39, 11–25.

Butler, P.M., 1987. Fossil insectivores from Laetoli. In: Leakey, M.D., Harris, J.M. (Eds.), Laetoli: A Pliocene Site in Northern Tanzania. Clarendon Press, Oxford, pp. 85–87.

Cadman, A., Rayner, R.J.R., 1989. Climatic change and the appearance of Australopithecus africanus in the Makapansgat sediments. Journal of Human Evolution 18, 107–113.

Cerling, T.E., 1992. Development of grasslands and savannas in East Africa during the Neogene. Palaeogeography, Palaeoclimatology, Palaeoecology 97, 241–247.

Cerling, T.E., Hay, R.L., O'Neil, J.R., 1977. Isotopic evidence for dramatic changes in East Africa during the Pleistocene. Nature 267, 137–138.

Cerling, T.E., Harris, J.M., Leakey, M.G., 1999. Browsing and grazing in elephants: the isotope record of modern and fossil proboscideans. Oecologia 120, 364–374.

Cerling, T.E., Harris, J.M., Leakey, M.G., 2003a. Isotope paleoecology of the Nawata and Nachukui Formations at Lothagam, Turkana Basin, Kenya. In: Leakey, M.G., Harris, J.M. (Eds.), Lothagam: The Dawn of Humanity in Eastern Africa. Columbia University Press, New York, pp. 605–624.

Cerling, T.E., Harris, J.M., Passey, B.H., 2003b. Diets of East African Bovidae based on stable isotope analysis. Journal of Mammalogy 84, 456–470.

Child, G.S., 1964. Some notes on the mammals of Kilimanjaro. Tanganyika Notes and Records 53, 77–89.

Child, G., Robbel, H., Hepburn, C.P., 1972. Observations on the biology of tsessebe, *Damaliscus lunatus lunatus*, in northern Botswana. Mammalia 36, 342–388.

Coe, M., McWilliam, N., Stone, G., Packer, M., 1999. Mkomazi: The Ecology, Biodiversity and Conservation of a Tanzanian Savanna. Royal Geographical Society (with The Institute of British Geographers), London.

Damuth, J.D., 1990. Problems in estimating body masses of archaic ungulates using dental measurements. In: Damuth, J.D., MacFadden, B.J. (Eds.), Body Size in Mammalian Paleobiology. Cambridge University Press, Cambridge, pp. 229–253.

Dankwa-Wiredu, B., Euler, D.L., 2002. Bushbuck (*Tragelaphus scriptus* Pallas) habitat in Molc National Park, northern Ghana. African Journal of Ecology 40, 35–41.

Dart, R.A., 1952. Faunal and climatic fluctuations in Makapansgat Valley: their relation to the geologic age and Promethean status of *Australopithecus*. In: Leakey, L.S.B., Cole, S. (Eds.), Proceedings of the 1st Pan African Congress on Prehistory, Nairobi, 1947, pp. 96–106.

Davidge, C., 1978. Ecology of baboons (*Papio ursinus*) at Cape Point. Zoologica Africana 13, 329–350.

Davies, C., 1987a. Fossil Pedetidae (Rodentia) from Laetoli. In: Leakey, M.D., Harris, J.M. (Eds.), Laetoli: A Pliocene Site in Northern Tanzania. Clarendon Press, Oxford, pp. 171–190.

Davies, C., 1987b. Note on the fossil Lagomorpha from Laetoli. In: Leakey, M.D., Harris, J.M. (Eds.), Laetoli: A Pliocene Site in Northern Tanzania. Clarendon Press, Oxford, pp. 190–193.

Denys, C., 1985. Palaeoenvironmental and palaeobiogeographical significance of the fossil rodent assemblages of Laetoli (Pliocene, Tanzania). Palaeogeography, Palaeoclimatology, Palaeoecology 52, 77–97.

Denys, C., 1987. Fossil rodents (other than Pedetidae) from Laetoli. In: Leakey, M.D., Harris, J.M.

(Eds.), Laetoli: A Pliocene Site in Northern Tanzania. Clarendon Press, Oxford, pp. 118–170.

Depew, L.A., 1983. Ecology and behaviour of baboons (*Papio anubis*) in the Shai Hills Game Production Reserve, Ghana. M.Sc. Thesis, Cape Coast University, Ghana.

Dieckmann, R.C., 1980. The ecology and breeding biology of the gemsbok *Oryx gazella gazella* (Linnaeus, 1758) in the Hester Malan Nature Reserve. M.Sc. Thesis, University of Pretoria, Pretoria.

Dodson, P., 1973. The significance of small bones in paleoecological interpretation. Contributions to Geology 12, 15–19.

Dörgeloh, W.G., 2000. Relative densities and habitat utilization of non-utilized, terrestrial gamebird populations in a natural savanna, South Africa. African Journal of Ecology 38, 31–37.

Drake, R.E., Curtis, G.H., 1979. Radioisotope date of the Laetolil Beds, the Hadar Formation and the Koobi Fora-Shungura Formations. American Journal of Physical Anthropology 50, 433–434.

Drake, R.E., Curtis, G.H., 1987. K-Ar Geochronology of the Laetoli fossil localities. In: Leakey, M.D., Harris, J.M. (Eds.), Laetoli: A Pliocene Site in Northern Tanzania. Clarendon Press, Oxford, pp. 48–52.

Dunbar, R.I.M., Dunbar, P., 1974. Ecological relations and niche separation between sympatric terrestrial primates in Ethiopia. Folia Primatologica 21, 36–60.

Eck, G.G., 1976. Cercopithecoidea from Omo Group deposits. In: Coppens, Y., Howell, F.C., Isaac, G.L., Leakey, R.E. (Eds.), Earliest Man and Environments in the Lake Rudolf Basin. University of Chicago Press, Chicago, pp. 332–344.

Eck, G.G., Jablonski, N.G., 1985. The skull of Theropithecus brumpti compared with those of other species of the genus Theropithecus. In: Coppens, Y., Howell, F.C. (Eds.), Les Faunes Plio-Pléistocènes de la basse vallée de l'Omo (Éthiopie). Éditions du Centre National de la Recherche Scientifique, Paris, pp. 18–122.

Eloff, F.C., 1964. On the predatory habits of lions and hyaenas. Koedoe 7, 105–112.

Fashing, P.J., 2001. Feeding ecology of guereza in the Kakamega Forest, Kenya: the importance of Moraceae fruit in their diet. International Journal of Primatology 22, 579–609.

Feibel, C.S., Harris, J.M., Brown, F.H., 1991. Palaeoenvironmental context for the late Neogene of the Turkana Basin. In: Harris, J.M. (Ed.), Koobi Fora Research Project, Volume 3: The Fossil Ungulates: Geology, Fossil Artiodactyls, and Paleoenvironments. Clarendon Press, Oxford, pp. 321–370.

Fernandez-Jalvo, Y., 1995. Small mammal taphonomy at La Trinchera de Atapuerca (Burgos, Spain). A remarkable example of taphonomic criteria used for stratigraphic correlations and palaeoenvironment interpretations. Palaeogeography, Palaeoclimatology, Palaeoecology 114, 167–195.

Fernandez-Jalvo, Y., 1996. Small mammal taphonomy and the Middle Pleistocene environments of Dolina, Northern Spain. Quaternary International 33, 21–34.

Fernandez-Jalvo, Y., Andrews, P., 1992. Small mammal taphonomy of Gran Dolina, Atapuerca (Burgos), Spain. Journal of Archaeological Science 19, 407–428.

Ferreira, N.A., Bigalke, R.C., 1987. Food selection by grey rhebuck in the Orange Free State. South African Journal of Wildlife Research 17, 123–127.

Flynn, J.J., 1986. Faunal provinces and the Simpson coefficient. Contributions to Geology., pp. 317–338. University of Wyoming, Laramie, WY, Special Paper 3.

Gagnon, M., Chew, A.E., 2000. Dietary preferences in extant African Bovidae. Journal of Mammalogy 81, 490–511.

Gaynor, D., 1994. Foraging and feeding behaviour of chacma baboons in a woodland habitat. Ph.D. Thesis, University of Natal, Natal.

Gentry, A.W., 1976. Bovidae of the Omo Group deposits. In: Coppens, Y., Howell, F.C., Isaac, G.L., Leakey, R.E. (Eds.), Earliest Man and Environments in the Lake Rudolf Basin. University of Chicago Press, Chicago, pp. 275–292.

Gentry, A.W., 1987. Pliocene Bovidae from Laetoli. In: Leakey, M.D., Harris, J.M. (Eds.), Laetoli: A Pliocene Site in Northern Tanzania. Clarendon Press, Oxford, pp. 378–408.

Goddard, J., 1968. Food preferences of two black rhinoceros populations. East African Wildlife Journal 6, 1–18.

Grafton, R.N., 1965. Food of the black-backed jackal: a preliminary report. Zoologica Africana 1, 41–53.

Gray, B.T., 1980. Environmental reconstruction of the Hadar Formation (Afar, Ethiopia). Ph.D. Dissertation, Case Western Reserve University, Cleveland, Ohio.

Grobler, J.H., Wilson, V.J., 1972. Food of the leopard *Panthera pardus* (Linn.) in the Rhodes Matopos National Park, Rhodeisa, as determined by faecal analysis. Arnoldia (Rhodesia) 5, 1–10.

Happold, D.C.D., 1987. The Mammals of Nigeria. Clarendon Press, Oxford.

Harris, J.M., 1983. Background to the study of the Koobi Fora fossil faunas. In: Leakey, M.G., Leakey, R.E. (Eds.), Koobi Fora Research Project, Volume II. Clarendon Press, Oxford, pp. 1–21.

Harris, J.M., 1987a. Summary. In: Leakey, M.D., Harris, J.M. (Eds.), Laetoli: A Pliocene Site in Northern Tanzania. Clarendon Press, Oxford, pp. 524–531.

Harris, J.M., 1987b. Fossil Suidae from Laetoli. In: Leakey, M.D., Harris, J.M. (Eds.), Laetoli: A Pliocene Site in Northern Tanzania. Clarendon Press, Oxford, pp. 349–358.

Harris, J.M., 1991. Koobi Fora Research Project, Volume 3: The Fossil Ungulates, Geology, Fossil Artiodactyls, and Palaeoenvironments. Clarendon Press, Oxford.

Harris, J.M., Cerling, T.E., 2002. Dietary adaptations of extant and Neogene African suids. Journal of Zoology. Lond. 256, 45–54.

Harris, J.M., White, T.D., 1979. Evolution of the Plio-Pleistocene African Suidae. Transactions of the American Philosophical Society 69, 1–128.

Harrison, T., 2002. The first record of fossil hominino from the Ndolanya Bedo, Laetoli, Tanzania. American Journal of Physical Anthropology 119, 83.

Harrison, T., 2005. Fossil Bird Eggs from the Pliocene of Laetoli, Tanzania: Their Taxonomic and Paleoecological Relationships. Journal of African Earth Sciences 41, 289–302.

Harrison, T., Msuya, C., 2005. Fossil struthinid eggshells from Laetoli, Tanzania: Taxonomic and Biostratigraphic Significance. Journal of African Earth Sciences 41, 303–315.

Hay, R.L., 1976. Geology of the Olduvai Gorge: A Study of Sedimentation in a Semiarid Basin. University of California Press, Berkeley.

Hay, R.L., 1978. Melilitite-carbonotite tuffs in the Laetolil Beds of Tanzania. Contributions to Mineralogy and Petrology 17, 255–274.

Hay, R.L., 1981. Paleoenvironment of the Laetolil Beds, Northern Tanzania. In: Rapp, G., Vondra, C.F. (Eds.), Hominid Sites: Their Geologic Settings. Westview Press, Boulder, pp. 7–24.

Hay, R.L., 1987. Geology of the Laetoli area. In: Leakey, M.D., Harris, J.M. (Eds.), Laetoli: A Pliocene Site in Northern Tanzania. Clarendon Press, Oxford, pp. 23–47.

Hay, R.L., Reader, R.J., 1982. The fossil footprints of Laetoli. Scientific American 246, 50–57.

Hill, R.A., Dunbar, R.I.M., 2002. Climatic determinants of diet and foraging behaviour in baboons. Evolutionary Ecology 16, 579–593.

Hoffman, R., 1988. The contribution of raptorial birds to patterning in small mammal assemblages. Paleobiology 14, 81–90.

Hooijer, D.A., 1987. Hipparion teeth from the Ndolanya Beds. In: Leakey, M.D., Harris, J.M. (Eds.), Laetoli: A Pliocene Site in Northern Tanzania. Clarendon Press, Oxford, pp. 312–315.

Janis, C.M., 1988. An estimation of tooth volume and hypsodonty indices in ungulate mammals, and the correlation of these factors with dietary preference. In: Russel, D.E., Santoro, J.-P., Sigogneau-Russell, D. (Eds.), Teeth Revisited: Proceedings of the VIIth International Symposium on Dental Morphology, Paris 1986. Memoirs of the Museé Nationale de l'Histoire Naturelle, Paris, pp. 367–387.

Janis, C.M., 1990. Correlation of cranial and dental variables with body size in ungulates and macropodoids. In: Damuth, J.D., MacFadden, B.J. (Eds.), Body Size in Mammalian Paleobiology. Cambridge University Press, Cambridge, pp. 255–300.

Joubert, S.C., 1976. The population ecology of the roan antelope, *Hippotragus equinus equinus* (Desmarest, 1804) in the Kruger National Park. D.Sc. Thesis, University of Pretoria, Pretoria.

Jungius, H., 1971. The Biology and Behaviour of the Reedbuck *Redunca arundinum* Boddaert, 1785 in the Kruger National Park. Mammalia Depicta. Paul Parey, Hamburg.

Kaiser, T., Bromage, T.G., Schrenk, F., 1995. Hominid Corridor Research Project update: new Pliocene fossil localities at Lake Manyara and putative oldest Early Stone Age occurrences at Laetoli (Upper Ndolanya Beds), northern Tanzania. Journal of Human Evolution 28, 117–120.

Kappelman, J., Plummer, T., Bishop, L., Duncan, A., Appleton, S., 1997. Bovids as indicators of Plio-Pleistocene paleoenvironments in East Africa. Journal of Human Evolution 32, 229–256.

Kay, R.F., 1984. On the use of anatomical features to infer foraging behavior in extinct primates. In: Rodman, P.S., Cant, I.G. (Eds.), Adaptations for Foraging in Nonhuman Primates. Columbia University Press, New York, pp. 21–53.

Kent, P.E., 1941. The recent history and Pleistocene deposits of the plateau north of Lake Eyasi, Tanganyika. Geological Magazine 78, 173–184.

Kingdon, J., 1974a. East African Mammals: An Atlas of Evolution in Africa, Volume I. The University of Chicago Press, Chicago.

Kingdon, J., 1974b. East African Mammals: An Atlas of Evolution in Africa, Volume IIA (Insectivores and Bats). The University of Chicago Press, Chicago.

Kingdon, J., 1974c. East African Mammals: An Atlas of Evolution in Africa, Volume IIB (Hares and Rodents). The University of Chicago Press, Chicago.

Kingdon, J., 1974d. East African Mammals: An Atlas of Evolution in Africa, Volume IIIA (Carnivores). The University of Chicago Press, Chicago.

Kingdon, J., 1974e. East African Mammals: An Atlas of Evolution in Africa, Volume IIIB (Large Mammals). The University of Chicago Press, Chicago.

Kingdon, J., 1974f. East African Mammals: An Atlas of Evolution in Africa, Volume IIIC (Bovids). The University of Chicago Press, Chicago.

Kingdon, J., 1974g. East African Mammals: An Atlas of Evolution in Africa, Volume IIID (Bovids). The University of Chicago Press, Chicago.

Kingdon, J., 1997. The Kingdon Field Guide to African Mammals. Academic Press, San Diego.

Korth, W.M., 1979. Taphonomy of microvertebrate fossil assemblages. Annals of the Carnegie Museum 48, 235–285.

Kovarovic, K., Andrews, P., Aiello, L., 2002. The palaeoecology of the Upper Ndolanya Beds at Laetoli, Tanzania. Journal of Human Evolution 43, 395–418.

Kummer, H., 1968. Social Organization of Hamadryas Baboons. Karger, Basel.

Kyauka, P.S., Ndessokia, P., 1990. A new hominid tooth from Laetoli, Tanzania. Journal of Human Evolution 19, 747–750.

Lamprey, H.F., 1962. The Tangarire Game Reserve. Tanganyika Notes and Records 60, 10–22.

Lamprey, H.F., 1963. Ecological separation of the large mammal species in the Tarangire Game Reserve, Tanganyika. East African Wildlife Journal 1, 63–92.

Leakey, M.D., 1981. Tracks and Tools. Philosophical Transactions of the Royal Society of London B 282, 95–102.

Leakey, M.D., 1987a. Introduction. In: Leakey, M.D., Harris, J.M. (Eds.), Laetoli: A Pliocene Site in Northern Tanzania. Clarendon Press, Oxford, pp. 1–22.

Leakey, M.D., 1987b. The hominid footprints: introduction. In: Leakey, M.D., Harris, J.M. (Eds.), Laetoli: A Pliocene Site in Northern Tanzania. Clarendon Press, Oxford, pp. 490–496.

Leakey, M.D., 1987c. The Laetoli hominid remains. In: Leakey, M.D., Harris, J.M. (Eds.), Laetoli: A Pliocene Site in Northern Tanzania. Clarendon Press, Oxford, pp. 108–109.

Leakey, M.D., Delson, E., 1987. Fossil Cercopithecidae from the Laetolil Beds. In: Leakey, M.D., Harris, J.M. (Eds.), Laetoli: A Pliocene Site in Northern Tanzania. Clarendon Press, Oxford, pp. 91–107.

Leakey, M.D., Harris, J.M. (Eds.), 1987. Laetoli: A Pliocene Site in Northern Tanzania. Oxford University Press, Oxford.

Leakey, M.D., Hay, R.L., 1979. Pliocene footprints in the Laetolil Beds at Laetoli, northern Tanzania. Nature 278, 317–323.

Leakcy, M.D., Hay, R.L., 1982. Les empreintes de pas fossils de Laetoli. Pour la Science 52, 28–37.

Leakey, M.D., Hay, R.L., Curtis, G.H., Drake, R.E., Jackes, M.K., White, T., 1976. Fossil hominids from the Laetolil Beds. Nature 262, 460–466.

Leakey, M.G., Harris, J.M., 2003. Lothagam: its significance and contributions. In: Leakey, M.G., Harris, J.M. (Eds.), Lothagam: The Dawn of Humanity in Eastern Africa. Columbia University Press, New York, pp. 625–660.

Leakey, M.G., Feibel, C.S., McDougall, I., Walker, A., 1995. New four-million-year-old hominid species from Kanapoi and Allia Bay, Kenya. Nature 376, 565–571.

Lewis, M.E., 1995. Plio-Pleistocene carnivoran guilds: implications for hominid paleoecology. Ph.D. Thesis, State University of New York, Stony Brook.

Maberly, C.T., 1950. The African bushpig – sagacious and intelligent. African Wildlife 4, 14–18.

Manega, P., 1993. Geochronology, geochemistry and isotopic study of the Plio-Pleistocene hominid sites and the Ngorongora volcanic highlands in northern Tanzania. Unpublished Ph.D. Dissertation, University of Colorado, Boulder.

Marean, C.W., 1989. Sabertooth cats and their relevance for early hominid diet and evolution. Journal of Human Evolution 18, 559–582.

McDougall, I., Feibel, C.S., 2003. Numerical age control for the Miocene-Pliocene succession at Lothagam, a hominoid-bearing sequence in the northern Kenya rift. In: Leakey, M.G., Harris, J.M. (Eds.), Lothagam: The Dawn of Humanity in Eastern Africa. Columbia University Press, New York, pp. 43–64.

McKee, J.K., 1999. The Autocatalytic Nature of Hominid Evolution in African Plio-Pleistocene Environments. In: Bromage, T.G., Schrenk, F. (Eds.), African Biogeography, Climate Change, and Human Evolution. Oxford University Press, Oxford, pp. 57–75.

Melton, D.A., 1976. The biology of aardvark (Tubulidentata Orycteropodidae). Mammal Review 6, 75–88.

Meylan, P.A., 1987. Fossil snakes from Laetoli. In: Leakey, M.D., Harris, J.M. (Eds.), Laetoli: A Pliocene Site in Northern Tanzania. Clarendon Press, Oxford, pp. 78–81.

Milstein, P. le S., 1971. The bushpig *Potamochoerus porcus* as a problem animal in South Africa. Report, Entomological Symposium, Pretoria. Mimeograph.

Mitchell, B.L., Uys, J.M.C., 1961. The problem of the lechwe (*Kobus leche*) on the Kafue Flats. Oryx 6, 171–183.

Mitchell, B.L., Shenton, J., Uys, J., 1965. Predation of large mammals in the Kafue National Park, Zambia. Zoologica Africana 1, 297–318.

Musiba, C., 1999. Laetoli Pliocene paleoecology: a reanalysis via morphological and behavioral approaches. Ph.D. Dissertation, University of Chicago, Chicago, Illinois.

Nakaya, H., 1994. Faunal change of late Miocene Africa and Eurasia: mammalian fauna from the Namurungule Formation, Samburu Hills, Northern Kenya. African Studies Monograph 20 (Suppl.), 1–112.

Ndessokia, P.N., 1990. The mammalian fauna and archaeology of the Ndolanya and Olpiro Beds, Laetoli, Tanzania. Ph.D. Dissertation, University of California, Berkeley.

Norton, G.W., Rhine, R.J., Wynn, G.W., Wynn, R.D., 1987. Baboon diet: a five-year study of stability and variability of the plant feeding and habitat of the yellow baboons (*Papio cynocephalus*) of Mikumi National Park, Tanzania. Folia Primatologica 48, 78–120.

Novellie, P.A., 1983. Feeding ecology of the kudu *Tragelaphus strepsiceros* (Pallas) in the Kruger National Park. D.Sc. Thesis, University of Pretoria, Pretoria.

Oates, J.F., 1994. The natural history of African Colobines. In: Davies, A.G., Oates, J.F. (Eds.), Colobine Monkeys: Their Ecology, Behaviour, and Evolution. Cambridge University Press, Cambridge, pp. 75–128.

Owen, R.E.A., 1970. Some observations on the sitatunga in Kenya. East African Wildlife Journal 8, 181–195.

Owen-Smith, N., 1973. The behavioural ecology of the white rhinoceros. Ph.D. Thesis, University of Wisconsin, Madison, Wisconsin.

Petter, G., 1987. Small carnivores (Viverridae, Mustelidae, Canidae) from Laetoli. In: Leakey, M.D., Harris, J.M. (Eds.), Laetoli: A Pliocene Site in Northern Tanzania. Clarendon Press, Oxford, pp. 194–234.

Pickering, R., 1964. Endulen. Quarter Degree Sheet 52. Geological Survey of Tanzania.

Pienaar, U. de V., 1969. Predatory-prey relationships amongst the larger mammals of the Kruger National Park. Koedoe 12, 108–187.

Player, I.C., Feely, J.M., 1960. A preliminary report on the square-lipped rhinoceros (*Ceratotherium simum simum*). Lammergeyer 1, 3–21.

Post, D.G., 1978. Feeding and ranging behaviour of the yellow baboon (*Papio cynocephalus*). Ph.D. Thesis, Yale University, New Haven, CT.

Rahm, U., 1966. Les mammifères de la forêt équatoriale de l'est du Congo. Musee Royal de l'Afrique Centrale Annales Serie 8, No. 149, 39–121.

Rasmussen, D.R., 1978. Environmental and behavioural correlates of changes in range use in a troop of yellow (*Papio cynocephalus*) and a troop of olive (*Papio anubis*) baboons. Ph.D. Thesis, University of California, Riverside.

Rautenbach, I.L., 1978a. A numerical re-appraisal of southern African biotic zones. Bulletin of the Carnegie Museum of Natural History 6, 175–187.

Rautenbach, I.L., 1978b. Ecological distribution of the mammals of the Transvaal (Vertebrata: Mammalia). Annals of the Transvaal Museum 31, 131–153.

Reed, K.E., 1996. The paleoecology of Makapansgat and other African Pliocene Hominid Localities. Ph.D. Dissertation, State University of New York, Stony Brook.

Reed, K.E., 1997. Early hominid evolution and ecological change through the African Plio-Pleistocene. Journal of Human Evolution 32, 289–322.

Reed, K.E., 1998. Using large mammal communities to examine ecological and taxonomic structure and predict vegetation in extant and extinct assemblages. Paleobiology 24, 384–408.

Robbins, L.M., 1987. Hominid footprints from Site G. In: Leakey, M.D., Harris, J.M. (Eds.), Laetoli: A Pliocene Site in Northern Tanzania. Clarendon Press, Oxford, pp. 497–502.

Sauer J.J., Skinner, J.D., Neitz, R., 1982. Seasonal utilisation of leaves by giraffes *Giraffa camelopardalis* and the relationship of the seasonal utilisation to the chemical composition of the leaves. South African Journal of Zoology 17, 210–219.

Schaller, G.B., 1968. Hunting behavior of the cheetah in the Serengeti National Park, Tanzania. East African Wildlife Journal 6, 95–100.

Sharman, M., 1981. Feeding, ranging and social organization of the guinea baboon. Ph.D. Thesis, Yale University, New Haven, CT.

Sheppe, W., Osborne, T., 1971. Patterns of use of a flood plains by Zambian mammals. Ecological Monographs 41, 179–205.

Shortridge, G.C., 1934. The Mammals of South West Africa, Volumes I and II. Heinemann, London.

Simpson, G.G., 1960. Notes on the measurement of faunal resemblance. American Journal of Science 258-A, 300–311.

Sinclair, A.R.E., 1977. The African Buffalo: A Study of Resource Limitation of Populations. University of Chicago Press, Chicago.

Skinner, J.D., Smithers, R.H., 1990. The Animals of the Southern African Subregion, 2 ed. University of Pretoria, Pretoria.

Skinner, J.D., Breytenbach, G.J., Maberly, C.T., 1976. Observations on the ecology and biology of the bushpig, *Potamochoerus porcus* Linn. in the northern Transvaal. South African Journal of Wildlife Research 6, 123–128.

Skinner, J.D., Davis, S., Ilani, G., 1980. Bone collecting by striped hyaenas, *Hyaena hyaena*, in Isreal. Palaentologica Africana 23, 99–104.

Smithers, R.H., 1971. The Mammals of Botswana. Trustees of the National Museum of Rhodesia, Salisbury.

Smithers, R.H., 1983. The Mammals of the Southern African Subregion. University of Pretoria, Pretoria.

Spencer, L.M., 1995. Antelopes and grasslands: reconstructing African hominid environments. Ph.D. Dissertation, State University of New York, Stony Brook.

Sponheimer, M., Reed, K.E., Lee-Thorp, J.A., 1999. Combining isotopic and ecomorphological data to refine bovid paleodietary reconstruction: a case study from the Makapansgat Limeworks hominin locality. Journal of Human Evolution 36, 705–718.

Sponheimer, M., Lee-Thorp, J.A., DeRuiter, D.J., Smith, J.M., van der Merwe, N.J., Reed, K., Grant, C.C., Ayliffe, L.K., Robinson, T.F., Heidelberger, C., Marcus, W., 2003. Diets of southern African Bovidae: stable isotope evidence. Journal of Mammalogy 84, 471–479.

Stewart, K.M., 2003. Fossil fish remains from Mio-Pliocene at Lothagam, Kenya. In: Leakey, M.G., Harris, J.M. (Eds.), Lothagam: The Dawn of Humanity in Eastern Africa. Columbia University Press, New York, pp. 75–111.

Stuart, C.T., 1977. The distribution, status, feeding and reproduction of carnivores of the Cape Province. Research Reports, Department of Nature and Environment, Conservation of Mammals 91–174.

Swynnerton, G.H., 1958. Fauna of the Serengeti National Park. Mammalia 22, 435–450.

Tinley, K.L., 1969. Dikdik, *Madoqua kirkii*, in South West Africa: notes on distribution, ecology and behaviour. Madoqua 1, 7–33.

Tuttle, R.H., 1985. Ape footprints and Laetoli impressions: a response to the SUNY claims. In: Tobias, P.V. (Ed.), Hominid Evolution: Past Present, and Future. Alan R. Liss, New York, pp. 129–133.

Tuttle, R.H., 1987. Kinesiological inferences and evolutionary implications from Laetoli bipedal trails G-1, G-2/3, and A. In: Leakey, M.D., Harris, J.M. (Eds.), Laetoli: A Pliocene Site in Northern Tanzania. Clarendon Press, Oxford, pp. 503–523.

Tuttle, R.H., 1990. The pitted pattern of Laetoli feet. Natural History 90, 60–65.

Tuttle, R.H., Webb, D., Tuttle, N., 1991. Laetoli footprint trails and the evolution of hominid bipedalism. In: Coppens, Y., Senut, B. (Eds.), Origine(s) de la bipédie chez les hominids. CNRS, Paris, pp. 203–218.

Tuttle, R.H., Webb, D., Tuttle, N., Baksh, M., 1992. Footprints and gaits of bipedal apes, bears, and barefoot peoples: perspectives on Pliocene tracks. In: Matano, S., Tuttle, R.H., Ishida, H., Goodman, M. (Eds.), Topics in Primatology, Volume 3: Evolution, Biology, Reproductive Endocrinology, and Viology. University of Tokyo Press, Tokyo, pp. 221–242.

Van Couvering, J.A.H., Van Couvering, J.A., 1976. Early Miocene mammal fossils from East Africa: aspects of geology, faunistics, and paleoecology. In: Isaac, G.L., McCown, E.R. (Eds.), Human Origins: Louis Leakey and the East African Evidence. Staples Press, Menlo Park, CA, pp. 155–207.

Van Valkenburgh, B., 1985. Locomotor diversity within past and present guilds of large predatory mammals. Paleobiology 11, 406–428.

Van Valkenburgh, B., 1988. Trophic diversity in past and present guilds of large predatory mammals. Paleobiology 14, 155–173.

Van Valkenburgh, B., 1990. Carnivore dental adaptations and diet: a study of trophic diversity within guilds. In: Gittleman, J.L. (Ed.), Carnivore Behavior, Ecology and Evolution. Cornell University Press, Ithaca, pp. 410–435.

Verdcourt, B., 1987. Mollusca from the Laetolil and Upper Ndolanya Beds. In: Leakey, M.D., Harris, J.M. (Eds.), Laetoli: A Pliocene Site in Northern Tanzania. Clarendon Press, Oxford, pp. 438–450.

Vesey-FitzGerald, D.F., 1964. Mammals of the Rukwa Valley. Tanganyika Notes and Records 62, 61–72.

Vincens, A., 1979. Paynologie, environnements actuels et plio-pléistocène à l'Est Turkana (Kenya). Ph.D. Dissertation, University d'Aix, Marseille.

Vrba, E.S., 1980. The significance of bovid remains as indicators of environment and prediction patterns. In: Behrensmeyer, A.K., Hill, A. (Eds.), Fossils in the Making: Vertebrate Taphonomy and Paleoecology. University of Chicago Press, Chicago, pp. 247–271.

Walker, A.C., 1987. Fossil Galaginae from Laetoli. In: Leakey, M.D., Harris, J.M. (Eds.), Laetoli: A Pliocene Site in Northern Tanzania. Clarendon Press, Oxford, pp. 88–91.

Watson, G.E., 1987. Pliocene bird fossils from Laetoli. In: Leakey, M.D., Harris, J.M. (Eds.), Laetoli: A Pliocene Site in Northern Tanzania. Clarendon Press, Oxford, pp. 82–83.

Wells, L.H., Cooke, H.S.B., 1956. Fossil bovidae from the Limeworks quarry, Makapansgat, Potgietersrust. Palaeontologica Africana 4, 1–55.

Werdelin, L., Lewis, M. E., 2001. A revision of the genus *Dinofelis* (Mammalia, Felidae). Zoological Journal Linnean Society 132, 147–258.

Wesselman, H.B., 1985. Fossil micromammals as indicators of climatic change about 2.4 Myr ago in the Omo valley, Ethiopia. South African Journal of Science 81, 260–261.

White, T.D., 1977. New fossil hominids from Laetoli, Tanzania. American Journal of Physical Anthropology 46, 197–230.

White, T.D., 1980. Additional fossil hominids from Laetoli, Tanzania: 1976–1979 specimens. American Journal of Physical Anthropology 53, 487–504.

White, T.D., 1981. Primitive hominid canine from Tanzania. Science 213, 348–349.

White, T.D., 1985. The hominids of Hadar and Laetoli: an element-by-element comparison of the dental samples. In: Delson, E. (Ed.), Ancestors: The Hard Evidence. Alan R. Liss, New York, pp. 138–152.

White, T.D., 1989. Evolutionary implications of Pliocene hominid footprints. Science 208, 175–176.

White, T.D., Suwa, G., 1987. Hominid footprints at Laetoli – facts and interpretations. American Journal of Physical Anthropology 72, 485–514.

Williamson, G.R., 1975. The condition and nutrition of elephant in Wankie National Park. Arnoldia (Rhodesia) 7, 1–20.

Wilson, V.J., 1966. Notes on the food and feeding habits of the common duiker, *Sylvicapra grimmia*, in eastern Zambia. Arnoldia (Rhodesia) 2, 1–19.

Wilson, V.J., 1969. The large mammals of the Rhodes Matopos National Park. Arnoldia (Rhodesia) 4, 1–32.

Zar, J.H., 1999. Biostatistical Analysis. Upper Saddle River. Prentice Hall, New Jersey.

# Appendix 1.  Summary of current interpretations of other fossil Plio-Pleistocene localities

### Nachukui Formation, Lothagam, Kenya (~5.0–<3.9 Ma)

There is a rich aquatic fauna, including crabs, fish, turtles, crocodiles, waterfowl, and hippopotamids (Leakey and Harris, 2003; Stewart, 2003). While the fish assemblage from the Apak Member (4.22–5 Ma) appears to be river adapted, the Kaiyumung Member (<3.9 Ma) has a predominantly lake fauna (Stewart, 2003; McDougall and Feibel, 2003). Evidence from oxygen isotope analyses of paleosols and mammalian tooth enamel indicate a mosaic of habitats (Cerling et al., 2003a). The Apak mammalian fauna points to a woodland habitat with abundant grassland nearby and the presence of a river, while the Kaiyumung assemblage suggests an open habitat with relative increase in grasslands and bushlands and the presence of a lake (Leakey and Harris, 2003).

### Lower Laetolil Beds, Laetoli, Tanzania (~4.6–3.8 Ma)

There has not been much attention given to the paleontology or paleoecology of the Lower Laetolil Beds. Harris (1987) described the presence of fish and crocodiles in the lower unit; however, there is only a single confirmed crocodile specimen in the Mary Leakey collections and no aquatic vertebrates have been recovered subsequently. Based on the similarity of the Lower Laetolil mammalian fauna to that of the Upper Laetolil Beds, Harris (1987) suggested that the two units had comparable environmental conditions, although the presence of aquatic vertebrates indicates that the lower unit had standing water.

### Kanapoi, Kenya (~4.2–3.9 Ma)

The mammalian fauna from Kanapoi indicates a dry woodland or bushland environment (Leakey et al., 1995). The primate fauna is dominated by Parapapio cf. ado, but it also includes colobines and Galago senegalensis. Bovids are dominated by Kobus and Aepyceros, species found near water and in edge habitats between grasslands and woodlands, respectively. The sediments were deposited by a large river, confirmed by the abundance of aquatic vertebrates (Leakey et al., 1995). The large river would have supported a wide gallery forest along the main river course (Leakey et al., 1995).

## Hadar, Ethiopia (~3.4–2.3 Ma)

There have been various alternative reconstructions of the paleoenvironment of Hadar ranging from an open grassland with humid conditions (Harris, 1991), a wooded to treeless savanna (Boaz, 1977), to an evergreen bushland with forest nearby (Bonnefille, 1983). Bonnefille et al. (1987, 2004) using palynological data, noted a change in habitat over time. During the Sidi Hakoma Member (3.4–3.22 Ma), there were elements of deciduous and evergreen forest or bushland, later replaced by a succession of montane forest and woodland. The habitat becomes more open with the Denen Dora Member (3.22–3.18 Ma), which is characterized by wet and dry grassland. At 2.9 Ma, evergreen bushland and montane forests reappeared, but conditions were not as humid as in the Sidi Hakoma Member. Reed (1997), using community structure and ecological diversity analyses, contends that there were no open, arid habitats before the Denen Dora Member. There is evidence of a lake with marshes in the early part of the Denen Dora Member that changed to floodplains and deltas later in the member (Aronson and Taieb, 1981). Combined with the faunal data, Reed (1997) concluded that during this period the environment was generally woodland, with forests around the margins of the lake, and edaphic grassland.

## Shungura Formation, Omo, Ethiopia (~3.5–1.3 Ma)

Fossil wood from the Shungura Formation shows that precipitation became more variable and lower in amount above Member C (2.95–2.6 Ma), causing more open and drought-resistant woodland–grassland communities to replace riverine forest communities (Eck and Jablonski, 1985). Pollen spectra also indicate that arboreal taxa were prevalent in Members B and C, but that they decrease after Member C (Bonnefille and DeChamps, 1983), although Bonnefille (1983) had noted the dominance of grassland during Member B. Micromammals in Members B and C indicate that there was a forest block with humid woodland–grassland, and some dry woodland–grassland (Wesselman, 1985). However, by Member F (2.35–2.33 Ma), the environment had shifted to a dry woodland–grassland and semi-arid steppe (Wesselman, 1985). The bovids indicate a change from a closed environment to one of a more open nature somewhere between Members B and G (Eck, 1976; Gentry, 1976). Reed (1997) ascertained that there was closed woodland with riverine forest and edaphic grasslands during Member B, but by Member C, and into Member F, habitats were dominated by bushland–woodland, even though riverine forest and edaphic grassland still existed. Recent study of the mammalian fauna in the Shungura Formation found that there was a steady decline of forest and closed woodland indicators after 3.2 Ma, while taxa indicating open woodland and grassland habitats increased moderately until after 2.5 Ma when they are more abundant that those associated with forests (Bobe et al., 2002). In his recent study of faunal change in the Shungura Formation, Alemseged (2003) suggests that while there is faunal composition change due to habitat change throughout the sequence, the most important faunal shift occurs during Member G at around 2.3 Ma when grasslands become an important part of the paleolandscape.

## Usno Formation, Omo, Ethiopia (3.36–3.0 Ma)

The paleoecology of the Usno Formation is less intensively studied than the Shungura Formation. Reed (1997) concluded that the environment was probably a closed habitat with bushland and thicket areas, as well as riverine forest and woodland.

## Koobi Fora Formation, Koobi Fora, Kenya (~4.0–1.3 Ma)

Evidence from stable isotopes (Cerling et al., 1977), palynology (Bonnefille, 1986a, b; Vincens, 1979), and faunal studies (Harris, 1983, 1987b) indicate that Koobi Fora was cooler and more humid during the Pliocene and early Pleistocene than at the present time, but it became progressively more arid throughout the sequence (Harris, 1983). Based on pollen and faunal data, the Tulu Bor Member (3.4–2.64 Ma) was probably a floodplain with gallery forest, while the Burgi Member (2.64–1.90 Ma) was closed woodland to the north becoming more open to the south (Harris, 1991). According to Reed (1997), the ecology during the Tulu Bor Member was scrub woodland on a riverine floodplain. The Burgi Member, in contrast, was open woodland with edaphic grassland and riparian woodland (Reed, 1997).

## Members 3, Limeworks Cave, Makapan Valley, South Africa (~3.2–2.7 Ma)

The habitat of Member 3 has been variously reconstructed as woodland (Vrba, 1980), forest (Cadman and Rayner, 1989), and open savanna with nearby bushland (Dart, 1952; Wells and Cooke, 1956). Reed (1997) has suggested that Member 3 was a mosaic habitat with riparian woodland, bushland, and edaphic grassland.

## Upper Ndolanya Beds, Laetoli, Tanzania (~2.5–2.7 Ma)

Analyses of the mammalian fauna from the Upper Ndolanya Beds, especially the equids and bovids, suggest an arid, grassland habitat. The equids are more hyposodont than those from the Laetolil Beds, and the bovid fauna is dominated by alcelaphines and antilopines (Harris, 1987). Recent analysis of the Upper Ndolanya large mammal fauna using ecological diversity analysis indicates that the paleohabitat was semi-arid bushland (Kovarovic et al., 2002).

# 12. Fauna, taphonomy, and ecology of the Plio-Pleistocene Chiwondo Beds, Northern Malawi

O. SANDROCK
*Department of Geology and Paleontology*
*Hessisches Landesmuseum Darmstadt, Friedensplatz 1, 64283 Darmstadt, Germany*
*sandrock@hlmd.de*


O. KULLMER
*Department of Paleoanthropology and Quaternary Paleontology*
*Forschungsinstitut Senckenberg, Senckenberganlage 25, 60325 Frankfurt am Main, Germany*
*ottmar.kullmer@senckenberg.de*


F. SCHRENK
*Department of Vertebrate Paleobiology*
*Johann Wolfgang Goethe-University, Siesmayerstrasse 70, 60054 Frankfurt am Main, Germany*
*schrenk@bio.uni-frankfurt.de*


Y.M. JUWAYEYI
*Department of Sociology and Anthropology*
*Long Island University, 1 University Plaza Brooklyn, New York, NY 11201-5372, USA*
*yusuf.juwayeyi@liu.edu*


T.G. BROMAGE
*Department of Biomaterials and Biomimetics*
*New York University College of Dentistry, 345 East 24th Street, New York, NY 10010, USA*
*tim.bromage@nyu.edu*

**Keywords:**   Human evolution, paleoecology, ecozones, Africa

## Abstract

The vertebrate fauna of the Chiwondo Beds in Northern Malawi is heavily biased towards the preservation of large terrestrial mammals, the majority being ungulates. The faunal diversity resembles an African short-grass plains assemblage. The taxonomic diversity is nevertheless low, emphasizing an incomplete fossil record. Based on modern bovid representation in African game parks, statistical tests show that the Chiwondo bovid assemblage consists of a mixture of species found in the Somali-Masai and the Zambezian ecozones. The composition of the terrestrial fauna is similar to Swartkrans 1 and the Upper Ndolanya Beds. The fossil assemblages can be assigned to three biostratigraphic time intervals that date from older than 4.0 Ma to less

315

*R. Bobe, Z. Alemseged, and A.K. Behrensmeyer (eds.) Hominin Environments in the East African*
*Pliocene: An Assessment of the Faunal Evidence, 315–332.*
© 2007 *Springer.*

than 1.5 Ma. The occurrence of *Paranthropus boisei* at a lake margin site in the Chiwondo Beds corresponds to robust australopithecine-bearing localities near Lake Turkana, Kenya. A case study showed that the investigated death assemblage on a delta plain in the Malema region was subject to heavy modification after deposition. This has affected the size distribution, the frequencies of skeletal elements, and thus the taxonomic composition. High-density skeletal elements such as molars and partial mandibles dominate the assemblage. The *Homo rudolfensis* locality at Uraha has a different faunal composition, the preservation in a paleosol points to a different taphonomic history and the Uraha area encompasses a longer time span.

## Introduction

Due to an interest in the paleobiogeography of southeastern Africa, the Hominid Corridor Research Project (HCRP) began its long-term study in the Malawi Rift in 1983. The research agenda is focused on the understanding of the ecological differences, migration events, and the origin and dispersion of Plio-Pleistocene faunas between eastern and southern Africa (Schrenk et al., 1993; Bromage et al., 1995; Kullmer et al., 1999).

Dixey (1927) first recognized the Chiwondo and Chitimwe Beds, which he attributed to the Pliocene and Pleistocene, respectively. J.D. Clark and colleagues' (Clark et al., 1966, 1970; Coryndon, 1966; Clark and Haynes, 1970; Mawby, 1970) explorations into the Chiwondo Beds of the northern Malawi Rift led to the first major contributions to the knowledge of the paleoecology and paleobiogeography of that area. Kaufulu et al. (1981) reinvestigated the deposits considering the high potential of the Chiwondo Beds for paleoanthropology.

## Geology

The sedimentology and geology was described by Ring and Betzler (1995) and Betzler and Ring (1995). Lake-beds and fluviatile deposits have been subdivided into five depositional units, which are bounded by unconformities (angular and erosional unconformities, paleosols) reflecting sedimentary breaks (Figure 1).

Unit 1 overlies the Mesozoic Dinosaur Beds with an angular unconformity. The sediments consist of reddish to grayish braided stream deposits. After a perennial lake was established around 4.5–4 Ma, the sedimentary system was lake dominated throughout Unit 2. A general flooding of the depositional area marks the lower limit of this unit.

The base of Unit 3 is marked by a change of the type in depositional system above an angular unconformity. Biostratigraphic data and facies repartition suggest that two subunits can be differentiated (Betzler and Ring, 1995).

A predominance of fluviatile processes was established in Unit 3A. Subunit 3A occurs throughout the Karonga-Chilumba area. Meandering rivers and minor lagoons developed in the proximal parts. Lake-ward, the sediment accumulated in the stream-mouth bars of deltas. Subunit 3B is restricted to the southern part of the Karonga-Chilumba area. It is a condensed section containing a series of calcimorphic paleosols. Pronounced lake regressions took place between 2.3–2 Ma and 1.6–1.5 Ma (= paleosol in Unit 3B), correlating with the Mbamba sequence of the offshore seismic record and the second magmatic pulse of the Rungwe volcanics, southern Tanzania (Ring and Betzler, 1995).

Deposition of Unit 4, which is bound to the southern part of the Karonga-Chilumba area, occurred following minor tectonic activity, allowing a transgression after 1.5 Ma. The lower part of Unit 4 consists of aeolian sands, while the upper part corresponds to a lake high stand, which is documented by open lake

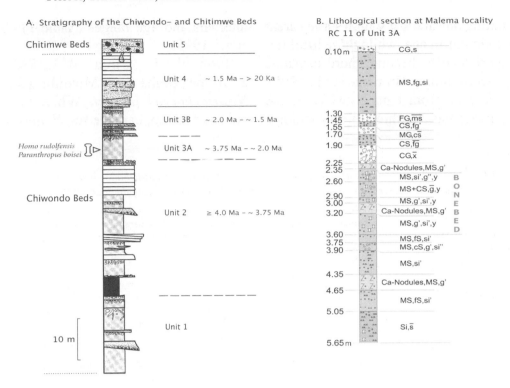

Figure 1. A. Generalized stratigraphic profile of Chiwondo and Chitimwe Beds Units 1–5 including their age ranges. Solid lines mark major unconformities; dashed lines mark minor unconformities. B. Lithological section of Malema hominin locality RC 11 of Unit 3A. The thickness refers to locality RC 11 only. The RC 11 bone bed extends from 2.35 to 3.6 m depth. Abbreviations: Si = silt, si = silty; fs = fine sand; ms = middle sand; cs = coarse sand; s = sandy; fg = fine gravel; mg = middle gravel; cg = coarse gravel; g = gravelly; x = stone; y = boulder; quantity of sediment fraction '– slight, " = very slight, — = strong.

limestones with abundant diatoms (Ring and Betzler, 1995). The alluvial fan deposits of the Chitimwe Beds, Unit 5, indicate a lake level lowering and are evidence of the changing sedimentary character.

## Biostratigraphy

The age of the Chiwondo Beds relies on faunal correlation with radiometrically dated biostratigraphic units in eastern Africa. The age range refers to radiometrically well-dated volcanic tuffs in the Koobi Fora and Shungura Formations as described by Brown et al. (1985), Brown and Feibel (1986), and Feibel et al. (1989).

Most of the Late Pliocene fossil localities are attributed to stratigraphic Unit 3A (e.g., Uraha, Mwimbi, Malema, and Mwenirondo localities), few are located in stratigraphic Unit 3B (Uraha), while some older Middle Pliocene localities occur in stratigraphic Unit 2 (Uraha and Mwimbi). The age of Unit 2 of the Chiwondo Beds ranges between >4 Ma and circa 3.75 Ma, the age of Unit 3A ranges between circa 3.75 and 2 Ma, the age of Unit 3B ranges between circa 2 and 1.5 Ma (Schrenk et al., 1993; Bromage et al., 1995). Stratigraphic Unit 3A is the longest time-aggregated deposit within the Chiwondo Beds. Short-term sedimentary unconformities exist, but none are significant enough to discriminate individual subunits. Nevertheless, localities approximating one another within survey areas tend to be similar in age, while those localities further apart tend to become more disparate in their biochronology (Bromage et al., 1995). Fossils recovered from Unit 2 and Units 3A and B correspond to three biochronologic

intervals and age ranges (Table 1) (Bromage et al., 1995).

Interval I. Circa 4–3.75 Ma (Unit 2 of the Uraha and Mwimbi areas): e.g., *Nyanzachoerus jaegeri*. White (1995) provides a first appearance for *N. jaegeri* (FAD)

Table 1. *List of identified mammal taxa at fossil areas of the Chiwondo Beds (modified after Bromage et al., 1995). "X" marks the presence of a taxon, the ones in parentheses refer to Kaufulu et al. (1981)*

| Stratigraphic unit 2 | Uraha | Mwimbi | Malema | Mwenirondo | Mwamberu | Sadala |
|---|---|---|---|---|---|---|
| *Nyanzachoerus jaegeri* | X | X | | | | |
| *Anancus* aff. *kenyensis* | X | | | | | |
| early *Loxodonta* sp. | | X | | | | |
| *Mammuthus subplanifrons* | X | | | | | |
| Stratigraphic Unit 3A | | | | | | |
| *Damaliscus* sp. | X | X | X | X | | |
| Alcelaphini, medium-sized | X | | X | | | |
| *Connochaetes* sp. | X | X | | | | |
| *Megalotragus* sp. | X | X | X | X | | |
| *Megalotragus kattwinkeli* | X | (X) | | (X) | | |
| *Gazella* sp. | X | X | X | | | |
| *Gazella* sp. aff. *vanhoepeni* | | | | | | (X) |
| Antilopini gen. indet. | X | | | X | | |
| *Tragelaphus* sp. | X | | X | | | |
| *Tragelaphus* cf. *angasi* | X | | | (X) | | |
| *Tragelaphus* cf. *strepsiceros* | X | X | | | | |
| Tragelaphini gen. indet. | | X | | | | |
| *Kobus* sp. | X | X | X | X | | |
| *Kobus* aff. *patulicornis* | | (X) | | | (X) | |
| Reduncini sp. | | | X | | | |
| *Oryx* aff. *gazella* | X | | | | | |
| *Hippotragus* sp. | X | | X | X | | |
| *Hippotragus* aff. *gigas* | | | X | | | |
| *Syncerus* sp. | X | X | X | X | | |
| *Ugandax.* sp. | | | X | X (X) | | |
| Bovini gen. indet. | | X | | | | |
| *Aepyceros* sp. | X (X) | X | X | X (X) | (X) | |
| *Madoqua* sp. | X | | | | | |
| Neotragini gen. indet. | | | X | | | |
| *Giraffa* aff. *pygmaea* | X | | X | | | |
| *Giraffa* aff. *stillei* | X | X | X | X | | |
| *Giraffa* aff. *jumae* | | | X | | | |
| *Sivatherium* sp. | | | X | | | |
| *Hippopotamus* sp. | X | X | X | X | | |
| *Metridiochoerus andrewsi* | X | X (X) | | | | |
| *Notochoerus euilus* | X | X | | X | | |
| *Notochoerus scotti* | X | X | X | X | | |
| early *Notochoerus scotti* | | | X | | | |
| *Notochoerus capensis* | | X (X) | | | (X) | |
| *Diceros bicornis* | X | | | | | |
| *Ceratotherium simum* | X | | X | X | | |
| Rhinoceratidae gen. indet. | | X | | | | |

(Continued)

*Table 1. List of identified mammal taxa at fossil areas of the Chiwondo Beds (modified after Bromage et al., 1995). "X" marks the presence of a taxon, the ones in parentheses refer to Kaufulu et al. (1981)—cont'd*

| Stratigraphic unit 2 | Uraha | Mwimbi | Malema | Mwenirondo | Mwamberu | Sadala |
|---|---|---|---|---|---|---|
| *Hipparion* sp. | X | X | X | X | | |
| *Elephas* sp. | X | X | | | | |
| *Elephas recki* | | | X | | | |
| *Elephas recki atavus* | X | | | | | |
| *Elephas recki shungurensis* | | | X | | | |
| *Deinotherium* sp. | X | | X | | | |
| *Parapapio* sp. | X | | X | X | | |
| *Theropithecus* sp. | | | X | | | |
| Papionini gen. indet. | | | | X | | |
| *Homo rudolfensis* | X | | | | | |
| *Paranthropus boisei* | | | X | | | |
| Stratigraphic Unit 3B | | | | | | |
| *Metridiochoerus compactus* | X | | | | | |

at 5.0–6.0 Ma at Lothagam 1C, with a quality score 2 (= date possibly actual FAD) and a last appearance (LAD) at 3.75 Ma in the eastern Central Awash Complex (with quality score 0 = arbitrary split of lineage). The VT-3 Tuff with an age of 3.75 Ma (White et al., 1993) occurs in the Aramis subunit of Kalb et al. (1982), where *N. jaegeri* was discovered (Kalb et al., 1982).

An age of older than 4 Ma for parts of Unit 2 is further indicated by the occurrence of *Anancus* aff. *kenyensis*. This taxon occurs in the Adu-Asa Formation (WoldeGabriel et al., 2001; Haile-Selassie et al., 2004), the Kuseralee Member of the lower Sagantole Formation (Kalb and Mebrate, 1993; Kalb, et al. 1995; Renne et al., 1999), and in the Mursi Formation (Beden, 1976) at an age of >4 to <4.15 Ma (Brown and Nash, 1976; Brown et al., 1985; Feibel et al., 1989).

An early *Loxodonta* sp. occurs at Mwimbi. *L. adaurora* is known from the Mursi Formation (Beden, 1987) and Hadar Formation (Kalb et al., 1982; Kalb, 1995). *Mammuthus subplanifrons* is identified at Uraha. Maglio (1973) also reported its occurrence in the Chiwondo Beds. *M. subplanifrons* from Kalb et al. (1982) Aramis subunit occurs between the VT-1 and CT tuffs and is dated to <4.1–3.75 Ma respectively (White et al., 1993).

Interval II. Between 3.75 and 1.8 Ma and younger (Unit 3A of Chiwondo Beds): e.g., *Notochoerus euilus*. The FAD of the taxon derives anagenetically and falls coincident with the LAD of *Nyanzachoerus jaegeri* at 3.75 Ma (with quality score 0 = arbitrary split of lineage) (White, 1995). A reliable LAD for *Notochoerus euilus* is 2.0 Ma (Shungura upper Member G) (Harris and White, 1979; Feibel et al., 1989) and for *Notochoerus scotti* an age of 1.8 Ma (Koobi Fora, just above the KBS tuff) (with quality score 3 = date probably actual LAD) (White, 1995).

The third molars from Interval II of the early *Notochoerus scotti* types are slightly more advanced than the early specimens from Shungura C, but less progressive than specimens from Shungura G, the correlation therefore suggests Shungura D–F age of 2.52–2.33 Ma (Feibel et al., 1989). Their occlusal length is shorter than advanced *N. scotti* specimens recognized from Shungura G. The number of lateral pillar pairs is relatively low (six pairs) indicating an early stage of *N. scotti*, since younger specimens tend to increase their occlusal length by adding lateral pillar pairs to the distal end of the tooth crown (Kullmer, 1999).

*Elephas recki shungurensis* specimens are identified in Chiwondo Beds Unit 3A. The taxon occurs in the upper Hadar and Matabaietu Formations between ~3.0 and ~2.0 Ma (Kalb, 1995) and from Shungura Member C to lower Member F (Beden, 1987), indicating an age of 2.85–~2.35 Ma.

Interval III. 1.6 Ma and younger (Unit 3B of the Uraha area): e.g., *Metridiochoerus compactus* is encountered from 1.6 Ma (correlation with Koobi Fora, just below the Okote Tuff) to 0.78 Ma (correlation with Olduvai Bed IV). So, Unit 3B has a correlation of 1.6–0.8 Ma.

Other taxa fall in age ranges I–III[(+)], II–III[(o)], or II[(*)] (the FADs and LADs refer to Behrensmeyer et al., 1997): *Giraffa pygmaea*[(+)] (FAD 4.35–LAD 1.39 Ma; E. Turkana), *Sivatherium marusium* [(+)] (FAD 4.10–1.33 Ma; W. and E. Turkana), *Giraffa stillei*[(o)] (FAD 3.50–LAD 1.39 Ma; E. Turkana), *Megalotragus isaaci*[(o)] (FAD 2.52–LAD 1.39 Ma; E. Turkana), *Ceratotherium simum*[(o)] (FAD 3.40–LAD 0.70 Ma; W. and E. Turkana, Omo), *Diceros bicornis*[(o)] (FAD 3.40–LAD 0.70 Ma; W. and E. Turkana, Omo), *Ugandax* sp. nov. WT[(*)] (FAD 3.36–LAD 1.88 Ma; W. Turkana), *Hippotragus gigas*[(*)] (FAD 3.36–LAD 1.6 Ma; W. and E. Turkana).

Most of the equids can be referred to species of *Hipparion* sp. One isolated molar of *Equus* sp. from northern localities derives from a layer of upper Unit 3A just below the contact of the Chiwondo and Chitimwe Beds (Bromage et al., 1995).

The *Homo rudolfensis* locality at Uraha is situated in Unit 3A within a ferruginous, calcimorphic paleosol, 5.5 m above an oncoid layer of Unit 2, and circa 6 m below the boundary of Unit 3 to 4 (Figure 7; Betzler and Ring, 1995). Equids that derive from the same lithologic unit as *H. rudolfensis* are referred to *Hipparion* sp. (Bromage et al., 1995). One equid molar fragment from Uraha may tentatively be assigned to *Equus* sp. (Ray Bernor, pers. comm.). It derives from the upper part of Unit 3A above the ferruginous paleosol-yielding *Homo rudolfensis*.

According to Table 1 and information given above, Uraha, Malema, Mwimbi, and Mwenirondo share certain taxa, which co-occur in biostratigraphic Unit 3A.

## Fauna

The sample of vertebrates recovered from Units 2 and 3 of the Chiwondo Beds consists of about thousand identifiable vertebrate specimens from areas between the towns of Karonga in the north and Chilumba in the south, a distance of over 70 km (Figure 2). Three major fossiliferous areas can be distinguished within the Chiwondo Beds: Uraha, Mwimbi, and Malema.

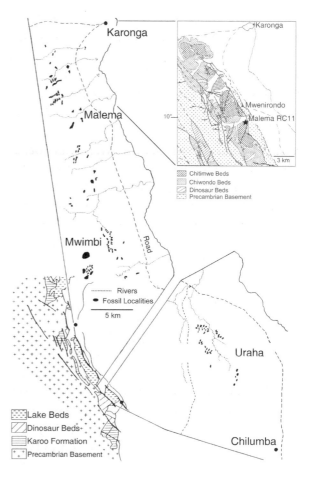

Figure 2. Topographical and geological setting of the Chiwondo Beds, Northern Malawi.

Most of the fossil specimens have been collected from the erosional surface by systematic surveys. Bovids, followed by equids, suids, hippopotamids, and giraffids, dominate the fauna of the Chiwondo Beds (Figure 3). The recovered skeletal elements are mostly isolated molars, mandible fragments, or high-density limb bones. Because of the very low fossil density in the Chiwondo Beds, only a few systematic excavations were conducted during years of fieldwork, until a fossil-rich horizon containing also *Paranthropus boisei* was discovered at Malema locality RC 11, which considerably enlarged the collection (Bromage and Schrenk, 1986; Sandrock et al., 1999).

Taking a closer look at the fauna of the Uraha, Malema, and Mwimbi regions, it is evident that bovids dominate the faunal assemblages (Figure 4a–c). Equids have about the same relative abundance at Uraha and Malema. However, there are some differences between the regions in their suid proportions: at Malema only early *Notochoerus scotti* specimens are found, whereas at Uraha remains of *Nyanzachoerus jaegeri*, *Notochoerus euilus*, *Notochoerus scotti*, *Metridiochoerus andrewsi*, *Metridiochoerus compactus* were discovered. The latter species occurs only at Uraha. Mwimbi is the only area where specimens attributable to *Notochoerus capensis* are found. Giraffes are more common in the

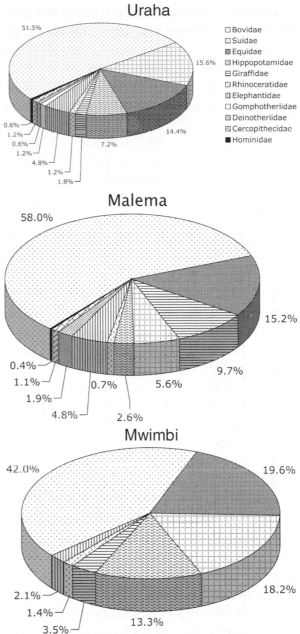

Figure 4a–c. The mammal fauna at Uraha (a), Malema (b), and Mwimbi (c) localities.

Malema region. Three size groups (i.e., large, similar to the extant species, medium size species, and a smaller one) can be distinguished at Malema and are tentatively attributed to *Giraffa* aff. *jumaea*, *Giraffa* aff. *stillei*, *Giraffa* aff. *pygmaea* respectively. *Sivatherium* sp. remains are represented only at Malema.

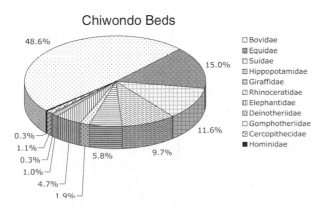

Figure 3. The mammal fauna of the Chiwondo Beds.

Bovid specimens are identified as *Syncerus* sp. or *Ugandax* sp. The *Ugandax* sp. specimens from Malema share morphological affinities with specimens of the Mwenirondo region that are curated at the University of California Museum of Paleontology (UCMP) (Kaufulu et al., 1981). Gentry and Gentry (1978) mentioned a right mandible with P$_4$-M$_3$ from Mwenirondo and identified it as *Syncerus* sp. Hippotragins are common at Malema; they may be referred to *Hippotragus* aff. *H. gigas*. Coryndon (1966) assigned two mandibles from Mwenirondo to *Oryx*, but Gentry and Gentry (1978) thought these specimens likely belonged to *Syncerus*. Gentry and Gentry (1978) attributed a broken right mandible with P$_4$-M$_3$ to *Hippotragus gigas*. Kaufulu et al. (1981) identified two size groups of hippotragins without further distinguishing them. Alcelaphins make up the majority of the bovid fauna. They comprise large-, medium- and small-sized specimens. The largest is *Megalotragus* sp., and the smallest *Damaliscus* sp. *Megalotragus* is recognized by its huge well-rounded lobes and its very simple central cavity. Especially the upper molars of *Megalotragus* sp. at Malema are extremely hypsodont.

As very rare elements, specimens of the cercopithecid *Parapapio* occur at northern and southern localities, while *Theropithecus* sp. is only found in the Malema region.

The Malawi Rift faunas include representatives from three geographically based assemblages: some species represent eastern African endemics, a smaller group consists of southern African endemics, and many species are shared between eastern and southern Africa. Interesting is the substantial occurrence of *Aepyceros* sp. at Uraha, which marks the southernmost distribution of this genus in the Late Pliocene.

It is striking that the abundance of mammalian families within the Chiwondo Beds differs significantly from that of East and South Africa fossil sites. This is especially highlighted by the apparent absence of micromammals and carnivores (Schrenk et al., 1995).

It is important to note, that the fossil sample of the main fossiliferous Unit 3A is likely spanning approximately 1.7 Ma, but this cannot be demonstrated based on the current taxonomic content. Only species like *E. recki shungurensis* and early *Notochoerus scotti* give some localities a more precise age range.

More fine-scaled stratigraphic subunits within Units 2 and 3A cannot be defined. Consequently, the fossil occurrences within these units might represent a longer time span, but adjacent localities within laterally extensive areas agree in their biochronologic range estimation (Bromage et al., 1995).

The faunal data from the excavation site of the Malema RC 11 bone bed diminishes large-scaled time averaging. The early *Notochoerus scotti* specimens at this locality give a more precise chronological control (Kullmer, 1999), making these deposits readily correlated with other East African strata between 2.6 and 2.4 Ma.

**Taphonomy**

Malema site RC 11 excavations were undertaken over several field seasons. The sedimentary sequences in Unit 3A at Malema can be referred to fine- to medium-grained fluviatile sediments of a delta plain. Due to a slow lake level lowering and increased sediment input by rivers, the fluviatile deposits became dominant after 3.7- to 2.0-Ma ago in the Malema sequence (Betzler and Ring, 1995). The emerging picture is a stable north–south directed land corridor bordered by the rift shoulder to the west and the lake to the east. Meandering rivers with minor lagoons developed proximally, lake-ward prograding river deltas with stream mouth bars developed (Betzler and Ring, 1995). The sedimentological and taphonomical study indicates that the Malema bone assemblage was deposited at the lake margin, probably on a delta plain (Sandrock, 1999) (see Figure 1).

Lacustrine transgressions were minor occurrences during this time and only inter-fingered with the delta front. Small-scaled lake bottom layers with gastropods, which can be found in almost the entire Malema valley, derive from minor transgressive phases of paleolake Malawi, while the overlying sand units represent fluviatile sequences. Geological sections indicate that this system consisted of quite shallow distributaries. These channels eroded into fine to middle sand units of previously formed overbank deposits (Sandrock, 1999).

Sudden and gradual abandonment of channels are evident by fine-grained channel fills and mixed-grained fills fining upward. Coarse bed-load sediments are entirely missing within the sequences. The angular to subangular rounding of the mineral components suggests a proximate source area for the sediment (Sandrock, 1999).

The features of the RC 11 case study may serve as a good example for the Chiwondo Beds assemblages as a whole. The observed taphonomic variables indicate a mixture of an attritional and hydraulically winnowed assemblage (Figure 5; after Behrensmeyer, 1991). The representation of the skeletal parts is strongly linked to the sedimentary environment. The most obvious feature is the dominance of high-density elements, especially teeth and half mandible fragments that fall into the Voorhies Group II–III range. The removal of the lighter skeletal elements of Voorhies Group I and the much greater amount of dental material than postcranial elements is evidence for hydraulic transport (Behrensmeyer, 1975, 1991). It is reasonable to conclude that more postcranial material existed at RC 11, which was subsequently destroyed and is now a part of the large fraction of indeterminate fragments. A hydraulic sorting is overprinted by the destruction of many skeletal elements, apparently biasing the assemblage to those that are dense and towards a classic channel lag deposit.

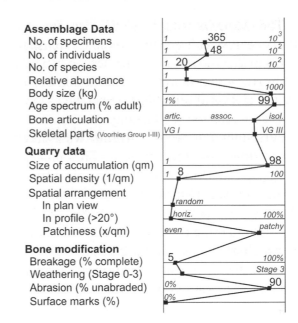

Figure 5. Taphonomic variables at Malema locality RC 11. The excavation area has a size of 22m × 10m; the fossiliferous layer has a thickness of 1.3 m. Regular numbers refer to the locality, italicized numbers are underlining the graphic.

The relative abundance was measured by using the MNI. Using the NISP would have overestimated the number of individuals due to unrecognized matches among the large amount of broken specimens and unrecognized association among specimens (Badgley, 1986).

The mixed features of an autochthonous-attritional assemblage and an allochthonous-sorted assemblage are the best analogues to describe site RC 11. Sorting, the dominance of Voorhies Groups II–III elements, the lack of association, lower stages of weathering, and overrepresentation of adults indicate allochthony. Autochthony is indicated by at least a few Voorhies Group I elements and a random orientation of the skeletal elements. Abrasion cannot really be inferred from the RC 11 assemblage, since it is heavily overprinted, if this was minimal it would also suggest autochthony. The same is true for associated skeletal elements that are obviously lacking at Malema.

The Malema site is similar to the late Miocene Tonopah quarry, Nevada, (Henshaw, 1942) in that the fossils are preserved in a relatively fine-grained matrix. The fossils consist of only teeth and limb bones. Skeletal associations and abrasion are lacking.

Persistent but weak currents were likely responsible for the deposit. The currents likely operated over a considerable time to create this lag deposit. In contrast, the early Pliocene Verdigre quarry of Voorhies (1969) was characterized by strong currents acting over a shorter time resulting in the fossil elements showing abrasion and no associations.

The Malema bone assemblage shows that *Hipparion* sp. and alcelaphins like *Megalotragus* sp. and *Damaliscus* sp. are the most common fossils from the lake margin area in Unit 3A. This leads to two potential scenarios: one scenario suggests that fluvial processes brought in all mammals representative of open habitats, being allochthonous in this sense. But this does not yet explain the scarcity of water-dependent reduncins in a deltaic deposit. The other scenario is that the Malema bone bed samples only the fauna of a microhabitat on the flat delta plains of the drier lake margin areas where rivers are an only minor constituent and water dependent taxa are few or absent.

The majority of the animals in the assemblage are adult animals, just 1% being juvenile. Due to their earlier destruction, the juvenile individuals of RC 11 probably share the same destiny as the micromammals. A bias against their survival potential could possibly be reworking in channels or leaching. Despite special sieving efforts, no micromammals were discovered. The lack of micromammals is striking since even in case of a "category 5 predator" (Dauphin et al., 1997) with felids or canids and the most destructive effects commonly observed to operate, at least a few bones should be preserved.

## Paleoecology

Terrestrial species range between 22% and 57% in 15 modern environments (Kovarovic et al., 2002). In contrast, the Chiwondo Beds sample contains nearly 90% terrestrial species – fossorial, scansorial, and (semi-) arboreal animals are absent. This places them towards modern tropical grassland or semi-arid bushland and to Swartkrans 1 and the Upper Ndolanya Beds, which also have high percentages of terrestrial taxa. In the Chiwondo Beds small animals are underrepresented or absent altogether. Again, in modern ecosystems, species of 1 g to 10 kg range between 53% and 82%: the African tropical grassland and bushland faunas exhibit the smallest percentage (Kovarovic et al., 2002), but the Malawi fauna corresponds neither to the observed weight pattern of Swartkrans 1 nor to the East African localities of Kanapoi, Aramis, Laetolil Beds, or the Ndolanya Beds.

The fauna is dominated by medium- to large-sized species, the latter being large bovids, proboscideans, and hippopotamuses. Consequently an overrepresentation of herbivorous species exists. Grazers and browsers make up the majority, but especially alcelaphin grazers dominate like in Swartkrans 1 and the Ndolanya Beds.

A correspondence analysis was applied to a contingency-table data set of modern bovid genera from 29 African game parks using data derived from Shipman and Harris (1988). Variable 1 consists of the game parks, and variable 2 of the bovid genera. In addition, the bovid genera of Malema and Uraha have been included (Figure 6, Table 2). The resulting cluster of points on two axes allow the interpretation of the relations between the variables. Distances between points within a variable do have a meaning, but proximity between points only *can* indicate a consistent association.

The results share strong resemblance to the existing phytochorions in Africa today

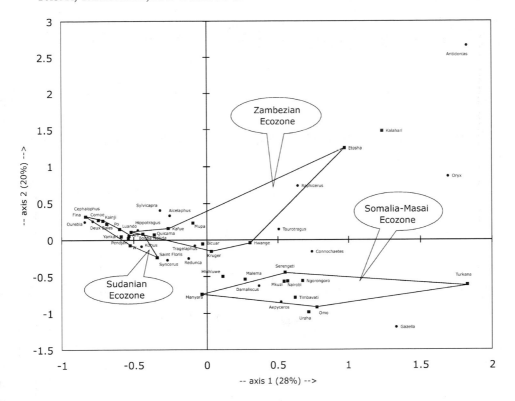

Figure 6. Correspondence analysis of game parks and bovid genera on axes 1 and 2. Chi-square distance on data = 876.6201 ~ degrees of freedom = 480, corresponding probability: 0.0000, limit Chi square for the chosen confidence range = 532.0753. Using this test one should reject the hypothesis of independence. Variable 1 (rows) consists of 29 African parks and Pliocene sites Uraha and Malema. Variable 2 (columns) consists of modern bovid genera living in the parks and bovid genera discovered at Uraha and Malema (variables 1 and 2 from Table 2).

(Sandrock, 1999). Axis 1 and 2 separate the Zambezian and Sudanian ecozones from the Somalia-Masai ecozone. Axis 1 shows increasing aridity to the right side of the plot towards the habitats of the Somalia-Masai ecoregion, which consists mainly of dry woodlands and scrub, with a gradation to grasslands (White, 1983). Turkana, Serengeti, Nairobi, Ngorongoro, Omo, and Manyara parks belong to the Somalia-Masai (SM) domain and plot separately from the Zambezian (Z) phytochorion including the parks Hwange, Kruger, Kafue, Quiçama, Luando, Etosha, Bicuar, and Mupa. All other parks but Hluhluwe, Mkuzi (both referring to the Tonga-Pondoland eco-

zone), Timbavati, and Kalahari belong to the Sudanian (S) phytochorion.

The Zambezian ecozone as a whole is close to the center of gravity of the multivariate cloud, which reflects that most bovids live in this ecozone in different proportions, while in the Somalia-Masai ecozone, tribes like Antilopini and Alcelaphini dominate the assemblages. The antilopin *Antidorcas* only occurs at Etosha and Kalahari game parks; today it is a southern African endemic. The antilopin *Gazella*, on the other hand, only occurs in the Somalia-Masai domain. Members of the cephalophins, neotragins, reduncins, and *Hippotragus* on the left side of the plot are clearly separated from

Table 2. Logarithmic abundances of bovid genera in 29 African parks plus Pliocene localities Malema and Uraha

| | Alcela. | Conno. | Dama. | Anti. | Gaz. | Aepy. | Tauro. | Tragel. | Kobus | Redun. | Syn. | Hippotr. | Oryx | Oureb. | Raphi. | Cephalo. | Sylvi. |
|---|---|---|---|---|---|---|---|---|---|---|---|---|---|---|---|---|---|
| Malema | 0 | 1.47 | 1.44 | 0 | 0.47 | 0.6 | 0 | 0.3 | 0.9 | 0 | 1.25 | 1.39 | 0 | 0 | 0.3 | 0 | 0 |
| Uraha | 0 | 1 | 1.11 | 0 | 1.14 | 1.07 | 0 | 0.84 | 0.3 | 0 | 0 | 0.69 | 0 | 0 | 0 | 0 | 0 |
| Kruger | 0 | 3.97 | 2.8 | 0 | 0 | 5.18 | 2.62 | 4.02 | 3.5 | 3.17 | 4.38 | 3.14 | 0 | 2 | 3.69 | 2.47 | 3 |
| Luando | 0 | 0 | 0 | 0 | 0 | 0 | 2.3 | 2.84 | 3.24 | 2.69 | 2.17 | 3.47 | 0 | 2.39 | 0 | 2 | 3 |
| Hwange | 0 | 3.39 | 2.11 | 0 | 0 | 3.9 | 3.25 | 3.56 | 3 | 2.39 | 4.11 | 3.39 | 2.07 | 0 | 3.47 | 0 | 3.3 |
| Nairobi | 3.03 | 2.4 | 0 | 0 | 2.92 | 2.8 | 1.76 | 1.3 | 1.95 | 1.17 | 0 | 0 | 0 | 0 | 0.3 | 0 | 0.3 |
| Turkana | 0 | 0 | 3.3 | 0 | 3.02 | 0 | 0 | 0 | 0 | 0 | 0 | 0 | 3.12 | 0 | 0 | 0 | 0 |
| Serengeti | 4.25 | 5.61 | 4.43 | 0 | 5.27 | 4.81 | 3.84 | 3.39 | 3.47 | 3.39 | 4.69 | 3.39 | 3.39 | 0 | 0 | 0 | 0 |
| Ngorongoro | 2 | 4.13 | 0 | 0 | 3.69 | 0 | 2.6 | 0 | 1.77 | 1.77 | 1.77 | 0 | 0 | 0 | 0 | 0 | 0 |
| Etosha | 2.77 | 3.6 | 0 | 4.07 | 0 | 0 | 2.69 | 3.3 | 0 | 0 | 0 | 2.47 | 3.65 | 0 | 2.69 | 0 | 2.39 |
| Bouba Ndjida | 3.84 | 0 | 2.15 | 0 | 0 | 0 | 2.97 | 2.92 | 3.14 | 3.75 | 3.24 | 3.63 | 0 | 4.06 | 0 | 1.7 | 3.73 |
| Quicama | 0 | 0 | 0 | 0 | 0 | 0 | 3.39 | 3.47 | 0 | 3.17 | 3.9 | 3.17 | 0 | 0 | 0 | 3.39 | 3 |
| Hluhluwe | 0 | 3.54 | 0 | 0 | 0 | 3.68 | 0 | 3.5 | 2.91 | 2.27 | 3.34 | 0 | 0 | 0 | 0 | 0 | 1.17 |
| Kafue | 1.95 | 2 | 0 | 0 | 0 | 0.3 | 0.3 | 1.6 | 1.89 | 2.07 | 1.92 | 1.49 | 0 | 1.69 | 1.6 | 1.47 | 1.6 |
| Pendjari | 3.94 | 0 | 2.77 | 0 | 0 | 0 | 0 | 0 | 4.27 | 2.91 | 3.54 | 3.7 | 0 | 3.89 | 0 | 0 | 3.39 |
| Comoe | 3.9 | 0 | 0 | 0 | 0 | 0 | 0 | 3 | 3.87 | 0 | 2.69 | 2 | 0 | 4.23 | 0 | 3.81 | 3.81 |
| Omo | 0 | 0 | 3.32 | 0 | 2.81 | 0 | 2.97 | 0.47 | 0 | 0 | 2.54 | 0 | 0 | 0 | 0 | 0 | 0 |
| Kalahari | 4.15 | 3.8 | 0 | 4.38 | 0 | 0 | 3.81 | 0 | 0 | 0 | 0 | 0 | 4.2 | 0 | 3.21 | 0 | 2.85 |
| Manyara | 0 | 0 | 0 | 0 | 0 | 2.84 | 0 | 1.39 | 1.17 | 1.6 | 3.17 | 0 | 0 | 0 | 0 | 0 | 0 |
| Bicuar | 0 | 2.69 | 0 | 0 | 0 | 2.17 | 2.39 | 2.3 | 2 | 1.69 | 2 | 2.69 | 0 | 1.69 | 1 | 0 | 2.69 |
| Mupa | 0 | 2.39 | 0 | 0 | 0 | 0 | 2.3 | 2.74 | 2.77 | 3 | 0 | 2.3 | 0 | 2.69 | 2.69 | 0 | 2.69 |
| Mkuzi | 0 | 3.14 | 0 | 0 | 0 | 3.97 | 0 | 2.74 | 0 | 1.83 | 0 | 0 | 0 | 0 | 0.77 | 0 | 0.47 |
| Arli | 2 | 0 | 1.95 | 0 | 0 | 0 | 0 | 1.9 | 3.07 | 1.74 | 1.81 | 3.27 | 0 | 3 | 0 | 0 | 3.33 |
| Deux Bales | 2.65 | 0 | 0 | 0 | 0 | 0 | 0 | 2.29 | 2.24 | 1.7 | 1.6 | 3.07 | 0 | 2.81 | 0 | 2.18 | 2.69 |
| Po | 2.73 | 0 | 0 | 0 | 0 | 0 | 0 | 2.03 | 1.99 | 2.27 | 2.39 | 2.89 | 0 | 2.7 | 0 | 0 | 2.68 |
| Saint Floris | 2.95 | 0 | 3.32 | 0 | 0 | 0 | 0 | 0 | 3.5 | 0 | 3.25 | 2.7 | 0 | 0 | 0 | 0 | 0 |
| Yankari | 2.82 | 0 | 0 | 0 | 0 | 0 | 0 | 0 | 2.81 | 0 | 2.82 | 2.55 | 0 | 0 | 0 | 0 | 0 |
| Fina | 3.47 | 0 | 0 | 0 | 0 | 0 | 0 | 0 | 3.1 | 0 | 0 | 3.59 | 0 | 3.66 | 0 | 0 | 0 |
| Kainji | 3.8 | 0 | 0 | 0 | 0 | 0 | 0 | 2.74 | 3.68 | 0 | 2.07 | 3.76 | 0 | 3.44 | 0 | 2.19 | 3.09 |
| W | 3 | 0 | 2.62 | 0 | 0 | 0 | 0 | 2.38 | 3.78 | 0 | 3.61 | 3.45 | 0 | 3.32 | 0 | 0 | 0 |
| Timbavati | 0 | 3.09 | 0 | 0 | 0 | 3.66 | 0 | 1.89 | 0 | 1.51 | 0 | 0 | 0 | 0 | 0 | 0 | 0 |

Alcela. = *Alcelaphus*, Conno. = *Connochaetes*, Dama. = *Damaliscus*, Anti. = *Antidorcas*, Gaz. = *Gazella*, Aepy. = *Aepyceros*, Tauro. = *Taurotragus*, Tragel. = *Tragelaphus*, Redun. = *Redunca*, Syn. = *Syncerus*, Hippotr. = *Hippotragus*, Oureb. = *Ourebia*, Raphi. = *Raphicerus*, Cephalo. = *Cephalophus*, Sylvi. = *Sylvicapra*.

antilopins and alcelaphins to the right side. The analysis clearly shows the association of open arid-adapted Alcelaphini–Antilopini and *Oryx* to the parks of the Somalia-Masai ecozone. These results coincide with previous analyses (Greenacre and Vrba, 1984; Shipman and Harris, 1988; Alemseged, 2003).

Due to the difficulties to identify all Chiwondo bovids to species level, the correspondence analysis was applied to generic level – for both the fossils and the modern taxa. Furthermore, it is not necessary in this case to work at the species level, because it is well known that species of Alcelaphini and Antilopini are living in open arid environments (Vrba, 1975, 1976, 1980; Shipman and Harris, 1988).

Based on bovid proportions, Late Pliocene Malema and Uraha seem to share more affinities with the arid grassland of the Somalia-Masai than with the Zambezian phytochorion. Today Malawi belongs to the Zambezian phytochorion.

In agreement with the site comparison presented above is another fact: the micromammals of the Upper Ndolanya Beds at Laetoli show a Somalia-Masai composition (Denys, 1999). This suggests that similar ecological conditions are represented in the two regions, disregarding the lack of small mammals in the Chiwondo Beds.

A second correspondence analysis refers to the 29 game parks, once again as variable 1, but this time variable 2 refers to modern bovid tribes plus the Chiwondo and Upper Ndolanya Beds' bovid tribes (Figure 7, Table 3; data for the Ndolanya Beds derive from Gentry, 1987). It shows a similar ecological dispersion to the previous graph, separating the Zambezian and Sudanian from the Somalia-Masai ecozone. The Chiwondo and Ndolanya Beds plot between the Somalia-Masai and Zambezian ecozones. The Ndolanya Beds plot closer to the arid parks of the Zambezian ecozone. This may be explained by the frequent occurrences of Neotragini and *Antidorcas* sp. at Etosha, Kalahari and the Ndolanya Beds.

Limitations of the analyses are, first, the use of only bovids, and second, the combined application of modern and fossil data, but due to the incomplete fossil record we believe it is both useful and important to gain this large-scaled ecological information.

A closer look at the bovid distribution of Malema and Uraha may reveal different habitats within the Chiwondo Beds. Although more antilopins were discovered in Uraha, these together with the alcelaphins suggest general grassland conditions. These tribes dominate in both areas. Nevertheless more Aepycerotini and Tragelaphini specimens were discovered in the Uraha region, which imply a closed/dry habitat for at least some parts of the environment, that was apparently less significant in the Malema area.

Shipman and Harris (1988, p. 375) state that there is a "strong taphonomic bias in habitat representation dictated by the mode of deposition." The lake margin sites (Olduvai) represent open to wet habitats, the riverine settings (Koobi Fora, W. Turkana, and Omo) sample closed/wet and closed/dry habitats, and the South African cave localities only sample the open arid spectra. Reed (1997) did not find strong support for *P. boisei* favoring dry or preferred closed habitats. Instead her analysis shows that *P. boisei* occupied habitats with abundant water and edaphic grasslands. Reed's study supports the interpretation by Behrensmeyer (1978) of *P. boisei* living in delta environments at Koobi Fora. This implies a preference for the vicinity of lake margins or rivers in the habitat of this early robust hominin. *Paranthropus boisei* KNM-ER-406 was recovered in a tributary on the delta plain (Behrensmeyer, 1978).

The sedimentary environment of the maxillary fragment RC-911 is thus similar to its East African analogue. Furthermore in both hominin localities the large alcelaphin *Megalotragus* dominates at the lake margin. The comparison with Koobi Fora shows that the Malema alcelaphins and *Hipparion* do not contradict a lake margin setting.

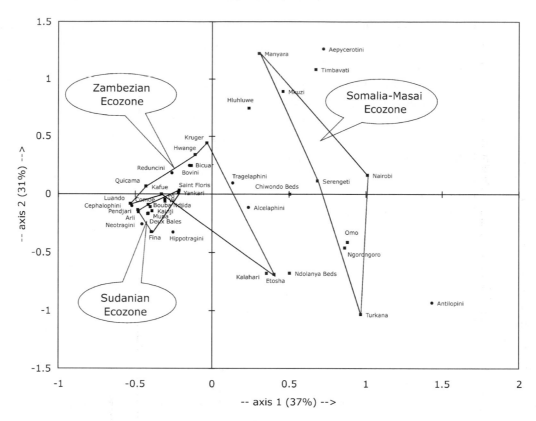

Figure 7. Correspondence analysis of game parks and bovid tribes on axes 1 and 2. Chi-square distance on data = 322.5253 ~ degrees of freedom = 240, corresponding probability = 0.0003, limit Chi square for the chosen confidence range = 277.1377. Using this test one should reject the hypothesis of independence. Variable 1 (rows) consists of 29 African parks and the Pliocene Chiwondo and Ndolanya Beds. Variable 2 (columns) consists of modern bovid tribes living in the parks and bovid tribes discovered in the Chiwondo- and Ndolanya Beds (variables 1 and 2 from Table 3).

## SUMMARY

The Plio-Pleistocene Chiwondo and Chitimwe Beds of the northern Malawi Rift are exposed in the Karonga-Chilumba region and can be referred to five depositional units. The age of the Chiwondo Beds relies on faunal correlation with radiometrically dated biostratigraphic units in eastern Africa.

Fauna from the Chiwondo Beds can be referred to three time intervals. The age of the fauna of interval I spans circa 4–3.75 Ma, the fauna of interval II ranges between circa 3.75 and 2 Ma, interval III encloses the time period of circa 2–1.5 Ma. The majority of the fauna derives of interval II, which derives from stratigraphic

Unit 3A, which was dominated by fluviatile processes and represents a time-aggregated deposit. More fine-scaled stratigraphic subunits cannot be defined. Consequently, the fossil occurrences within these units might represent a longer time span, but adjacent localities within laterally extensive areas agree in their biochronologic range estimation.

The hominins *Homo rudolfensis* and *Paranthropus boisei* refer to time interval II of Unit 3A. More precise temporal information for both hominin localities can be given based on the presence of early *Notochoerus scotti* specimens, which indicate an age of circa 2.3–2.5 Ma.

The Chiwondo Beds assemblages are taphonomically altered. A case study at the *P. boisei*

*Table 3. Logarithmic abundances of bovid tribes in 29 African parks plus Pliocene Chiwondo Beds and Ndolanya Beds*

| | Alcelaphini | Antilopini | Aepycerotini | Tragelaphini | Reduncini | Bovini | Hippotragini | Neotragini | Cephalophini |
|---|---|---|---|---|---|---|---|---|---|
| Ndolanya Beds | 1.81 | 1.6 | 0 | 0.95 | 0.3 | 0.3 | 0.95 | 1.56 | 0 |
| Chiwondo Beds | 2.03 | 1.5 | 1.27 | 1.14 | 1.17 | 1.39 | 1.51 | 0.6 | 0 |
| Kruger | 4 | 0 | 5.18 | 4.04 | 3.67 | 4.38 | 3.14 | 3.7 | 3.11 |
| Luando | 0 | 0 | 0 | 2.95 | 3.35 | 2.17 | 3.47 | 2.39 | 3.04 |
| Hwange | 3.41 | 0 | 3.9 | 3.73 | 3.09 | 4.11 | 3.41 | 3.47 | 3.3 |
| Nairobi | 3.12 | 2.92 | 2.82 | 1.89 | 2.02 | 0 | 0 | 0.3 | 0.3 |
| Turkana | 3.3 | 3.02 | 0 | 0 | 0 | 0 | 3.12 | 0 | 0 |
| Serengeti | 5.65 | 5.27 | 4.81 | 3.97 | 3.74 | 4.68 | 3.69 | 0 | 0 |
| Ngorongoro | 4.13 | 3.69 | 0 | 2.6 | 2.07 | 1.77 | 0 | 0 | 0 |
| Etosha | 4.22 | 4.07 | 0 | 3.39 | 0 | 3.24 | 3.68 | 2.69 | 2.39 |
| Bouba Ndjida | 3.85 | 0 | 0 | 3.25 | 3.84 | 3.9 | 3.63 | 4.06 | 3.73 |
| Quicama | 0 | 0 | 0 | 3.74 | 3.17 | 3.9 | 3.17 | 0 | 3.54 |
| Hluhluwe | 3.54 | 0 | 3.68 | 3.5 | 3 | 3.34 | 0 | 0 | 1.17 |
| Kafue | 2.27 | 0 | 0.3 | 1.62 | 2.29 | 1.92 | 1.49 | 1.95 | 1.84 |
| Pendjari | 3.97 | 0 | 0 | 0 | 4.29 | 3.54 | 3.7 | 3.89 | 3.39 |
| Comoe | 3.9 | 0 | 0 | 3 | 3.87 | 2.68 | 2 | 4.23 | 4.11 |
| Omo | 3.32 | 2.81 | 0 | 2.97 | 0 | 2.54 | 0 | 0 | 0 |
| Kalahari | 4.31 | 4.38 | 0 | 3.81 | 0 | 0 | 4.2 | 3.21 | 2.85 |
| Manyara | 0 | 0 | 2.84 | 1.39 | 1.74 | 3.17 | 0 | 1.77 | 0 |
| Bicuar | 2.69 | 0 | 2.17 | 2.65 | 2.17 | 2 | 2.69 | 1.77 | 2.69 |
| Mupa | 2.39 | 0 | 0 | 2.87 | 3.2 | 0 | 2.3 | 3 | 2.69 |
| Mkuzi | 3.14 | 0 | 3.97 | 2.74 | 1.83 | 0 | 0 | 0.77 | 0.6 |
| Arli | 2.27 | 0 | 0 | 1.9 | 3.09 | 1.81 | 3.27 | 3 | 3.33 |
| Deux Bales | 2.65 | 0 | 0 | 2.29 | 2.35 | 1.6 | 3.07 | 2.81 | 2.81 |
| Po | 2.73 | 0 | 0 | 2.03 | 2.45 | 2.39 | 2.89 | 0 | 2.68 |
| Saint Floris | 3.32 | 0 | 0 | 0 | 3.5 | 3.25 | 2.7 | 0 | 0 |
| Yankari | 2.82 | 0 | 0 | 0 | 2.81 | 2.82 | 2.55 | 0 | 0 |
| Fina | 3.47 | 0 | 0 | 0 | 3.1 | 0 | 3.59 | 3.66 | 0 |
| Kainji | 3.8 | 0 | 0 | 2.74 | 3.68 | 2.07 | 3.76 | 3.44 | 3.15 |
| W | 3.15 | 0 | 0 | 2.38 | 3.78 | 3.61 | 3.45 | 3.32 | 0 |
| Timbavati | 3.09 | 0 | 3.66 | 1.89 | 1.51 | 0 | 0 | 0 | 0 |

locality of Malema RC 11 indicates a mixture of an attritional and hydraulically winnowed assemblage. The sediments of the Malema bone assemblage consist of fine- to medium-grained fluviatile sediments that accumulated at the lake margin, likely on a delta plain. Isolated molars, half mandible fragments, and high-density postcranial elements dominate at the site. The fauna is dominated by large terrestrial herbivores. Juvenile specimens and small mammals are underrepresented. Large bovids, proboscideans, and hippopotamuses rule the Chiwondo Beds assemblages. Especially alcelaphin grazers dominate like in Swartkrans 1 and the Ndolanya Beds, pointing towards grass–bushland habitats.

Throughout the Chiwondo Beds the tribes Alcelaphini and Antilopini make up the majority of the bovid assemblage. Species of Aepycerotini and Tragelaphini, that are adapted to a closed dry habitat, are more common in the Uraha region than in the northern deposits. Correspondence analysis of African bovid generic abundances shows that that the Chiwondo Beds bovid assemblage show more affinities to the more arid Somalia-Masai ecozone, than to the Zambezian ecozone to which the Malawi Rift belongs today.

Interpretation of the paleoenvironments represented by the Chiwondo Beds is limited by the incomplete preservation of the fossils, the limited number of localities, and the discontinuous stratigraphic sequence. But, its geographic position between the classic East and South African vertebrate fossil sites and the recovery of two hominins demonstrate its important potential for the understanding of faunal distribution in Africa during the Late Pliocene.

## Acknowledgments

The authors wish to thank R. Bobe, Z. Alemseged, and A.K. Behrensmeyer for the invitation to the symposium "Hominid Environments and Paleoecology in the East African Pliocene: An Assessment of the Faunal Evidence," which was held during the AAPA meeting in 2003 at Phoenix, AZ. We thank the editors, John Harris and two anonymous reviewers for critical comments on earlier versions of this manuscript. Fieldwork in Malawi was conducted by permission of the Malawi Government under the auspices of the Department of Antiquities and funded by DFG.

## References

Alemseged, Z., 2003. An integrated approach to taphonomy and faunal change in the Shungura Formation (Ethiopia) and its implication for hominid evolution. Journal of Human Evolution 44, 451–478.

Badgley, C., 1986. Counting individuals in mammalian fossil assemblages from fluvial environments. Palaios 1, 328–338.

Beden, M., 1976. Proboscideans from Omo group deposits. In: Coppens, Y., Howell, F.C., Isaac, G.Ll., Leakey, R.E.F. (Eds.), Earliest Man and Environments in the Lake Rudolf Basin. University of Chicago Press, Chicago, pp. 193–208.

Beden, M., 1985. Les proboscidiens des grands gisements à hominidés Plio-Pléistocènes d'Afrique Orientale. In: (Colloque Int.) L'environnement des Hominidés au Plio-Pléistocène, Foundation Singer-Polignac, Masson, Paris, pp. 21–44.

Beden, M., 1987. Les faunes Plio- Pléistocène de la basse vallée de 'omo (Éthiopie) Tome 2, Les éléphantidés (Mammalia: Proboscidea). Cahiers de paléontologie. CNRS, Paris.

Behrensmeyer, A.K., 1975. The taphonomy and paleoecology of Plio-Pleistocene vertebrate assemblages east of Lake Rudolf, Kenya. Bulletin of the Museum of Comparative Zoology 146(10), 473–578.

Behrensmeyer, A.K., 1978. The habitat of Plio-Pleistocene hominids in East Africa: taphonomic and microstratigraphic evidence. In: Jolly, C.J. (Ed.), Early Hominids of Africa. Duckworth, London, pp. 165–190.

Behrensmeyer, A.K., 1991. Terrestrial vertebrate accumulations. In: Allison, P.A., Briggs, D.E.G. (Eds.), Taphonomy, Releasing the Data Locked in the Fossil Record. Plenum Press, New York and London, pp. 291–335.

Behrensmeyer, A.K., Todd, N.E., Potts, R., McBrinn, G.E., 1997. Late Pliocene faunal turnover in the Turkana Basin, Kenya and Ethiopia. Science 278, 1589–1594.

Betzler, C., Ring, U., 1995. Sedimentology of the Malawi Rift: facies and stratigraphy of the Chiwondo Beds, northern Malawi. Journal of Human Evolution 28, 23–35.

Bromage, T.G., Schrenk, F., 1986. A cercopithecoid tooth from the Pliocene in Malawi. Journal of Human Evolution 15, 497–500.

Bromage, T.G., Schrenk, F., Juwayeyi, Y.M., 1995. Palaeobiogeography of the Malawi Rift: age and vertebrate paleontology of the Chiwondo Beds, northern Malawi. Journal of Human Evolution 28, 37–59.

Brown, F.H., Feibel, C.S., 1986. Revision of lithostratigraphic nomenclature in the Koobi Fora region, Kenya. Journal of the Geological Society of London 143, 297–310.

Brown, F.H., Nash, W.P., 1976. Radiometric dating and tuff mineralogy of Omo Group deposits. In: Coppens, Y., Howell, F.C., Isaac, G.Ll., Leakey, R.E.F. (Eds.), Earliest Man and Environments in the Lake Rudolf Basin. University of Chicago Press, Chicago, pp. 50–63.

Brown, F.H., McDougall, I., Davies, T., Maier, R., 1985. An integrated Plio-Pleistocene chronology for the Turkana Basin. In: Delson, E. (Ed.), Ancestors: The Hard Evidence. Alan R. Liss, New York, pp. 82–90.

Clark, J.D., Haynes, C.V., 1970. An elephant butchery site at Mwanganda's Village, Karonga, Malawi and its relevance of Palaeolithic Archaeology. World Archaeology 1, 390–411.

Clark, J.D., Stephens, E.A., Coryndon, S.C., 1966. Pleistocene fossiliferous lake beds of the Malawi (Nyasa) Rift: a preliminary report. American Anthropologist 68(2), 46–49.

Clark, J.D., Haynes, C.V., Mawby, J.E., Gautier, A., 1970. Interim report on palaeo-anthropological investigations in the Lake Malawi Rift. Quaternaria 13, 305–354.

Coryndon, S.C., 1966. Preliminary report on some fossils from the Chiwondo Beds of the Karonga District, Malawi. American Anthropology 68, 59–66.

Dauphin, Y., Denys, C., Kowalski, K., 1997. Analysis of accumulations of rodent remains: role of the chemical composition of skeletal elements. Neues Jahrbuch für Geologie und Paläontologie Abhandlungen 203, 295–315.

Denys, C., 1999. Of mice and men: evolution in East and South Africa during Plio-Pleistocene times. In: Bromage, T.G., Schrenk, F. (Eds.), African Climate Change, Biogeography and Hominid Evolution. Oxford University Press, Oxford, pp. 226–252.

Dixey, F., 1927. The Tertiary and post-Tertiary lacustrine sediments of the Nyasan Rift valley. Quarterly Journal of Geological Society of London 83, 432–447.

Feibel, C.S., Brown, F.H., McDougall, I., 1989. Stratigraphic context of fossil hominids from the Omo Group deposits: Northern Turkana basin, Kenya and Ethiopia. American Journal of Physical Anthropology 78, 595–622.

Gentry, A.W., 1987. Pliocene Bovidae from Laetoli. In: Leakey, M.D., Harris, J.M. (Eds.), Laetoli: A Pliocene Site in Northern Tanzania. Clarendon Press, Oxford, pp. 378–408.

Gentry, A.W., Gentry, A., 1978. The Bovidae (Mammalia) of Olduvai Gorge, Tanzania. Part I and II. Bulletin of the British Museum (Natural History) (Geology) 29, 289–446; 30, 1–83.

Greenacre, M.J., Vrba, E.S., 1984. A correspondence analysis of biological census data. Ecology 65, 984–997.

Haile-Selassie, Y., WoldeGabriel, G., White, T.D., Bernor, R.L., Degusta, D., Renne, P.R., Hart, W.K., Vrba, E., Ambrose, S., Howell, F.C., 2004. Mio-Pliocene mammals from the Middle Awash, Ethiopia. Geobios 37(4), 536–552.

Harris, J.M., White, T.D., 1979. Evolution of the Plio-Pleistocene African Suidae. Transactions of the American Philosophical Society 69, 1–128.

Henshaw, P.C., 1942. A Tertiary mammalian fauna from the San Antonio Mountains near Tonopah, Nevada. Carnegie Institution of Washington, Contribution to Paleontology Publication 530, 77–168.

Kalb, J.E., 1995. Fossil elephantoids, Awash paleo-lake basins, and the Afar triple junction, Ethiopia. Palaeogeography, Palaeoclimatology, Palaeoecology 114, 357–368.

Kalb, J.E., Mebrate, A., 1993. Fossil elephantoids from the hominid-bearing Awash Group, Middle Awash Valley, Afar depression, Ethiopia. Transactions of the American Philosophical Society 83(1), 1–110.

Kalb, J.E., Jolly, C.J., Mebrate, A., Tebedge, S., Smart, C., Oswald, E.B., Cramer, D., Whitehead, P., Wood, C.B., Conroy, G.C., Adefris, T., Sperling, L., Kana, B., 1982. Fossil mammals and artefacts from the Middle Awash Valley, Ethiopia. Nature 298, 25–29.

Kaufulu, Z.M., Vrba, E.S., White, T.D., 1981. Age of the Chiwondo Beds, northern Malawi. Annals of the Transvaal Museum 33, 1–8.

Kovarovic, K., Andrews, P., Aiello, L., 2002. The palaeoecology of the Upper Ndolanya Beds at Laetoli, Tanzania. Journal of Human Evolution 43, 395–418.

Kullmer, O., 1999. Evolution of African Plio-Pleistocene suids (Suidae: Artiodactyla) based on tooth pattern analysis. In: Schrenk F., Gruber, G. (Eds.), Kaupia, Darmstädter Beiträge zur Naturgeschichte, Heft 9, Current Research 2, Plio-Pleistocene Mammalian Evolution. Darmstadt, pp. 1–34.

Kullmer, O., Sandrock, O., Abel, R., Schrenk, F., Bromage, T.G., Juwayeyi, Y.M., 1999. The first *Paranthropus* from the Malawi Rift. Journal of Human Evolution 37, 121–127.

Maglio, V.J., 1973. Origin and evolution of the Elephantidae. Transactions of the American Philosophical Society 63(3), 1–26.

Mawby, J.E., 1970. Fossil vertebrates from northern Malawi: preliminary report. Quaternaria 13, 319–323.

Reed, K., 1997. Early hominid evolution and ecological change through the African Plio-Pleistocene. Journal of Human Evolution 32, 289–322.

Renne, P., WoldeGabriel, G., Hart, W.K., Heiken, G., White, T., 1999. Chronostratigraphy of the Miocene–Pliocene Sagantole Formation, Middle Awash Valley, Afar Rift, Ethiopia. Geological Society of America Bulletin 111(6), 869–885.

Ring, U., Betzler, C., 1995. Geology of the Malawi Rift: kinematic and tectonosedimentary background to the Chiwondo Beds, northern Malawi. Journal of Human Evolution 28, 7–21.

Sandrock, O., 1999. Taphonomy and paleoecology of the Malema hominid site, Northern Malawi. Ph.D. Dissertation, University of Mainz.

Sandrock, O., Dauphin, Y., Kullmer, O., Abel, R., Schrenk, F., Denys, C., 1999. Malema: preliminary taphonomic analysis of an African hominid locality. Comptes Rendus de l Academie des Sciences Paris 328, 133–139.

Schrenk, F., Bromage, T.G., Betzler, C.G., Ring, U., Juwayeyi, Y.M., 1993. Oldest Homo and Pliocene biogeography of the Malawi Rift. Nature 365, 833–836.

Schrenk, F., Bromage, T.G., Gorthner, A., Sandrock, O., 1995. Paleoecology of the Malawi Rift: ver-tebrate and invertebrate faunal contexts of the Chiwondo Beds, northern Malawi. Journal of Human Evolution 28, 59–70.

Shipman, P., Harris, J.M., 1988. Habitat preference and paleoecology of *Australopithecus boisei* in Eastern Africa. In: Grine, F.E. (Ed.), Evolutionary History of the "Robust" Australopithecines. Aldine de Gruyter, New York, pp. 343–381.

Voorhies, M.R., 1969. Taphonomy and Population Dynamics of an Early Pliocene Vertebrate Fauna, Knox Country, Nebraska. Contribution to Geology Special Paper 1.

Vrba, E.S., 1975. Some evidence of chronology and palaeoecology of Sterkfontein, Swartkrans and Kromdraai from the fossil Bovidae. Nature 254, 301–304.

Vrba, E.S., 1976. The fossil Bovidae of Sterkfontein, Swartkrans and Kromdraai. Transvaal Museum Memoir 21, 1–166.

Vrba, E.S., 1980. The significance of bovid remains as indicators of environment and predation patterns. In: Behrensmeyer, A.K., Hill, A.P. (Eds.), Fossils in the Making. University of Chicago Press, Chicago, pp. 247–271.

White, F., 1983. The vegetation of Africa, a descriptive memoir to accompany the UNESCO/AETFAT/UNSO Vegetation Map of Africa (3 Plates, Northwestern Africa, Northeastern Africa, and Southern Africa, 1:5,000,000). UNESCO, Paris.

White, T.D., 1995. African omnivores: global climatic change and plio-pleistocene hominids and suids. In: Vrba, E.S., Denton, G.H., Partridge, T.C., Burckle, L.H. (Eds.), Paleoclimate and Evolution, with Emphasis on Human Origins. Yale University Press, New Haven and London, pp. 369–384.

White, T.D., Suwa, G., Hart, W.K., Walter, R.C., WoldeGabriel, G., de Heinzelin, J. deClark, J.D., Asfaw, B., Vrba, E.S., 1993. New discoveries of Australopithecus at Maka in Ethiopia. Nature 366, 261–265.

WoldeGabriel, G., Haile-Selassie, Y., Renne, P.R., Hart, W.K., Ambrose, S.H., Asfaw, B., Heiken, G., White, T., 2001. Geology and paleontology of the Late Miocene Middle Awash Valley, Afar rift, Ethiopia. Nature 412, 175–178.

# Finale and future

Investigating faunal evidence for hominin paleoecology in East Africa

A.K. BEHRENSMEYER
*Department of Paleobiology*
*Smithsonian Institution*
*P.O. Box 37012, NHB MRC 121*
*Washington, DC 20013-7012, USA*
*behrensa@si.edu*


Z. ALEMSEGED
*Department of Human Evolution*
*Max Planck Institute for Evolutionary Anthropology*
*Deutscher Platz 6*
*04103 Leipzig, Germany*
*zeray@eva.mpg.de*


R. BOBE
*Department of Anthropology*
*The University of Georgia*
*Athens, GA 30602, USA*
*renebobe@uga.edu*

## Introduction

The East African Plio-Pleistocene fossil record is important both for understanding diversification, extinction, and ancestor–descendent relationships among extinct and living African mammals and for investigating the interaction of environmental change and evolution. The 2004 Workshop on Faunal Evidence for Hominin Paleoecology at the Smithsonian Institution focused on the second of these research themes, bringing together a multi-national group of 44 professionals and students for in-depth discussion about how the East African mammalian faunal record can be used to reconstruct the changing paleoeco-logical context of the hominins during their time of major diversification from late Miocene to early Pleistocene. Many of the contributions to this volume originated as papers presented at a 2003 American Association of Physical Anthropologists symposium in Tempe, Arizona, organized by Bobe, Alemseged, and Behrensmeyer. These contributions evolved as a consequence of the discussions at the Smithsonian workshop, taking on broader issues and adapting to a wide range of ideas and concerns expressed by the participants.

Three over-arching topics guided the workshop discussions and provide underlying themes for papers in this volume. These

*R. Bobe, Z. Alemseged, and A.K. Behrensmeyer (eds.) Hominin Environments in the East African*
*Pliocene: An Assessment of the Faunal Evidence, 333–345.*
© 2007 *Springer.*

include: (1) key paleoecological and paleoenvironmental questions in human evolution, (2) methodological approaches to the collection and analysis of fossil data in relation to hominin paleoecology, and (3) strategies for storing, retrieving, and sharing the expanding paleontological information that presently resides in many different electronic databases. The ideas and collaborations generated during the workshop not only helped to shape the papers in this volume around these themes, they provided a forum for ideas and recommendations that extend well beyond the topics covered in these papers. In this concluding chapter of the volume, we provide commentary on the workshop presentations and share highlights from the wide-ranging discussions, many of which raise important issues not covered in the volume articles that should be taken into account in future field and museum research. For instance, the workshop discussion groups developed a list of protocols for documenting field surveying and collection procedures (Theme 2 below) and provided important insights on issues regarding database access and intellectual property rights (Theme 3). We include text citations and notes to help readers understand how volume contributors incorporated and expanded upon some of these issues in their papers. Suggestions and recommendations to guide future research in faunal analysis are included at the end of each section.

## Theme 1: Major Questions in Hominin Paleoecology

The workshop led off with four invited speakers, Andrew Hill, Bernard Wood, Richard Potts, and Nina Jablonski, who highlighted how their own thinking about key questions in hominin paleoecology has changed over time. They discussed: (1) late Miocene hominoids and the emergence of hominins (Andrew Hill), (2) the radiation and diversification of hominin species in the Pliocene (Bernard Wood), (3) cultural and biological changes in the Pleistocene (Richard Potts), and (4) the relevance of non-hominin primates to hominin paleoecology (Nina Jablonski). Hill pointed out that the earliest hominins have been considered "ecological apes" (Andrews, 1995), and that we can document that several ape species lived in the African Rift Valley during the time when hominins first appeared. The major question is, however, which hominoid species evolved into the first hominin? Hill suggested that the human ancestor must have been an "anthropomorphic ape" that lived during the Miocene, probably before 6.5 to 7.0 Ma, given what is now known of the earliest hominins. When and where did this evolutionary transition take place? Hill noted the fact that only 0.1% of the African continent is represented by fossil localities, providing perspective on the scale of our knowledge versus the huge area of diverse habitats where early hominins might have evolved (see Hill's Preface to this volume for further discussion).

Wood outlined two views on hominin diversity: the "simple" philosophy, which assumes a succession of species with low diversity over time, and the "complex" philosophy, in which the hominin clade has often been represented by several species at any given time. He stressed that hominin diversity would be better understood if we could first define the most critical and data-rich periods in hominin evolution and suggested that we look at appearances of key hominin features during these periods. The critical time periods should include ~4.2 Ma (the appearance of megadont hominins such as *Australopithecus anamensis*) (Leakey et al., 1995), ~2.8 Ma (the appearance of the "robust australopithecine" group), ~2.2 Ma (when the transition from *Paranthropus aethiopicus* to *P. boisei* took place (Suwa et al., 1996), with the latter showing molarized premolars and smaller incisors), and ~2.0 Ma (when human-like body proportions first appear). Correlating major morphological changes such as these with well-documented paleoecological and paleoenvironmental events could shed light on our understanding of the role of ecological factors in shaping hominin evolution.

For the later part of the human evolutionary story, Potts emphasized that faunal analysis can provide evidence about where hominins lived, how they interacted with their environments, and spatial/temporal changes in these parameters. Evidence from fossil mammals can, for instance, address questions such as whether the evolution of faunal elements, or faunas in general, was faster during periods of environmental stability or periods of change. To answer this question we need to expand our range of sites and compare different basins and regions across time and space. Faunal evidence can also tell us about past vegetation. Potts pointed to the need to document the expansion of human ecological boundary conditions, i.e., their levels of tolerance for different habitats, climates, and terrains. In about 100 Kyr during the early Pleistocene, hominins expanded their geographic range immensely, from Africa to Dmanisi, Java, and China (Swisher et al., 1994; Gabunia et al., 2000; Zhu et al., 2004). But what was happening in Africa at that time? Were there changes in how hominins interacted with the African faunas, or environmental shifts that encouraged hominin emigration from Africa? In his paper for this volume, Potts provides an in-depth review of environmental hypotheses about hominin evolution and an agenda for future contributions from the faunal record (Potts, 2007), and the contribution by Bobe et al. (2007) demonstrates the potential of the inter-basin comparisons of faunal evolution.

Jablonski discussed a wide range of important general issues in faunal analysis. Documentation of context at the specimen level is critical. How does the catalogued fauna relate to the actual fossil fauna from a particular time and place? Small mammals (e.g., rodents, small primates, and carnivores) are often a neglected component of paleo-

ecological analysis, yet they provide much finer-grained evidence for environmental parameters than large mammals (Frost, 2007; Reed, 2007). Adequate documentation of geographic continuity versus variability of fossil mammal taxa is elusive to absent in faunal analysis (but see Cooke, 2007). There remain many "taxonomic equivalence" problems in inter-basin comparisons. Successive "chronospecies" that have morphoclines must be clearly distinguished from morphospecies that are characterized by obvious morphological breaks. Misuse and mixing of taxonomic ranks (tribes, genera, species) can bias diversity estimates and make comparisons among localities difficult. Subtle morphological variation often goes unnoticed, so we may be missing pockets of endemism. Communities generally are not "locked" in species composition through time; they are perpetually in flux and taxonomically ephemeral; we can expect "non-analogue" communities in the past and should beware of over-using ecological uniformitarianism. It is also risky extrapolating habitat preferences from recent to fossil members of the same tribe or genus, as in the example of fossil colobines, some of which may have been more terrestrial than any of their modern counterparts (see also Frost, 2007).

Discussions following the presentations focused on the terminology of fauna-based paleoecology. Clear definitions are important to avoid confusion between paleoenvironments – the physical habitats – and paleoecology, which encompasses many more variables including characteristics of the co-occurring plant and animal species (communities) and habitat structure (see Behrensmeyer et al., 2007). Studies that deal with interactions among organisms, or interactions between organisms and their environment, must be included in paleoecology. The term paleoenvironment is more appropriate in discussions of climate, substrate, physical and chemical setting, topography, and lateral variation across the landscape in these variables. However, there is a continuum between these two concepts, so the demarcation may not always be clear.

It is also important to ask to what extent actualistic (i.e., modern analogue) approaches are justified for the questions at hand. We need to compile and compare actualistic studies from different areas to get at underlying variables, which can then serve as more powerful analogues for the past (see Reed, 2007). The traditional way of applying actualistic approaches could be misleading if researchers look only for the ways in which ancient ecosystems resemble modern ones, thereby missing the ways in which these ecosystems may be unique and without modern analogues (Tooby and DeVore, 1987). This then brings us to an important point agreed to by all discussants: the need for more interaction with ecologists working on modern ecosystems. For example, knowing more about the behavioral ecology of modern great apes and distilling general features of these varying ecologies could provide insights into the paleoecology of early hominins. Environmentally linked behaviors are currently under study and we need to pay more attention to these. Paleoecological theory-building and model-building (testable hypotheses) need to be strengthened through interaction with biologists and ecologists. Given that paleoecologists work on fossils, which only sample a fraction of the paleoecosystem, our data are not often readily comparable and testable using models that are based on information that comes only from modern ecosystems. Thus, we should clearly identify which paleoecological questions are appropriate given the data we have at hand (Behrensmeyer et al., 2007).

The study of hominin paleoecology includes the interactions that our ancestors had with other animals and their environments. We would like to know more about how these hominins behaved, more about their physiology and ecological niches. We should first attempt to understand what the key hominin adaptations are, to guide what we investigate in the associated mammalian faunas. It is also critical to focus on dynamic habitat reconstructions, not static ones, by incorporating changes that may have occurred over short and long periods of time (Potts, 2007). Our research would benefit from the use of multiple sites (e.g., contemporaneous faunas in different basins) for comparisons and tests of hypotheses (Behrensmeyer, 2006; Behrensmeyer et al., 2007), and from focus on both short- and long-range space–time variation in faunal composition.

The reconstruction of possible habitat preferences of hominin species is important and must involve alternative, testable hypotheses. Understanding hominin diets should be given high research priority because diet provides a direct link to environment, ecology, and adaptation. A related issue is the appearance of meat in the hominin diet and its role in shaping human behavior. Does meat consumption lead to interaction with a wider or a narrower range of other animal species, relative to herbivory? Can we tell which hominin species were associated with which behaviors? It will be important to develop more specific hypotheses regarding alternative hominin dietary strategies, supported by increased knowledge of adaptations in other mammalian species and groups that were more abundant than hominins in the Plio-Pleistocene ecosystems (Frost, 2007; Lewis and Werdelin, 2007; Potts, 2007).

## SPECIFIC QUESTIONS FOR FUTURE RESEARCH

(1) How can we develop ways to recognize biotic forcing (competition and interaction) in human evolution versus environmental forcing? How do we differentiate between regional and global events?

(2) How can we map biogeographic and stratigraphic variation in hominins and their adaptability, e.g., evidence for their ability to transcend environmental changes within a basin (versus habitat-specific adaptations and sensitivity to environmental shifts)?

(3) What environments are associated with early forms of bipedalism?

(4) What were the centers of endemism, speciation, sources, and sinks of faunas?

(5) How can we best define and investigate different scales of paleoenvironmental analysis?

(6) Hominins were not initially a keystone species: they were uncommon and ecologically marginal. When did this begin to change?

## Theme 2: Methodologies in Paleofaunal Analysis

Temporal and spatial scales are major issues in paleofaunal analysis because paleontological sites (i.e., fossiliferous areas with multiple localities), localities, and lists of species vary in how much time and space they represent; one may reflect the ecology of a few years, another a time-averaged sample of 1000s to 100,000s of years. Obviously, spatial/temporal resolution of faunal samples affects their diversity and taxonomic abundances, hence the validity of time-specific and through-time comparisons among localities and regions. This topic is considered in detail in the volume contribution by Behrensmeyer et al. (2007).

Paleoecological change through time can be examined in areas that provide relatively rich faunal samples from multiple levels within a long stratigraphic sequence (e.g., Alemseged et al., 2007; Bobe et al., 2007; Sandrock et al., 2007). The paleoecology of specific time intervals can also be studied either locally (Musiba et al., 2007; Su and Harrison, 2007)

or over a wide geographic area (Bobe et al., 2007) depending on how well fossil-producing sites can be correlated in time and controlled for taphonomic and collecting biases. For both types of paleoecological goals, it is desirable to minimize the amount of time-averaging represented by each locality and faunal list as much as possible in order to obtain the best approximation of an ecologically contemporaneous fauna. Examples of paleolandscape sites whose faunas have relatively high time resolution include Peninj (one widespread 30-cm unit with fossils) (Domínguez-Rodrigo et al., 2001), Olorgesailie UM1paleosol (estimated at ~500 years) (Potts et al., 1999), Olduvai Lower Bed II (~50 Kyr) (Blumenschine et al., 2003), Hadar BKT-2, and the Laetoli Upper Units (Musiba et al., 2007; Su and Harrison, 2007). To these could be added many archeological and paleontological quarry samples, in which single excavated levels may be 10–30-cm thick and likely represent periods of days or years to, at most, centuries. Many different types of evidence can be integrated at such sites, such as macro- and microfaunas, pedogenic features, stable isotope analysis, clay mineral analysis, and the spatial distribution of taphonomic features. There appears to be a pattern of patchiness of faunal remains, even in the laterally extensive paleolandscape samples, which may indicate short-term, highly localized, favorable circumstances for the accumulation and burial of faunal remains. In the future, it will be important to focus on what these patches mean taphonomically, sedimentologically, and paleoecologically. They may not indicate evidence for a concentration of living animals in a favorable habitat, but instead a taphonomic window of opportunity for permanent bone preservation (e.g., a low, wet area where bones were quickly buried).

Sedimentological evidence comes primarily from times when sediments were aggrading (building up), but it is during the pauses (hiatuses) in deposition that ecosystems develop on a land surface, soils form, and bones and artifacts accumulate. Thus, the faunal record may represent times with less direct geological evidence for environmental variables, except when bones were incorporated into soils that record paleoenvironmental conditions via pedogenic features and stable isotope signals in soil carbonates. It is important to realize that sedimentological evidence can be temporally out of phase with faunal evidence, representing finer scales and coarser scales of stratigraphic and temporal resolution as well as different time intervals than associated fossil and archeological remains.

A number of modern East African ecosystems have been used for comparison with Plio-Pleistocene faunas and taphonomy, including Parc Nationale de Virunga (Tappen, 1995), Serengeti (Blumenschine, 1989; Reed, 2007), Amboseli (Behrensmeyer, 1993; Faith and Behrensmeyer, 2006), Tsavo (Domínguez-Rodrigo, 1996), and Laikipia (Pobiner and Blumenschine, 2003). It is clear, however, that we need more examples of mixed- to closed-vegetation habitats, as well as information on how cosmopolitan versus habitat-specific different modern species are (e.g., *Crocuta*). Some habitat types that existed in the past may now be extinct, i.e., have no modern analogues. There also may be behavioral clues in the fossil record that we are missing; for instance, the proportions of carnivore-modified skeletal remains found in landscape assemblages or especially in fossil den accumulations might indicate different levels of inter- or intra-specific competition (Faith and Behrensmeyer, 2006). Also, it appears that there are diverse carnivores in some paleontological localities but not in others, and we do not yet have a basis for understanding this from the study of modern analogues. To tease out underlying patterns in the structure of faunal communities, we should use faunas from other continents (not just Africa) to expand our sample. It would be informative to document in greater detail the shift from more to less diverse ungulates and carnivores (see Lewis

and Werdelin, 2007) around 1.8 Ma, comparing sites with and without hominins, including Africa, China, and Europe. This could test the hypothesis of hominins as emerging competitors in the carnivore guild. It is important to remember the non-mammalian carnivores, such as crocodiles, and also large raptors, were potentially important as predators and/or competitors for hominins.

Workshop participants agreed that we need to build more bridges to ecological and evolutionary research in order to enhance opportunities for paleoanthropology students to learn from these fields. It was suggested that we try to generate studies (e.g., doctoral dissertations or postdoctoral projects) that would serve both ecological and paleoecological/taphonomic purposes, e.g., "variability in lion behavioral ecology and its importance to conservation biology," in which we could encourage the development of better understanding of the impact of such behavior on bone assemblages.

Micromammals usually come from a small area of outcrop, and also in life can represent habitat patches on the order of 1–5 km$^2$; therefore they provide a different level of spatial resolution for paleoecological reconstruction than macromammals (Andrews, 1990; Reed, 2007). Microfaunas have been documented in South Africa (Avery, 2003; Matthews et al., 2005), that reflect changes in species ranges and/or habitat preferences; these suggest the possibility of "non-analogue" Pleistocene mammalian paleocommunities in Africa, as have been proposed for North American faunas (Graham, 1992; Lyons, 2003). If the time/space scale issues can be dealt with, comparisons of macro- and micropaleofaunas could provide different but complementary views of the same paleoecosystem.

A major topic of discussion revolved around the need for standardized field data collection, i.e., a general checklist of what to record during field collecting of fossil vertebrates (Eck, 2007) in addition to the standard information about field number, place, and stratigraphy. Too much is never written down about what was and was not collected, making it difficult or impossible to use collections for some types of analysis, especially those relating to paleoecological questions where it may be critical to know collection strategies and goals in order to understand biases in the abundance of different taxa (Alemseged et al., 2007; Bobe et al., 2007; Eck, 2007). It would be very helpful to have general repositories for such information; e.g., on websites devoted to the museum collections themselves. One participant suggested that any field collector should plan for his/her field notes and collection information to be comprehensible to people at least 50 years into the future, and it is also crucial to leave these in a secure repository where they will be accessible to coming generations.

Suggestions for a standard field collection note-taking checklist:

(1) Description of the terrain and limits of a locality – meticulous, detailed note taking, everyday.
(2) Resolution of stratigraphic position, issues in tracing of marker beds.
(3) Type of preservation of fossils, degree of surface fragmentation.
(4) Goals of collecting, sampling strategy, who was in charge of decisions about what to collect or what not to collect, basis for these decisions.
(5) Areas searched that do and do not have fossils ("null collecting events").
(6) Name of recorder for the team; best to use one person for consistency's sake.

The ideal is to do the geology first, then the fossil sampling. However, this is rarely possible because most successful grant proposals target fossil discovery and collection. All research projects, however, should address the issue of documenting geological context as a critical goal, and collecting should be restricted until the geological framework is in place. Another important issue is how to define a fossil locality targeted for collection. The best strategy is to go for the smallest area or stratigraphic unit that is producing the fossils and document that; such localities can always be grouped, but a large one that is actually a composite of different patches or levels cannot be resolved more finely after the collecting party has left the field. Usually paleontologists target large survey areas and then spend more time in places where there are fossils; these become their localities. Sometimes drawing locality boundaries can be a challenge if fossils are widely dispersed. It is always imperative to document the place on the map where the fossils occur in addition to an initial assessment of stratigraphic level, so that the latter can be checked later based on the exact original location of the fossils on the ground.

The above list is preliminary and should be further developed, with input from a range of paleoanthropologists as well as organizations that face similar field documentation problems. NAGPRA (Native American Graves Protection and Repatriation Act) and other organizations that oversee archeological surveys in the US have guidelines for types of data to record in field surveys (e.g., www.kshs.org/resource/siteformhome.htm), and modern biodiversity surveys could also be a source of information/models. High-resolution GIS imagery is required for NSF standards; $15\,m^2$ pixel size is too large; at $1\,m^2$ pixel it is possible to get enough resolution regarding a fossil's position on the ground. GIS provides a good answer to spatial documentation, but there are caveats; the technology alone can give a false sense of precision. Ground examination may be the only way to determine whether a fossil lies above or below a critical stratigraphic boundary, even if it is precisely located using satellite coordinates. It is also essential to be sure that one is using the same GIS datum (e.g., ARC 1960 versus WGS1984 can result in a 100 m or more difference in position). Redundancy is always a good strategy, with air photo documentation in addition to GPS readings. A camera on a kite or balloon can provide higher-resolution spatial documentation for a rich patch of fossils. For consistency's sake, one person on a team should be assigned the responsibility for collecting GIS information on fossils and other spatial variables.

The strength of African museums will be critical to the future of research and collections. There must be trained people who can respond to public reports of "funny bones" from the general population and assess their paleontological importance. African museum staff must have increased access to funds for curation and collections management, as well as for developing new paleontological sites. Most of the current external funding is for fieldwork and collecting, leaving an imbalance of funding resources once the fossils are deposited at the museum. Whose responsibility are these collections? This is an international problem, and there need to be agreements between institutions about funding for maintaining and databasing these collections into the foreseeable future. The National Science Foundation (US) recognizes the need to support collections resulting from fieldwork and has worked with the Department of Antiquities in Ethiopia to further this goal, but this problem needs to be addressed with increased levels of funding, training, and international cooperation.

## Theme 3: Paleofaunal Databases

Databases compiling large bodies of faunal information are becoming indispensable tools for paleobiological analyses, as demonstrated in the volume chapters by Frost (2007), Bobe et al. (2007), Alemseged et al. (2007), and Eck (2007). There are many platforms and programs available for designing and maintaining such databases. The workshop participants agreed that it is not necessary to centralize the diversity of paleontological information avail-

able now and in the future. Instead, it appears that a distributed network of linked databases is the best option for future paleontological and paleoecological research. The databases discussed during the workshop focus on East Africa, but we should think about a future "Pan-African Paleofaunal Data Network" – expanding the scope from East Africa to the whole continent.

The consensus was against recommending any particular software platform; instead we should focus on making sure that the data are downloadable from program to program, and all users should know exactly what the fields mean. This could take the form of a data dictionary and thesaurus. For example, we often use the terms "locality" and "site" interchangeably, but these can have different meanings for different researchers – in this volume, "site" usually refers to an area with many localities, and "locality" is a specific place where fossils occur. It would be easy to look up a word or field name with a data dictionary, with citations to different ways these terms may be used. Minimum standards for paleontological database fields are usually very similar from worker to worker. Key fields include:

(1) Accession number
(2) Basic information on taxonomy
(3) Geological context – formation, member, stratum, age
(4) Geography of the locality or site
(5) References or citations to published work
(6) Person responsible for the identification
(7) Links to additional information from field notes, geochronological context, etc.

The Society of Vertebrate Paleontology website (http://www.vertpaleo.org/) provides a general checklist for such information, and the Database Manual of the Evolution of Terrestrial Ecosystems Program (ETE) offers a comprehensive listing of field types and allowed values (Damuth et al., 1997 http://www.nmnh.si.edu/ete/, with data on African sites and faunas available via The Paleobiology Database at http://paleodb.org/cgi-bin/bridge.pl).

There was lively discussion at the workshop about the issue of public access to databases and the concerns of individuals and museums about "losing control" of their data. It was generally agreed by the participants that providing public access to previously published information is important, and there is no good reason not to make such databases available. The people, institutions, and funding agencies that have supported data compilation and quality control should have a say in how and when the data are made public. The position of the National Science Foundation is that all data and databases resulting from work funded by the foundation must become public access within a reasonable period of time. It was suggested that an up-front "subscriber" page, to be filled in before access to the database is allowed, would be one way to keep track of who is using the database. The programming and maintenance of such a page will be an issue for institutions without sufficient infrastructure and software expertise, however. Although high-resolution images of specimens should be controlled, low-resolution images can go up on the web.

Obviously, museums will continue to control access to their specimens and the irreplaceable "ground-truth" data they represent. Published tables of measurements and drawings or images may be copyrighted and cannot be freely distributed on the web (e.g., as scans of material published in journals or books); there was a question about whether previously published measurements themselves could be reproduced and made available as part of a public access database if cited appropriately – this question was left unanswered. Although naïve use of a public database is a risk, there are checks and balances in the peer-review process regarding scientific publication, and the group felt overall that positive aspects of making basic data compilations available to the public outweigh the risks, particularly for educational purposes and general public awareness of paleontological resources.

The workshop participants came up with a series of recommendations regarding paleontological databases:

Recommendation 1: Databases derived from published data should be accessible to the public and would help to advance educational goals and public awareness of the value of collections. There are valid reasons for unpublished data to be restricted until issues such as control of access, individual researcher protection, and institutional protection are resolved. Much time and effort goes into assembling faunal data, and the researchers involved should control these databases until they have published analyses, with agreed-upon time limits for how long data can be restricted. Many scientific publications now require that the primary data from which analyses were derived must also be published concurrently with the analyses and interpretations on journal websites.

The accepted policy within the scientific community is that data should be made generally available only after publication. However, it is not always clear when the data have been officially "published." This can be an issue when a fossil is announced (e.g., in *Science* or *Nature*) a long time before it is carefully described in a monograph. Matters of this kind must be worked out with the museum staff and the researchers involved. Web publication is not yet taken as seriously in paleontology as it is in genetics but is likely to become more of a concern in the near future.

Data collected using public funding should be made available to the public. In the world of genomics, for example, data collected using NSF or other government funding must be posted within 24h after it is compiled and published. However, NSF recognizes that there is variability in what is appropriate for different areas of science. One of the sources to be consulted in this regard is the Federal Geographic Data Committee guidelines (http://www.fgdc. gov/standards/), which also includes rules about the Freedom of Information Act.

Recommendation 2: Restricted databases can have a "uses and practices" statement that users would read and agree to before obtaining access to that database. This would be equivalent to what is required by many software licenses and some museums for collections access (e.g., American Museum of Natural History, New York), applying both to the primary collections and also to data derived from them. For public databases, a voluntary sign-in form (and the user database derived from this) may help to monitor how the public is using the database. Truly public databases are available freely to all users, and those making the data public agree that the benefits outweigh the potential misuses of such access.

It is important to remember that ease of access keeps a database growing and dynamic (e.g., the NOW Database: http://www.helsinki.fi/science/now/). Although researchers may have misgivings about making their data available, there are many benefits in sharing data and promoting feedback, including quality control from interaction with the scientific community.

Recommendation 3: A public or research database must be credited in any publication that uses the data, including the website address, and the peer-review process should be charged with controlling this requirement.

Recommendation 4: Some data fields should always be restricted to protect fossils and fossil localities, e.g., locations of fossils and localities or sites, and locations of specimens within museums.

Recommendation 5: Future workshops can bring together working groups on more specific topics, perhaps with a smaller core group and including other specialists representing different disciplines. Ideas for focal topics for different workshops include faunal database development and sharing, planning for the future of African fossil collections, the ecology and behavior of carnivores, bovids, and primates with implications for interpreting the fossil record, with invitations to include ecologists and other biologists as well as paleontologists.

## Conclusion

The contributions in this volume provide examples and guidelines that should lead to more thorough and rigorous use of East African paleontological data in the analysis of hominin paleoecology. Databases are great resources for organizing the enormous body of faunal information that has resulted from decades of intensive field and museum work. However, the impact of analyses that use this information is often limited by problems with data comparability from site to site and region to region. The experts, students, and young professionals that contributed to the symposium, workshop, and this volume examined a range of existing faunal databases, exchanged data and ideas, and discussed data standardization. They have laid the groundwork for enhanced collaborations and comparative research on collections-based faunal analysis in the study of hominin evolution and initiated discussion on data sharing and intellectual property rights with regard to databases developed by individuals and institutions. These are positive developments that should help to generate more funding, fossils, and integrative research in future African paleofaunal studies.

## Acknowledgments

We would like to thank all of the participants in the 2003 AAPA symposium and the 2004 Smithsonian workshop for freely sharing their ideas and experiences relating to late Cenozoic faunas. The workshop was funded by the National Science Foundation (BCS 0422048) and by the Smithsonian Institution's Evolution of Terrestrial Ecosystems (ETE) Program and Human Origins Program. This is ETE contribution #168.

# References

Alemseged, Z., Bobe, R., Geraads, D., 2007. Comparability of fossil data and its significance for the interpretation of hominin environments: a case study in the lower Omo valley, Ethiopia. In: Bobe, R., Alemseged, Z., Behrensmeyer, A.K. (Eds.), Hominin Environments in the East African Pliocene: an Assessment of the Faunal Evidence. Springer, Dordrecht.

Andrews, P., 1990. Owls, Caves and Fossils. The Natural History Museum, London.

Andrews, P., 1995a. Ecological apes and ancestors. Nature 376, 555–556.

Andrews, P., 1995b. Mammals as palaeoecological indicators. Acta Zoologica Cracoviensia 38(1), 59–72.

Avery, D.M., 2003. Early and Middle Pleistocene environments and hominid biogeography; micromammalian evidence from Kabwe, Twin Rivers and Mumbwa Caves in central Zambia. Palaeogeography, Palaeoclimatology, Palaeoecology 189, 55–69.

Behrensmeyer, A.K., 1993. The bones of Amboseli: Bone assemblages and ecological change in a modern African ecosystem. National Geographic Research 9(4), 402–421.

Behrensmeyer, A.K., 2006. Climate change and human evolution. Science 311, 476–478.

Behrensmeyer, A.K., Bobe, R., Alemseged, Z., 2007. Approaches to the analysis of faunal change during the East African Pliocene. In: Bobe, R., Alemseged, Z., Behrensmeyer, A.K. (Eds.), Hominin Environments in the East African Pliocene: an Assessment of the Faunal Evidence. Springer, Dordrecht.

Blumenschine, R.J. 1989. A landscape taphonomic model of the scale of prehistoric scavenging opportunities. Journal of Human Evolution 18, 345–371.

Blumenschine, R.J., Peters, C.R., Masao, F.T., Clarke, R.J., Deino, A.L., Hay, R.L., Swisher, C.C.III, Stanistreet, I.G., Ashley, G.M., McHenry, L.J., Sikes, N.E., vander Merwe, N.J., Tactikos, J.C., Cushing, A.E., Deocampo, D.M., Njau, J.K., Ebert, J.I., 2003. Late Pliocene homo and hominid land use from Western Olduvai Gorge, Tanzania. Science 299(5610), 1217–1221.

Bobe, R., Behrensmeyer, A.K., Eck, G.G., Harris, J.M., 2007. Patterns of abundance and diversity in late Cenozoic bovids from the Turkana and Hadar Basins, Kenya and Ethiopia. In: Bobe, R., Alemseged, Z., Behrensmeyer, A.K. (Eds.), Hominin Environments in the East African Pliocene: an Assessment of the Faunal Evidence. Springer, Dordrecht.

Cooke, H.B.S., 2007. Stratigraphic variation in Suidae from the Shungura Formation and some coeval deposits. In: Bobe, R., Alemseged, Z., Behrensmeyer, A.K. (Eds.), Hominin Environments in the East African Pliocene: an Assessment of the Faunal Evidence. Springer, Dordrecht.

Damuth, J.D., 1997. ETE Database Manual. Evolution of Terrestrial Ecosystems Consortium, National Museum of natural History, Smithsonian Institution, Washington, D.C., 250p.

Domínguez-Rodrigo, M., 1996. A landscape study of bone conservation in the Galana and Kulalu (Kenya) ecosystem. Origini 20, 17–38.

Domínguez-Rodrigo, M., Lopez-Saez, J.A., Vincens, A., Alcalá, L., Luque, L., Serrallonga, J., 2001. Fossil pollen from the Upper Humbu Formation of Peninj (Tanzania): hominid adaptation to a dry open Plio-Pleistocene savanna environment. Journal of Human Evolution 40, 151–157.

Eck, G.G., 2007. The effects of collection strategy and effort on faunal recovery: a case study of the American and French collections from the Shungura Formation, Ethiopia. In: Bobe, R., Alemseged, Z., Behrensmeyer, A.K. (Eds.), Hominin Environments in the East African Pliocene: an Assessment of the Faunal Evidence. Springer, Dordrecht.

Faith, J.T., Behrensmeyer, A.K., 2006. Changing patterns of carnivore modification in a landscape bone assemblage, Amboseli Park, Kenya. Journal of Archaeological Science 33, 1718–1733.

Frost, S.R., 2007. African Pliocene and Pleistocene cercopithecid evolution and global climatic change. In: Bobe, R., Alemseged, Z., Behrensmeyer, A.K. (Eds.), Hominin Environments in the East African Pliocene: an Assessment of the Faunal Evidence. Springer, Dordrecht.

Gabunia, L., Vekua, A., Lordkipanidze, D., Swisher, C.C. III, Ferring, R., Justus, A., Nioradze, M., Tvalchrelidze, M., Antón, S.C., Bosinski, G., Jöris, O., de Lumley, M.-A., Majsuradze, G., Mouskhelishvili, A., 2000. Earliest Pleistocene hominid cranial remains from Dmanisi, Republic of Georgia: taxonomy, geological setting, and age. Science 288(5468), 1019–1025.

Graham, R.W., 1992. Late Pleistocene faunal changes as a guide to understanding effects of greenhouse warming on the mammalian fauna of North America. In: Peters, R.L., Lovejoy, T.E. (Eds.), Global Warming and Biological Diversity. Yale University Press, New Haven, pp. 76–87.

Leakey, M.G., Feibel, C.S., McDougall, I., Walker, A., 1995. New four-million-year-old hominid species from Kanapoi and Allia Bay, Kenya. Nature 376, 565–571.

Lewis, M.E., Werdelin, L., 2007. Patterns of change in the Plio-Pleistocene carnivorans of eastern Africa: implications for hominin evolution. In: Bobe, R., Alemseged, Z., Behrensmeyer, A.K. (Eds.), Hominin Environments in the East African Pliocene: an Assessment of the Faunal Evidence. Springer, Dordrecht.

Lyons, S.K., 2003. A quantitative assessment of the range shifts of Pleistocene mammals. Journal of Mammalogy (Special Feature) 84, 385–402.

Matthews, T., Denys, C., Parkington, J.E., 2005. The palaeoecology of the micromammals from the late middle Pleistocene site of Hoedjiespunt 1 (Cape Province, South Africa). Journal of Human Evolution 49, 432–451.

Musiba, C., Magori, C., Stoller, M., Stein, T., Branting, S., Vogt, M., Tuttle, R., Hallgrímsson, B., Killindo, S., Mizambwa, F., Ndunguru, F., Mabulla, A., 2007. Taphonomy and paleoecological context of the Upper Laetolil Beds (Localities 8 and 9), Laetoli in northern Tanzania. In: Bobe, R., Alemseged, Z., Behrensmeyer, A.K. (Eds.), Hominin Environments in the East African Pliocene: an Assessment of the Faunal Evidence. Springer, Dordrecht.

Pobiner, B.L., Blumenschine, R.J., 2003. A taphonomic perspective on Oldowan hominid encroachment on the carnivoran paleoguild. Journal of Taphonomy 1, 115–141.

Potts, R., 2007. Environmental hypotheses of Pliocene human evolution. In: Bobe, R., Alemseged, Z., Behrensmeyer, A.K. (Eds.), Hominin Environments in the East African Pliocene: an Assessment of the Faunal Evidence. Springer, Dordrecht.

Potts, R., Behrensmeyer, A.K., Ditchfield, P., 1999. Paleolandscape variation and early Pleistocene hominid activities: Members 1 and 7, Olorgesailie Formation. Journal of Human Evolution 37, 747–788.

Reed, D., 2007. Serengeti micromammals and their implications for Olduvai paleoenvironments. In: Bobe, R., Alemseged, Z., Behrensmeyer, A.K. (Eds.), Hominin Environments in the East African Pliocene: an Assessment of the Faunal Evidence. Springer, Dordrecht.

Sandrock, O., Kullmer, O., Schrenk, F., Juwayeyi, Y.M., Bromage, T.G., 2007. Fauna, taphonomy and ecology of the Plio-Pleistocene Chiwondo Beds, Northern Malawi. In: Bobe, R., Alemseged, Z., Behrensmeyer, A.K. (Eds.), Hominin Environments in the East African Pliocene: an Assessment of the Faunal Evidence. Springer, Dordrecht.

Su, D., Harrison, T., 2007. The paleoecology of the Upper Laetolil Beds at Laetoli: a reconsideration of the large mammal evidence. In: Bobe, R., Alemseged, Z., Behrensmeyer, A.K. (Eds.), Hominin Environments in the East African Pliocene: an Assessment of the Faunal Evidence. Springer, Dordrecht.

Suwa, G., White, T.D., Howell, F.C., 1996. Mandibular postcanine dentition from the Shungura Formation, Ethiopia: crown morphology, taxonomic allocations, and Plio-Pleistocene hominid evolution. American Journal of Physical Anthropology 101(2), 247–282.

Swisher, C.C., Curtis, G.H., Jacob, T., Getty, A.G., Suprijo, A., Widiasmoro, 1994. Age of the earliest known hominids in Java, Indonesia. Science 263, 1118–1121.

Tappen, M., 1995. Savanna ecology and natural bone deposition: implications for early hominid site formation, hunting, and scavenging. Current Anthropology 36, 223–260.

Tooby, J., DeVore, I. (Eds.), 1987. The Reconstruction of Hominid Behavioral Evolution through Strategic Modeling. The Evolution of Human Behavior: Primate Models. SUNY Press, Albany.

Vrba, E.S., 1999. Habitat theory in relation to the evolution in African Neogene biota and hominids. In: Bromage, T.G., Schrenk, F. (Eds.), African Biogeography, Climate Change, and Human

Evolution. Oxford University Press, Oxford, pp. 19–34.

Zhu, R.X., Hoffman, K.A., Potts, R., Deng, C.L., Pan, Y.X., Guo, B., Shi, C.D., Guo, Z.T., Yuan, B.Y.,

Hou, Y.M., Huang, W.W., 2004. New evidence on the earliest human presence at high northern latitudes in northeast Asia. Nature 431(7008), 559–562.

# Index

**Taxonomic terms**

These terms appear in the *text* of the various chapters; additional taxa not listed here may be found in the chapters' tables and figures.

# Index

**Geologic-geographic names**

These terms appear in the *text* of the various chapters; additional terms not listed here may be found in the chapters' tables and figures

# Other terms

**A**

Acheulean, 43
Actualistic, 218, 268, 336
Adaptability, 36–39, 337
Adaptation, 2, 5, 9, 13–15, 17–19, 30, 35–39, 51
Aeolian (eolian), 262, 264, 283, 284, 316
Allochthonous (allochthony), 323, 324
Allopatry, 51
Aquatic, 11, 228, 288, 294, 311, 942
Arboreal, 28, 38, 232, 233, 248, 281, 288, 290, 294, 296,
    298, 300, 302, 312, 324
Arid, 32–35, 37, 93, 139, 140, 144, 145, 150,
    151, 229
Aridity, 32–34, 43, 151, 249, 325
Artifacts, 29, 36, 83, 125, 131, 147, 219, 338
Autecology, 4–6, 13, 228
Autochthonous, 8

**B**

Biocoenosis, 218, 243
Biogeography, 2, 4, 5, 11, 28, 241
Biostratigraphy (biostratigraphic), 5, 7, 11, 19, 316–320
Bioturbation, 10, 11, 262, 264, 272, 275
Bipedalism, 337
Bipedality, 28, 37
Brain, 28, 98
Browsing, 18, 222, 303
Bunodont, 125
Bushland, 139, 144, 150, 151, 224, 226–228, 235, 236,
    238, 245, 246, 280–282, 286, 287, 299, 301, 311–313,
    324, 330

**C**

$C_3$, 18, 32
$C_4$, 17, 18, 32
Carbon isotopes, 281
Carcass, 29, 30, 78, 79, 82, 83, 87–93, 95, 275
Chronospecies, 335
Climate, 2–5, 8, 10–12, 16, 31–35, 37–39, 44
Cluster analysis, 41, 42, 66, 68, 139, 145, 150
Community, 4, 5, 10, 13, 16, 39, 40, 42, 52, 160
Competition, 35, 36, 39, 42, 52, 78, 79, 83, 88, 93–97, 192,
    261, 338

Coprocoenosis (coprocoenoses), 218, 219, 221, 228, 235,
    248, 249
Correspondence analysis, 41, 42, 138–145, 150, 170–172,
    174, 175, 230, 243, 245, 246, 249, 250, 324, 325, 327,
    330
Cyclicity, 16, 32, 33

**D**

Database, 6, 19, 136, 152, 165, 178, 179, 222, 334, 340–342
Diet, 5, 6, 17, 18, 28, 30, 31, 78, 83, 92, 95, 112, 116, 125,
    126, 218, 243, 293, 298, 336
Dispersal, 39, 42, 270
Diversity, 5, 6, 9, 26, 28, 37, 44, 72, 81, 83, 93, 94, 97, 125,
    131, 136, 140

**E**

Ecological diversity analysis, 281, 283, 287–290,
    296, 313
Ecomorphology (ecomorphological, ecomorphic), 2, 4–6, 9,
    13, 14, 139, 160, 177, 246, 287, 293, 303
Ecotone (Ecotonal), 8–11, 282
Ecovariable, 288, 291, 292, 296, 298–300, 302
Ecozone, 325, 327, 330
Edaphic, 125, 150, 222, 237, 312, 327
Encephalization, 28
Eurybiomic, 71
Evaporation, 31
Exfoliation, 267, 271, 273, 274
Extinction, 4, 5, 12, 26, 35–6, 38, 52, 71, 83, 93, 95–98,
    160, 333

**F**

Fluvial, 10, 14, 15, 18, 40, 42, 69, 148, 159, 161, 165, 171,
    177, 184, 186, 189, 238, 324
Food, 28–31, 37, 42, 79, 92, 95, 97
Foraging, 6, 30, 31, 79, 82, 95
Forest, 9, 10, 34, 35, 37, 70, 89, 91, 125, 218, 224, 226–230,
    233, 235, 236
Fossorial, 288, 290, 292, 294, 324

**G**

GIS, 339, 340
Glaciation (glacial), 31, 32, 124, 217

353

# Vertebrate Paleobiology and Paleoanthropology

## Published and forthcoming titles: